Wireless Digital Mobile Robotics – A College Curriculum Perspective

By

Troy Mack, Ph.D

Table of Contents

Preface

Robotics extends across many different disciplines, particularly in digital communication systems. Digital communications is rapidly outpacing its predecessor analog communications. It encompasses many fields in engineering – digital signal processing, digital control systems, digital image processing, and so on. This work touches several of these areas relative to mobile robotics, which allows this book to be used at different educational levels from undergraduate to graduate work.

In an undergraduate program, this book can be used as an advanced course in digital systems with application to robotics. Relative to digital systems, some concepts covered are basic electrical components, number systems, combinational logic circuits, and sequential logic circuits. The basic electrical components in Chapter 2 provide the practical knowledge needed when dealing with resistors, inductors, capacitors, diodes, transistors, etc. Number systems in Chapter 3 cover the conversion from one number system to another, including two's compliments numbers and binary number addition. The combinational logic circuit principles of Chapter 4 deal with the basic gates, digital theorems, design of digital circuits and their simplification using K-maps, multiplexers, demultiplexers, encoders, decoders, arithmetic circuits, etc., with application to VHDL. The sequential logic circuits in Chapter 5 consist of latches, flip flops, counters, registers, memory devices, design of sequential logic circuits, state machines and more, with application to VHDL. In this first edition, the student is required to have some basic knowledge of digital design kits used to construct robots. Basic knowledge on the use of Altera Quartus or similar digital design tools is also assumed. In a robotic curriculum, this book would also be an excellent resource for a pre-robotic and robotic course in digital systems as part of the VHDL language in Chapter 7, labs in Chapter 9, and with the various projects from other Chapters. The students can construct the circuits provided in conjunction with the given VHDL codes using Quartus. The provided circuits consists of data acquisition circuits such as analog to digital converters, digital to analog converters, sensors, etc. Students have enjoyed bringing the digital circuits to life using robotics.

Within an undergraduate program, another important aspect of this book is electronic communications and wireless applications. I have taught these courses over many terms at DeVry University and have used many of the step-by-step examples as labs. The projects typically require the latter half of a fifteen-week term to complete with three or more students collaborating in a group. The first half of the term focused on a few examples that lead up the familiarity of the internal circuitry of the projects using wireless transmitter/receiver pairs while the second half focused on integrating the internal digital circuitry with external analog circuitry. Examples of the internal digital circuitry include the asynchronous transmitter (txuart), in Section 9.6, which converts parallel data to serial data and the asynchronous receiver (rxuart), in Section 9.7 and 9.8, which converts serial data back to parallel data. Examples of the external circuitry include photo sensors whose outputs are converted to digital using a comparator. The various external circuits are discussed in Sections 9.9 through 9.14. The WyJen Technologies transmitters, receivers, and transceivers are easily adaptable to wireless applications requiring serial transmission and are introduced in Chapter 10. Furthermore,

Chapter 11 provides several experiments setting up and testing these wireless devices. Chapter 12, on transmission lines and antennas, provides the principles to maximize signal transmission by optimizing the antenna circuitry. Finally, mobile robotic projects, PLD-bot and PIC-bot, involving programmable logic devices and PIC microcontrollers respectively, are noted in Chapter 15.

A digital or industrial controls class would also benefit from the labs and projects in this book, specifically those given in Chapter 9 and Chapter 14, respectively. Robots are rich in data acquisition systems including various sensors. Thus, the focus is to understand, at a low level, digital circuits and how they control mobile robotic control systems. Furthermore, this book provides an excellent practical guide for implementing data acquisition systems and interfacing. It uses a simple analog to digital converter (ADC) that interfaces to a programmable logic device as discussed in Chapter 9. VHDL code is provided to demonstrate how a programmable logic device (PLD) sends the appropriate control signals to capture the ADC data. The data is then sent back to the outside world using a digital to analog converter (DAC).

Robotics is also rich in signal processing. For example, a robot that recognizes speech using VHDL would serve as an excellent graduate level project. Chapter 16 discusses a speaker recognition module in VHDL. An extension of this Chapter would be to use a digital signal processor (DSP) for speech recognition. The implementation of DSPs in robotic systems is a technology that is also widely being pursued. DSPs are further introduced in Appendix A, Appendix B, and Appendix C. Other DSP books which I used for teaching signal processing courses are listed in the reference. Other advanced topics to consider, such as data communication projects, are provided in Chapter 17.

Finally, a CD with most VHDL codes used in the examples is available separately from this edition of the book. The files and projects on the CD match the names specified in the text. Altera Maxplus II and Altera Quartus II are the software used for creating, simulating, and programming Altera programmable logic devices used in this work. Some of the Maxplus files were converted to Quartus since Maxplus is no longer being supported by Altera. Quartus is used to program a variety of Altera programmable logics devices including field programmable logic gate arrays.

Acknowledgements

It is difficult to write a technical book without the support from others. I wish to thank my Rockwell Collins Colleague, Fred Studenburg, a communications manager, for his recommendations and edits on several Chapters. I thank Eugene Mack for his edits to the book cover and the introductions. I want to thank my wife Celia Mack for her review and suggestions on several Chapters.

I am grateful for the many DeVry students who have helped to make this work possible. I am also thankful for WyJen Technologies Incorporated for the design and manufacturing of several radio frequency modules used in several wireless projects in this work. I am thankful for my family, Celia, Chaya, Landen, and Asya who have all been quite patient with me during this work. I dedicate this work to my dear mother Lillie Mack who passed away on June 15, 2010. I am grateful for Jerome who was always at her side through the years while I was away from home. We loved her very much. Most of all, I thank the creator who gave me the strength to complete this work.

Troy Mack

1. Introduction

This book deals with the technologies related to digital communications, controls, and signal processing systems. The following Sections will discuss the organization of this book and how to use it for mobile robotic applications.

1.1. Organization of Chapters

Chapter One is the Introduction.

Chapter Two gives an overview of some basic electronic components.

Chapter Three discusses some basic number systems.

Chapter Four discusses basic combinational logic circuits.

Chapter Five discusses basic sequential logic circuits.

Chapter Six discusses basic communication systems.

Chapter Seven provides a basic introduction to VHDL.

Chapter Eight deals with the description of some digital hardware kits that can be used for VHDL projects in this book.

Chapter Nine discusses the top level and low level VHDL modules for the robot. VHDL codes are provided in the text.

Chapter Ten introduces the WyJen Technologies Transmitters/Receivers used in some of the experiments and projects.

Chapter Eleven deals with WyJen Transmitter and Receiver Experiments.

Chapter Twelve gives basic concepts in antenna design and transmission lines. In general, the vertical and horizontal antennas are discussed. Basic transmission line characteristics such as matching, standing wave ratio, and reflections are discussed.

Chapter Thirteen deal with the introduction of the Project Descriptions for signal processing, controls, and communications.

Chapter Fourteen gives projects with application to Control Electronics and Robotics.

Chapter Fifteen gives projects with application to Communication Electronics and Robotics.

Chapter Sixteen gives projects with application to Signal Processing Electronics and Robotics.

Chapter Seventeen introduces Data Communications Projects.

Chapter Eighteen introduces Digital Filtering using PLDs or DSPs.

Chapter Nineteen discusses DC motors instead of servo motors in the robotic projects.

Appendix A introduces Texas Instrument Digital Signal Processors.

Appendix B discusses the Texas Instrument DSPs and Reference Frame Works.

Appendix C introduces FIR filter design and the Texas Instrument TMS320C6713 DSP.

Appendix D gives several distributors of robotic components and where to purchase parts and supplies for the projects.

The References provide reading sources on where to find more details on some subjects in this book.

The Index provides a quick way to get to pages in this book based on specific key words.

1.2. Required Hardware and Software

The robotic projects require the use of digital hardware such as programmable logic devices (PLD) that can be easily programmed. The PLDs used in this book are manufactured by Altera and reside on the DeVry University Electronic System on Chip (ESOC) development board. While the DeVry ESOC is used here, any programmable platform can be adapted to for the robotic projects in this text. Field programmable logic gate arrays (FPGAs) can also be used, but will have to be reprogrammed when the power is turn off. Of course, this can be eliminated if the FPGA has a supporting PROM device. The FPGA resides on the Altera DE2 board. For the ESOC designs, the reference clock was a 4MHz clock from which all frequencies were derived. The servo motors are controlled using the selected frequencies derived from the 4MHz reference. If a different reference is used, just test as specified in the robotic project sections to get the correct forward, reverse, or stop frequencies for each servo.

1.3. Example Curriculum for Mobile Robotics

Table 1.1 gives an example of a Fifteen-week Communication System curriculum with mobile Robotic applications using the Sections of this book. Ideal classes normally met three times a week: two lectures at two hours each and one lab at two hours. Lectures taught the concepts and theory while labs provided hands on experience performing experiments that enforced lecture concepts with application to robotic projects. The first week involved an introduction of the class, projects, robotic team assignments, and acquiring the necessary components for the wireless mobile robot. Teams were selected and weekly reports required from each group. There were a range of 3 to 5 member teams. Based on class size, that resulted in 3 to 6 teams in total. For more advanced or graduate level courses, the number per team can be smaller. The first week lectures involved the basic communication system, communication mathematics, etc., as stated in

Table 1.1. Each week builds on the previous week. The second week assumes that members of each team have acquired the kits and components necessary for the first lab, which involves wireless transmitters/receivers. By the third week, team members of each group should know and assign roles for the project:

1.) RF Engineer - RF Transmitter Design
2.) Digital Engineer - txuart digital design using VHDL
3.) RF Engineer - RF Receiver Design
4.) Digital Engineer - rxuart digital design using VHDL
5.) Antenna Engineer - RX and TX Antenna design
6.) Mechanical Engineer - Robot platforms and housing, etc.
7.) Sensor Engineer - Sensor 1 design
8.) Sensor Engineer - Sensor 2 design
9.) Sensor Engineer - Sensor 3 design
10.) Technical Writer - Final Power Point presentation
11.) Technical Writer - Final written report
12.) Technical Writer - Marketing

Each student, regardless of the designated group, should perform individual labs up to week 7 as indicated in Table 1.1 for week 1 through 7. In parallel, team members should be responsible for their assigned roles up to week 7 and through week 14. Most importantly, the group leader, typically chosen by team members must have good people skills, be dependable, etc. The team leader should organize and lead all activities of the team for their respective robot project. The expectation was that groups would have all the parts in their possession to start building the mobile robotic by week 8. That requires the circuit designs to be completed and parts ordered in advance of week 8. Some students found it advantageous to work on the robot design and tests after individual labs were completed. For example, teams would often brainstorm the robot platform, design it, start construction, order servo motors, brainstorm the features and corresponding sensors for their robot, etc. By week 8, the team members were expected to be done with individual labs and began working 100% of their job related tasks. Some team members took on multiple jobs if needed. Prior to or during week 8, students started constructing the base of their robots and adding servos. The PLD platform (or PIC platform, or DSP platform, etc) was mounted to the base top side along with the battery pack. The servo motors were mounted to the base. Once the servos were mounted, the students implemented the VHDL code given in Section 9.9 (on page 370). The servo VHDL code was then compiled and loaded into the PLD device. Interfacing the PLD to the left and right servos were done according to the schematic, etc., given in Figure 9.36 on page 375, or Figure 15.13 on page 481. Two PLD platforms were needed if a robot base and its remote control unit were desired. The top level VHDL code of the robot base PLD was similar to Figure 9.2 on page 317. The remote control unit's PLD top level VHDL code was similar to that shown in Figure 9.3 on page 318. Manually (manual mode), the servos were tested first using the robot's base VHDL code. The remote control unit was then be integrated to control the servos remotely (remote mode). In the following weeks (9-14), two or more sensors and supporting circuitry were added and tested to control the servos. These included the compass sensor circuitry in Section 9.10 on page 376, the infrared sensor circuitry in Section 9.11 on page 379, the tape tracking sensor circuitry in Section 9.12 on page 384, the track light sensor circuitry in Section 9.13 on page 387, and the speaker identification sensor in Section 9.14 on page 390. Finally, week 15 ended with a professional styled project presentation and invitation of students and staff. The students presented in business attire.

In summary, the first half of the 15-week session was geared toward learning the essential communication systems for the wireless robot. For a communication system curriculum, it was important to start digital communications in weeks 3 or 4 after the basic communication concepts (see Section 6, on page 185). Lectures for weeks 1 to 3 of the communication system curriculum covered intro to communication systems (CS), CS block diagrams, Intro to Fourier series, time and frequency domain analysis, and oscillator concepts and circuits. Week 4 lectures started with digital communications systems, i.e., ASK, FSK, PSK, QPSK, 8PSK, QAM, etc., bit rate, symbol rate, information theory, etc. For the labs, the second half of the term involved using the concepts learned from lectures and labs during weeks 1-7 to design and build the mobile robot. The curriculum assumed that the student has prior class experience with combinational logic circuits (see Section 4, on page 68), sequential logic circuits (see Section 5, on page 115), basic VHDL (see Section 7, on page 260), and electronics and component interfacing. The robotic projects are categorized into the following main areas:

1. Control electronics using sensors – See Section 14 on page 452.
2. Communication Electronics using wireless systems – see Section 15 on page 465.
3. Signal processing electronics and data acquisition systems using sensors and signal processing – see Section 16 on page 523.

Table 1.2 demonstrates an example of an 8-week communication system curriculum with mobile Robotic applications. It is similar to the 15-week curriculum given in Table 1.1 but is performed twice as fast. It is suggested that VHDL code be provided to the student to make this compressed curriculum flow smoother. It is also suggested that kits be made available as part of the curriculum materials to eliminate the risk involved in ordering parts or shipping delays. In the scenario described, a kit containing the TX/RX pair was available at the bookstore. The student used their PLD or PIC platform from a previous digital systems class. Team members and their leads should be made aware of the time constraints and act accordingly to get electronic components, mechanical parts, and other materials, in order to finish projects on time. Appendix D, Robotic Electronic Component Stores, on page 577 provides a list of where to get robotic components, etc. For the accelerated 8-week curriculum, the requirement to use 3 sensors could also be reduced to 1 sensor circuit, for example, track tape sensor, etc.

Week Number	Subject: This Book.	Lab
1	See Section 6.1 on page 185: The basic communication system - block diagram, Digital transmission, digital radio, Power, communication system modes, Electromagnetic spectrum, etc. Communication math: Time domain and frequency domain of sinusoidals: addition, subtraction, and multiplication of sinusoidals, phase, phasors, basic transistors and circuits; Fourier series, linear and nonlinear mixing. For digital circuits and design refer to Chapters 3, 4, and 5. Assign Project Teams!	For the labs you need to purchase the 315MHz or 433MHz TX/RX set from the book store. Their info is below. Transceiver, TX, and RX Part Numbers: **315 MHz Transceiver Set** 2 GTR315M_M13A **315MHz TX and RX Set** 1 GTX315M-3V-A00S06A-M27A 1 GRX315M-3V-A00S12B-M07A **433MHz TX and RX Set** 1 GTX433M-3V-A00S06A-M27A 1 GRX433M-3V-A00S12B-M07A Lab Assignment 1. Do A or B depending on what RX/TX you have. A. WyJen Transceiver, Transmitter and Receiver Simplex Test For the simplex TX to RX set do Section 11.1 on page 407. B. Transceiver Simplex Test For the transceiver set do Section 11.2 on page 410. Groups need to initiate some of the following: 1. Research robot project requirements. 2. Download datasheet of TX and RX on web. 3. Allocate jobs to various team members. 4. Determine group leader.

Table 1.1. Communication Systems 15 week curriculum with applications to mobile Robotics.

2	Continue week 1.	Lab Assignment 2: VHDL txuart Transmitter and VHDL rxuart receiver. Do Section 9.6.1on page 348 for the VHDL code for the txuart and Section 9.7 on page 351 for the VHDL code for the rxuart. Compile the VHDL codes. txuart will be used for parallel to serial data conversion of your robotic project while rxuart will be used for the serial to parallel conversion of your project. **Lab report due next Lab.** Use the following Lab report format: Lab Report Requirement document on how to write lab reports.
3	See Section 6.5 on page 200: Feedback, Colpritts, Hartley, Voltage controlled oscillator, Phase Locked Loop, etc.	Lab Assignment 3: Preparing digital data for wireless transmission. Do Section 11.3.1 on page 412, "Preparing digital data for wireless transmission - from txuart to transmitter. **Due Week 4.** Project: Robotics. Week 3 status report due Groups need to initiate the following: 1. Determine group leader. 2. Order the following parts: Servos, etc. 3. Need to start building the Robot's platform. 4. Allocate jobs to various team members. **Final Project Job assignments:** 1.) RF Engineer - RF Transmitter Design 2.) Digital Engineer - txuart digital design using VHDL 3.) RF Engineer - RF Receiver Design 4.) Digital Engineer - rxuart digital design using VHDL 5.) Antenna Engineer - RX and TX Antenna design 6.) Mechanical Engineer - Robot platforms and housing, etc. 7.)Sensor Engineer - Sensor 1 design 8.) Sensor Engineer - Sensor 2 design 9.) Sensor Engineer - Sensor 3 design 10.) Technical Writing - Final Power Point presentation 11.) Technical Writer - Final written report 12.) Technical Writer - Status Reports **Status Report Format**: Each status report should show Term Week Number, Group members, and should explain what each group member contributed to the project up to the date the status report is due. 1. Project supporting documentation – this book. 2.) HS-300 Servo modification for continuous rotation, see http://www.robotstore.com, 3.) For the wireless transmitters and receivers shop online here: www.wyjen.com

Table 1.1. Continued.

4	See Section 6.7 on page 232: What is Digital Radio and Digital Transmission Information Capacity, Amplitude Shift Keying TX and RX Frequency Shift Keying TX and RX.	Lab Assignment 4: Sending data from txuart to rxuart over a wireless link. Do "Sending data from the txuart to a transmitter and over a wireless link to a receiver and the rxuart", Section 11.3.2 on page 416. **Due Week 5.**
5	See Section 6.7 on page 232: Phase Shift Keying TX and RX. Quadrature amplitude modulation. Start Digital Transmission.	Lab Assignment 5: Do "A Simple Simplex Communication System using a 315/433MHz Transmitter, Encoder, Decoder, and Receiver Module Pair", Section 11.6 on page 425.
6	PWM, PPM, PAM, and PCM. See Section: 6.6.9 on page 231: Time division multiplexing or TDM. Frequency division multiplexing or FDM.	Lab Assignment 5 continued. **Project Report Format and Project Notebook:** See the <u>project report requirement document</u> for the required format of the project written report. **Week 6 status report due.**
7	See Section 6.6.1 on page 204: Amplitude modulation transmission and Concepts: time and frequency domain, modulation index, etc. Amplitude modulation circuits. AM Transmitters: high level and low level.	Lab Assignment 5 continued.
8	See Section 6.6.5 on page 210: Amplitude modulation reception and concepts. Tuned radio frequency receiver or TRF. Superherodyne receiver. Amplitude reception circuits.	Important rules to know: FCC rules regarding wireless devices, see sections 15.231 and 15.249., http://www.access.gpo.gov/nara/cfr/waisidx_01/47cfr15_01.html Other helpful guidelines: <u>Where to Go for Regulations Concerning Short-Range Devices (SRD)</u>, http://www.maxim-ic.com/appnotes.cfm/appnote_number/1772/CMP/BTDT Project: Robotics. Week 8 status report due.
9	See Section 6.6.7 on page 224: Frequency modulation and phase modulation: time	Project: Robotics, Week 9 status report due.
10	See Section 6.6.7 on page 224: Frequency domain of FM signals using Bessel functions: narrow band and wide band. FM TXs and circuits.	Project: Robotics, Week 10 status report due.

Table 1.1. Continued.

11	See Section 6.6.8 on page 228: FM receiver concepts. FM receiver circuits or frequency discriminators.	Project: Robotics, Week 11 status report due.
12	FM stereo.	Project: Robotics. Week 12 status report due. Initial written report (1st draft) and Power Point presentation (1st draft). Accept Email or CD-ROM.
13	Transmission lines. Relationship of transmission line wavelength, velocity factor, and frequency. Characteristic impedance. Incident and reflected waves. Standing waves and reflection coefficient. Transmission line input impedance. Wave propagations, ground, sky, and space wave.	Project: Robotics.
14	See Section 12on page 432: Vertical and Horizontal Antennas. Antenna radiation Patterns, antenna power, antenna gain, radiation resistance, antenna impedance. Antenna loading.	Project: Robotics, Final Project Power Point presentations and Project Notebooks due. **Group Presentation Time Slots: TBA.** Everyone must participate! Dress professionally and be prepared - other faculty may be present.
15		No Lab.

Table 1.1. Continued.

Week Number	Subject: This Book.	Lab
1	See Section 6.1 on page 185: The basic communication system - block diagram, Digital transmission, digital radio, Power, communication system modes, Electromagnetic spectrum, etc. Communication math: Time domain and frequency domain of sinusoidals: addition, subtraction, and multiplication of sinusoidals, phase, phasors, basic transistors and circuits; Fourier series, linear and nonlinear mixing, Assign Project Teams.	For the labs you need to purchase the 315MHz or 433MHz TX/RX set from the book store. Their info is below. Transceiver, TX, and RX Part Numbers: **315 MHz Transceiver Set** 2 GTR315M_M13A **315MHz TX and RX Set** 1 GTX315M-3V-A00S06A-M27A 1 GRX315M-3V-A00S12B-M07A **433MHz TX and RX Set** 1 GTX433M-3V-A00S06A-M27A 1 GRX433M-3V-A00S12B-M07A Lab Assignment 1. Do A or B depending on what RX/TX you have. A. WyJen Transceiver, Transmitter and Receiver Simplex Test For the simplex TX to RX set do Section 11.1 on page 407. B. Transceiver Simplex Test For the transceiver set do Section 11.2 on page 410. Groups need to initiate some of the following: 1. Research robot project requirements. 2. Download datasheet of TX and RX on web. 3. Allocate jobs to various team members. 4. Determine group leader. Lab Assignment 1: VHDL txuart Transmitter and VHDL rxuart receiver. Do Section 9.6.1 on page 348 for the VHDL code for the txuart and Section 9.7 on page 351 for the VHDL code for the rxuart. Compile the VHDL codes. txuart will be used for parallel to serial data conversion of your robotic project while rxuart will be used for the serial to parallel conversion of your project. **Lab report due next Lab.** Use the following Lab report format: Lab Report Requirement document on how to write lab reports.

Table 1.2. Communication Systems 8 week curriculum with applications to mobile Robotics.

| 2 | See Section 6.5 on page 200: Feedback, Colpritts, Hartley, Voltage controlled oscillator, Phase Locked Loop, etc.

See Section 6.7 on page 232: What is Digital Radio and Digital Transmission, Information Capacity, Amplitude Shift Keying TX and RX, Frequency Shift Keying TX and RX. | Project: Robotics. Week 3 status report due.
Groups need to initiate the following:
1. Determine group leader.
2. Order the following parts: Servos, etc.
3. Need to start building the Robot's platform.
4. Allocate jobs to various team members.

Final Project Job assignments:
1.) RF Engineer - RF Transmitter Design
2.) Digital Engineer - txuart digital design using VHDL
3.) RF Engineer - RF Receiver Design
4.) Digital Engineer - rxuart digital design using VHDL
5.) Antenna Engineer - RX and TX Antenna design
6.) Mechanical Engineer - Robot platforms and housing, etc.
7.)Sensor Engineer - Sensor 1 design
8.) Sensor Engineer - Sensor 2 design
9.) Sensor Engineer - Sensor 3 design
10.) Technical Writing - Final Power Point presentation
11.) Technical Writer - Final written report
12.) Technical Writer - Status Reports

Status Report Format: Each status report should show Term Week Number, Group members, and should explain what each group member contributed to the project up to the date the status report is due.
1. Project supporting documentation – this book.
2.) HS-300 Servo modification for continuous rotation, see http://www.robotstore.com,
3.) For the wireless transmitters and receivers shop online here: www.wyjen.com

Lab Assignment 2: Sending data from txuart to rxuart over a wireless link. Do "Sending data from the txuart to a transmitter and over a wireless link to a receiver and the rxuart", Section 11.3.2 on page 416. |

Table 1.2. Continued.

3	See Section 6.7 on page 232: Phase Shift Keying TX and RX. Quadrature amplitude modulation. Start Digital Transmission. Chapter 10 & 11: PWM, PPM, PAM, and PCM. Time division multiplexing or TDM. Frequency division multiplexing or FDM.	Lab Assignment 3: Do "A Simple Simplex Communication System using a 315/433MHz Transmitter, Encoder, Decoder, and Receiver Module Pair", Section 11.6 on page 425. **Project Report Format and Project Notebook:** See the project report requirement document for the required format of the project written report. Week 2 status report due.
4	See Section 6.6.1on page 204: Amplitude modulation transmission and Concepts: time and frequency domain, modulation index, etc. Amplitude modulation circuits. AM Transmitters: high level and low level. See Section 6.6.5 on page 210: Amplitude modulation reception and concepts. Tuned radio frequency receiver or TRF. Superherodyne receiver. Amplitude reception circuits.	Lab Assignment 3 continued. Important rules to know: FCC rules regarding wireless devices, see sections 15.231 and 15.249., http://www.access.gpo.gov/nara/cfr/waisidx_01/47cfr15_01.html Other helpful guidelines: Where to Go for Regulations Concerning Short-Range Devices (SRD), http://www.maxim-ic.com/appnotes.cfm/appnote_number/1772/CMP/BTDT
5	See Section 6.6.7 on page 224: Frequency modulation and phase modulation: time domain, modulation index, etc. Frequency domain of FM signals using Bessel functions: narrow band and wide band. FM TXs and circuits.	Project: Robotics, Week 4 status report due.

Table 1.2. Continued.

6	See Section 6.6.8 on page 228: FM receiver concepts. FM receiver circuits or frequency discriminators. FM stereo.	Project: Robotics. Week 5 status report due. Initial written report (1st draft) and Power Point presentation (1st draft). Accept Email or CD-ROM.
7	Transmission lines. Relationship of transmission line wavelength, velocity factor, and frequency. Characteristic impedance. Incident and reflected waves. Standing waves and reflection coefficient. Transmission line input impedance. Wave propagations, ground, sky, and space wave. See Section 12 on page 432: Vertical and Horizontal Antennas. Antenna radiation Patterns, antenna power, antenna gain, radiation resistance, antenna impedance. Antenna loading. Other antennas.	Project: Robotics, Final Project Power Point presentations and Project Notebooks due. **Group Presentation Time Slots: TBA.** Everyone must participate! Dress professionally and be prepared. Invite faculty and friends.
8		No Lab.

Table 1.2. Continued.

Week	Subject: Book - Telecommunications	Lab
1	See Section 6.7: Review of communication systems and math. Phasor diagram and relative phase. Digital modulation Techniques: ASK, FSK, PSK, and QAM.	Lab Assignment 1: Micro phone, ADC, and PLD platform. Do all of Section 9.2 starting on page 321. Build A/D circuit.
2	See Sections 6.4, 6.7, and 11.5: Basic PCM system block diagram. Pulse modulation techniques: PCM, PPM, PAM, PWM. A/D process: Quantization, sampling (flattop and natural), flattop and natural sampling circuit, and encoding into bits. Line codes: NRZ, RZ, Manchester, AMI - bipolar and unipolar. **Project group assignments.**	Lab Assignment 1 continued: Micro phone, ADC, and PLD platform. Build VHDL code and compile. Connect ADC circuit to PLD platform.
3	A/Ds and DACs: and encoding into bits. Dynamic range definition, Dynamic range of DACs, bipolar and unipolar A/D and DAC concepts, codec and combo chips.	Complete Lab Assignment 1. Start Lab Assignment 2: PLD Platform, DAC, and Speaker experiment. Do all of Section 9.3 starting on page 330. Build DAC circuit. Group Tasks: Groups need to acquire wireless TX/RXs.
4	PCM block diagram: analog companding and digital companding, ulaw, Alaw, line coding.	Lab Assignment 2: PLD Platform, DAC, and Speaker experiment. Do all of Section 9.3 starting on page 330. Build DAC circuit.
5	See Section 6.7: Basic modem block diagram, Baud rate vs. bit rate, basic FSK modem, modem classifications, Bell system and Bell system modems.	Lab Assignment 2: PLD platform, DAC, and Speaker experiment. Build VHDL circuit and compile, connect DAC circuit to PLD platform.
6	Table of Bell System modems, Bell System modem 103, the ITU, FSK TX and FSK RX, details of Bell 103 modem using FSK TX/FSK RX, and QPSK TX. Start serial interfaces.	Lab Assignment 3: Encoding and Decoding: VHDL, PLDE platform, TX UART and RX UART. For speech need to replace current ADC0804 with ADC0801 and current DAC0808 with DAC0830.

Table 1.3. Data Communications 15 week curriculum with applications to wireless phones.

7	The Open Systems Interconnection (OSI) seven layer model, serial/parallel interfaces (RS-232, RS422,UART, USRT, USAR) terminals and data sources.	Project: Wireless telephone. Group Assignments starts: Interface Transmitter, Receiver, or Transceiver to PLD Platform for wireless telephones.
8	Data communication network: terminals, LCU, data modems, printer, host, servers, etc. Serial methods: RS232, Firewire, USB, Ethernet, etc. Hoika and Tomasi details of TX UART and RX UART. UART data format, RS232 interface and voltage levels.	Project: Wireless telephone
9	Parity, UART chips, 8251A, 16550A. The RS 232 specification, electrical, functional, and mechanical	Project: Wireless telephone
10	Telecommunication Topics.	Project: Wireless telephone
11	Telecommunication Topics.	Project: Wireless telephone
12	Telecommunication Topics.	Project: Wireless telephone
13	Telecommunication Topics.	Project: Wireless telephone
14	Telecommunication Topics.	Project due.
15		No Lab.

Table 1.3. Continued.

2. Basic Electronic Components

In this Chapter it is important to know a few basic concepts before moving forward. These can always be revisited when the need arises. Basic Electronics include electronic component understanding such as resistor color codes, resistor types, LED anode and cathode, inductors, how to read capacitors, diodes, NPN and PNP transistors, etc. Let's start with resistors in the next Section.

2.1. Resistors

Carbon resistors have color codes. These are described in Section 2.1.1. Resistors also have standard values as described in Section 2.1.2.

2.1.1.   Carbon Resistor Color Code

The resistor color code for the carbon resistor is shown below in Figure 2.1. One statement that is often used to remember the color code is the following: **B**ig (black) **B**oys (brown) **R**ace (red) **O**ur (orange) **Y**oung (yellow) **G**irls (green) **B**ut (blue) **V**iolet (violet) **G**enerally (gray) **W**ins (white).

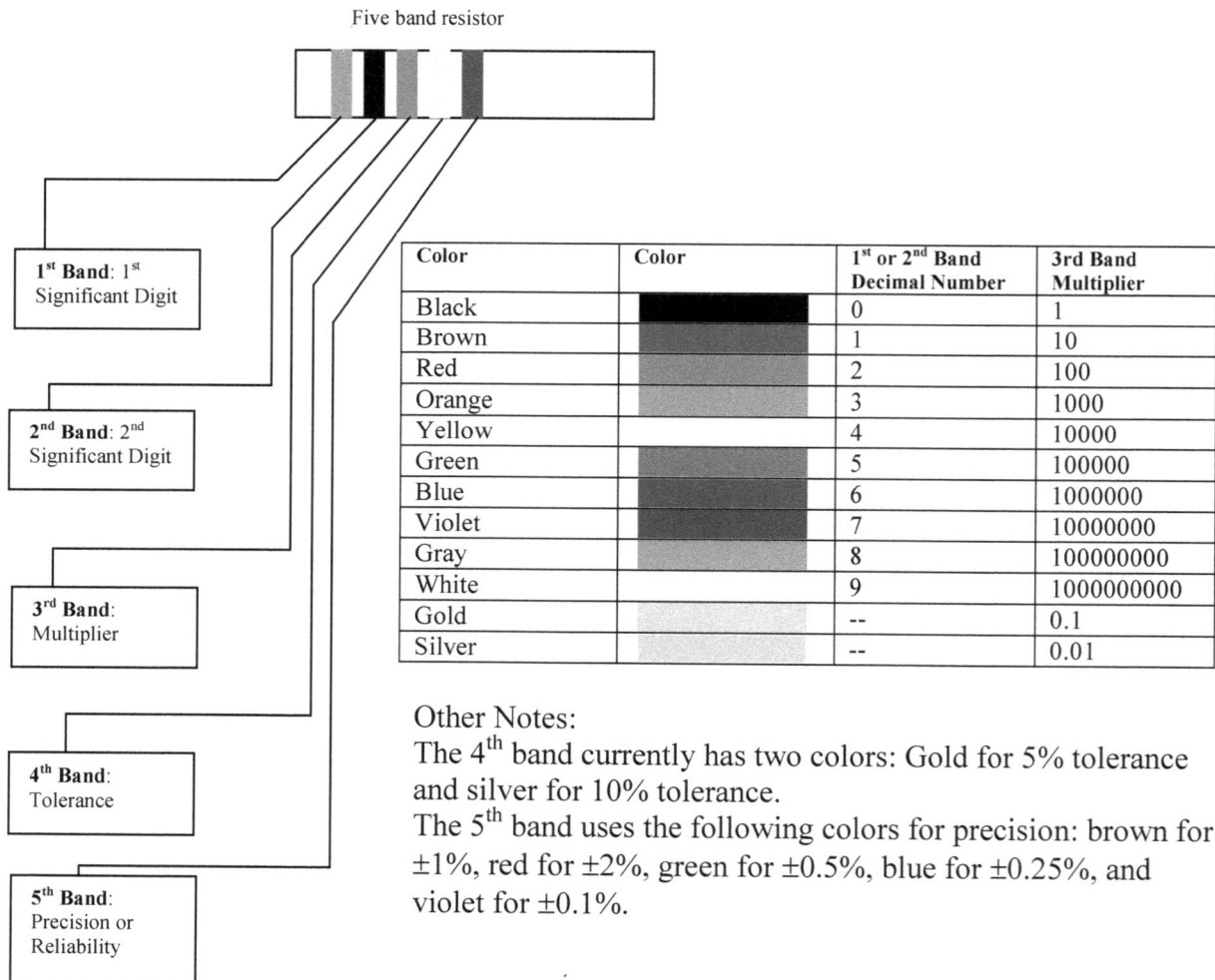

Five band resistor

| 1st **Band**: 1st Significant Digit |
| 2nd **Band**: 2nd Significant Digit |
| 3rd **Band**: Multiplier |
| 4th **Band**: Tolerance |
| 5th **Band**: Precision or Reliability |

Color	Color	1st or 2nd Band Decimal Number	3rd Band Multiplier
Black		0	1
Brown		1	10
Red		2	100
Orange		3	1000
Yellow		4	10000
Green		5	100000
Blue		6	1000000
Violet		7	10000000
Gray		8	100000000
White		9	1000000000
Gold		--	0.1
Silver		--	0.01

Other Notes:
The 4th band currently has two colors: Gold for 5% tolerance and silver for 10% tolerance.
The 5th band uses the following colors for precision: brown for $\pm1\%$, red for $\pm2\%$, green for $\pm0.5\%$, blue for $\pm0.25\%$, and violet for $\pm0.1\%$.

Figure 2.1. Resistor color code for carbon resistor.

2.1.2. Standard Resistor Values

When selecting resistor values for a given circuit select practical values close to theoretical values. For practical purposes, select resistors with decade multiples of the values listed in Table 2.1 or Table 2.2 for ±1% or ±10% tolerances, respectively. Examples of standard resistance values with ±10% tolerances and decade multiples of column two are already shown in Table 2.3.

10	10.2	10.5	10.7	11.0	11.3	11.5	11.8	12.1	12.4	12.7	13.0
13.3	13.7	14.0	14.3	14.7	15.0	15.4	15.8	16.2	16.5	16.9	17.4
17.8	18.2	18.7	19.1	19.6	20.0	20.5	21.0	21.5	22.1	22.6	23.2
23.7	24.3	24.9	25.5	26.1	26.7	27.4	28.0	28.7	29.4	30.1	30.9
31.6	32.4	33.2	34.0	34.8	35.7	36.5	37.4	38.3	39.2	40.2	41.2
42.2	43.2	44.2	45.3	46.4	47.5	48.7	49.9	51.1	52.3	53.6	54.9
56.2	57.6	59.0	60.4	61.9	63.4	64.9	66.5	68.1	69.8	71.5	73.2
75.0	76.8	78.7	80.6	82.5	84.5	86.6	88.7	90.9	93.1	95.3	97.6

Table 2.1. Standard Resistor Values at ±1% . Select decade multiples of these values.

10	12	15	18	22	27	33	39	47	56	68	82

Table 2.2. Standard Resistor Values at ±10%. Select decade multiples of these values.

	Column 2					
1.0	10	100	1.0K	10K	100K	1.0M
1.1	11	110	1.1K	11K	110K	1.1M
1.2	12	120	1.2K	12K	120K	1.2M
1.3	13	130	1.3K	13K	130K	1.3M
1.5	15	150	1.5K	15K	150K	1.5M
1.6	16	160	1.6K	16K	160K	1.6M
1.8	18	180	1.8K	18K	180K	1.8M
2.0	20	200	2.0K	20K	200K	2.0M
2.2	22	220	2.2K	22K	220K	2.2M
2.4	24	240	2.4K	24K	240K	2.4M
2.7	27	270	2.7K	27K	270K	2.7M
3.0	30	300	3.0K	30K	300K	3.0M
3.3	33	330	3.3K	33K	330K	3.3M
3.6	36	360	3.6K	36K	360K	3.6M
3.9	39	390	3.9K	39K	390K	3.9M
4.3	43	430	4.3K	43K	430K	4.3M
4.7	47	470	4.7K	47K	470K	4.7M
5.1	51	510	5.1K	51K	510K	5.1M
5.6	56	560	5.6K	56K	560K	5.6M
6.2	62	620	6.2K	62K	620K	6.2M
6.8	68	680	6.8K	68K	680K	6.8M
7.5	75	750	7.5K	75K	750K	7.5M
8.2	82	820	8.2K	82K	820K	8.2M
9.1	91	910	9.1K	91K	910K	9.1M

Table 2.3. Standard Resistor Values at ±5%. Decade multiples of second column already selected. For example, decade multiples of 10 are 1, 10, 100, 1000, 10000, 100000, and 1000000 ohms.

2.2. **The LED**

The LED is called a **L**ight **E**mitting **D**iode, thus the name LED. The LED consists of an anode and cathod as shown in Figure 2.2. The arrow represents light coming from the LED when it is on. The LED comes on when the anode is more positive that the cathode. That is, when the voltage applied to the anode is greater than the voltage applied to the cathod. **The anode is typically represented by the longest side while the cathode is the shortest lead. The cathode is also indicated by the flat lead.**

Anode Lead   Cathode Lead

Figure 2.2. LED representation showing anode and cathode.

2.3. **Capacitors**

Capacitors come in various shapes, sizes, and electrical characteristics. As illustrated in the capacitor family tree in Figure 2.3 the basic capacitor types are:
   A. Electrostatic
   B. Electrolytic

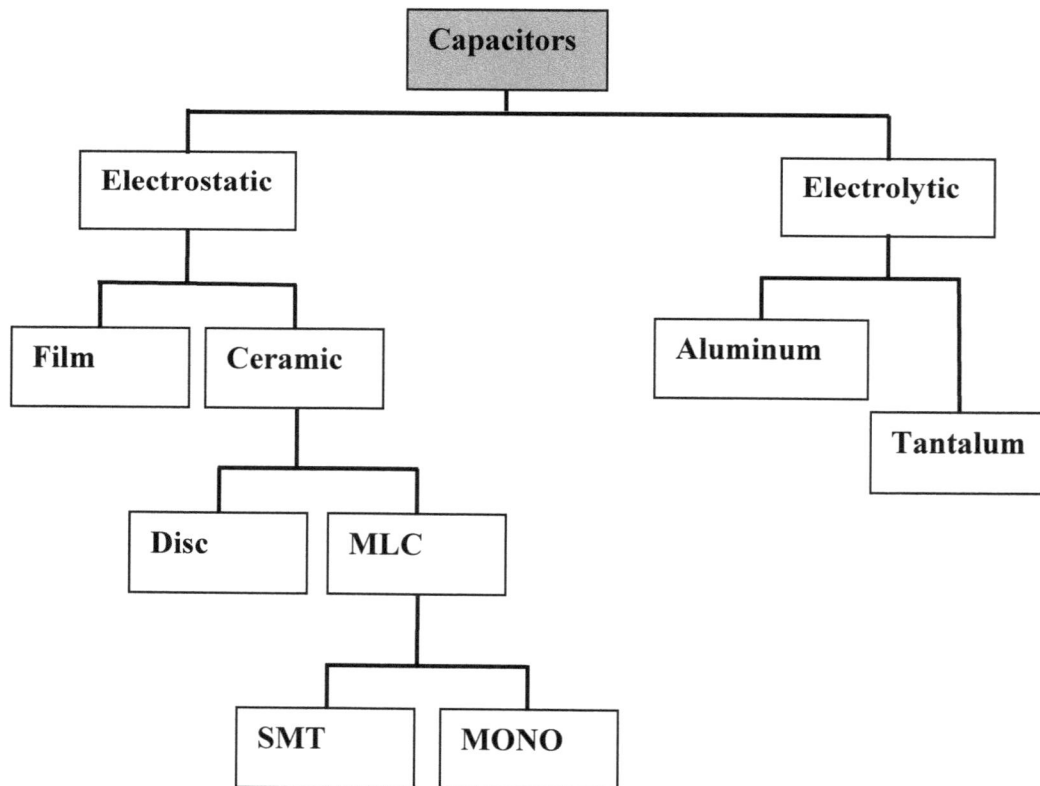

Figure 2.3. Capacitor Family Tree.

Let's talk about the most popular capacitors – ceramic capacitors. Ceramic capacitors belong to the family of electrostatic capacitors. They have the following characteristics:
  A. They are both leaded and surface mount style
  B. Most are fixed capacitance values, that is, they are not changed by the user
  C. They are non-polar
  D. They dominate the lower range of capacitance values, see Section 2.3.1 on capacitor values
  E. They are the most widely used

Ceramic capacitors include:
  A. Multilayer ceramic capacitors (MLC)
  B. Disc ceramic capacitors
Multiple layers construction results in multiple increase in capacitance surface area. An increase in surface area means an increase in capacitance as given in EQ. 2.1 In Section 2.3.1. For example, one layer results in 1X the capacitance while 5 layers results in 5X the capacitance.

The Disc ceramic capacitor is shaped like a disk as shown in Figure 2.5.

MLC capacitors include:
  A. Monolithic (mono) capacitors
  B. Surface mount (SMT) capacitors
Monolithic ceramic capacitors are formed from a single structure. The ceramic material acts as a dielectric and encapsulant of the basic element. Electrodes are buried within the ceramic and exits both ends. The ends are covered by silver glass compound. The number of electrodes is 20 or 30 in commercial practice but can be 60 or 80 for higher values of capacitance.

  Surface mount capacitors come in several sizes. The sizes are: 0402 (1005), 0603 (1608), 0805 (2012), 1206 (3216), 1210(3225), 1812(4532), and 2225(5764), from smallest to the largest size. These are the English sizes with the metric sizes enclosed in parenthesis.

Other types of capacitors are the electrolytic type. That is, they are made of an electrolyte. One type of electrolytic type is a conductive salt in solvent with aluminum electrodes that use a thin oxidation membrane. These are commonly used in applications that involve timing circuits, ripple filters. They are not very accurate so are not suitable for high frequency applications. The most common electrolytic type is the polarized capacitor. When selecting an electrolytic capacitor, one rule of thumb is to select one that is twice the supply voltage. For example, if the supply voltage is 8V, choose a 16V type. Also, be aware that they will explode if the working voltage is exceeded or the polarity is reversed. The other type of electrolytic capacitor is the tantalum type, made of tantalum pentoxide with electrodes made of tantalum material. These capacitors have excellent temperature and frequency characteristics. The tantalum has popular use in analog signal systems because of the lack of current spike noise.

2.3.1. Capacitor Construction

A capacitor can be created by placing two plates in parallel with some material between the plates and connecting wires on each plate. The capacitance of a capacior is measured in Farads. The capacitance of a capacitor can be calculated using the following formula:

$$C = \frac{8.85 * k * A}{10^{12} s}$$ **EQ. 2.1**

where A is the area of the plates facing each other in inches, k is the dielectric strength of the material between the plates, s is the separation between plates in meters, and C is the capacitance in μF. See Figure 2.4 for the general construction of a capacitor.

Some capacitors have a code written on them wich represents the capacitance in picoFarads (pF). The code typically consists of four characters. The representation of each character for a ceramic disk capacitor is described in Figure 2.5. The first two digits represent the actual digits in the capacitance value, in this case 10. The 3rd digit represents the multiplier exponent using 10 as the base. The number 3 gives a multiplier, n, where n represents the exponent in 10^n. For example, n=3 => 10^3=1000. The next character is a letter representing the tolerance.

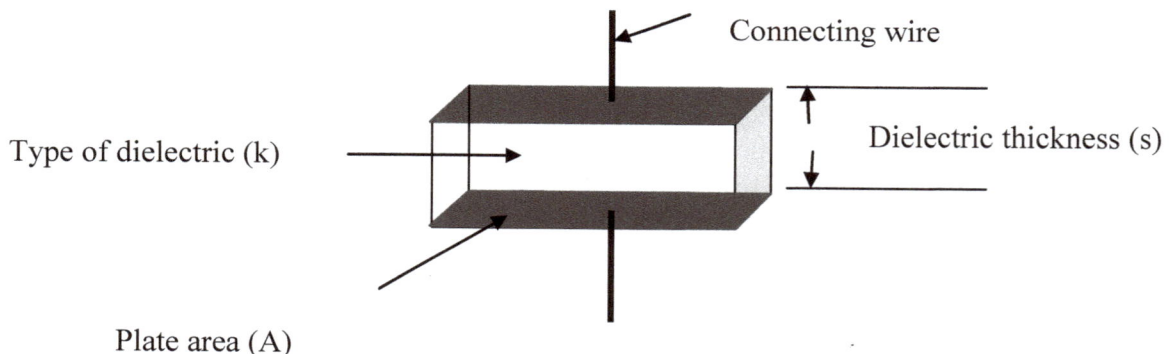

Connecting wire

Dielectric thickness (s)

Type of dielectric (k)

Plate area (A)

Figure 2.4. Construction of a capacitor.

Number	Multiplier
0	1
1	10
2	100
3	1000
4	10000

1st digit

2nd digit

Tolerance

Multiplier

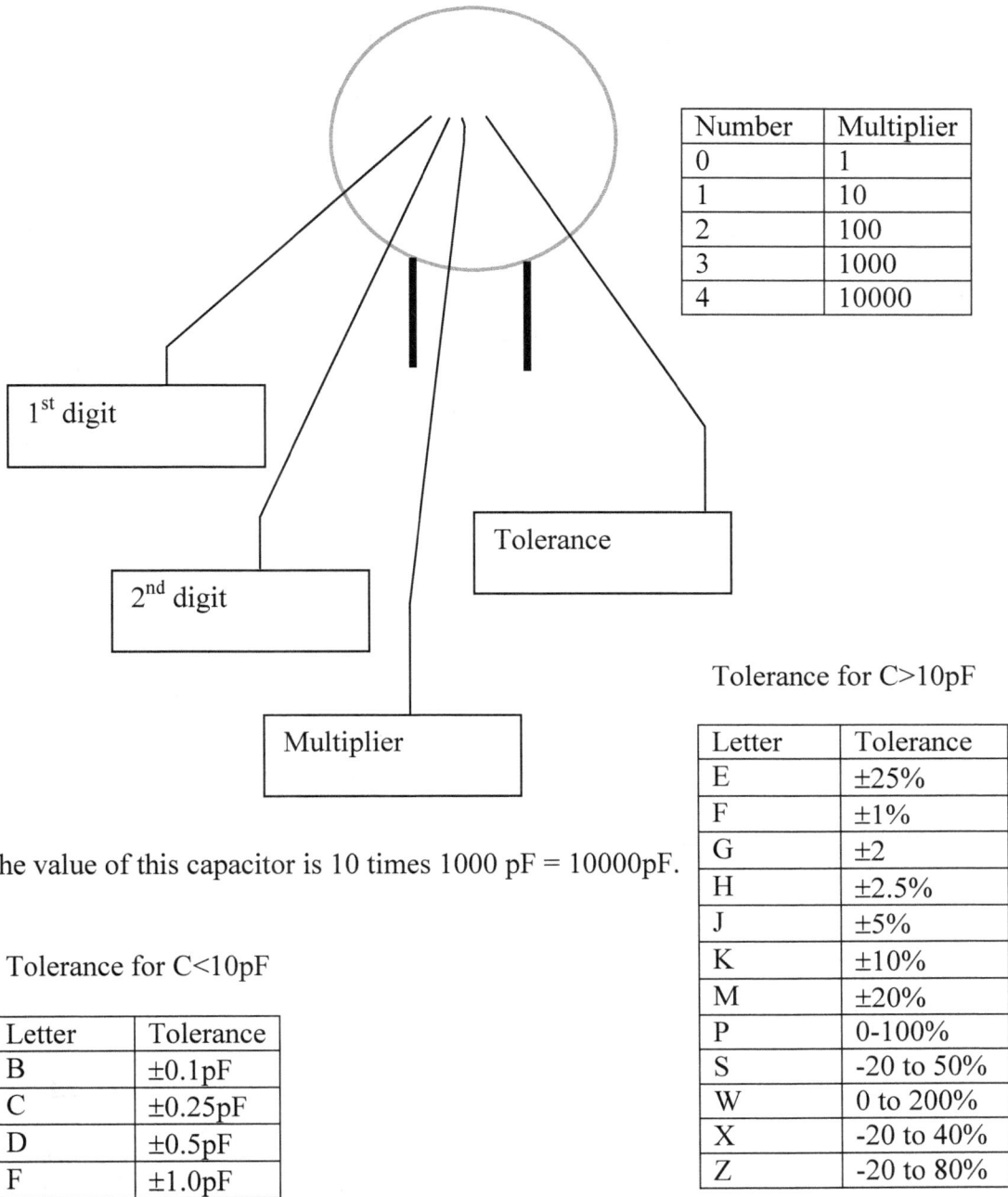

The value of this capacitor is 10 times 1000 pF = 10000pF.

Tolerance for C>10pF

Letter	Tolerance
E	±25%
F	±1%
G	±2
H	±2.5%
J	±5%
K	±10%
M	±20%
P	0-100%
S	-20 to 50%
W	0 to 200%
X	-20 to 40%
Z	-20 to 80%

Tolerance for C<10pF

Letter	Tolerance
B	±0.1pF
C	±0.25pF
D	±0.5pF
F	±1.0pF

Figure 2.5. Four character capacitor code for ceramic capacitors.

2.3.2. Standard Capacitor Values

Similar to standard resistor values there are also standard capacitor values. When selecting capacitor values for a given circuit select practical values close to theoretical values. Per Electronic Industries Alliance 575 (EIA-575) & RS 460 standard values take on the decade multiples of the following base numbers:

10 12 15 18 22 27 33 39 47 56 68 82

For example, for practical purposes, examples of standard capacitance values at ±10% are shown in the **Table 2.4**. There are also nonstandard values shown in-between and below the standard values here:

standard →10   12   15   18   22   27   33   39   47   56   68   82
              11   13   16   20   24   30   36   43   51   62   75   91← non-standard

The in-between values are considered odd non-preferred values and as such are typically not stocked and should be discouraged from being selected.

1.0pF	10pF	100pF	1000pF	.010µF	.10µF	1.0µF	10µF
1.2pF	12pF	120pF	1200pF	.012µF	.12µF	1.2µF	
1.5pF	15pF	150pF	1500pF	.015µF	.15µF	1.5µF	
1.8pF	18pF	180pF	1800pF	.018µF	.18µF	1.8µF	
2.2pF	22pF	220pF	2200pF	.022µF	.22µF	2.2µF	22µF
2.7pF	27pF	270pF	2700pF	.027µF	.27µF	2.7µF	
3.3pF	33pF	330pF	3300pF	.033µF	.33µF	3.3µF	33µF
3.9pF	39pF	390pF	3900pF	.039µF	.39µF	3.9µF	
4.7pF	47pF	470pF	4700pF	.047µF	.47µF	4.7µF	47µF
5.6pF	56pF	560pF	5600pF	.056µF	.56µF	5.6µF	
6.8pF	68pF	680pF	6800pF	.068µF	.68µF	6.8µF	
8.2pF	82pF	820pF	8200pF	.082µF	.82µF	8.2µF	

Table 2.4. Decade multiples of Standard capacitor values at ±10%. pF = pico Farad =1 x 10^{-12} F, nF = nano-Farad = 1 x 10^{-9} F, uF = micro-Farad 1 x 10^{-6} F. For example, 1000pF=1nF, 1,000,000pF=1000nF=1uF.

2.4. Inductors

The inductance of an inductor is measured in Henrys. It can be calculated by the following formula:

$$L = 12.57 \times 10^{-7} \times \frac{\mu N^2 A}{\ell}$$ EQ. 2.2

where L is the inductance in Henrys, µ is the relative permeability of the coil (= 1 for air), N is the number of turns of the coil, and A is the cross sectional area of the coil in meters. See Figure 2.6 for the general construction of an inductor.

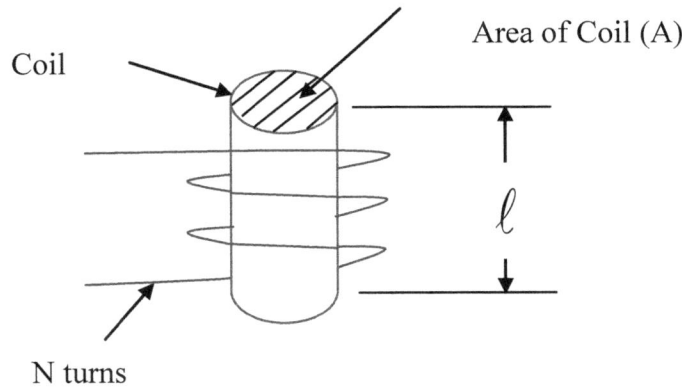

Figure 2.6. Construction of an inductor.

2.5. Potentiometers

A potentiometer is a variable resistor that functions as a voltage divider. The potentiometer converts displacement to a voltage. A displacement occurs when the movable part of the potentiometer moves away from its original position. The displacement can be linear or angular depending of the physical makeup of the pot. Displacement causes a change in resistance which causes a change in voltage or current. The popular usage of the term potentiometer (or pot for short) describes an electrical device which has an adjustable resistance controlled by the user. Typically, the pot is a three-terminal resistor with a sliding contact in the center (the wiper). If all three terminals are used, it can act as a variable voltage divider. If only two terminals are used (one side and the wiper), it acts as a variable resistor. One major disadvantage is that of corrosion or wearing of the sliding contact, especially if it is kept in one position. The schematic symbol of the pot is shown in Figure 2.7.

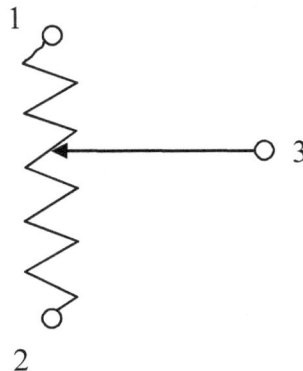

Figure 2.7. Schematic symbol of the pot (3 terminals).

There are many different types of potentiometers:
1. Low power types
2. High power types
3. Digital controlled types

Low power types are those used for low power or low current applications. Examples of Low power trimmer pots are shown in Figure 2.8. They are typically adjusted infrequency such as when they are installed. Most of the trimmer pots in Figure 2.8 can be installed on a printed circuit board (PCB).

Figure 2.8. Trimmer potentiometers.

High power pots are those that are used in high powered or high current applications such as an air conditioning unit. An example of a high power pot is the rheostat. Rheostats are usually used as variable resistors rather than variable voltage dividers.

Digitally Controlled Potentiometers can be used in analogue signal processing circuits where they replace potentiometers. These types of pots allow fine adjustments to be made to the circuit by software, instead of a mechanical adjustment.

The equivalent circuit for a pot with a resistive load is shown in Figure 2.9.

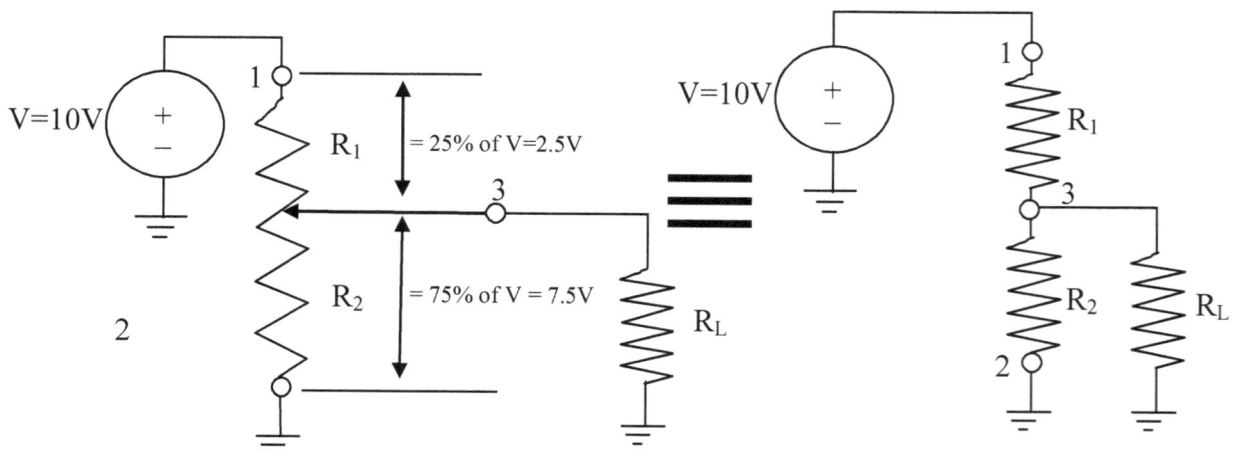

Figure 2.9. A potentiometer with a resistive load, illustrating equivalent fixed resistors.

The voltage across the resistive load is calculated by:

$$V_L = \frac{R_2 \parallel R_L}{R_1 + R_2 \parallel R_L} V$$

$$= \frac{R_2 R_L}{R_1 R_L + R_2 R_L + R_1 R_2} V$$

EQ. 2.3

2.6. Diodes

The typical diode symbol is shown in Figure 2.10. The anode is n-type material while the cathod is p-type material. The p-type and n-type material are metallurgically joined together.

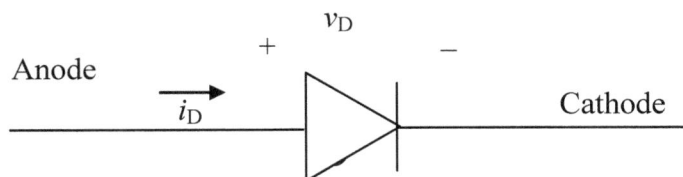

Figure 2.10. Typical Diode symbol.

The diode is forward biased, or "on", when the anode voltage is greater than the cathod voltage or $v_D \geq 0$, $i_D \geq 0$. Otherwise, the diode is reversed biased or "off".

The diode current i_D is represented by

$$i_D = I \left(e^{v_D/V_T} - 1 \right) A$$

EQ. 2.4

where

$V_T \equiv kT/q$ (V)
$v_D \equiv$ diode terminal voltage, V
$I_o \equiv$ saturation current, A
$T \equiv$ absolute temperature, K
$k \equiv$ Boltzmann's constant (1.38×10^{-23} J/K)
$q \equiv$ charge of an electron (1.6×10^{-19} C)

The graphical relationship between v_D and i_D are shown in Figure 2.11 and Figure 2.12 for the Silicon and Germanium diodes respectively. For large negative values of v_D the reverse current increases significantly. The diode voltage at which this occurs is referred to as avalanche breakdown, $-V_R$.

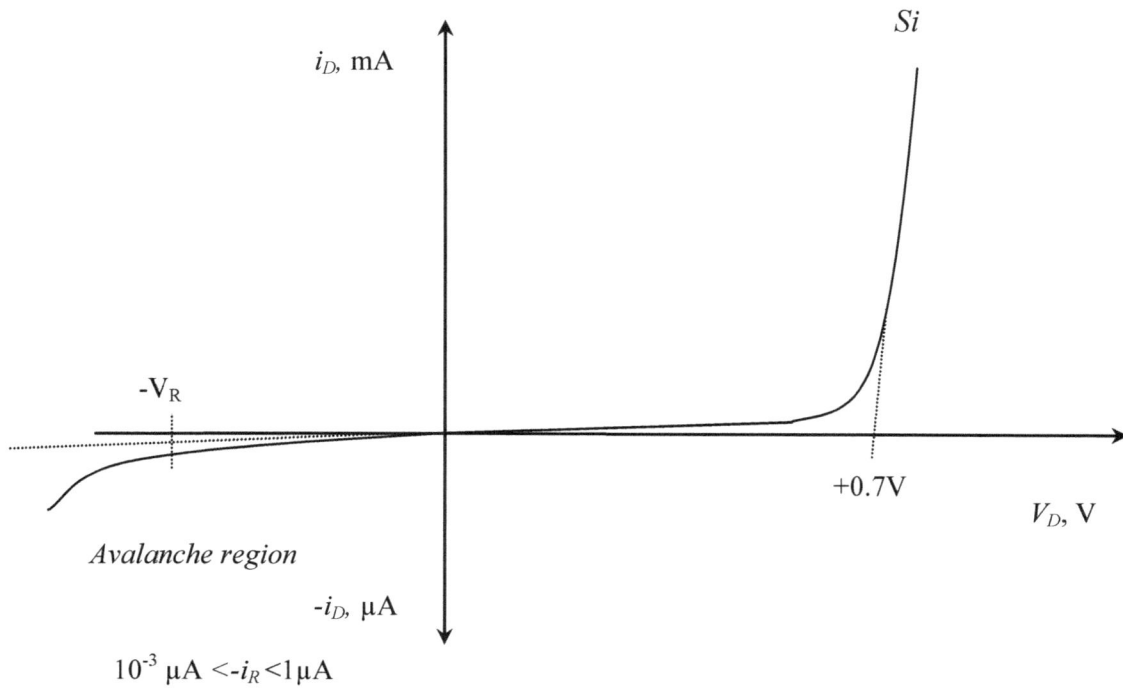

Figure 2.11. Typical Silicon (Si) diode electrical characteristics.

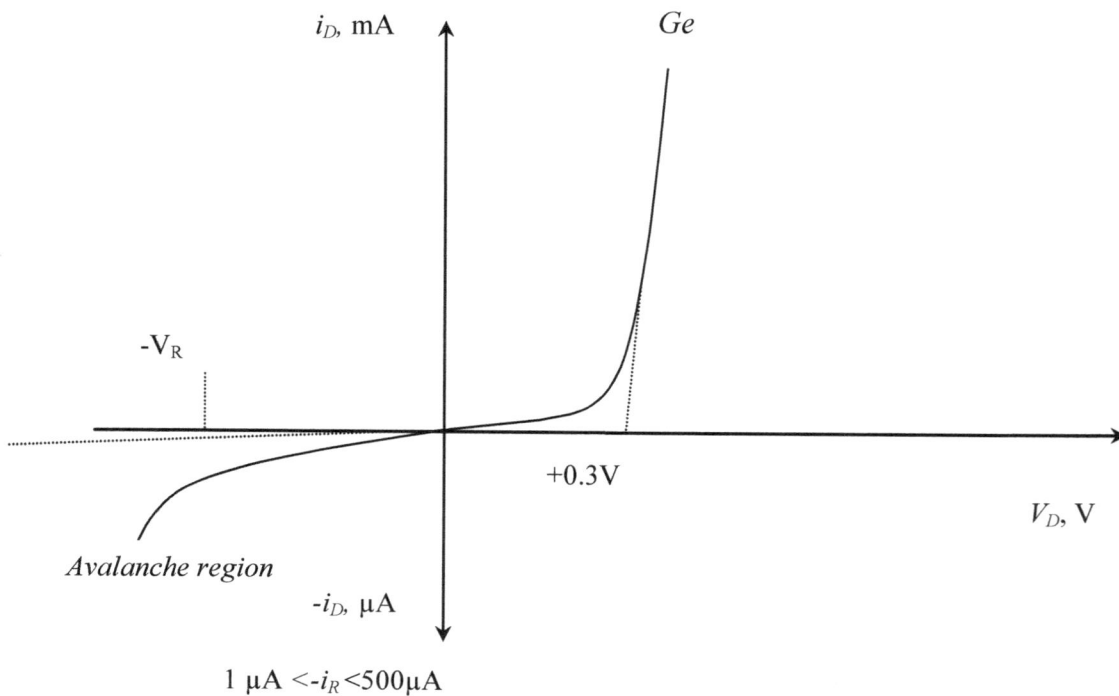

Figure 2.12. Typical Germanium (Ge) diode electrical characteristics.

2.7. **Bipolar Transistors**

Bipolar transistors have two standard types, NPN and PNP. These letters refer to the layers of semiconductor material used to make the transistor. The most popular transistors used today are NPN because they are the easiest type to make from silicon. NPN transistors are the best place to begin for someone new to transistors. The bipolar junction transistor (BJT) is a three terminal device. It has leads labeled as the base (B), emitter (E), and collector (C) as shown for the typical NPN and PNP symbols in Figure 2.13. The internal representation of NPN and PNP transistors is shown in Figure 2.14. As shown, the internal representation consists of alternating layers of n and p semiconductor materials joined metallurgically.

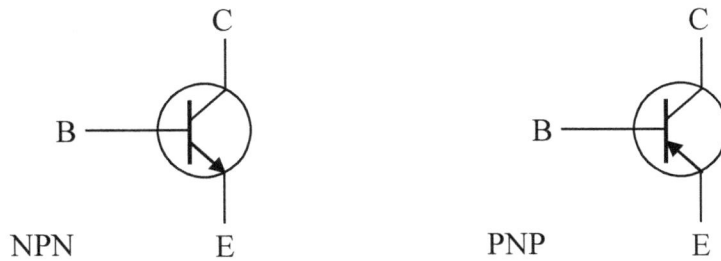

Figure 2.13. Typical Symbols for NPN and PNP transistors.

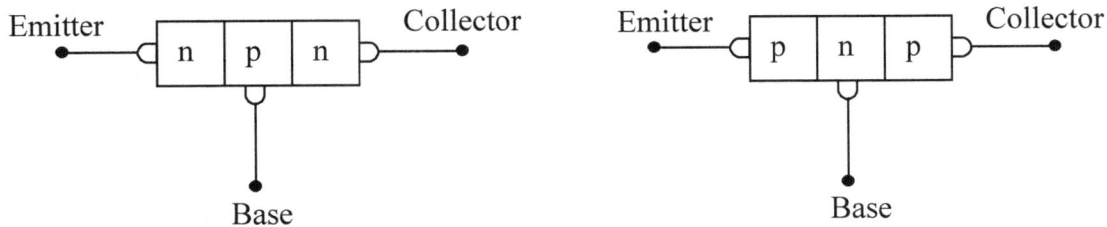

Figure 2.14. Internal representation of NPN and PNP transistors.

The **field effect transistor** (FET) is another type of transistor. The FET has a different circuit symbol and different electrical properties and is not covered here.

2.7.1. **Transistor Soldering**

Transistors can be damaged by heat when soldering. For the soldering novice use a heat sink clipped to the lead between the joint and the transistor body. A crocodile clip, such as the one shown Figure 2.15 or similar can be used as a heat sink.

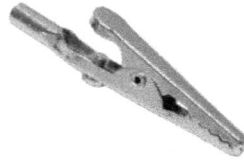

Figure 2.15. A crocodile clip.

2.7.2. Transistor Testing

One way to test a transistor is to use a multimeter. Each pair of leads can be checked for conduction. Configure a digital multimeter to diode test and or an analogue multimeter to a low resistance range. Using Figure 2.16, test each pair of leads as follows:

1. The base-emitter (BE) junction should act like a diode and conduct.
2. The emitter-base (EB) junction should act like a diode and not conduct.
3. The base-collector (BC) junction should act like a diode and conduct.
4. The collector-base (CB) junction should act like a diode and not conduct.
5. The collector-emitter (CE) should not conduct.
6. The emitter-collector (EC) should not conduct.

The same principle holds for the PNP transistor, except the diodes are in reverse.

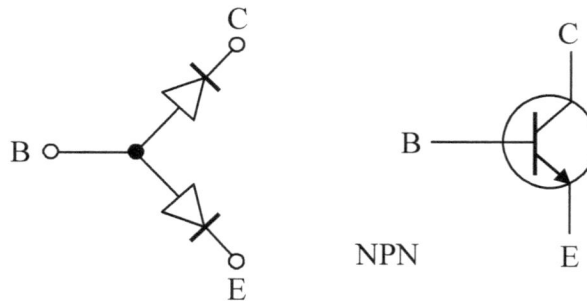

Figure 2.16. Diode representation of NPN transistor.

2.8. Bipolar Transistors and Terminal Characteristics

The notation of the transistor terminal voltages and currents are given in Table 2.5.

Value	Variable	Subscript	Examples
total instantaneous	lowercase	uppercase	i_C, v_{BE}
DC	uppercase	uppercase	I_C, V_{BE}
quiescent point	uppercase	uppercase plus Q	I_{BQ}, V_{BEQ}
ac instantaneous	lowercase	lowercase	i_c, v_{be}
RMS	uppercase	lowercase	I_c, V_{be}
maximum sinusoidal	uppercase	lowercase plus m	I_{cm}, V_{bem}

Table 2.5. Transistor current, voltage, etc., notations.

The two PN junctions of the bipolar junction transistor (BJT) can be independently biased, resulting in four possible operating modes of the transistor:

      A. saturation
      B. cutoff
      C. inverse
      D. linear or active.

These modes are indicated in Table 2.6 for the BJT NPN or PNP transistor.

Mode #	Emitter Base Bias	Collector Base Bias	Operating mode
1	forward	forward	saturation
2	reverse	reverse	cutoff
3	reverse	forward	inverse
4	forward	reverse	Linear or active

Table 2.6. BJT Terminal characteristics.

Saturation represents the operation of the transistor with maximum current flow as shown in Figure 2.21 and Figure 2.28 for the NPN and PNP transistors respectively. The transistor acts like a closed switch from collector to emitter terminals. Si based transistors have the following saturation characteristics:

$$|v_{CE}| \approx 0.2V \qquad\qquad \text{EQ. 2.5}$$

and

$$|v_{BC}| \approx 0.5V \qquad\qquad \text{EQ. 2.6}$$

Cutoff represents a region of operation where the transistor acts like an open switch. In this mode of operation only leakage current flows. Thus,

$$i_C = I_{CEO} \approx 0A \qquad\qquad \text{EQ. 2.7}$$

for the collector-base junction and

$$i_C = I_{CBO} \approx 0A \qquad\qquad \text{EQ. 2.8}$$

for the collector-emitter junction.

The **inverse** mode is an inefficient active mode with the transistor emitter and collector interchanged. Therefore, this mode is not used much.

The **linear** or **active** region of operation describes the behavior of the transistor as it operates in the region to the right of saturation and above cutoff as shown in Figure 2.21 and Figure 2.28 for the NPN and PNP transistors respectively. In the active region the terminal currents are nearly linear. In the active mode DC currents are defined as:

$$I_E = I_C + I_B$$

EQ. 2.9

and constants of proportionality are defined by:

$$\alpha \equiv \frac{I_C - I_{CBO}}{I_E}$$

EQ. 2.10

and

$$\beta \equiv \frac{\alpha}{1-\alpha} \equiv \frac{I_C - I_{CEO}}{I_B}$$

EQ. 2.11

where the leakage currents are related by:

$$I_{CEO} = (\beta + 1)I_{CBO}$$

EQ. 2.12

2.9. Common Emitter Input and Output Characteristics

A common emitter connection is a two port transistor configuration where the base shares a common point with the input and output terminals as shown in Figure 2.17. The practical analysis of the common emitter input variables is based on I_B and V_{BE} with V_{CE} held constant as shown in Figure 2.20. Similarly, the output characteristics are based on I_C and V_{CE} with constant values of I_B as shown in Figure 2.21.

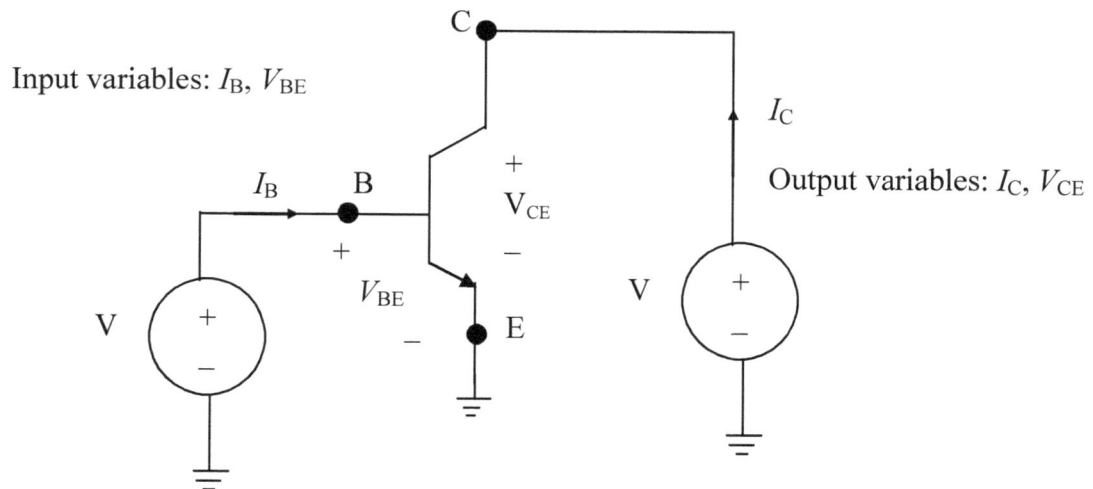

Figure 2.17. Common emitter circuit configuration for NPN BJT.

Two methods can used to analyze BJT circuits: Large signal analysis and small signal analysis. Large signal analysis or DC analysis is used to find the operating point of the BJT. It consists of replacing capacitors by opens, inductors by shorts, AC voltage sources with opens, and AC current sources with opens. Small signal analysis is used to determine the characteristics of the BJT circuit to sinusoidal inputs. Capacitors are replaced by shorts, inductors by opens, DC voltage sources with shorts, DC current sources with opens, and lastly, replace transistor by its AC model.

Popular methods for small signal analysis are A.) hybrid-π model B.) h parameter model C.) and the r_e parameter model. These are approximation models for the silicon BJT. Two will be shown here. The hybrid-π model for a bipolar transistor is shown in Figure 2.18 while the r_e model is shown in Figure 2.19B. The DC model is shown in Figure 2.19A.

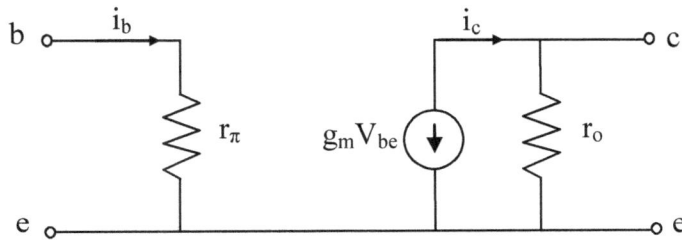

Figure 2.18. Hybrid-π model for the BJT.

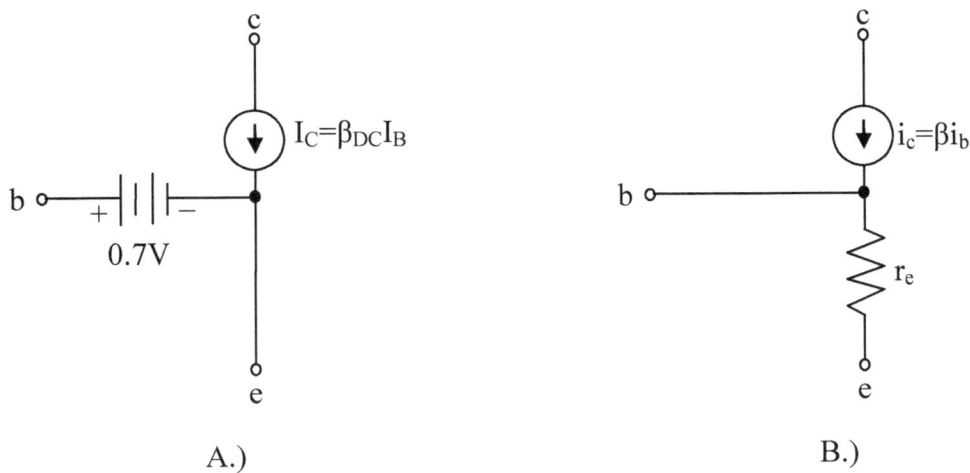

Figure 2.19. A.) DC model of NPN BJT. B.) AC analysis - r_e model of NPN and PNP BJT.

As shown in Figure 2.19A, in the active region the BJT is a current amplifier, that is

$$I_C = \beta_{DC} I_B$$ **EQ. 2.13**

and for AC r_e model,

$$i_c = \beta I_b$$ **EQ. 2.14**

As indicated in Table 2.6, the active region is when the collector base region is reversed biased and the emitter base junction is forward biased. β can be found graphically by looking at the family of curves for I_B given in Figure 2.21. These curves are normally provided in the BJT manufacturer's datasheet. Therefore, since the BJT input vs. output relationship depends on three variables, the input I_B is held constant for a given collector current verses the collector-to-emitter voltage, V_{CE}. As shown in Figure 2.21, for a given base current, I_B, the collector currents (I_C) are constant with a slight slope. For example, for a base current of 45uA, the collector current is about 5mA. In this case, $\beta_{DC} = I_C / I_B = 5mA/45uA = 111$. The static DC beta ($\beta_{DC}$) is typically close to the dynamic AC beta (β). Beta is normally kept constant to simplify circuit analysis but is actually temperature sensitive. Beta can vary widely between manufacturing facilities and should be kept in mind when designing BJT circuits. In reference to the h parameter model, h_{FE} and h_{fe} are approximately the same as β_{DC} and β, respectively.

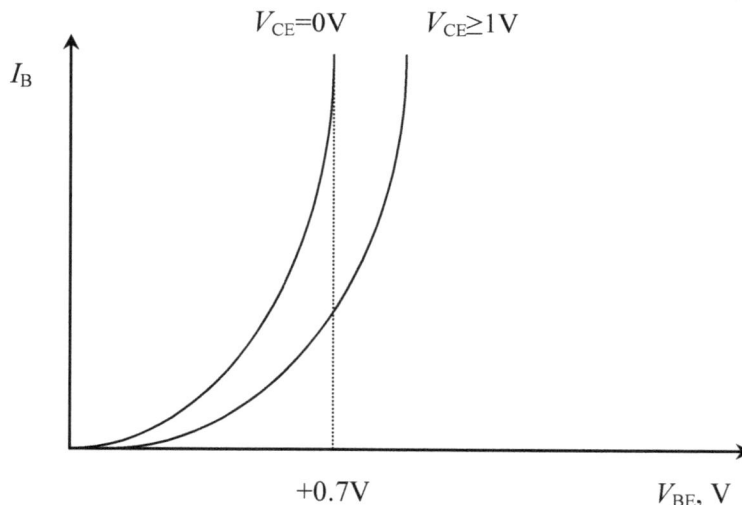

Figure 2.20. Common emitter configuration input characteristics for NPN BJT.

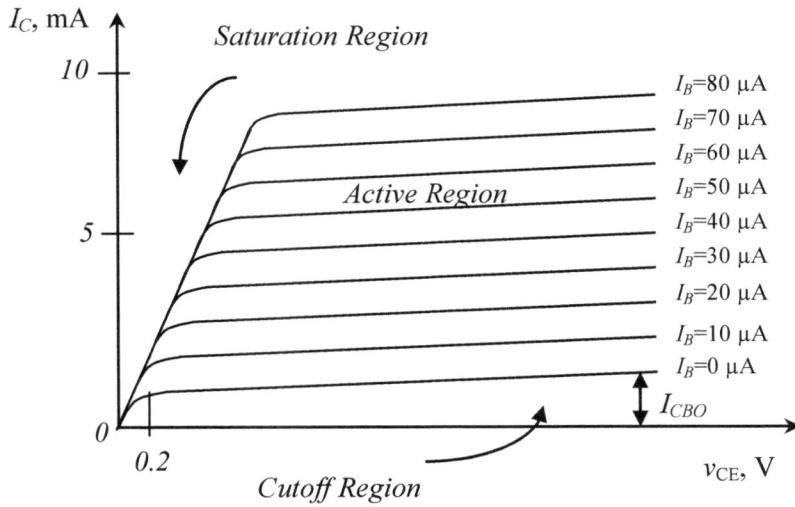

Figure 2.21. Common emitter configuration output characteristics for NPN BJT.

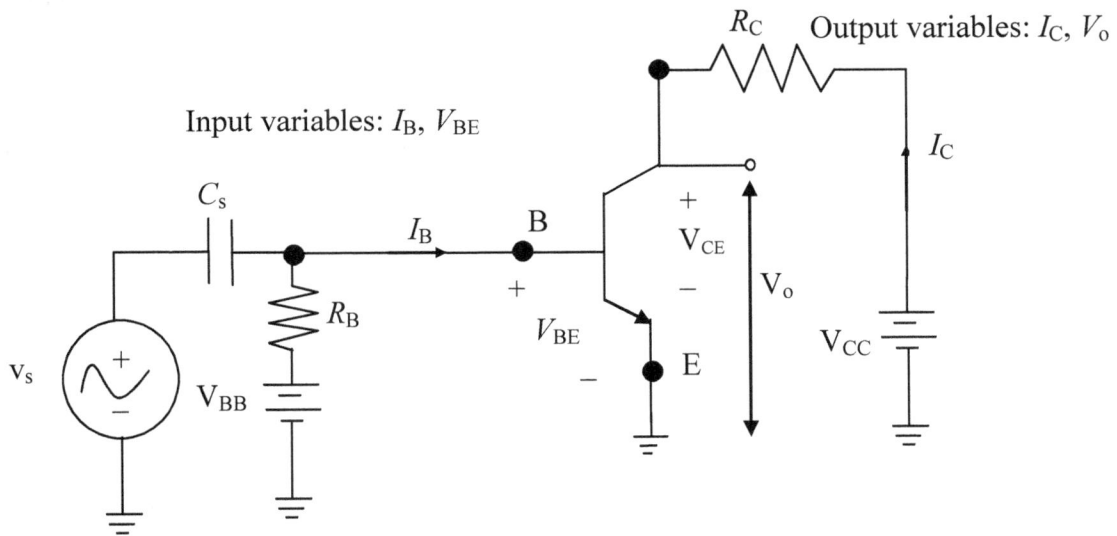

Figure 2.22. Common emitter transistor circuit.

In Figure 2.19B, the internal AC emitter resistance, r_e, is

$$r_e = \frac{26mV}{I_E}$$ **EQ. 2.15**

where the DC emitter current, I_E, has been determined in a prior DC circuit analysis of the circuit.

For a DC and AC circuit analysis, let's use the common emitter circuit in Figure 2.22. The equivalent DC circuit configuration is shown in Figure 2.23A while the equivalent AC circuit is shown in Figure 2.23B. Figure 2.23B uses the hybrid-π model. Observing Figure 2.23B, the base voltage, V, is given by

$$V = \frac{r_\pi}{r_\pi + R_B} V_s$$ **EQ. 2.16**

where the base emitter junction resistor, r_π, is given by

$$r_\pi = \frac{kT}{q} \frac{\beta}{I_C} = \frac{0.026\beta}{I_C}$$ **EQ. 2.17**

where β is the transistor base current (I_B) to collector current (I_C) gain, I_C is the collector DC bias current, q is the charge on an electron, k is the Boltzmann's constant, and T is the temperature. At room temperature T=290K and (kT)/q=0.026V.

Furthermore, in Figure 2.23B, the current gain is related to the transistor transconductance, g_m, by

$$g_m r_\pi = \beta$$ **EQ. 2.18**

Rearranging gives

$$g_m = \frac{qI_C}{kT} \approx 40I_C$$ **EQ. 2.19**

In Figure 2.23B, the collector to emitter resistance is denoted by r_o. Typically, it has a value of 15,000Ω. The output voltage is given by

$$v = -g_m v \left(\frac{r_o R_c}{r_o + R_c} \right) = -g_m (v_s) \left(\frac{r_o R_c}{r_o + R_c} \right)$$ **EQ. 2.20**

where we have assumed R_s=0 for simplicity. Now, rearranging gives the voltage gain as follows

$$A_v = \frac{V_o}{V_s} = -g_m\left(\frac{r_o R_c}{r_o + R_c}\right)$$ **EQ. 2.21**

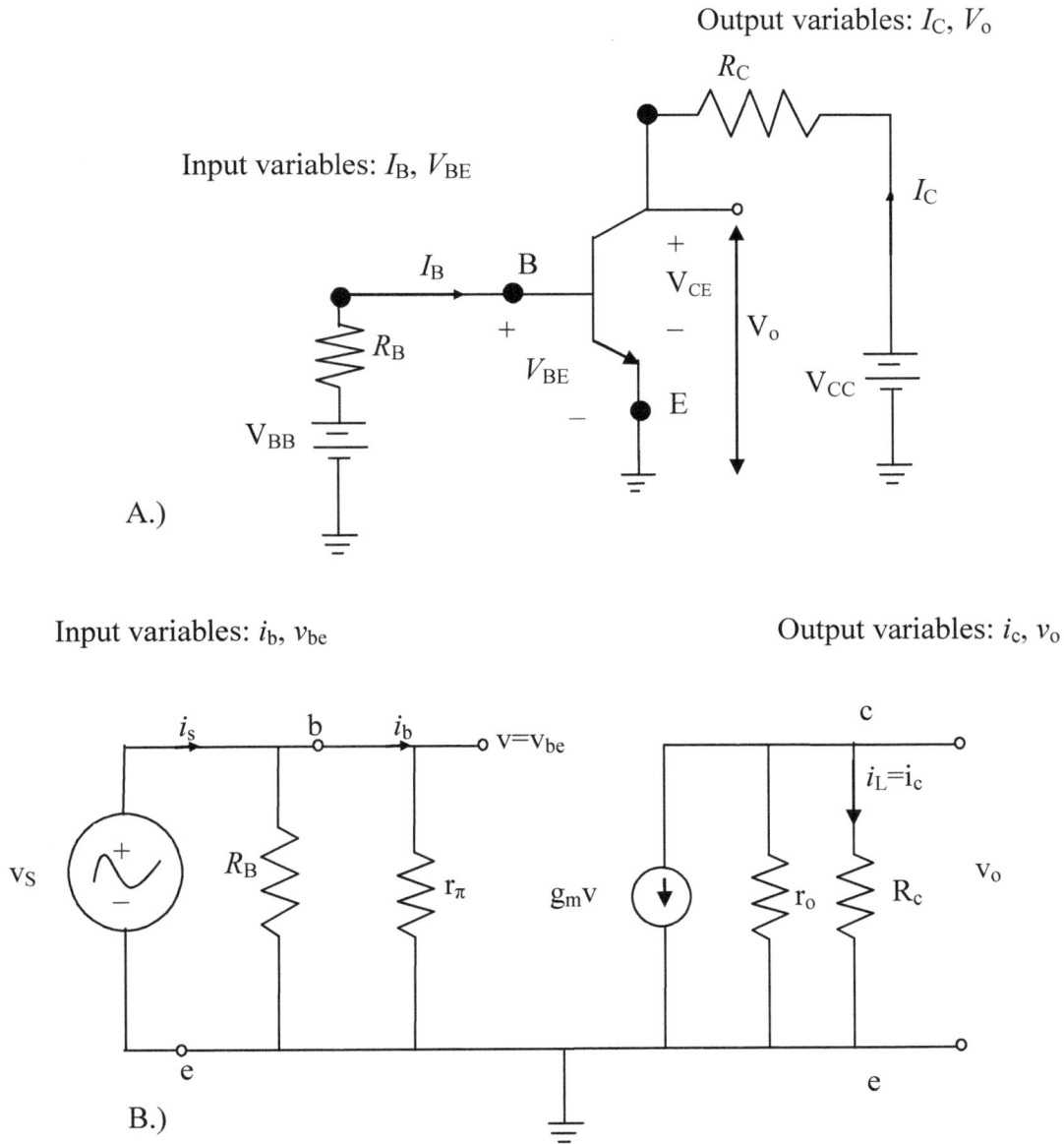

Output variables: I_C, V_o

Input variables: I_B, V_{BE}

A.)

Input variables: i_b, v_{be}

Output variables: i_c, v_o

B.)

Figure 2.23. Common Emitter: A.) DC equivalent circuit of Figure 2.22 . B.) AC equivalent circuit of Figure 2.22.

The phase shift of midfrequency voltage gain of the common emitter amplifier is 180°. The current gain is the load current, i_L, divided by the input current, i_s, given by

$$A_i = \frac{i_L}{i_S} = \frac{\left(\dfrac{r_o}{r_o + R_L}\right)\left(-g_m\left(v_S\right)\right)}{\dfrac{v_S}{\dfrac{r_\pi R_B}{r_\pi + R_B}}} = -g_m\left(\frac{r_o}{r_o + R_L}\right)\left(\frac{r_\pi + R_B}{r_\pi R_B}\right)$$

EQ. 2.22

In Figure 2.23B, the input impedance is given by

$$Z_i = \frac{v_S}{i_S} = \frac{r_\pi R_B}{r_\pi + R_B}$$

EQ. 2.23

The source resistance, R_s, has not been included in Figure 2.23B, and is not used in determining the input impedance of the circuit. The output impedance is determined by

$$Z_o = \frac{v_o}{i_o} = r_o$$

EQ. 2.24

The load resistance, $R_L = R_c$, is usually not used in determining the output impedance of the circuit.

Figure 2.24 represents a more advanced circuit for the common emitter amplifier.

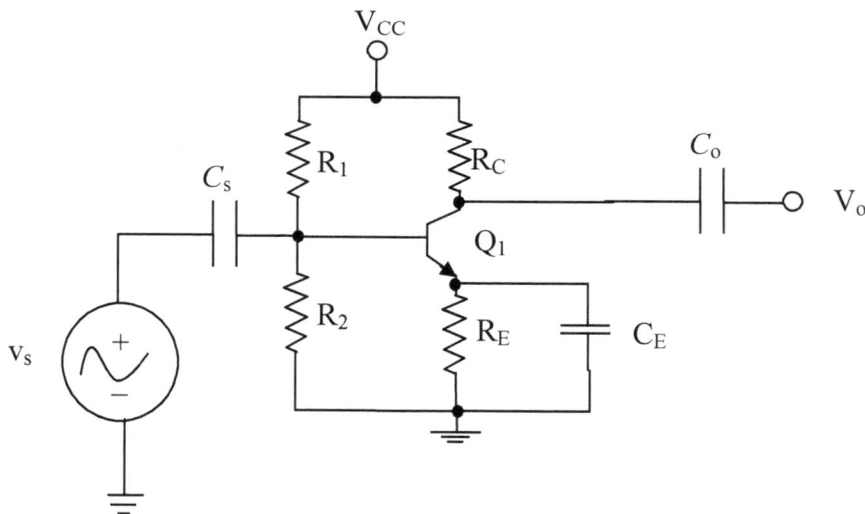

Figure 2.24. Common emitter amplifier.

The equivalent AC circuit for the common emitter amplifier in Figure 2.24 has a voltage gain of

$$A_v = \frac{-R_c}{r_e}$$ **EQ. 2.25**

The equivalent AC circuit for the common emitter amplifier in Figure 2.25 with the additional emitter resistor R_{E1} has a voltage gain of

$$A_v = \frac{-R_c}{r_e + R_{E1}}$$ **EQ. 2.26**

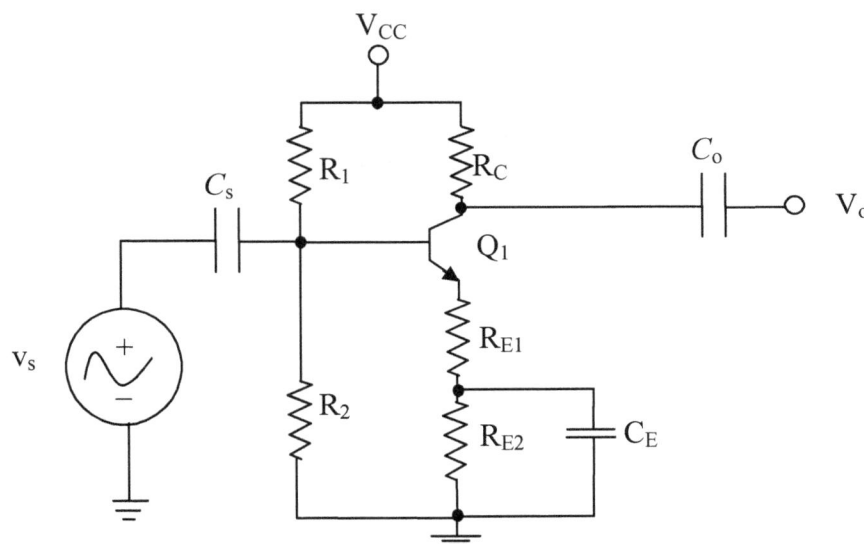

Figure 2.25. Common emitter amplifier.

2.10.   Common Base Input and Output Characteristics

A common base connection is a two port transistor configuration where the base shares a common point with the input and output terminals.

The practical DC analysis of the common base input variables is based on I_E and V_{EB} with V_{CB} held constant as shown in Figure 2.27. Similarly, the output characteristics are based on I_C and V_{CB} with constant values of I_E as shown in Figure 2.28.

For an AC analysis involving the common base the same small signal equivalent circuit model in Figure 2.18 can also be used for the common base and common collector amplifiers.

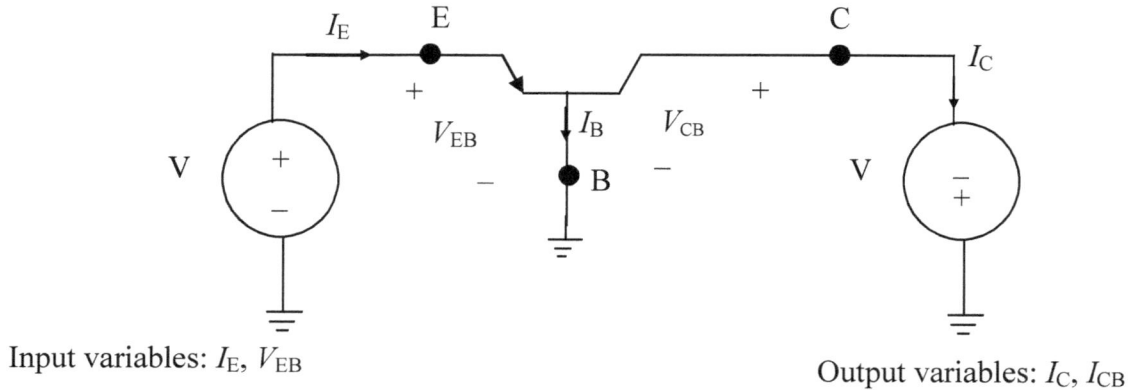

Input variables: I_E, V_{EB}

Output variables: I_C, I_{CB}

Figure 2.26. Common base configuration for PNP BJT.

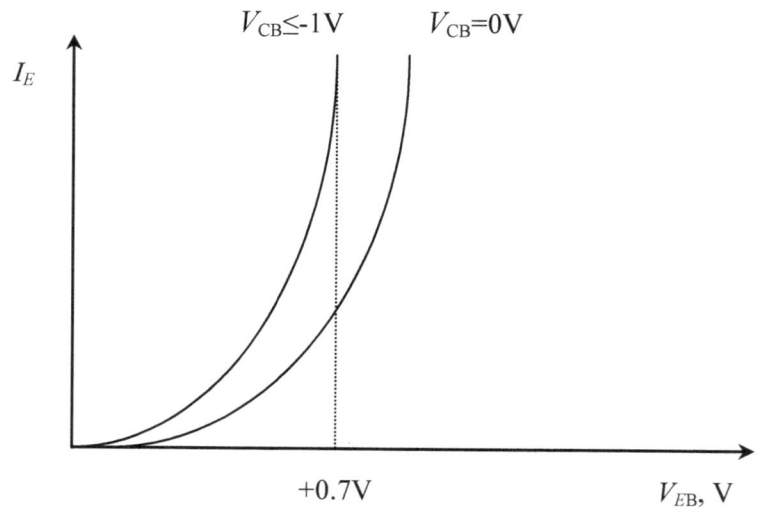

Figure 2.27. Common base configuration input characteristics for PNP BJT.

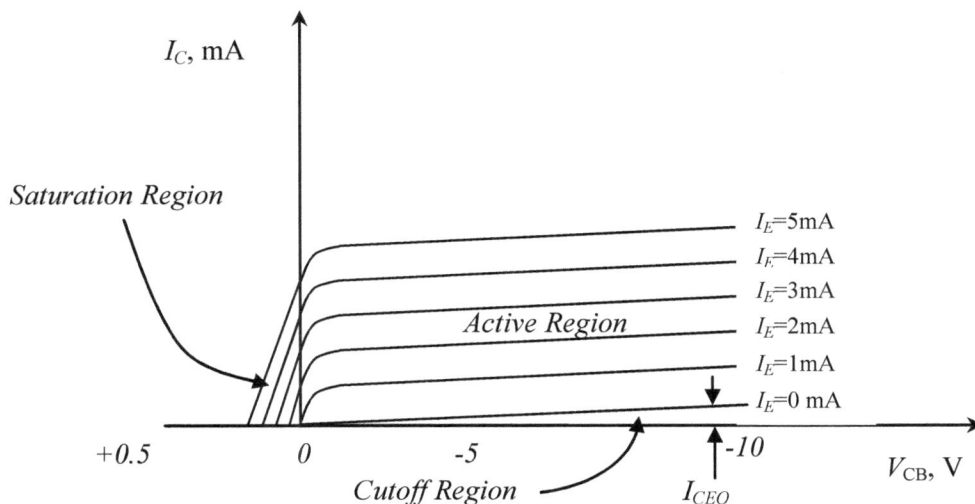

Figure 2.28. Common base configuration output characteristics for PNP BJT.

2.11. **Operational Amplifiers**

The operational amplifier (OPAMP), shown in Figure 2.29, was originally used as a basic building block in analog computers. It was used to implement mathematical operations such as integration, differentiation, addition, sign changing, and scaling. Therefore, it was given the name operational amplifier.

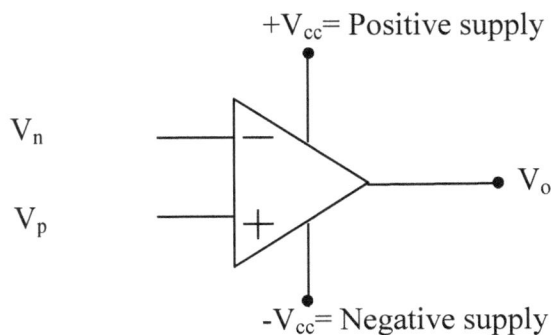

Figure 2.29. Operational Amplifier.

2.11.1. OP Amp Input vs. Output Characteristics

The input vs. output voltage characteristics for the OPAMP is shown in Figure 2.30. The output voltage is a function of the difference between the input voltages, $V_p - V_n$. The output voltage is given by

$$V_o = A(V_p - V_n)$$

EQ. 2.27

Saturation occurs when $V_o = V_{CC}$ or

$$V_{CC} = A(V_p - V_n)$$

$$\Rightarrow \frac{V_{CC}}{A} = (V_p - V_n)$$

EQ. 2.28

Thus, to keep the OPAMP in the linear region requires

$$(V_p - V_n) \leq \frac{V_{CC}}{A} \quad \text{or} \quad (V_p - V_n) \geq \frac{-V_{CC}}{A}$$

EQ. 2.29

Typically, the OPAMP has the following characteristics
   A. A is rarely less than 10,000
   B. $-V_{CC}$ or $+V_{CC}$ seldom exceeds 20V.

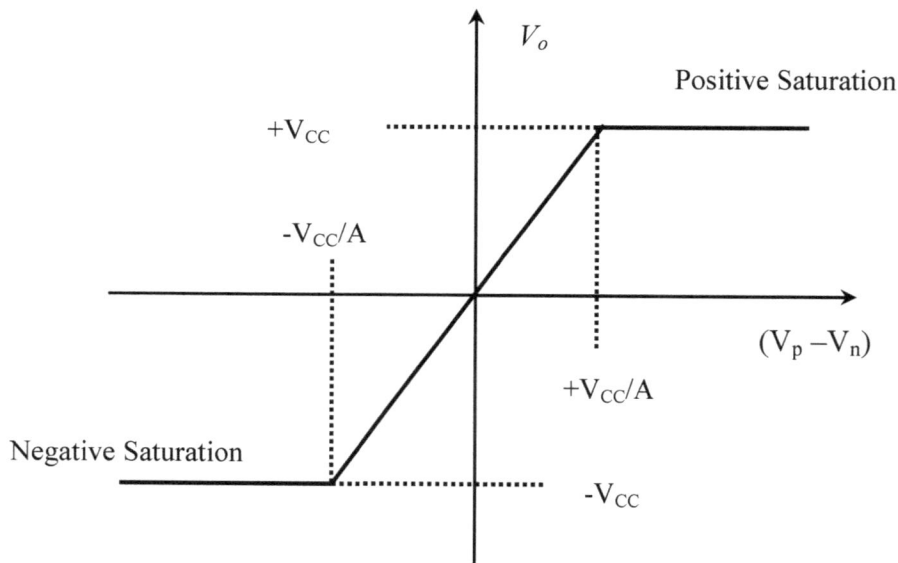

Figure 2.30. OP Amp input vs. output characteristics.

Therefore, for worse case, $V_p - V_c = V_{CC}/A = 20/10000 = 2mV$. Typically, node voltages are much larger than 2mV so a voltage difference of 2mV is essentially the same. Thus,

$$V_p \cong V_n \qquad \qquad \text{EQ. 2.30}$$

This condition keeps the OPAMP in its linear region.

2.11.2. OP Amp Input Resistance

Typically, the input resistance is seen by the input terminals is very large, about $1M\Omega$ or more. Ideally, the input resistance is infinite resulting in the following conditions

$$i_p \cong i_n \qquad \qquad \text{EQ. 2.31}$$

2.11.3. Comparator OPAMP

Figure 2.31 and Figure 2.32 illustrate the OP amp comparator. The OPAMP comparator is a OPAMP that is operating in the open loop mode. The output of the comparator will be positive or negative depending on whether the inverting input is more negative than the none inverting input.

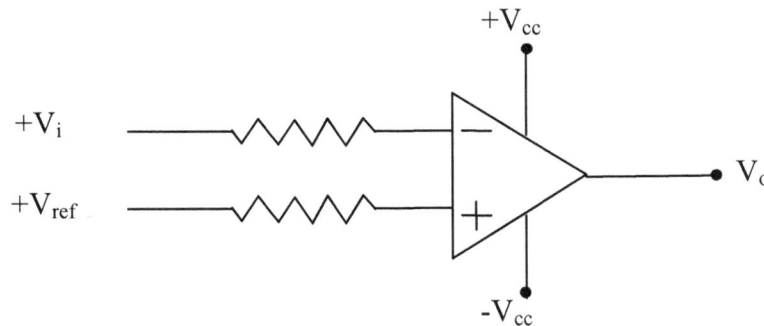

Figure 2.31. OPAMP comparator.

In Figure 2.31, the output of the comparator is given by

$$V_o = A(V_{ref} - V_i) \qquad \qquad \text{EQ. 2.32}$$

Thus, if $V_i > V_{ref}$, then $V_o = -V_{cc}$, otherwise $V_o = V_{cc}$.

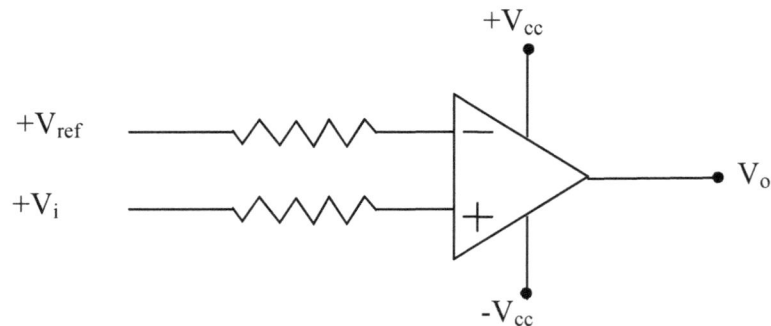

Figure 2.32. Comparator OPAMP.

In Figure 2.32, the output of the comparator is given by

$$V_o = A(V_i - V_{ref})$$

EQ. 2.33

Thus, if $V_i > V_{ref}$, then $V_o = +V_{cc}$, otherwise $V_o = -V_{cc}$.

In summary, the output of the comparator has two states
   A. positive saturation ($+V_{cc}$)
   B. negative saturation ($-V_{cc}$)

2.11.4. The Inverting OPAMP
The inverting OPAMP is shown in Figure 2.33.

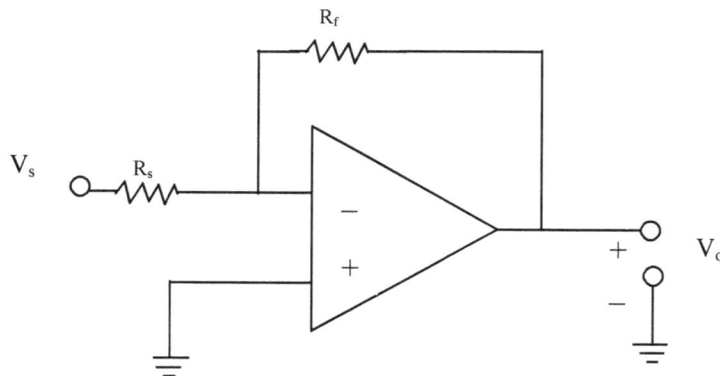

Figure 2.33. The inverting OPAMP.

The output voltage of the inverting amplifier in Figure 2.33 is given by

$$V_o = -\frac{R_f}{R_s} V_s$$

EQ. 2.34

The output of the inverting amplifier is an inverted scaled version of the input voltage.

2.11.5. The Summing OPAMP

The summing OPAMP is shown in Figure 2.34. The output voltage is given by

$$V_o = -\left(\frac{R_f}{R_1} V_1 + \frac{R_f}{R_2} V_2\right)$$

EQ. 2.35

If $R_1 = R_2 = R$, then

$$V_o = -\left(\frac{R_f}{R} V_1 + \frac{R_f}{R} V_2\right) = -\frac{R_f}{R}\left(V_1 + V_2\right)$$

EQ. 2.36

The output of the summing OPAMP is an inverted scaled sum of the input voltages. Correspondingly, for n input voltages with equivalent input resistors, the output voltage is

$$V_o = -\frac{R_f}{R}\left(V_1 + V_2 + \cdots + V_n\right)$$

EQ. 2.37

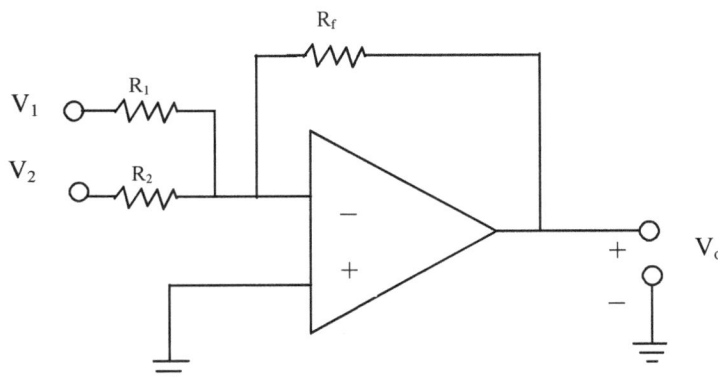

Figure 2.34. The summing OPAMP.

2.11.6. The Differential OPAMP

The differential OPAMP is shown in Figure 2.35. The output voltage is given by

$$V_o = -\frac{R}{Z_c} V_{in}$$

EQ. 2.38

$$= -(RCj\omega)V_{in}$$

The differential OPAMP input vs. output characteristics for various curves is shown in Figure 2.36, Figure 2.37, and Figure 2.38.

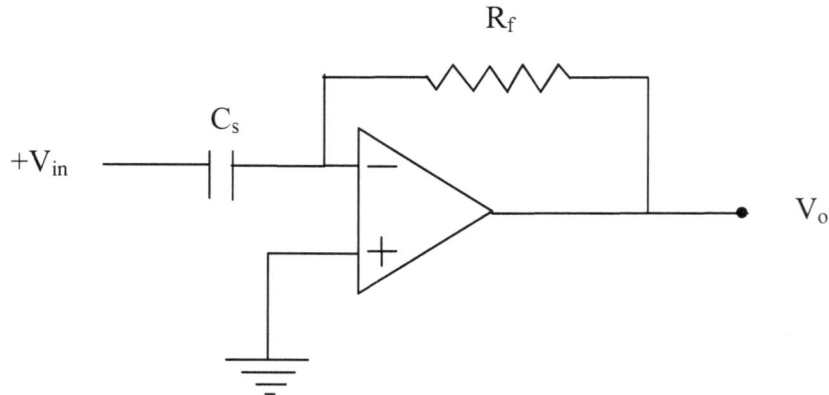

Figure 2.35. The differential OPAMP.

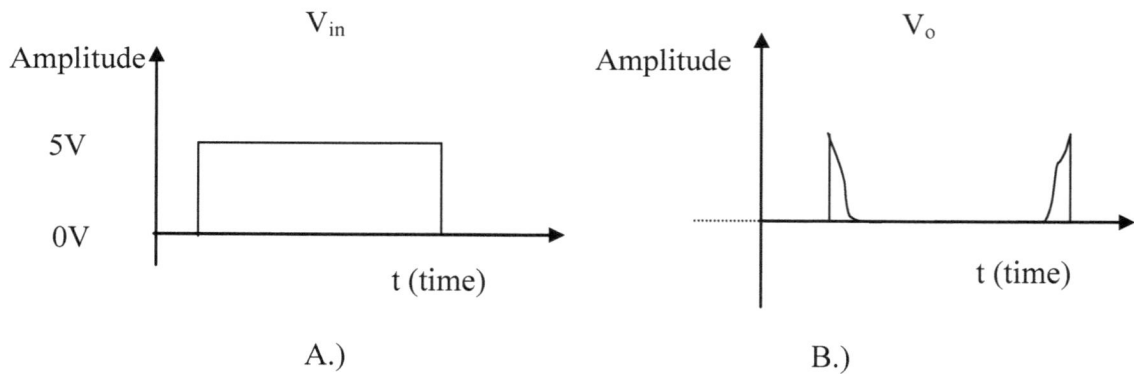

Figure 2.36. A.) Differentiator input step function. B.) Differentiator output.

Amplitude

Amplitude

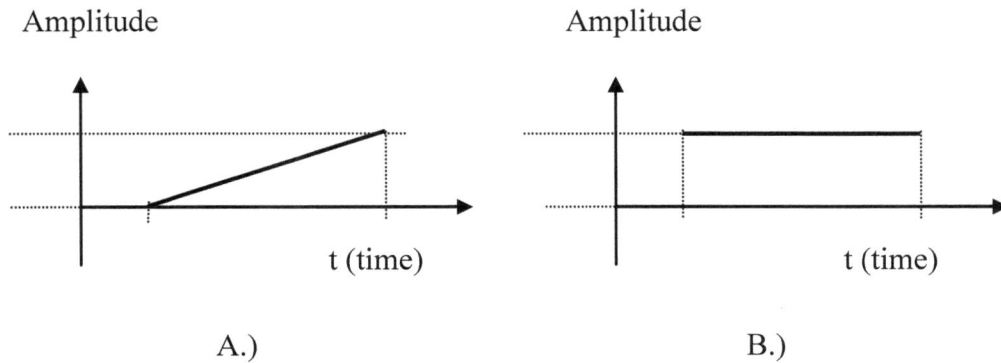

t (time)

t (time)

A.)

B.)

Figure 2.37. A.) Differentiator input ramp function. B.) Differentiator output.

Amplitude

Amplitude

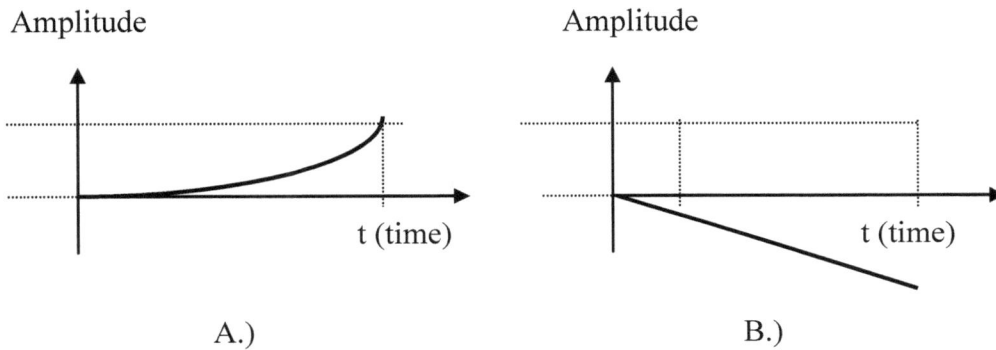

t (time)

t (time)

A.)

B.)

Figure 2.38. A.) Differentiator input square function. B.) Differentiator output.

2.11.1. The Integrator OPAMP

The integrator OPAMP is shown in Figure 2.39. The output voltage is given by

EQ. 2.39

$$V_o = -\frac{Z_f}{R} V_{in}$$

$$= \left(\frac{-j}{\omega RC}\right) V_{in}$$

The integrator generates an output proportional to the integral of the input. For the special case where the input is a horizontal line between two points, the output equation is

$$V_o = -\frac{1}{R_s C_f} V_{in} t \qquad\qquad \textbf{EQ. 2.40}$$

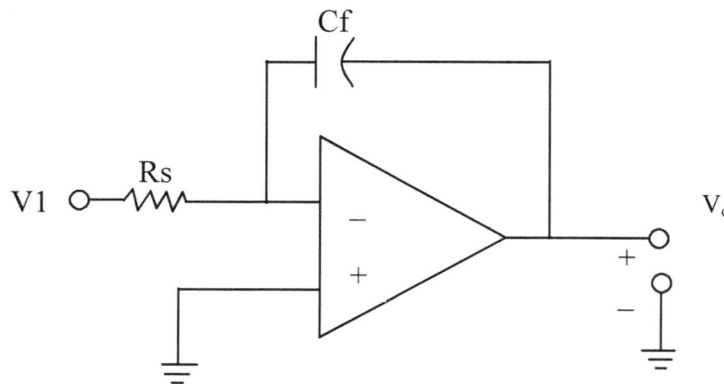

Figure 2.39. The Integrator OPAMP.

2.11.1. The Difference OPAMP

The difference OPAMP is shown in Figure 2.40. The output voltage is given by

$$V_o = \frac{R_f}{R_a}\left(V_b - V_a\right) \qquad\qquad \textbf{EQ. 2.41}$$

where $R_d = R_f$ and $R_a = R_c$ must be met, otherwise, the equation for V_o will be different and is not shown here. The output of the difference OPAMP is proportional to the difference between two input voltages. A difference OPAMP amplifies the difference signals and rejects common mode signals. For example, noise is found in all electric signals, but is suppressed by the difference amplifier. Difference mode is the difference between the 2 input voltages, that is

$$V_{dm} = V_b - V_a \qquad\qquad \textbf{EQ. 2.42}$$

and common mode is the average of the input voltages

$$V_{cm} = \frac{V_b + V_a}{2} \qquad\qquad \textbf{EQ. 2.43}$$

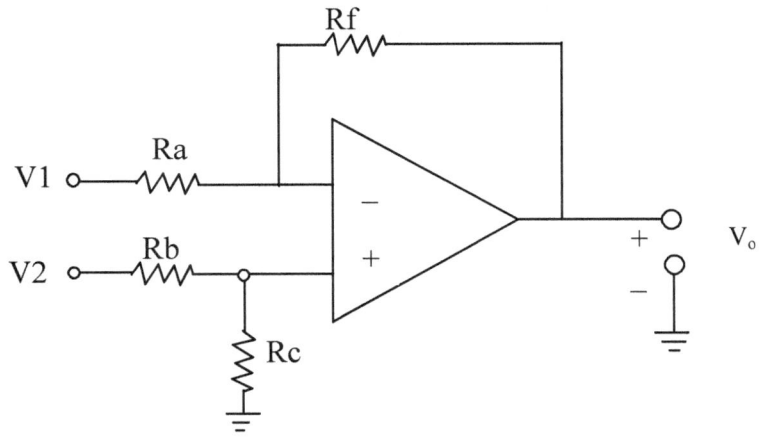

Figure 2.40. The Difference OPAMP.

PROBLEMS

1. Find the two closest practical values at 5% tolerance for each of the following calculated resistor values:

A. 6.9 ohms
B. 610 ohms
C. 10,600 ohms
D. 2,800,000 ohms

2. Find the closest, next higher, practical value for each of the following calculated capacitor values:

A. 1.9 pF
B. 2.3 uF
C. 107 pF
D. 7300pF

3. A capacitor has a temperature coefficient of 30PPM. What is the expected change in capacitance at 125 degrees Celsius?

4. A capacitor has a temperature coefficient of 10PPM. What is the expected change in capacitance at 85 degrees Celsius?

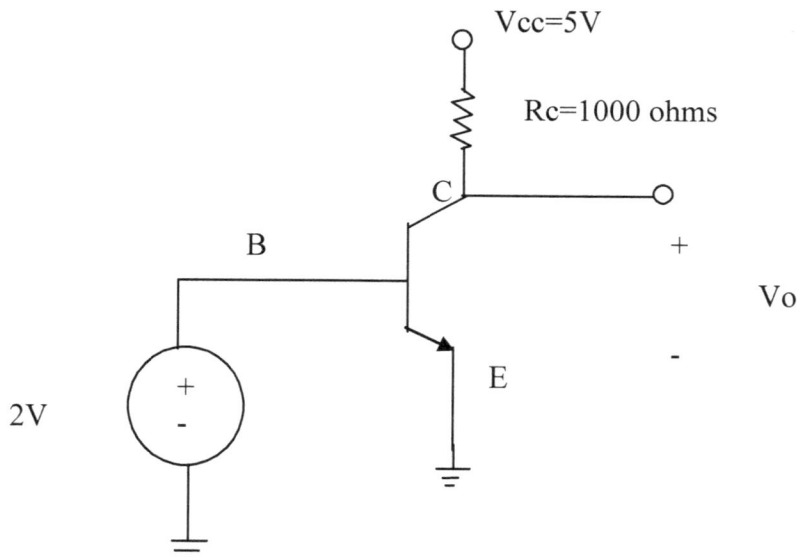

Figure 2.41.

5. The transistor in Figure 2.41 is in which of the following states:
A. Active Region   B. Cutoff   C. Saturated   D. None

6. The transistor in Figure 2.41 is configured as:
A. Common Collector   B. Common Emitter   C. Common Base   D. None

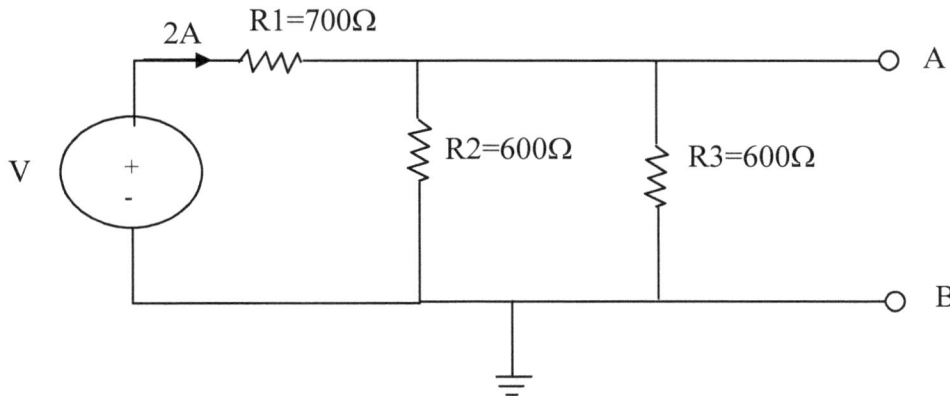

Figure 2.42.

7. The total resistance seen by the source of the circuit in Figure 2.42 is:
A.) 1300 ohms B.) 1900 ohms C.) 300 ohms D.) 1000 ohms

8. The current through R3 in Figure 2.42 is:
A.) 2A   B.) 4A   C.) 1A   D. 0 A

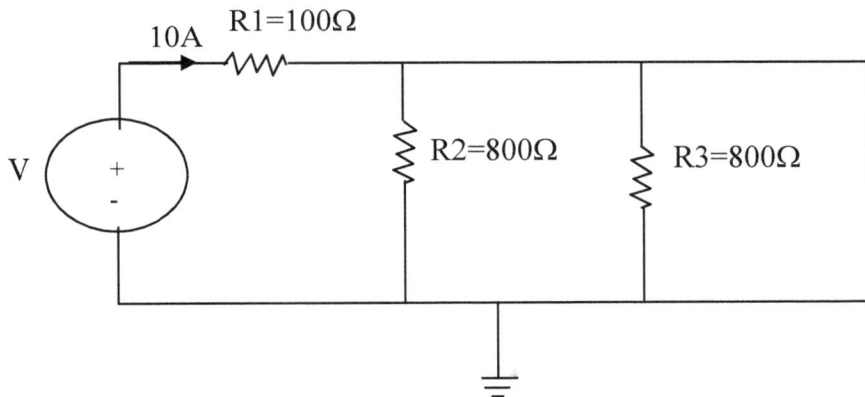

Figure 2.43.

9. The current through R3 in Figure 2.43 is:
A.) 10A   B.) 5A   C.) 8A   D.) 0 A   E.) None

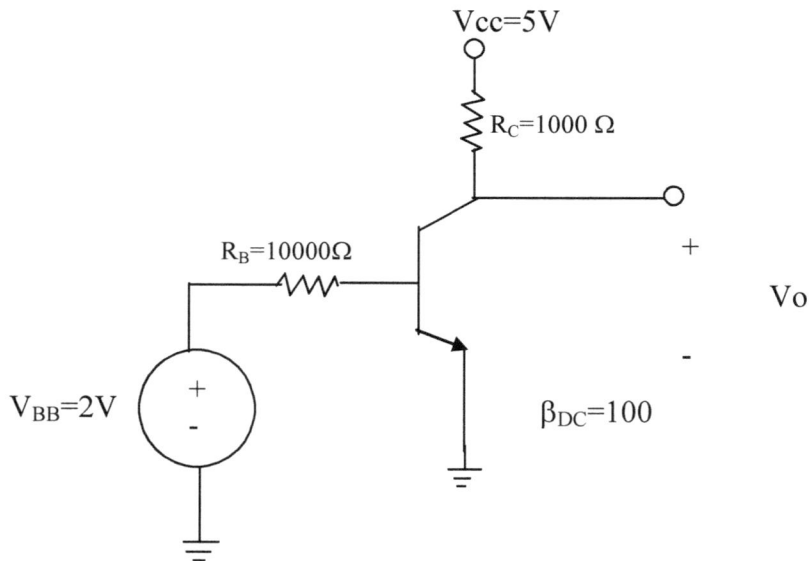

Figure 2.44.

10. Find the following using the circuit in Figure 2.44:
   a.) I_B
   b.) I_C
   c.) I_E
   d.) V_c

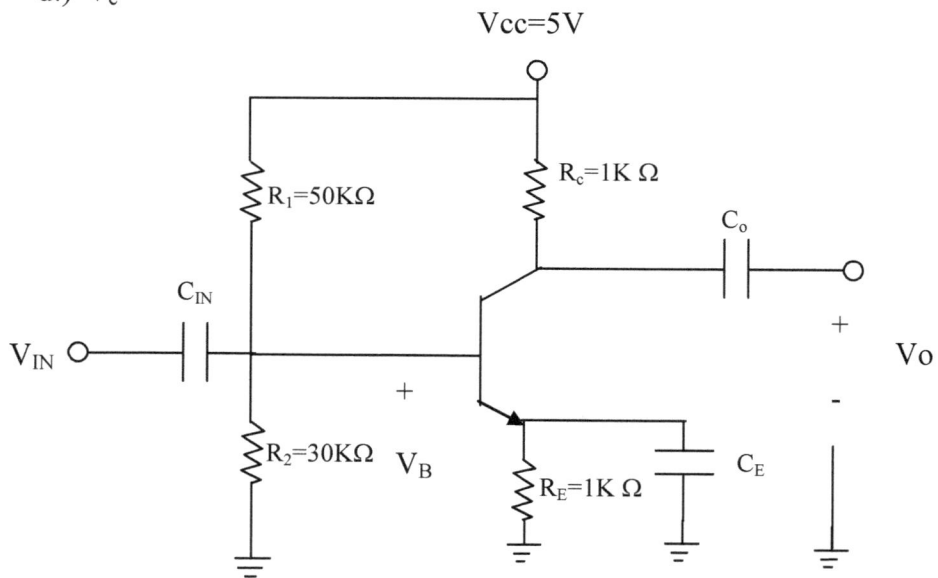

Figure 2.45.

11. Find the following using the circuit in Figure 2.45:
    a.) the bias voltage V_B
    b.) the voltage gain if $r_e = 21.277\Omega$
    c.) the voltage gain without the capacitor C_E

12. Derive the expression for the voltage gain of the circuit in Figure 2.45. Use the r_e parameter model. Show all work.

13. Derive the expression for the voltage gain of the circuit in Figure 2.25. Use the r_e parameter model. Show all work.

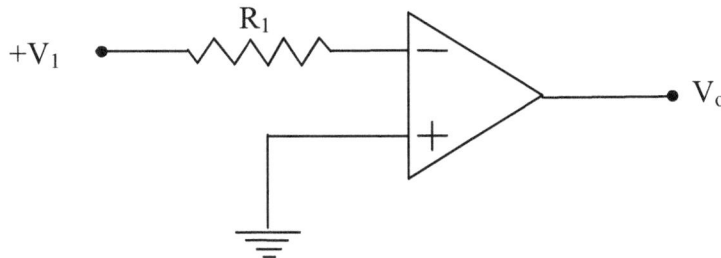

Figure 2.46.

14.) Using Figure 2.46, which one of the following is true?
A.) The output is positively saturated
B.) The output is negatively saturated
C.) The output is a high impedance

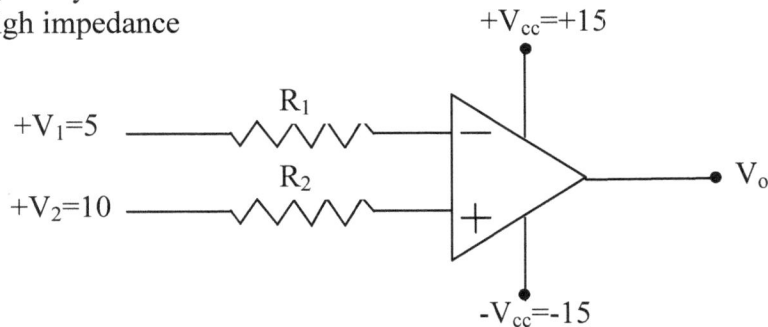

Figure 2.47.

15.) The circuit in Figure 2.47 is:
A.) a comparator
B.) difference amplifier
C.) none of the above

16.) Using Figure 2.47, which one of the following is true?
A.) The output is positively saturated at $\approx 15V$
B.) The output is negatively saturated at $\approx -15V$
C.) The output is a high impedance
D.) None of the Above

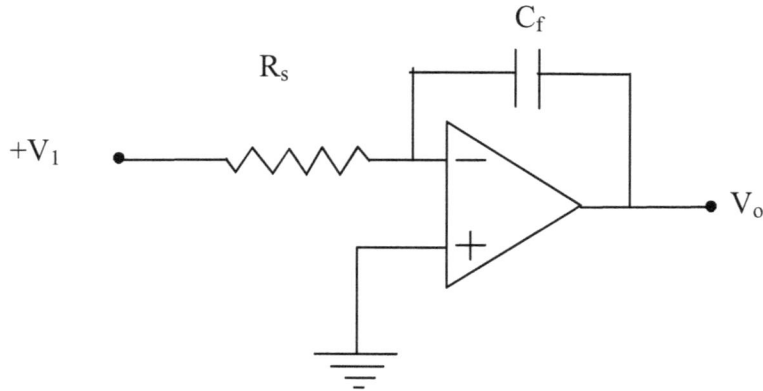

Figure 2.48.

17.) The circuit in Figure 2.48 is:
A.) an inverting amplifier
B.) noninverting amplifier
C.) a differentiator
D.) an integrator

18.) The circuit in Figure 2.48 has an output V_o:
A.) that is proportional to the derivative of the input V_1
B.) that is proportional to the input V_1
C.) that is proportional to the integral of the input V_1
D.) that is equal to the input V_1

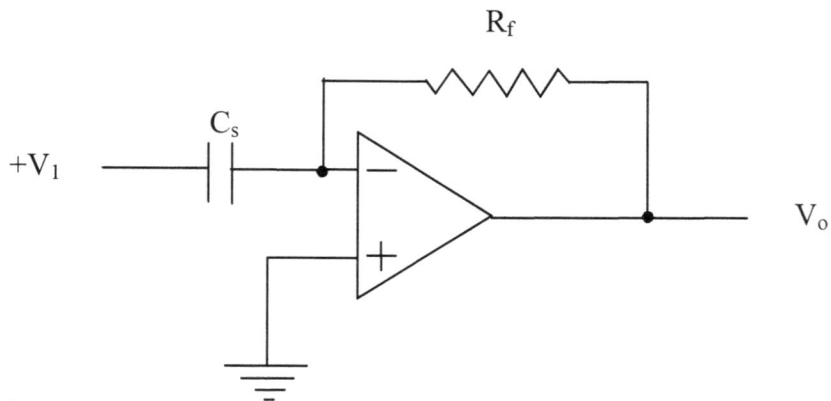

Figure 2.49.

19.) The circuit in Figure 2.49 is:
A.) an integral controller
B.) noninverting amplifier
C.) a differentiator
D.) an integrator

20.) The circuit in Figure 2.49 has an output V_o:
A.) that is proportional to the derivative of the input V_1
B.) that is proportional to the input V_1
C.) that is proportional to the integral of the input V_1
D.) that is equal to the input V_1

21. A differentiator is a an operational amplifier circuit with an output that is
       A.) the integral of its output
       B.) the derivative of its output
       C.) directly equal to its input
       C.) the derivative of its input

22. The circuit in Figure 2.51B is a(n)
       A.) integrator
       B.) differentiator
       C.) differential amplifier
       D.) summing amplifier

23. In Figure 2.51A, if Rs=1000 ohms, R_f=10000 ohms, and the input voltage is V_1 shown in Figure 2.50,

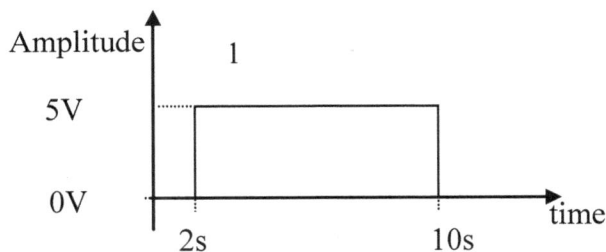

Figure 2.50.

what is the output Vo?

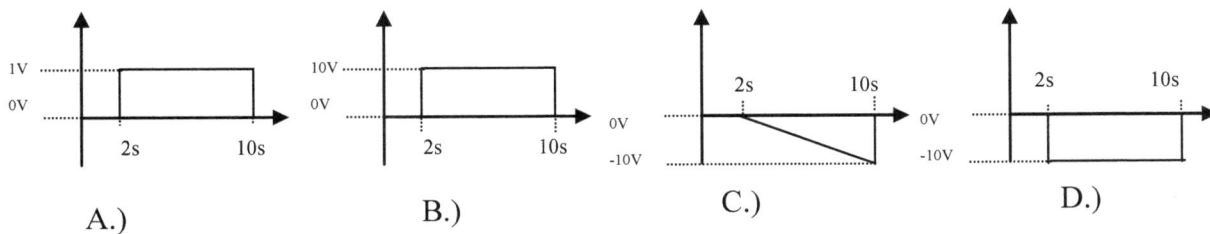

A.)   B.)   C.)   D.)

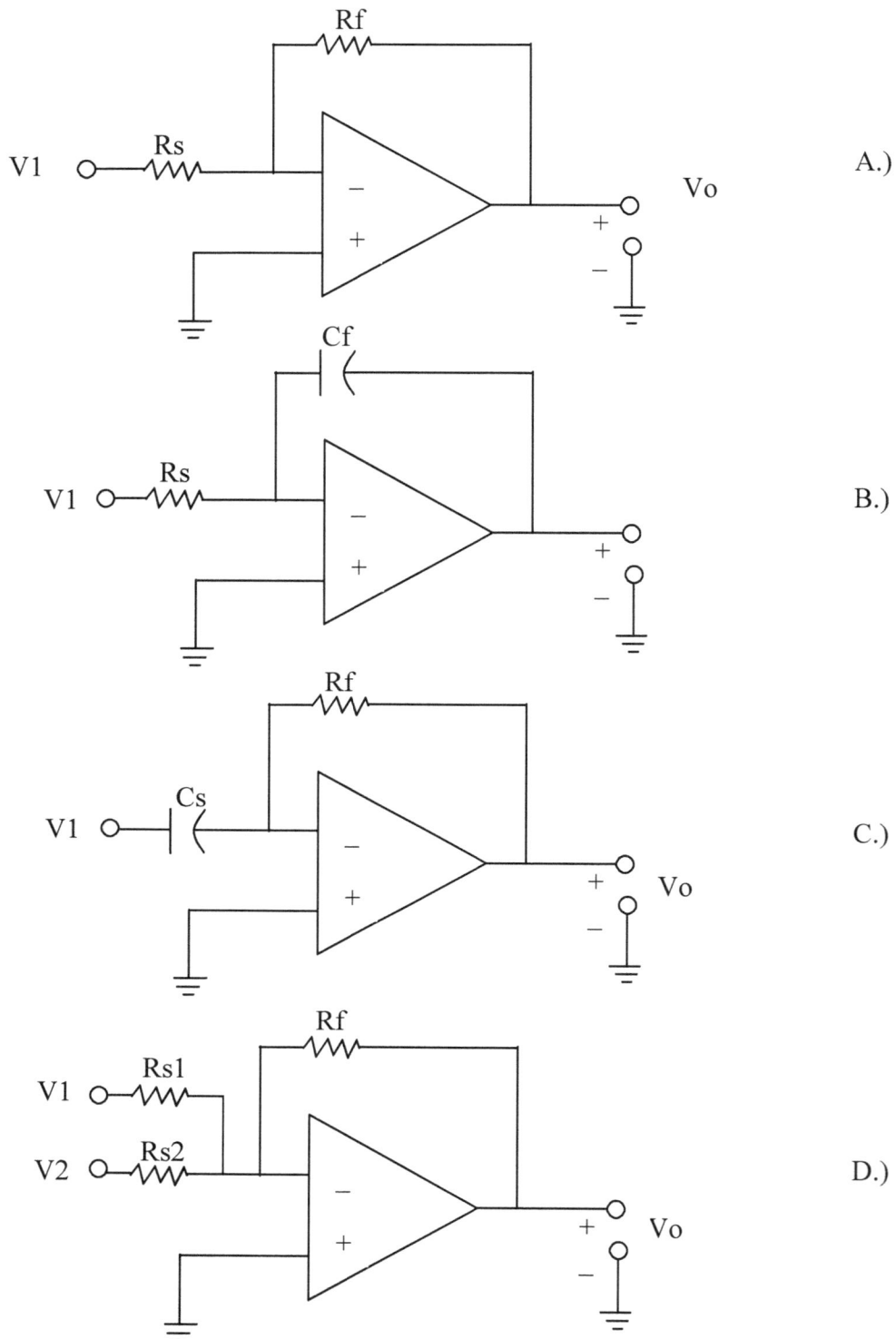

Figure 2.51. OPAMPs.

24. In Figure 2.51B, if the input voltage is V1 shown in Figure 2.50, what is the output voltage Vo?

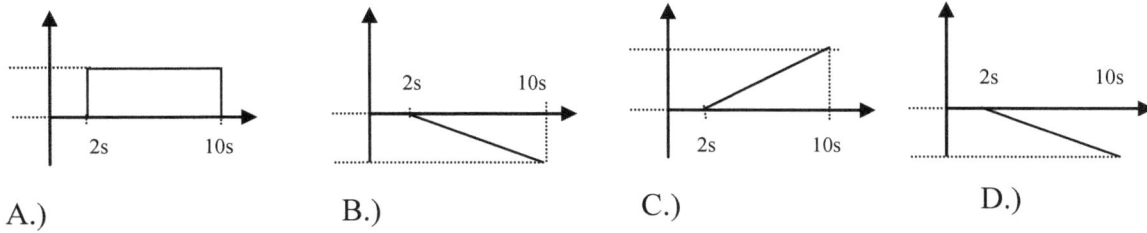

A.)   B.)   C.)   D.)

25. In Figure 2.51C, if the input voltage is V1 shown in Figure 2.50, what is the output voltage Vo?

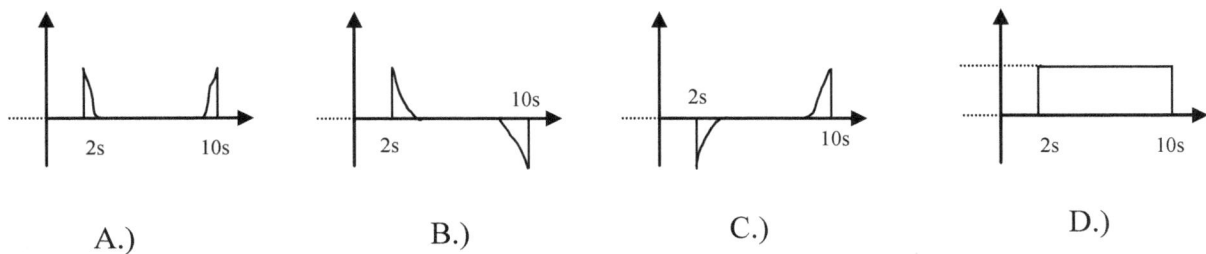

A.)   B.)   C.)   D.)

26. Consider Figure 2.51C. If the input voltage V1 is a ramp like function as shown in Figure 4.48, then the output voltage Vo is

        A.) a ramp like voltage
        B.) zero
        C.) a DC voltage
        D.) parabolic like voltage

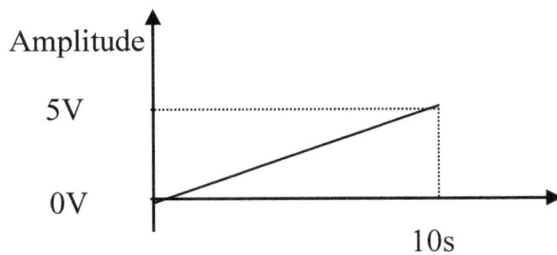

Figure 2.52

27. Consider Figure 2.51A. If $R_s=R_f$ the gain will be equal to

        A.) 0
        B.) -1
        C.) 1
        D.) Vo

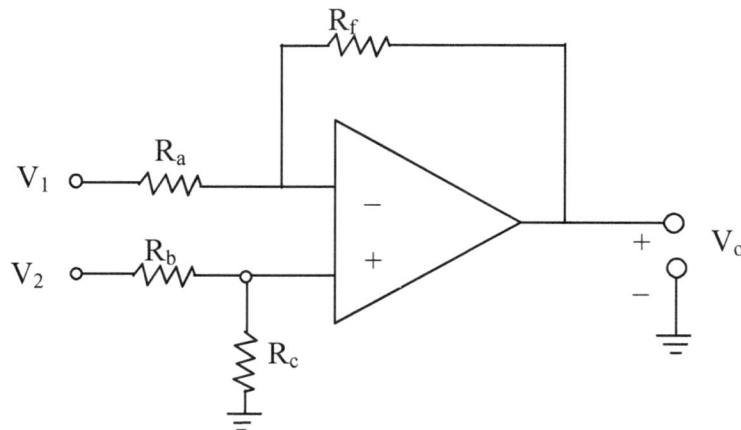

Figure 2.53.

28. The circuit in Figure 2.53 is a(n)
    A.) difference amplifier
    B.) instrumentation amplifier
    C.) subtractor
    D.) summer

29. Consider the circuit in Figure 2.53. If R_f =R_c= 1000 ohms and R_a=R_b = 1000 ohms, the gain of the system is
    A.) -1
    B.) –2
    C.) 2
    D.) 1

30. Consider the circuit in Figure 2.53. If R_f=R_c= 1000 ohms and R_a=R_b = 100 ohms, the gain of the system is
    A.) -10
    B.) –100
    C.) 10
    D.) 100

31. Consider the circuit in Figure 2.53. If R_f =R_c= 1000 ohms, R_a=R_b = 100 ohms, V_1= 5 volts, and V_2 = 10 volts, the output voltage V_o is
    A.) -50
    B.) -100
    C.) 50
    D.) 500

3. Number Systems

Some common number systems are shown in Table 3.1. These include decimal, binary, octal, and hexadecimal number systems.

Decimal	Binary	Octal	Hexadecimal
0	0000	0	0
1	0001	1	1
2	0010	2	2
3	0011	3	3
4	0100	4	4
5	0101	5	5
6	0110	6	6
7	0111	7	7
8	1000	10	8
9	1001	11	9
10	1010	12	A
11	1011	13	B
12	1100	14	C
13	1101	15	D
14	1110	16	E
15	1111	17	F

Table 3.1. Common number systems.

The **base** or **radix** is used to distinguish one number system from another. The decimal number system has the base or radix 10. The base indicates the number of digits that will be used. A subscript is usually used to indicate the number systems being used. The notation 100_{10} indicates a value of 100 with base 10. Therefore, the decimal number system is being used. It has decimal value 100. Also, 100_{10} has three digits. The notation 100_2 indicates a value of 100 with base 2. Therefore, the binary number system with three digits is being used. The decimal value of $100_2 = 1 \times 2^2 + 0 \times 2^1 + 0 \times 2^0 = .4 + 0 + 0 = 4_{10}$. The notation 100_{16} indicates a value of 100 with base 16. Therefore, the hexadecimal number system with three digits is being used. The decimal value of $100_{16} = 1 \times 16^2 + 0 \times 16^1 + 0 \times 16^0 = 256 + 0 + 0 = 256_{10}$. So, the position of a number in a number system determines the weight of that digit. So, using 100_{16} from above, the number in position 2 starting from the right, 1, has weight 16^2; the number in position 1 starting from the right, 0, has weight 16^1; the number in position 0 starting from the right, 0, has weight 16^0. So, the decimal equivalent of the number is obtained by summing the multiplication of the digit (when applicable, converted to its decimal value) and its corresponding weight in the number. Mathematically, speaking,

$$D = a^{(n-1)} \times w^{(n-1)} + \ldots + a^1 \times w^1 + a^0 \times w^0$$

EQ. 3.1

where D is the decimal sum, $a^{(n-1)}$, a^1, and a^0 are the digits of the nonfractional number system, and $w^{(n-1)}$, w^1, w^0 are the weights of the number system starting from the right most position.

EXAMPLE 3.1

Convert 1010_2 to a decimal number.

Solution:
$$1010_2 = 1 \times 2^3 + 0 \times 2^2 + 1 \times 2^1 + 0 \times 2^0$$
$$= 1 \times 8 + 0 \times 4 + 1 \times 2 + 0 \times 1$$
$$= 8 + 0 + 2 + 0$$
$$= 10_{10}$$

EXAMPLE 3.2

Convert 1111_2 to a decimal number.

Solution:
$$1111_2 = 1 \times 2^3 + 1 \times 2^2 + 1 \times 2^1 + 1 \times 2^0$$
$$= 1 \times 8 + 1 \times 4 + 1 \times 2 + 1 \times 1$$
$$= 8 + 4 + 2 + 1$$
$$= 15_{10}$$

EXAMPLE 3.3

Convert $1AF_{16}$ to a decimal number.

Solution:
$$1AF_{16} = 1 \times 16^2 + A \times 16^1 + F \times 16^0$$
$$= 1 \times 256 + 10 \times 16 + 15 \times 1$$
$$= 256 + 160 + 15$$
$$= 431_{10}$$

3.1. From Binary Fractional Numbers to Decimal Fractional Numbers

In the previous Section we went from whole binary numbers to whole decimal numbers. In this Section we will go a step further by adding the fractional part. Binary fractions are treated the same way as decimal fractions, for example, the decimal number 4.851_{10} is represented as follows:

Solution:

$$4.851_{10} = 4 \times 10^0 + 8 \times 10^{-1} + 5 \times 10^{-2} + 1 \times 10^{-3}$$

$$= 4 \times 1 + 8 \times \frac{1}{10} + 5 \times \frac{1}{100} + 1 \times \frac{1}{1000}$$

$$= 4 + .8 + .05 + .001$$

$$= 4.851_{10}$$

EXAMPLE 3.4

Convert 1110.101_2 to decimal.

Solution:

$$1110.101_2 = 1 \times 2^3 + 1 \times 2^2 + 1 \times 2^1 + 1 \times 2^0. \ 1 \times 2^{-1} + 0 \times 2^{-2} + 1 \times 2^{-3}$$

$$= 1 \times 2^3 + 1 \times 2^2 + 1 \times 2^1 + 1 \times 2^0. \ 1 \times 2^{-1} + 0 \times 2^{-2} + 1 \times 2^{-3}$$

$$= 1 \times 8 + 1 \times 4 + 1 \times 2 + 1 \times 1. \ 1 \times .5 + 0 + 1 \times .125$$

$$= 8 + 4 + 2 + 0 + .5 + 0 + .125$$

$$= 14.625_{10}$$

3.2. From Decimal Numbers to Binary Numbers

There are two ways to convert from a decimal number to a binary number:
  A. Sum of powers of 2
  B. Repeated division by 2

Sum of powers of 2 is done by adding up powers of 2 using Table 3.2 or similar. A similar Table can be generated for the base 2^{-n}, 16^{-n} systems, where n=1,2,3, etc., or 2^n, 16^n, where n=0, 1,2,3, …

Powers of 2	2^8	2^7	2^6	2^5	2^4	2^3	2^2	2^1	2^0
Place Value	256	128	64	32	16	8	4	2	1

Table 3.2. Table of place values for powers of 2.

EXAMPLE 3.5

Use sum of powers of 2 to convert 58_{10} to binary.

Solution:

Using Table 3.2, choose powers of 2 with the greatest place value less than 58. This is the 32s place. Set this bit to 1 as shown below. Then add the remaining bits, setting them to 1, until the total value adds to 58.

128	64	32	16	8	4	2	1
0	0	1	1	1	0	1	0

Thus, $58_{10} = 111010_2$

EXAMPLE 3.6

Use repeated division by 2 to convert 58_{10} to binary.

Solution:

Repeated division of 58_{10} by 2 gives the following:

$$
\begin{array}{lll}
58 \div 2 = 29 & \text{rem} \quad 0 & \text{LSB} \\
29 \div 2 = 14 & \text{rem} \quad 1 & \\
14 \div 2 = 7 & \text{rem} \quad 0 & \\
7 \div 2 = 3 & \text{rem} \quad 1 & \\
3 \div 2 = 1 & \text{rem} \quad 1 & \\
1 \div 2 = 0 & \text{rem} \quad 1 & \text{MSB}
\end{array}
$$

where the dot above the LSB bit (0 in this case) is the decimal point and rem ≡ remainder. For example, the rem for 58÷2=29 is obtained by multiplying 29 by 2 to get 58 and subtracting 58-58=0 remainder, the rem for 29÷2=14 is obtained by multiplying 14 by 2 to get 28 and subtracting 29-28=1 remainder, the rem for 14÷2=7 is obtained by multiplying 7 by 2 to get 14 and subtracting 14-14=0 remainder, the rem for 7÷2=3 is obtained by multiplying 3 by 2 to get 6 and subtracting 7-6=1 remainder 1, and so on.

Thus, $58_{10} = 111010.0_2$

3.3. **From Decimal Fractions to Binary Fractions**
Decimal fractions can be converted to binary fractions by repeated multiplication by 2.

EXAMPLE 3.7

Use repeated multiplication by 2 to convert $.375_{10}$ to binary.

Solution:

Repeated multiplication of .375$_{10}$ by 2 gives the following:

$$\bullet$$

$$375\times2 \;\; =0.75 \;\; \rightarrow \;\; 0 \qquad \text{MSB}$$
$$.75\times2 \;\; =1.5 \;\; \rightarrow \;\; 1$$
$$.5\times2 \;\; =1.0 \;\; \rightarrow \;\; 1 \qquad \text{LSB}$$

The first multiplication is 375×2=0.75. Take the 0 or 1 from the left of the decimal point to be the LSB bit of the binary number. The second multiplication is obtained by taking the decimal portion (discarding the integer portion) of the previous multiplication and multiplying it by 2 to get .75×2=1.5, then take the 0 or 1 from the left of the decimal point, 1, to be the next bit of the binary number, and so on. Thus, the final result is:

$$0.375_{10} = 0.110_{2}$$

For decimal numbers that have an integer part and a fractional part, e.g., 5.252$_{10}$, perform the following steps:
A. Convert the nonfractional part to binary using repeated division by 2.
B. Convert the fractional part to binary using repeated multiplication by 2.
C. Combine the results of step A and B.

3.4. From Decimal Numbers to Two's Compliment Numbers

Up until now we have considered only positive numbers. We must also consider negative numbers since microprocessors process positive and negative numbers. The 2's compliment method allows you to represent the sign as well as the magnitude of a positive or a negative number. Consider the 8 bit binary number representation in Figure 3.1.

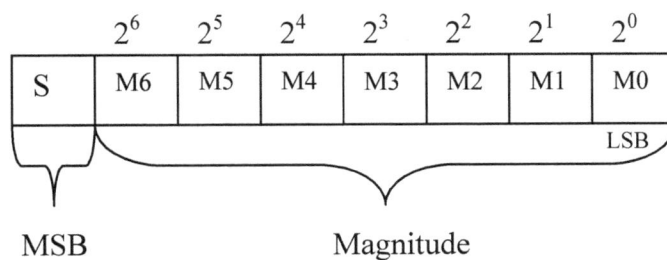

Figure 3.1. 8 bit 2's compliment number format.

In Figure 3.1, S is the sign bit. It is the most significant bit (MSB) while 2^{0} is the least significant bit (LSB). S is 0 for a positive number or 1 for a negative number. M0, M1, M2, … collectively, is the binary representation of the magnitude starting with the LSB bit 0, M0, then bit 1, M1, etc. The procedure to convert a decimal number to the 2's compliment n bit format is as follows:

1. Separate sign and magnitude of the decimal number
2. If the number is positive, convert the magnitude of decimal number to its n-1 bit binary number, put 0 in the S location, and stop here and let the binary number be the 2's compliment, otherwise if it is negative go to the next step.
3. Convert the magnitude of the decimal number to its n-1 bit binary number and put 1 in the S (MSB) bit position since the sign is negative.
4. Convert the binary number magnitude to its 1's compliment. The 1's compliment is inverting each 0 to get a 1 and inverting each 1 to get a zero.
5. Convert the 1's compliment to the 2's compliment. The 2's compliment is obtained by adding a 1 to the 1's compliment.

EXAMPLE 3.8

Convert -1 to its 2's compliment using an 8 bit format.

Solution:

The procedure to convert -1 to the 2's compliment format is as follows:

1. Separate sign and magnitude of -1
2. The number is not positive, go to the next step.
3. Convert the magnitude, 1, of the decimal number to its 7 bit binary number and put a 1 in the MSB bit position since the sign is negative. The result is:

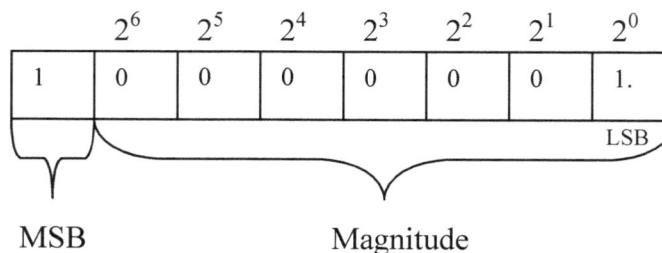

	2^6	2^5	2^4	2^3	2^2	2^1	2^0
1	0	0	0	0	0	0	1.
							LSB

MSB   Magnitude

4. Convert the binary number to its 1's compliment as follows:

	2^6	2^5	2^4	2^3	2^2	2^1	2^0	
1	1	1	1	1	1	1	0.	

LSB

MSB   Magnitude

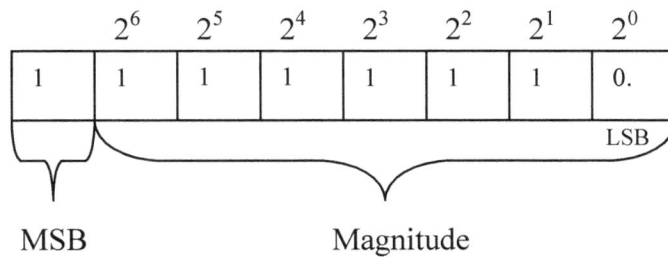

5. Convert the 1's compliment to the 2's compliment:

	2^6	2^5	2^4	2^3	2^2	2^1	2^0
1	1	1	1	1	1	1	0.
+ 0	0	0	0	0	0	0	1.
= 1	1	1	1	1	1	1	1.

3.5. From Two's Compliment Numbers to Decimal Numbers

The procedure to convert a 2's compliment n bit number to a decimal number is as follows:

1. Separate the sign bit (S) from the magnitude part of the 2's compliment : 0 => positive decimal number, 1=> negative decimal number.
2. If S=0, convert the magnitude part of the two's compliment number to its decimal number and stop here. This is the decimal number representing the two's compliment. Otherwise if S=1 go to the next step.
3. Take the 1's compliment of the magnitude part
4. Add 1 to the 1's compliment number to get the raw binary number
5. Convert the raw binary number to a decimal number

EXAMPLE 3.9

Find the decimal number of the 2's compliment number, 11111000.

Solution:

1. Separate the sign bit, S, from the magnitude of the 2's compliment number, in this case the MSB is a 1 => S=1 is a negative number. The remaining number is the magnitude portion as follows:

	1	1	1	1	0	0	0.

2. Take the 1's compliment of the magnitude portion

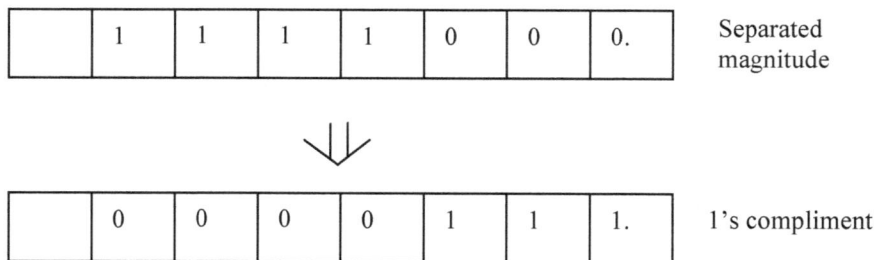

	1	1	1	1	0	0	0.	Separated magnitude

$$\Downarrow$$

	0	0	0	0	1	1	1.	1's compliment

3. Add 1 to the 1's compliment number result in step 2.

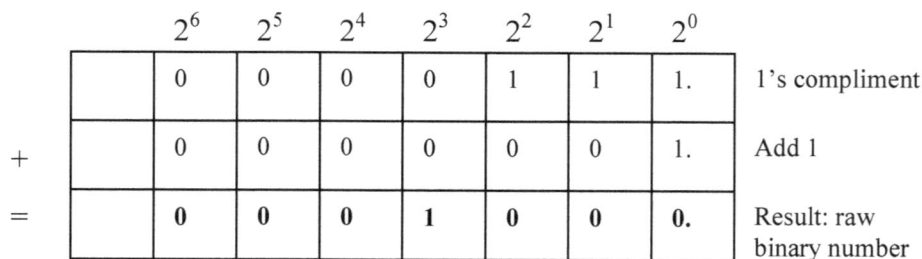

	2^6	2^5	2^4	2^3	2^2	2^1	2^0	
	0	0	0	0	1	1	1.	1's compliment
+	0	0	0	0	0	0	1.	Add 1
=	**0**	**0**	**0**	**1**	**0**	**0**	**0.**	Result: raw binary number

4. Convert the resulting binary number in step 3 to decimal.

$\therefore 0001000_2 = 8_{10}$, but from step 1 this is a negative number, so the final result is $= -8_{10}$,

3.6. **Binary Arithmetic**
The rules for binary addition are indicated below for rules 1, 2, 3, and 4. For more on binary arithmetic, see Section 4.12 on page 104 for designing digital circuitry for binary addition, etc.

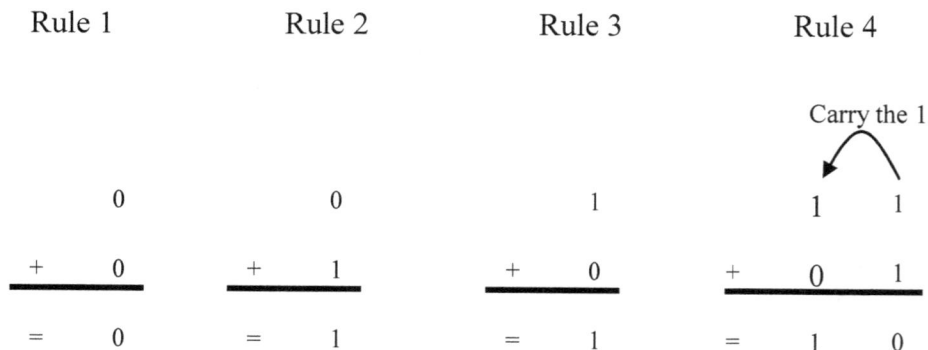

Rule 1	Rule 2	Rule 3	Rule 4
			Carry the 1
0	0	1	1 1
+ 0	+ 1	+ 0	+ 0 1
= 0	= 1	= 1	= 1 0

The rules are consistent with an XOR gate. That is, when the two binary numbers to be added are the same the result is 0, otherwise, when they are different the result is 1. However, for rule 4, when both are 1, the result is 0, but carry the 1 to the next column.

EXAMPLE 3.10

Perform the following decimal addition 4+2 in binary form.

Solution:

$$
\begin{array}{cccccc}
4 & & 0 & 1 & 0 & 0. \\
+ \quad 2 & \Rightarrow \quad + & 0 & 0 & 1 & 0. \\
\hline
6 & & 0 & 1 & 1 & 0. \\
\end{array}
$$

EXAMPLE 3.11

Perform the following decimal addition 6+3 in binary form. Show your work.

Solution:

Carry the 1

$$
\begin{array}{cccccc}
 & & 1 & 1 & & \\
6 & & 0 & 1 & 1 & 0. \\
+ \quad 3 & \Rightarrow \quad + & 0 & 0 & 1 & 1. \\
\hline
9 & & 1 & 0 & 0 & 1. \\
\end{array}
$$

PROBLEMS – Number systems

1. The hexadecimal number system uses the base 8.
   A.) True   B.) False

2. The binary number system uses the base 10.
   A.) True   B.) False

3. The octal number system uses the base 16.
   A.) True   B.) False

4. The decimal number system uses the base 2.
   A.) True   B.) False

5. The base represents the number of alphabets in a number system.
   A.) True   B.) False

6. The binary number system consists of 2 symbols.
   A.) True   B.) False

7. The octal number system consists of 7 symbols.
   A.) True   B.) False

8. The hexadecimal number system consists of 15 symbols.
   A.) True   B.) False

9. The decimal number system consists of 9 symbols.
   A.) True   B.) False

10. To convert a fractional decimal number to a fractional binary number you must perform which of the following:
   A.) A repeated multiplication process where the integers are repeatedly multiplied by 2 until the product becomes 1.00.
   B.) A repeated division process where the quotients are repeatedly divided by 16 until the quotient becomes 0 with a remainder of 1.
   C.) A repeated division process where the quotients are repeatedly divided by 2 until the quotient becomes 0 with a remainder of 1
   D.) A repeated multiplication process where the fractional parts are repeatedly multiplied by 2 until the product becomes 1.00.

11. Given the binary number 0111_2 the least significant bit (LSB) is 0.
   A.) True   B.) False

12. Given the binary number 10101_2 the most significant bit has a weight or place value of

   A.) 16   B.) 8   C.) 2   D.) 1

13. The decimal equivalent of 10001_2 is

   A.) 17_{10} B.) 16_{10} C.) 41_{10} D.) 12_{10}

14. The decimal number equivalent of 1000010_2 is

   A.) $10^6 + 10^1$
   B.) $2^6 + 2^0$
   C.) $2^6 + 2^1$
   D.) $10^6 + 10^0$

15. The decimal equivalent of $.11001_2$ is

   A.) $2^0 + 2^{-1} + 2^0$
   B.) $2^{-0} + 2^{-1} + 2^{-4}$
   C.) $2^{-1} + 2^{-2} + 2^{-5}$
   D.) $2^0 + 2^{-1} + 2^{-5}$

16. The hexadecimal number 01_{16} is equivalent to

   A.) 00000001_2
   B.) 00001111_2
   C.) 10000000_2
   D.) 11111111_2

17. Consider the decimal number 7312_{10}. It is equivalent to which of the following

   A.) $7 \times 2^3 + 3 \times 2^2 + 1 \times 2^1 + 2 \times 2^0$
   B.) $7 \times 10^3 + 3 \times 10^2 + 1 \times 10^1 + 2 \times 10^0$
   C.) $7 \times 16^3 + 3 \times 16^2 + 1 \times 16^1 + 2 \times 16^0$
   D.) $7 \times 8^3 + 3 \times 8^2 + 1 \times 8^1 + 2 \times 8^0$

18. The decimal number 15_{10} is equivalent to

   A.) 10000_2
   B.) 11110_2
   C.) 10110_2
   D.) 01111_2

19. The decimal number 11_{10} is equivalent to

   A.) $0F_{16}$
   B.) $0C_{16}$
   C.) $0B_{16}$
   D.) $0E_{16}$

20. The binary number 00011010111110001_2 is equivalent to
     A.) $9AF1_{16}$
     B.) $9FA1_{16}$
     C.) $1AF9_{16}$
     D.) $1FA_{16}$

21. The decimal equivalent of FAB1 is
     A.) $Fx16^0 + Ax16^1 + Bx16^2 + 1x16^3$
     B.) $Fx2^0 + Ax2^1 + Bx2^2 + 1x2^3$
     C.) $1x2^0 + Bx2^1 + Ax2^2 + Fx2^3$
     D.) $1x16^0 + Bx16^1 + Ax16^2 + Fx16^3$

22. The decimal number 16_{10} is equivalent to
     A.) 10_{16}
     B.) $1F_{16}$
     C.) $F1_{16}$
     D.) 11_{16}

4. Combinational Logic Circuits

Digital electronics is categorized into two general parts:
    1   Combinational logic circuits, and
    2   Sequential logic circuits.

Combinational logic circuits are created from three basic digital elements:
1. and gate
2. or gate
3. not gate

These three gates are the foundation of combinational logic circuits and will be discussed in this chapter. This Chapter shall also provide the basic concepts for understanding the design and implementation of combinational logic circuits such as encoders, decoders, multiplexers, demultiplexers, and arithmetic circuits.

Sequential logic circuits are created from one digital element called the SR latch. The SR latch is the foundation of sequential logic circuits and will be discussed in the next chapter.

4.1. The NOT Gate

The NOT symbol is shown in Figure 4.2.

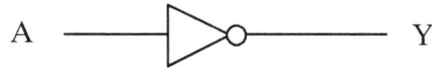

Figure 4.1. The symbol for the NOT gate.

The Boolean expression for the NOT symbol in Figure 4.2 is:

$$Y = \overline{A}$$

                        EQ. 4.1

The truth Table for the NOT gate is shown in Table 4.1.

A	Y
0	1
1	0

Table 4.1. NOT Gate Truth Table.

The VHDL expression for the NOT gate is:

```
Y <= not A;
```

4.2. **Boolean Multiplication - The AND Gate**

The AND symbol is shown in Figure 4.2.

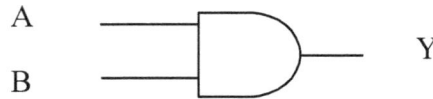

Figure 4.2. The symbol for the AND gate.

The Boolean expression for the AND symbol in Figure 4.2 is:

Y=AB

EQ. 4.2

Also expressed as:

$Y=A \cdot B$

EQ. 4.3

B	A	Y
0	0	0
0	1	0
1	0	0
1	1	1

Table 4.2. AND Gate Truth Table.

Boolean multiplication is equivalent to the AND operation. In a **product term** each literal or individual variable is separated by the AND operation. Examples are:

1. Y=AB
2. Y=ABC
3. $Y = \overline{AB}$

Examples that are not product terms are:

1. Y=A+BC
2. Y=AB+C

The VHDL expression for the AND gate is:

```
Y <= A and B;
```

4.3. Boolean Addition - The OR gate

The OR symbol is shown in Figure 4.3:

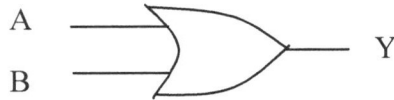

Figure 4.3. The symbol for the OR gate.

The Boolean expression for the OR symbol is:

Y=A+B   **EQ. 4.4**

B	A	Y
0	0	0
0	1	1
1	0	1
1	1	1

Table 4.3. OR Gate Truth Table.

Boolean addition is equivalent to the OR operation. In Boolean algebra, a **sum term** is a Boolean expression where each Boolean variable use OR operation, +, between them with no AND operation. Examples of sum terms are:

1. Y= A+B
2. Y=A+B+C
3. Y=A+B+C+D
4. $Y = \overline{A} + B$

Examples of sum terms that are not sum terms are:

1. Y=AB
2. Y=AB+C

The VHDL expression for the OR gate is:

```
Y <= A or B;
```

4.4. **Boolean Laws**

Boolean laws allow simplification of Boolean expressions. The laws are based Boolean operations relative to the following:
1. NOT Gate
2. AND Gate
3. OR Gate
4. Communicative properties
5. Associative properties
6. Distributive properties
7. DeMorgan's theorems

4.4.1. NOT Gate Laws

NOT laws are shown in Table 4.4.

NOT Law	NOT Gate Representation
$\bar{0} = 1$	
$\bar{1} = 0$	
$\bar{\bar{x}} = x$	

Table 4.4. NOT Gate Laws.

4.4.2. AND Gate Laws

The AND laws are listed in Table 4.5.

AND Law	AND Gate Representation
$x \cdot 0 = 0$	X 0 ⟩ 0
$x \cdot 1 = x$	X 1 ⟩ X
$x \cdot x = x$	X X ⟩ X
$x \cdot \overline{x} = 0$	X $\overline{x}$ ⟩ 0

Table 4.5. AND Gate Laws.

4.4.3. OR Gate Laws

The OR laws are listed in Table 4.6.

OR Law	OR Gate Representation
$x + 0 = x$	
$x + 1 = 1$	
$x + x = x$	
$x + \overline{x} = 1$	

Table 4.6. OR Gate Laws.

4.5. **Boolean Properties and Rules**

A mathematical operation is communicative if it can be applied to its operand in any order and still yield the same result. A mathematical operation is associative if its operands can be grouped in any order without affecting the results. The distributive property allows the AND property to multiply throughout its expression. Some common properties and rules are given in Table 4.7.

Applicable Property	Rules
Communicative	Rule 1: $xy = yx$
	Rule 2: $x + y = y + x$
Associative	Rule 3: $(xy)z = x(yz) = (xz)y$
	Rule 4: $x + y + z = x + (y + z) = (x + y) + z$
Distributive	Rule 5: $x(y + z) = xy + xz$
	Rule 6: $\begin{aligned}(x + y)(u + z) &= (x + y)u + (x + y)z \\ &= xu + yu + xz + yz\end{aligned}$
----	Rule 7: $x + xy = x$
	Rule 8: $x + \overline{x}y = x + y$
	Rule 9: $(x + y)(x + z) = x + yz$

Table 4.7. Common Properties and Rules.

EXAMPLE 4.1

Show that A+AB = A

$$A + AB = A(1 + B)$$
$$= A \cdot 1$$
$$= A$$

EXAMPLE 4.2

Show that A+ $\overline{A}$ B = A+B

$$A + \overline{A}B = A(A + AB) + \overline{A}B$$
$$= AA + AB + \overline{A}B$$
$$= AA + AB + \overline{A}A + \overline{A}B$$
$$= A(A + B) + \overline{A}(A + B)$$
$$= (A + \overline{A})(A + B)$$
$$= A + B$$

EXAMPLE 4.3

Show that $(A+B)(A+C)=A+BC$

$$(A+B)(A+C)= AA+AC+BA+BC$$
$$= A+AC+BA+BC$$
$$= A(1+C)+BA+BC$$
$$= A+BA+BC$$
$$= A(1+B)+BC$$
$$= A+BC$$

EXAMPLE 4.4

Simplify $AB+A(B+C)+B(B+C)$

$$AB+A(B+C)+B(B+C)= AB+AB+AC+BB+BC$$
$$= AB+AC+B+BC$$
$$= AB+AC+B$$
$$= (AB+B)+AC$$
$$= B+AC$$

EXAMPLE 4.5

Simplify $\left[A\overline{B}(C+BD)+\overline{A}\,\overline{B}\right]C$

$$\left[A\overline{B}(C+BD)+\overline{A}\,\overline{B}\right]C = \left[A\overline{B}C+A\overline{B}BD+\overline{A}\,\overline{B}\right]C$$
$$= A\overline{B}CC+A\overline{B}BDC+\overline{A}\,\overline{B}C$$
$$= A\overline{B}C+0+\overline{A}\,\overline{B}C$$
$$= \overline{B}C(A+\overline{A})$$
$$= \overline{B}C(1)$$
$$= \overline{B}C$$

4.6. **DeMorgan's Theorem**

DeMorgan's theorem states that the following:

$$\overline{x \cdot y} = \overline{x} + \overline{y}$$

EQ. 4.5

and

$$\overline{x + y} = \overline{x} \cdot \overline{y}$$

EQ. 4.6

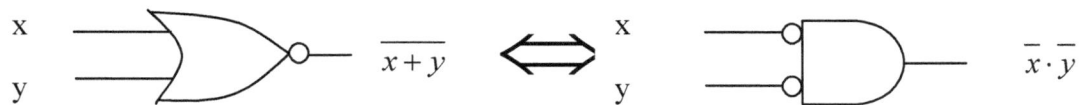

EXAMPLE 4.6

Use DeMorgan's Theorem to simplify $\overline{AB + AC} + \overline{A}\,\overline{B}\,C$

$$\overline{AB + AC} + \overline{A}\,\overline{B}\,C = \overline{\overline{(\overline{A} + \overline{B})(\overline{A} + \overline{C})}} + \overline{A}\,\overline{B}\,C$$
$$= (\overline{A} + \overline{B})(\overline{A} + \overline{C}) + \overline{A}\,\overline{B}\,C$$
$$= \overline{A} + \overline{B}\,\overline{C} + \overline{A}\,\overline{B}\,C$$
$$= A(1 + \overline{B}\,C) + \overline{B}\,\overline{C}$$
$$= A + \overline{B}\,\overline{C}$$

4.7. **Digital Circuit to Boolean Expression**

A digital circuit can be represented by a Boolean equation. The Boolean equation can then be used to derive the truth table of the digital circuit.

EXAMPLE 4.7

Find the Boolean expression and truth table of the following circuit.

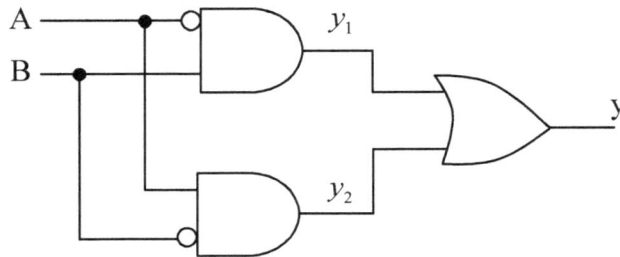

Figure 4.4. EXAMPLE 4.7 digital circuit.

Boolean Expressions:

$y = y_1 + y_2$

$y_1 = \overline{A}B$

$y_2 = A\overline{B}$

$\therefore \quad y = y_1 + y_2 = \overline{A}B + A\overline{B}$

Truth Table:

B	A	y
0	0	$y = \overline{A}B + A\overline{B} = \overline{0}0 + 0\overline{0} = 0 + 0 = 0$
0	1	$y = \overline{A}B + A\overline{B} = \overline{1}0 + 1\overline{0} = 0 + 1 = 1$
1	0	$y = \overline{A}B + A\overline{B} = \overline{0}1 + 0\overline{1} = 1 + 0 = 1$
1	1	$y = \overline{A}B + A\overline{B} = \overline{1}1 + 1\overline{1} = 0 + 0 = 0$

Table 4.8. EXAMPLE 4.4 truth table.

VHDL Expression:
Y<= ((not B) and A) or (B and (not A));

The circuit in Figure 4.4 is equivalent to the following simplified circuit:

Figure 4.5. The XOR Gate.

4.8. Deriving the Boolean Expression from the Truth Table

A typical design problem involves the following steps:

1. Determine the truth Table

2. Derive the Boolean Expression from the truth Table using the sum of products (SOP) or product of sum (POS) method.

   Standard SOP examples: $y = \overline{A}\,\overline{B}C + ABC + A\overline{B}C$, $y = \overline{A}B + A\overline{B}$

   Standard POS examples: $y = (\overline{A} + \overline{B} + \overline{C})(A + B + C)$, $y = (\overline{A} + \overline{B})(A + B)$

3. Finally, deriving the AND/OR digital circuit for the SOP Boolean expression or the OR/AND digital circuit from the POS Boolean expression.

4.8.1. AND/OR Circuits

An example of an AND/OR logic circuit is shown below:

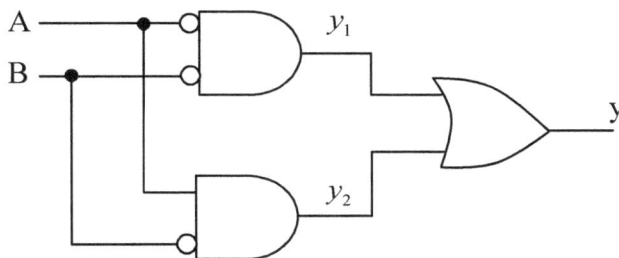

Figure 4.6. OR/AND Circuit.

It has the AND/OR Boolean expression given by:

$$y = \overline{A}\,\overline{B} + A\overline{B}$$

4.8.2. OR/AND Circuits

An example of an OR/AND logic circuit is shown below:

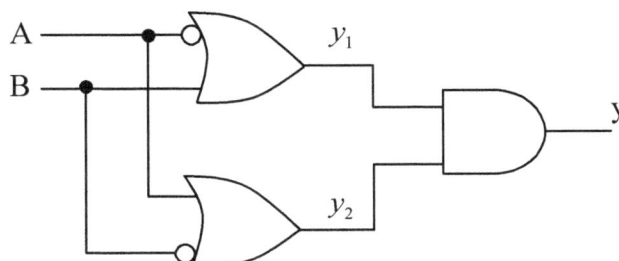

Figure 4.7. AND/OR Circuit.

It has the OR/AND Boolean expression given by:

$$y = \left(\overline{A}+B\right)\left(A+\overline{B}\right)$$

4.8.3. Sum of Products (SOP)

The SOP Boolean expression can be derived from the truth table as follows:
1. Select each row in the truth table where there is a 1 in the output column
2. Develop a term for the Boolean expression using the AND of the inputs. This is done for each row with a 1 in the output. Variables of the inputs that are high (1) are represented by no overbars while those that are low(0) are represented with an overbar.

EXAMPLE 4.8

A logic designer determined the following truth table for a digital circuit. Determine the :

1. Boolean expression using the SOP method,

2. and the corresponding digital circuit inside the box of Figure 4.8.

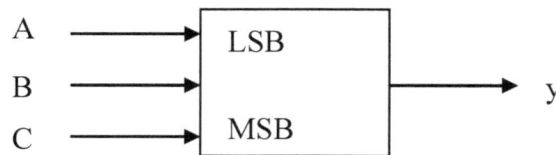

Figure 4.8. EXAMPLE 4.8 digital circuit.

C	B	A	y	
0	0	0	0	
0	0	1	0	
0	1	0	0	
0	1	1	1	y_1 term
1	0	0	0	
1	0	1	0	
1	1	0	1	y_2 term
1	1	1	0	

Table 4.9. Given truth table.

Solution - SOP Boolean expression:

$y_1 : 011 \Rightarrow \overline{C}BA$
$y_2 : 110 \Rightarrow CB\overline{A}$

$$\therefore \qquad y = y_1+y_2 = \overline{C}BA + CB\overline{A}$$

Solution - Digital Circuit:
The digital circuit for EXAMPLE 4.8 is given below in Figure 4.9.

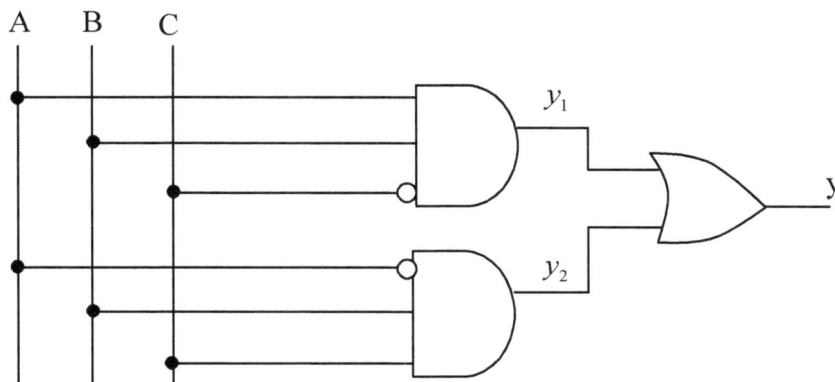

Figure 4.9. EXAMPLE 4.8 digital circuit.

4.9. Simplification Using Karnaugh Maps (K-Maps)

As seen previous sections Boolean laws can be used to simplify Boolean expressions. However, Karnaugh map, one of the easiest ways to simplify Boolean expressions can also be used. The Karnaugh map is a graphical method to simplify Boolean expressions using Boolean Theorems and laws. It is referred to as K-map. To simplify Boolean expressions using K-maps you must perform the following:

1. Determine the minterm Boolean expression (SOP) from the truth table. This is the unsimplified Boolean expression. This will be explained in more details in the following steps.
2. Determine the K-map from the truth table inputs, e.g., if there are 4 rows in the truth table, the K-map will consist of 4 boxes or cells, 8 rows => 8 cells, 16 rows => 16 cells, etc. Each box in the K-map must represent an input combination. Some examples are shown below.

B	A	y
0	0	1
0	1	1
1	0	0
1	1	0

$\Rightarrow$

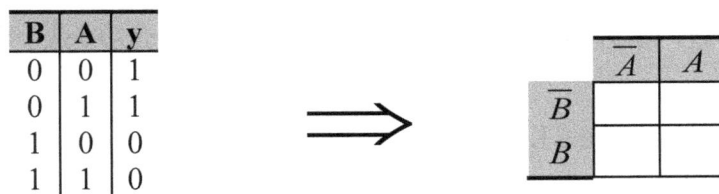

Figure 4.10. Relationship of 4 row truth table to 4 cell K-map.

C	B	A	y
0	0	0	0
0	0	1	1
0	1	0	1
0	1	1	1
1	0	0	0
1	0	1	1
1	1	0	0
1	1	1	1

	$\overline{A}$	A
$\overline{C}\,\overline{B}$		
$\overline{C}B$		
CB		
$C\overline{B}$		

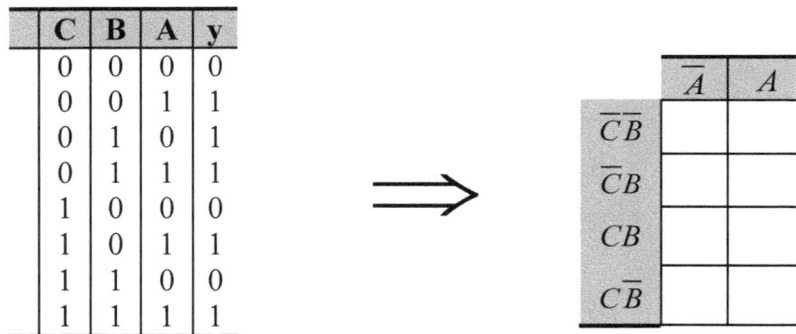

Figure 4.11. Relationship of 8 row truth table to 8 cell K-map.

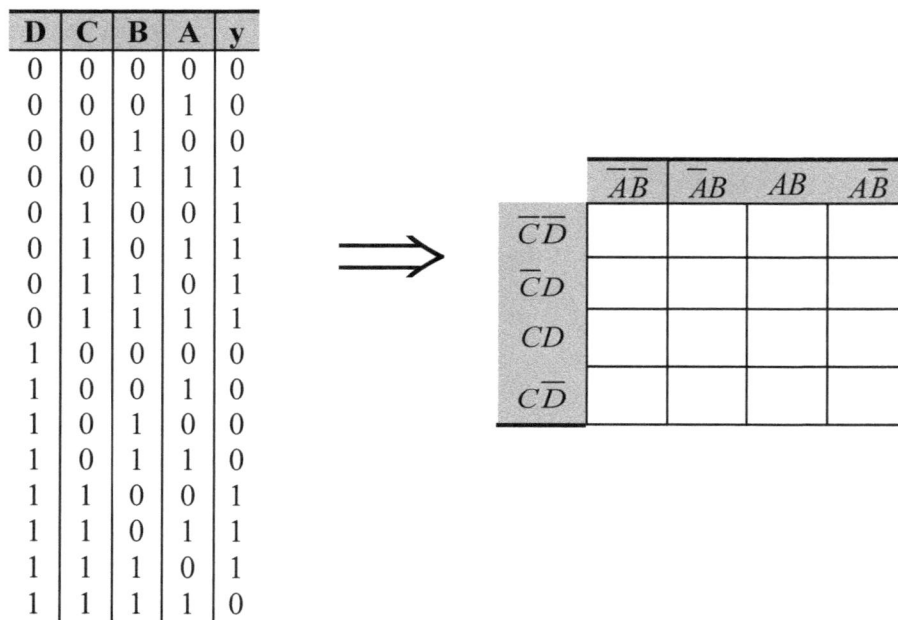

D	C	B	A	y
0	0	0	0	0
0	0	0	1	0
0	0	1	0	0
0	0	1	1	1
0	1	0	0	1
0	1	0	1	1
0	1	1	0	1
0	1	1	1	1
1	0	0	0	0
1	0	0	1	0
1	0	1	0	0
1	0	1	1	0
1	1	0	0	1
1	1	0	1	1
1	1	1	0	1
1	1	1	1	0

	$\overline{A}\,\overline{B}$	$\overline{A}B$	AB	$A\overline{B}$
$\overline{C}\,\overline{D}$				
$\overline{C}D$				
CD				
$C\overline{D}$				

Figure 4.12. Relationship of 16 row truth table to 16 cell K-map.

3. Enter the ones in the K-map corresponding to each minterm in the Boolean expression.

4. Loop adjacent groups of two, four, or eight 1s together.

5. Eliminate variables - when a variable and its complement are within a loop that variable is eliminated

6. OR the remaining variables. This is the simplified Boolean expression.

EXAMPLE 4.9

Find the simplified SOP (also called minterm) Boolean expression using the given truth table.

B	A	y	
0	0	0	
0	1	1	y_1
1	0	1	y_2
1	1	1	y_3

Figure 4.13. EXAMPLE 4.9 - Truth table.

Solution:

Step 1 - Unsimplified Boolean expression:

B	A	y		minterm
0	0	0		
0	1	1	y_1	$\overline{B}A$
1	0	1	y_2	$B\overline{A}$
1	1	1	y_3	BA

Figure 4.14. EXAMPLE 4.9 - Truth table minterms.

The unsimplified Boolean expression is:

$$y = y_1 + y_2 + y_3 = \overline{B}A + B\overline{A} + BA$$

Step 2 - Determine K-map:

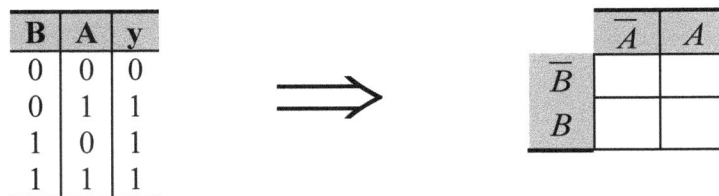

B	A	y
0	0	0
0	1	1
1	0	1
1	1	1

$\Longrightarrow$

	$\overline{A}$	A
$\overline{B}$		
B		

Figure 4.15. EXAMPLE 4.9 - Truth table to K-map.

Step 3 – Enter 1s into K-map:

B	A	y
0	0	0
0	1	1
1	0	1
1	1	1

$\Longrightarrow$

	$\overline{A}$	A
$\overline{B}$	0	1
B	1	1

Figure 4.16. EXAMPLE **4.9** - Fill K-map in.

Step 4 – Loop adjacent groups of 2, 4, or 8 ones:

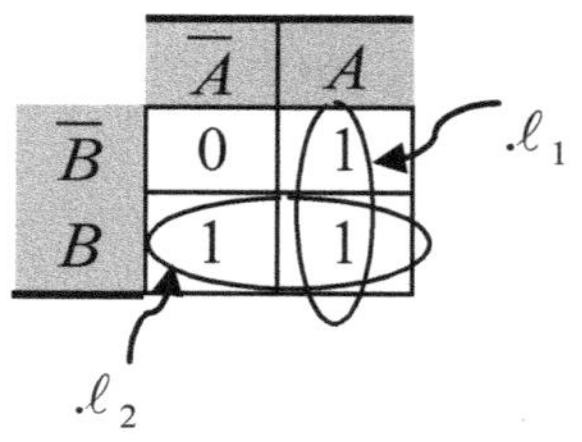

	$\overline{A}$	A
$\overline{B}$	0	1	$.\ell_1$
B	1	1

$.\ell_2$

Figure 4.17. EXAMPLE **4.9** - Loop adjacent 1s in K-map.

Step 5 – Eliminate variables that appear with their complement in a loop and save the others:

$\ell_1 = A$

$\ell_2 = B$

Step 6 – OR the remaining variables:

Therefore, the simplified Boolean expression is:
$y = \ell_1 + \ell_2 = A + B$

EXAMPLE 4.10

Find the simplified SOP (also call minterm) Boolean expression using the given truth table.

C	B	A	y
0	0	0	0
0	0	1	1
0	1	0	1
0	1	1	1
1	0	0	0
1	0	1	1
1	1	0	0
1	1	1	1

Figure 4.18. EXAMPLE 4.10 - Truth table.

Solution:

Step 1 - Unsimplified Boolean expression:

C	B	A	y		
0	0	0	0		
0	0	1	1	y_1	$\overline{C}\,\overline{B}A$
0	1	0	1	y_2	$\overline{C}B\overline{A}$
0	1	1	1	y_3	$\overline{C}BA$
1	0	0	0		
1	0	1	1	y_3	$C\overline{B}A$
1	1	0	0		
1	1	1	1	y_5	CBA

Figure 4.19. EXAMPLE 4.10 - Truth table minterms.

The unsimplified Boolean expression is:

$$y = y_1 + y_2 + y_3 = \overline{B}A + B\overline{A} + BA$$

Step 2 - Determine K-map:

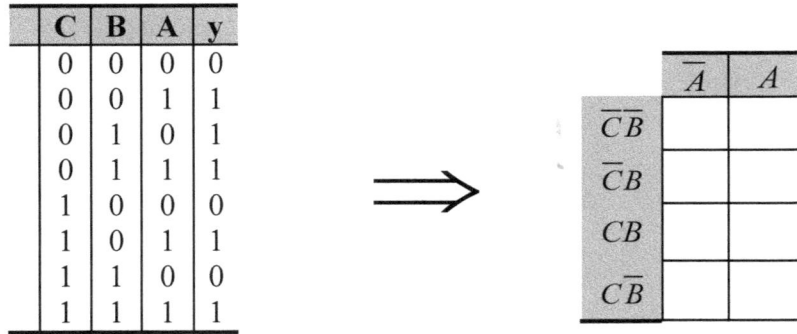

C	B	A	y
0	0	0	0
0	0	1	1
0	1	0	1
0	1	1	1
1	0	0	0
1	0	1	1
1	1	0	0
1	1	1	1

$\Longrightarrow$

	$\overline{A}$	A
$\overline{C}\,\overline{B}$		
$\overline{C}B$		
CB		
$C\overline{B}$		

Figure 4.20. EXAMPLE 4.10 - Truth table to K-map.

Step 3 – Enter 1s into K-map:

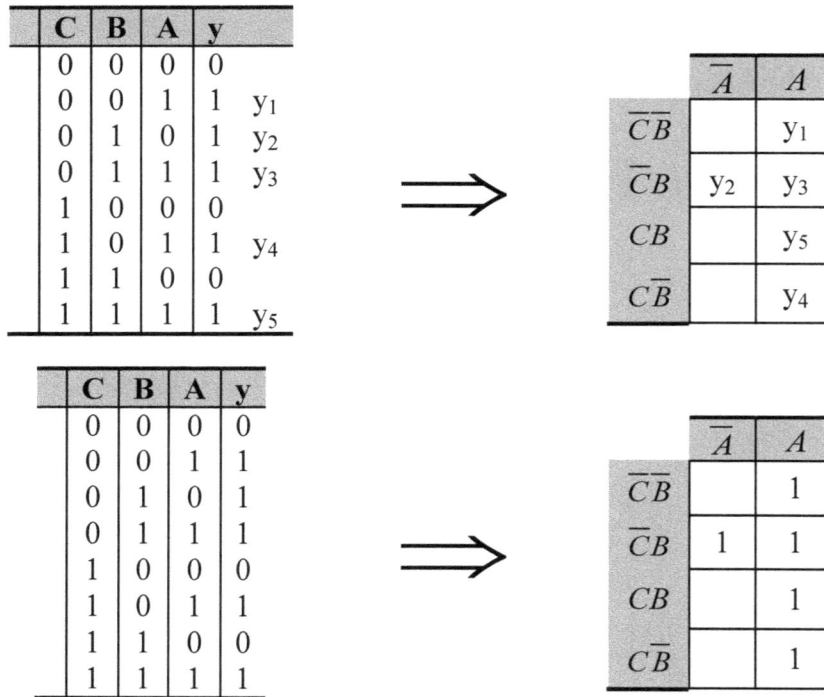

C	B	A	y	
0	0	0	0	
0	0	1	1	y_1
0	1	0	1	y_2
0	1	1	1	y_3
1	0	0	0	
1	0	1	1	y_4
1	1	0	0	
1	1	1	1	y_5

$\Longrightarrow$

	$\overline{A}$	A
$\overline{C}\,\overline{B}$		y_1
$\overline{C}B$	y_2	y_3
CB		y_5
$C\overline{B}$		y_4

C	B	A	y
0	0	0	0
0	0	1	1
0	1	0	1
0	1	1	1
1	0	0	0
1	0	1	1
1	1	0	0
1	1	1	1

$\Longrightarrow$

	$\overline{A}$	A
$\overline{C}\,\overline{B}$		1
$\overline{C}B$	1	1
CB		1
$C\overline{B}$		1

Figure 4.21. EXAMPLE 4.10 - Fill K-map in.

Step 4 – Loop adjacent groups of 2, 4, or 8 ones:

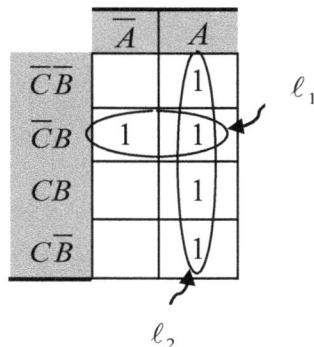

Figure 4.22. EXAMPLE 4.10 - loop adjacent 1s in K-map.

Step 5 – Eliminate variables that appear with their complement in a loop and save the others:

$$\ell_1 = \overline{C}B$$

$$\ell_2 = A$$

Step 6 – OR the remaining variables:

Therefore, the simplified Boolean expression is:

$$y = \ell_1 + \ell_2 = \overline{C}B + A$$

4.10.   Encoders and Decoders

An encoder/decoder application is shown in Figure 4.23. An encoder is used to translate from the language of humans to the language of machines (string of bit, ones and zeros). The decoder is used to translate from the language of machines to the language of humans, 0, 1, 2, 3, …., 9, a, b, c, d, …, z, +,-,÷, *, etc. The language of machines consists of binary coded decimal (BCD), octal, etc. BCD is a code in which decimal numbers 0 to 9 is represented by a 4 bit binary number. BCD is discussed in Chapter 3 on page 56. Computers must use these codes since they are built of digital circuits and systems which process only 1s and 0s.

The seven segment LED in Figure 4.23 receives the decoded message from the decoder. The decoder is controlled by the central processing unit (CPU). The seven segment LED can be configured as common anode or common cathode as shown in Figure 4.25 and Figure 4.26, respectively. Each segment of the seven segment LED can be represented as a Light emitting diode (LED) and resistor combination as shown in Figure 4.24.

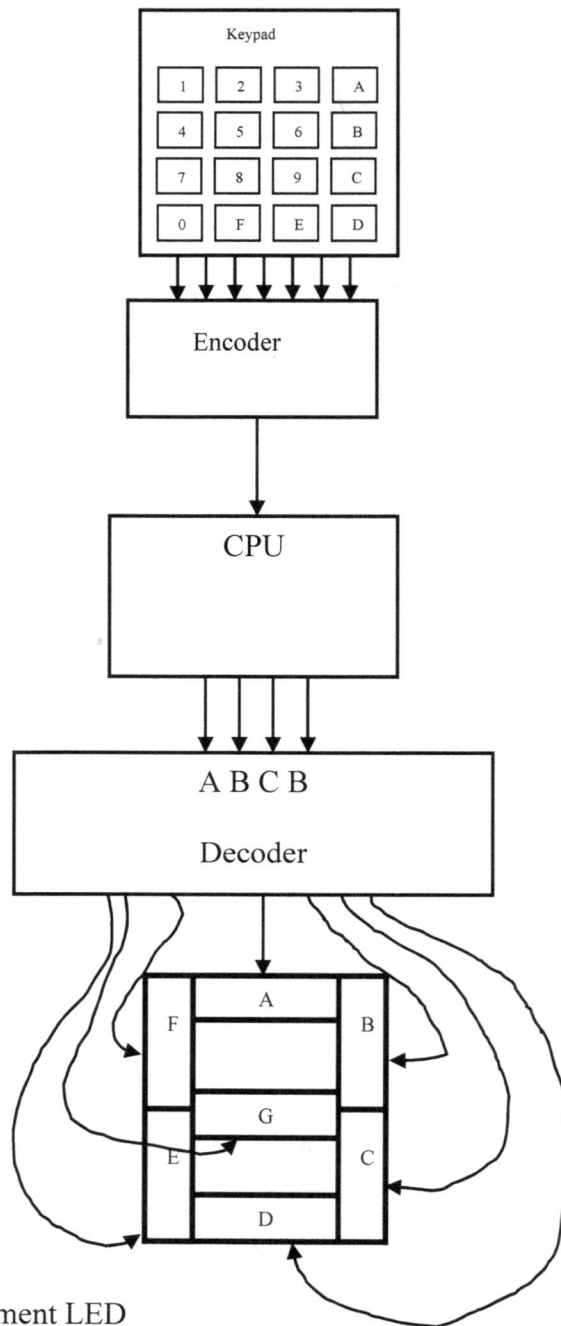

Figure 4.23. Example of encoder/decoder application.

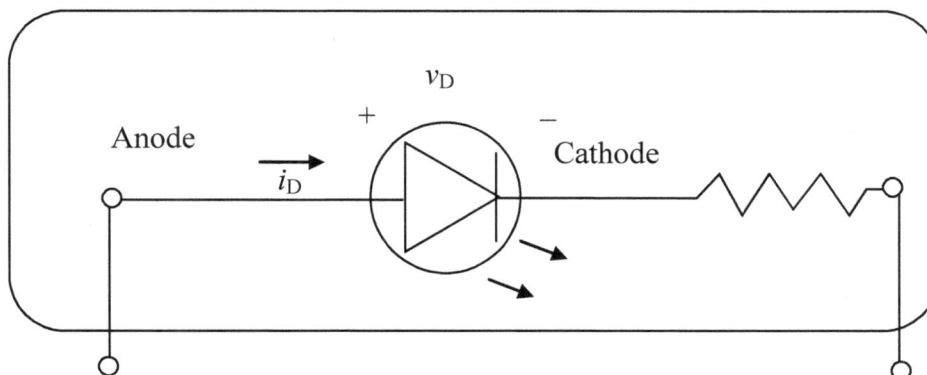

Figure 4.24. Each segment of a seven segment LED.

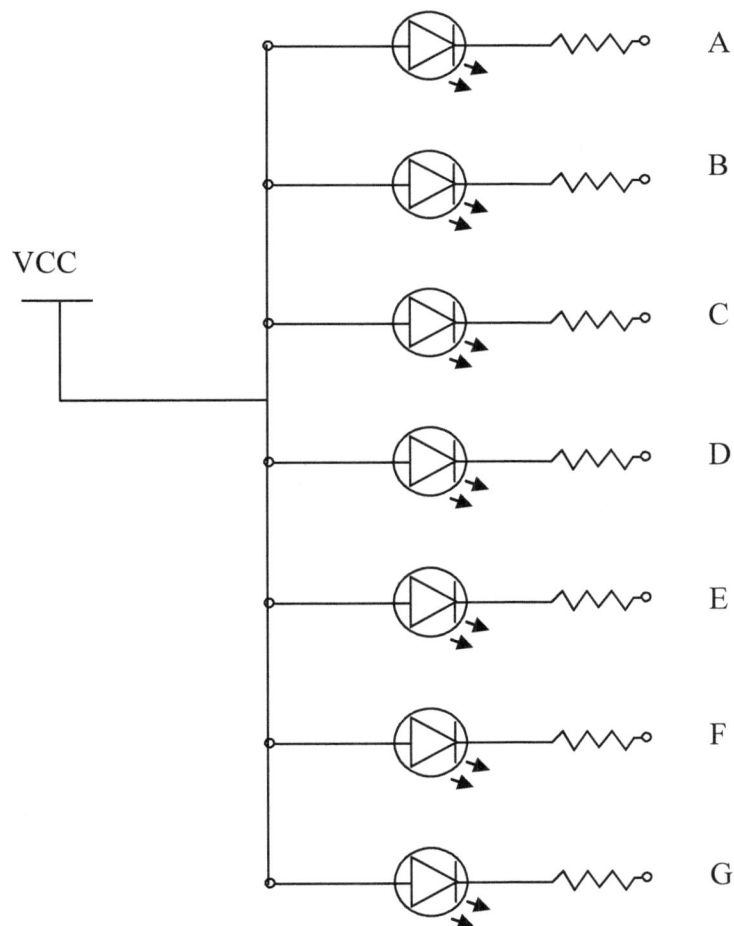

Figure 4.25. Seven segment LED common anode configuration.

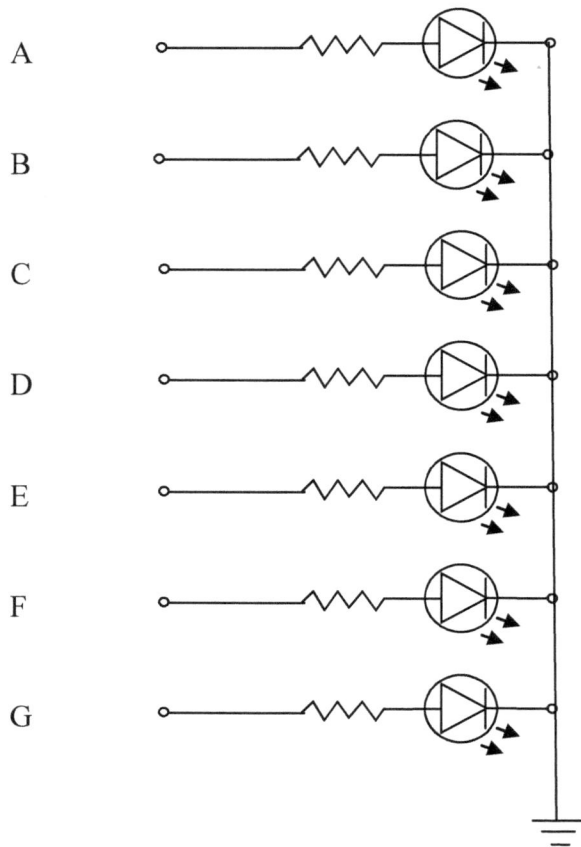

Figure 4.26. Seven segment LED common anode configuration.

4.10.1. n-Line-to-m-Line Decoder

Generally, a decoder with n inputs and m outputs is be represented by

$$m = 2^n$$

EQ. 4.7

Therefore, a decoder can be called a n-line-to-m-line decoder as shown in Figure 4.27.

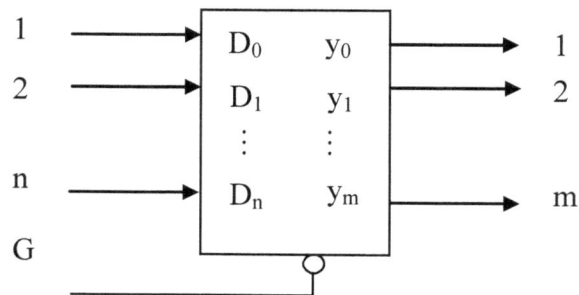

Figure 4.27. n-line-to-m-line decoder.

The general function of a decoder is to activate one or more outputs when a particular input state is detected. The simplest decoder is a single logic gate, e.g., an AND or a NAND. A **n-line-to-m-line decoder** (or simply n x m decoder) has n active high data inputs D_0, D_1,...., and D_n and m active high data outputs y_0, y_1, ..., and y_m. The input G (active low) is used to enable or disable the decoder.

EXAMPLE 4.11

Determine the number of outputs for a **n-line-to-m-line decoder** where n =2. Then draw the symbol.

$$m = 2^2 = 4$$

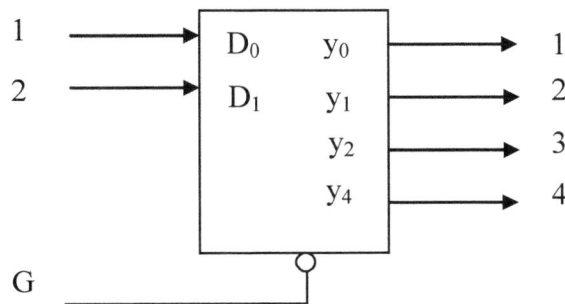

The truth table for the **n-line-to-m-line decoder** given in EXAMPLE 4.11 is given as follows:

G	D_1	D_0	y_0	y_1	y_2	y_3
0	0	0	1	0	0	0
0	0	1	0	1	0	0
0	1	0	0	0	1	0
0	1	1	0	0	0	1
1	X	X	0	0	0	0

Table 4.10. Truth table of 2-line-to-4-line decoder.

As indicated, when the decoder is enabled, G=0, only one output is available at a time. In this case, the outputs are active high. The Truth Table in Table 4.10 can be used to derive the Boolean equations and the corresponding digital circuit. X is a **don't care** condition. That is, when the decoder is disabled, G=1, a 0, 1, etc., on the inputs will not affect the outputs. The Boolean equations using SOPs are given as follows:

$$y_0 = \overline{G}\,\overline{D_1}\,\overline{D_0}$$

$$y_1 = \overline{G}\,\overline{D_1}\,D_0$$

$$y_2 = \overline{G}\,D_1\,\overline{D_0}$$

$$y_3 = \overline{G}\,D_1\,D_0$$

Using the equations above, the digital circuit can now be derived as shown in Figure 4.28.

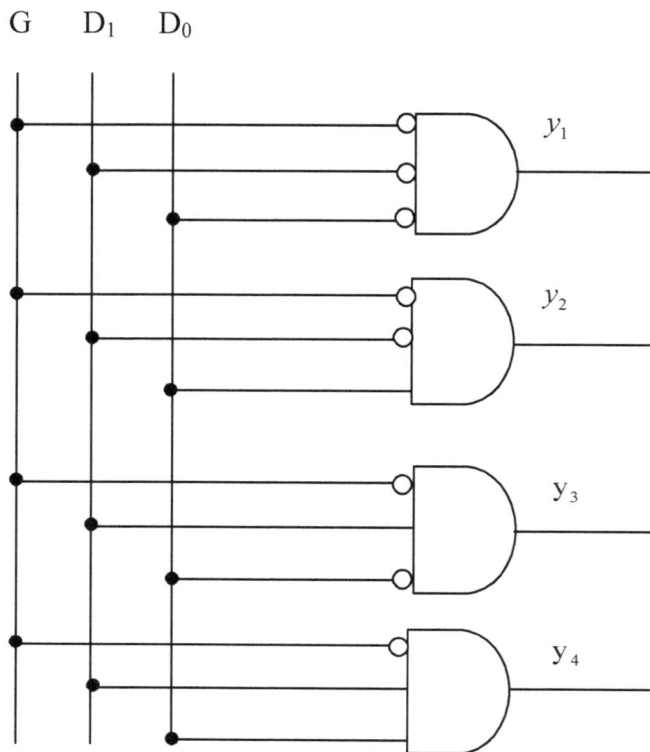

Figure 4.28. 2-line-to-4-line decoder circuit.

The VHDL code for the decoder in Table 4.10 is given below as follows:

```
with D select
    Y <=        '1' when "00",
                '1' when "01",
                '1' when "10",
                '1' when "11",
                '0' when others;
```

The VHDL code for the decoder implements the selected signal assignment statement which is explained in Section 7.5.3 on page 308.

4.10.2. Multiple Input Single Output Decoders

In general, a decoder activates one or more outputs when a specific set of inputs are detected. In particular,

A. the outputs, y, can be any of the following: 0, 1, X

B. can use minterms or maxterms to derive the digital circuit
C. when y is 0 the output is active low; when y is 1 the output is active high

A decoder with one output is a n-line-to-1-line decoder circuit.

EXAMPLE 4.12

Design a 4-line-to-1-line decoder digital circuit with an active low output that can detect 0011 as indicated in Table 4.11.

D_3	D_2	D_1	D_0	y
0	0	0	0	0
0	0	0	1	0
0	0	1	0	0
0	0	1	1	1
0	1	0	0	0
0	1	0	1	0
0	1	1	0	0
0	1	1	1	0
1	0	0	0	0
1	0	0	1	0
1	0	1	0	0
1	0	1	1	0
1	1	0	0	0
1	1	0	1	0
1	1	1	0	0
1	1	1	1	0

Table 4.11. Truth table of 4-line-to-1-line decoder.

Solution:

The Boolean equation of y is derived using the SOPs and DeMorgan's Theorem to give

$$y = D_3 + D_2 + \overline{D_1} + \overline{D_0} = \overline{\overline{D_3}\,\overline{D_2}\,D_1\,D_0}$$

4.10.3. Address Decoder

One application of a decoder is for enabling or disabling a block of memory as shown in Figure 4.29. The last two address bits are connected to a 2x4 decoder. The outputs of the decoder connects to the G input on four 16K RAMs.

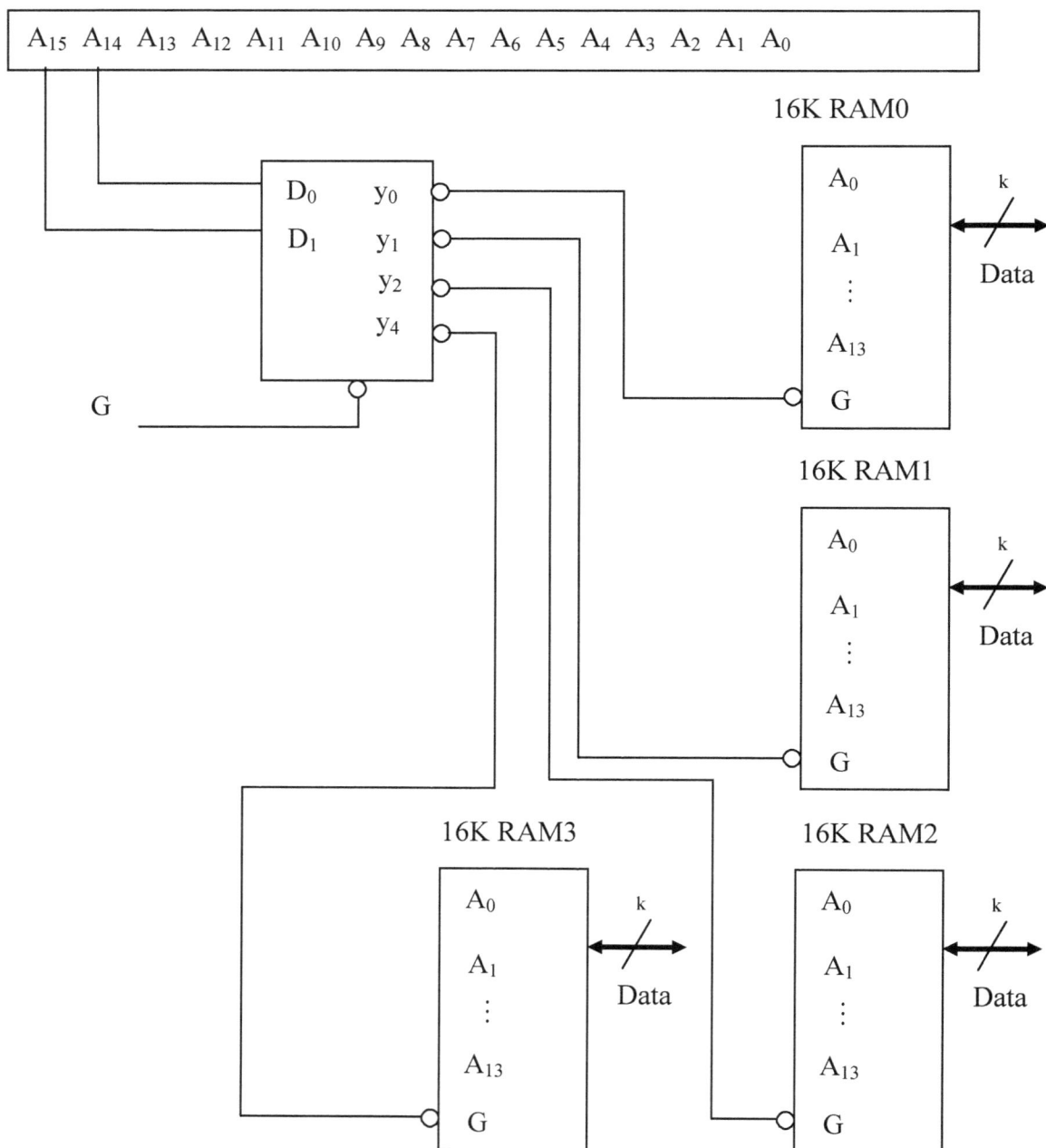

Figure 4.29. Decoder (2x4) used as an address decoder. Note: R/W are not indicated and address lines are not connected for simplicity.

In Figure 4.29, one decoder output is low at a given time which means only one block of memory can be active at any point in time. For example, A_{15} A_{14} $= 10 => y_2 = 0$ (active low), $y_0 = 1$, $y_1 = 1$, and $y_3 = 1$. Thus, RAM2 will be enabled and RAM0, RAM1, and RAM3 will be disabled. . When G $=1$, the decoder's outputs will be high resulting in no memory blocks being selected.

4.10.4. Encoder

An encoder performs the opposite of a decoder. As an example, let's look at the Truth Table for a 3 bit encoder represented in Table 4.12. This encoder results in the following Boolean equations:

$$Y_0 = D_7 + D_5 + D_3 + D_1$$

$$Y_1 = D_7 + D_6 + D_3 + D_2$$

$$Y_2 = D_7 + D_6 + D_5 + D_4$$

This encoder requires at least one input to be high at a time in order to get the right outputs results. For example, if D_3 and D_6 are high at the same time the output is 111 instead of the expected 011 or 110. This problem is resolved by using a priority encoder with Truth Table given in Table 4.13. The resulting equations are now:

$$Y_0 = D_7 + \overline{D}_6 D_5 + \overline{D}_6 \overline{D}_4 D_3 + \overline{D}_6 \overline{D}_4 \overline{D}_2 D_1$$

$$Y_1 = D_7 + D_6 + \overline{D}_5 \overline{D}_4 D_3 + \overline{D}_5 \overline{D}_4 D_2$$

$$Y_2 = D_7 + D_6 + D_5 + D_4$$

D_7	D_6	D_5	D_4	D_3	D_2	D_1	Y_2	Y_1	Y_0
0	0	0	0	0	0	0	0	0	0
0	0	0	0	0	0	1	0	0	1
0	0	0	0	0	1	0	0	1	0
0	0	0	0	1	0	0	0	1	1
0	0	0	1	0	0	0	1	0	0
0	0	1	0	0	0	0	1	0	1
0	1	0	0	0	0	0	1	1	0
1	0	0	0	0	0	0	1	1	1

Table 4.12. Truth table of 3 bit encoder.

D_7	D_6	D_5	D_4	D_3	D_2	D_1	Y_2	Y_1	Y_0
0	0	0	0	0	0	0	0	0	0
0	0	0	0	0	0	1	0	0	1
0	0	0	0	0	1	X	0	1	0
0	0	0	0	1	X	X	0	1	1
0	0	0	1	X	X	X	1	0	0
0	0	1	X	X	X	X	1	0	1
0	1	X	X	X	X	X	1	1	0
1	X	X	X	X	X	X	1	1	1

Table 4.13. Truth table of 3 bit priority encoder.

4.11.   Multiplexers/Demultiplexers

This Section covers basic concepts of multiplexers and demultiplexers necessary for understanding the robotic projects in this book.

4.11.1. Multiplexers m to 1/1 to m Demultiplexers

A multiplexer is represented by the symbol in Figure 4.30A. A **multiplexer** is a data selector that allows data to flow from one of several multiplexer input channels, D_0, D_1,...., and D_m, to an output, Z. The selection of the multiplexer inputs is controlled its select inputs, S_0,..., S_n. The **demultiplexer** performs the reverse of a multiplexer as shown in Figure 4.30B. It passes a binary signal from a single input to one of several demultiplexer output channels, D_0, D_1,...., and D_m. The selection of the demultiplexer outputs is controlled by its select inputs, S_0,..., S_n. The multiplexer is also referred to as a MUX while the demultiplexer is referred to as a DEMUX.

Multiplexer   Demultiplexer

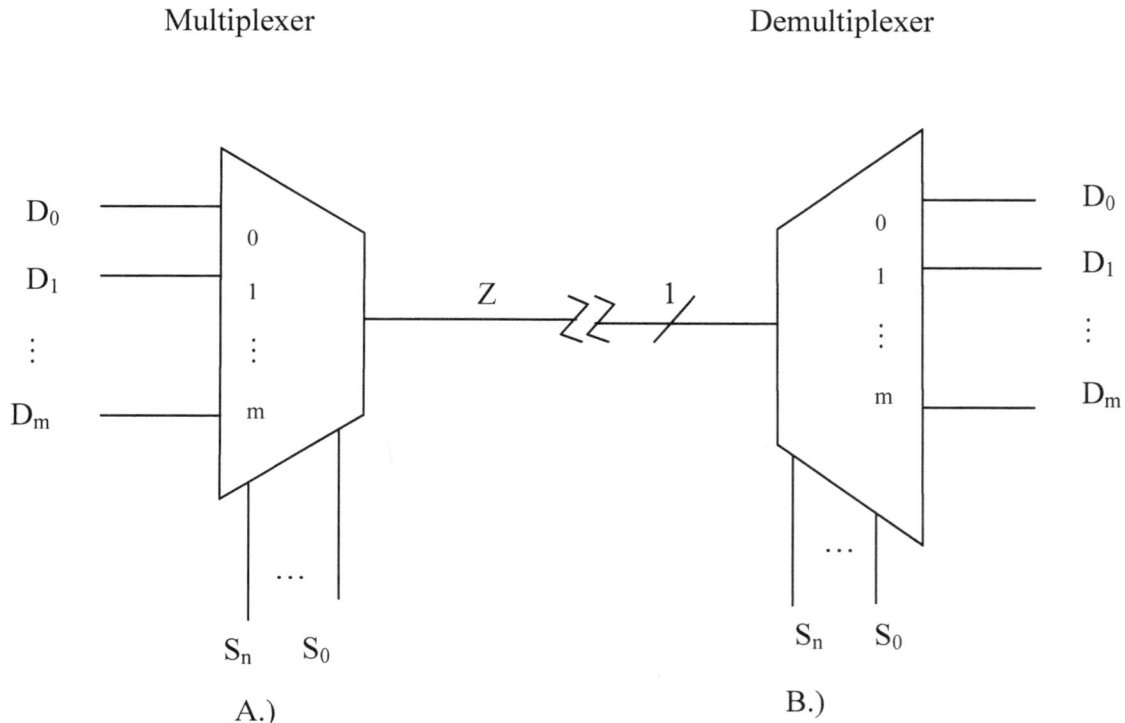

Figure 4.30. A.) m to 1 multiplexer. B.) 1 to m demultiplexer.

The number of select inputs, n, is related to the number of multiplexer inputs (m) or demultiplexer outputs (m) by

$$m = 2^n$$

EQ. 4.8

The Truth table of a 4 to 1 MUX is given in Table 4.14. The Truth Table of a 8 to 1 MUX is given in Table 4.15. The Truth Table of a 16 to 1 MUX is given in Table 4.16.

S_1	S_0	Z
0	0	D_0
0	1	D_1
1	0	D_2
1	1	D_3

Table 4.14. Truth table of 4 to 1 MUX.

S_2	S_1	S_0	Z
0	0	0	D_0
0	0	1	D_1
0	1	0	D_2
0	1	1	D_3
1	0	0	D_4
1	0	1	D_5
1	1	0	D_6
1	1	1	D_7

Table 4.15. Truth table of 8 to 1 MUX.

S_3	S_2	S_1	S_0	Z
0	0	0	0	D_0
0	0	0	1	D_1
0	0	1	0	D_2
0	0	1	1	D_3
0	1	0	0	D_4
0	1	0	1	D_5
0	1	1	0	D_6
0	1	1	1	D_7
1	0	0	0	D_8
1	0	0	1	D_9
1	0	1	0	D_{10}
1	0	1	1	D_{11}
1	1	0	0	D_{12}
1	1	0	1	D_{13}
1	1	1	0	D_{14}
1	1	1	1	D_{15}

Table 4.16. Truth table of 16 to 1 MUX.

EXAMPLE 4.13

Design a 4 to 1 MUX with no enable. Determine the Boolean equation. Then draw the digital circuit representing the Boolean equation.

Solution: The Truth Table of a 4 to 1 MUX is given in Table 4.17.

S_1	S_0	Z	Z Variable	Minterms
0	0	D_0	Z_0	$\overline{S}_1\overline{S}_0D_0$
0	1	D_1	Z_1	$\overline{S}_1S_0D_0$
1	0	D_2	Z_2	$S_1\overline{S}_0D_0$
1	1	D_3	Z_3	$S_1S_0D_0$

Table 4.17. Truth table of 4 to 1 MUX.

The Boolean equation is derived by using the SOPs and is given by:

$$Z = Z_0 + Z_1 + Z_2 + Z_3$$
$$= \overline{S}_1\overline{S}_0D_0 + \overline{S}_1S_0D_1 + S_1\overline{S}_0D_2 + S_1S_0D_3$$

Therefore, the digital circuit is shown in Figure 4.31.

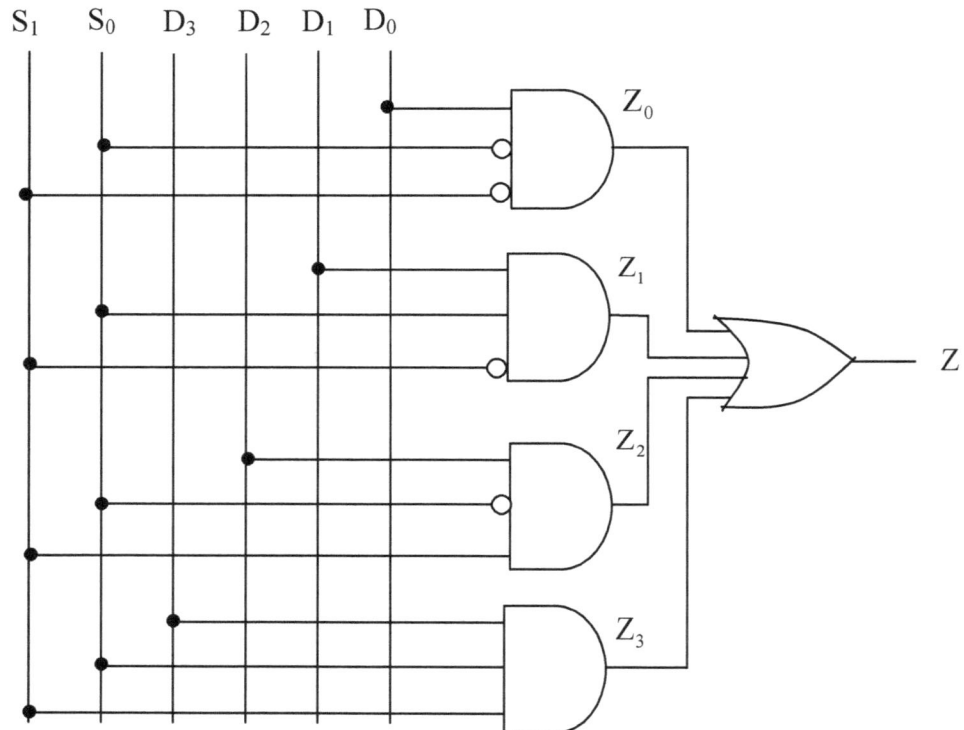

Figure 4.31. 4 to 1 MUX circuit.

The VHDL code of the 4 to 1 MUX in Table 4.17 is given below as follows:

```
entity mux4to1 is
port(D0, D1, D2, D3: IN BIT;
     S: IN BIT_VECTOR(1 downto 0);
     Z: OUT BIT);
end mux4to1;

architecture mux4to1 of mux is
begin

with S select
     Y <=      D0 when "00",
               D1 when "01",
               D2 when "10",
               D3 when "11";

end mux4to1;
```

The VHDL code for the 4 to 1 MUX implements the selected signal assignment statement which is explained in Section 7.5.3 on page 308.

Another method to implement the VHDL code of the 4 to 1 MUX in Table 4.17 is to use the case statement as follows:

```
entity mux4to1 is
port(D0, D1, D2, D3: IN BIT;
     S: IN BIT_VECTOR(1 downto 0);
     Z: OUT BIT);
end mux4to1;

architecture mux4to1 of mux is
begin
process(s)
  case S is
     when "00" => y <= D0;
     when "01" => y <= D1;
     when "10" => y <= D2;
     when "11" => y <= D3;
  end case;
end process;
end mux4to1;
```

The VHDL case statement is explained in Section 7.3.3 on page 293.

A demultiplexer can be implemented using a decoder. The truth Table of the 1 to 4 demultiplexer is given in Table 4.18.

S_1	S_0	D_0	D_1	D_2	D_3
0	0	D	0	0	0
0	1	0	D	0	0
1	0	0	0	D	0
1	1	0	0	0	D

Table 4.18. Truth table of 4 to 1 DEMUX.

Using the minterms for D_0, D_1, D_2, and D_3 in Table 4.18, the Boolean equations of the 4 to 1 DEMUX is given as follows:

$$D_0 = \overline{S_1}\,\overline{S_0}\,D$$
$$D_1 = \overline{S_1}S_0 D$$
$$D_2 = S_1\overline{S_0}D$$
$$D_3 = S_1 S_0 D$$

The digital circuit representing these equations is given in Figure 4.32.

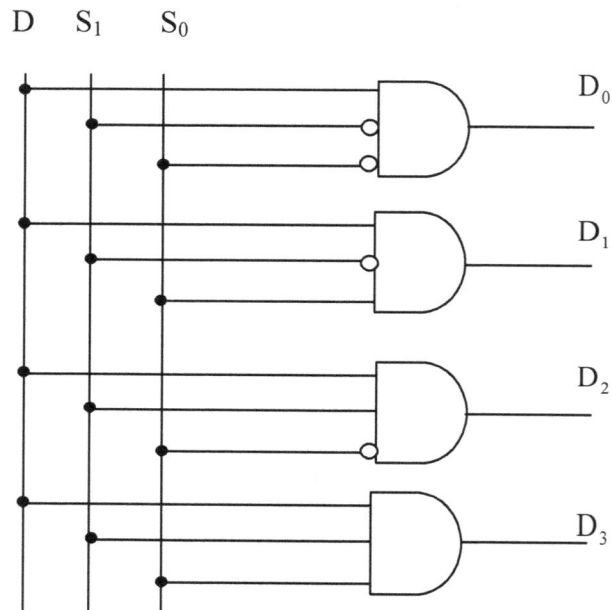

Figure 4.32. 4 to 1 DEMUX using a 2-line-to-4-line decoder circuit.

4.11.1. Multiplexers m to N bit bus/ N to m Demultiplexers

In the previous Section multiplexers with single bit inputs to a single bit output were discussed. In this Section we shall see that multiplexers can be used to select and pass one of several input buses to an output bus and vice versa using a demultiplexer. **Double subscript notation** is used where the first subscript represents the channel number and second subscript is the kth bit number from 0 to N-1. N is the total number of bits per bus. The m to 1 N bit bus multiplexer is shown in Figure 4.33A. The m to 1 N bit bus **multiplexer** is a data selector which allows data to flow from one of several input N bit bus channels, D_{0k}, D_{1k},...., and D_{mk}, to an N bit bus output, where k=0, 1, 2,..., N-1. The selection of the multiplexer input buses is controlled by the select inputs, S_0,..., S_n. The N bit bus **demultiplexer** performs the reverse of a multiplexer as shown in Figure 4.33B. It passes an N bit bus from a single input bus to one of several N bit bus output channels, D_{0k}, D_{1k},...., and D_{mk}, where k=0, 1, 2,..., N-1. The selection of the demultiplexer N bit bus outputs is controlled by the select inputs, S_0,..., S_n.

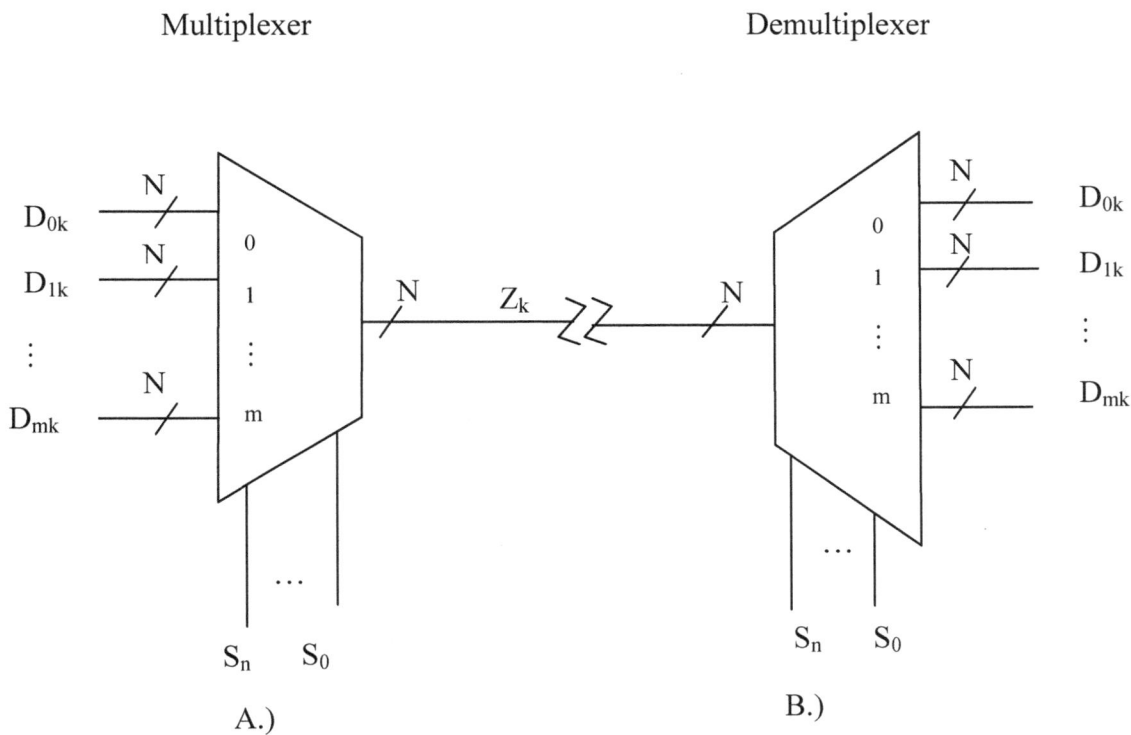

Figure 4.33. A.) m to 1 N bit bus multiplexer. B.) 1 to m N bit bus demultiplexer. k=0,1,2,....,N-1.

The Truth Table of a 4 to 1 4 bit bus MUX is given in Table 4.19. The Truth Table of a 8 to 1 4 bit bus MUX is given in Table 4.20. The Truth Table of a 16 to 1 2 bit bus MUX is given in Table 4.21.

S_1	S_0	$Z_3 \, Z_2 \, Z_1 \, Z_0$
0	0	$D_{03}D_{02}D_{01}D_{00}$
0	1	$D_{13}D_{12}D_{11}D_{10}$
1	0	$D_{23}D_{22}D_{21}D_{20}$
1	1	$D_{33}D_{32}D_{31}D_{30}$

Table 4.19. Truth table of 4 to 1 4 bit bus MUX.

S_2	S_1	S_0	$Z_3 \, Z_2 \, Z_1 \, Z_0$
0	0	0	$D_{03}D_{02}D_{01}D_{00}$
0	0	1	$D_{13}D_{12}D_{11}D_{10}$
0	1	0	$D_{23}D_{22}D_{21}D_{20}$
0	1	1	$D_{33}D_{32}D_{31}D_{30}$
1	0	0	$D_{43}D_{42}D_{41}D_{40}$
1	0	1	$D_{53}D_{52}D_{51}D_{50}$
1	1	0	$D_{53}D_{62}D_{61}D_{60}$
1	0	0	$D_{73}D_{72}D_{71}D_{70}$

Table 4.20. Truth table of 8 to 1 4 bit bus MUX.

D_3	D_2	D_1	D_0	$Z_1 \, Z_0$
0	0	0	0	$D_{01}D_{00}$
0	0	0	1	$D_{11}D_{10}$
0	0	1	0	$D_{21}D_{20}$
0	0	1	1	$D_{31}D_{30}$
0	1	0	0	$D_{41}D_{40}$
0	1	0	1	$D_{51}D_{50}$
0	1	1	0	$D_{61}D_{60}$
0	1	1	1	$D_{71}D_{70}$
1	0	0	0	$D_{81}D_{80}$
1	0	0	1	$D_{91}D_{90}$
1	0	1	0	$D_{101}D_{100}$
1	0	1	1	$D_{111}D_{110}$
1	1	0	0	$D_{121}D_{120}$
1	1	0	1	$D_{131}D_{130}$
1	1	1	0	$D_{141}D_{140}$
1	1	1	1	$D_{151}D_{150}$

Table 4.21. Truth table of 16 to 1 2 bit bus MUX.

EXAMPLE 4.14

Design a 2 to 1 4 bit bus MUX with no enable. Determine the Boolean equations. Draw the digital symbol representing the 2 to 1 4 bit bus MUX.

Solution: The Truth Table of a 2 to 1 4 bit bus MUX is given in Table 4.22.

S	$Z_3 Z_2 Z_1 Z_0$	Z_3 Minterm	Z_2 Minterm	Z_1 Minterm	Z_0 Minterm
0	$D_{03}D_{02}D_{01}D_{00}$	$\overline{S}D_{03}$	$\overline{S}D_{02}$	$\overline{S}D_{01}$	$\overline{S}D_{00}$
1	$D_{13}D_{12}D_{11}D_{10}$	SD_{13}	SD_{12}	SD_{11}	SD_{10}

Table 4.22. Truth table of 2 to 1 4 bit bus MUX.

Thus, N=4 and k=0,1,2 and 3. The Boolean equation for the output Z_k is derived by using SOP given by:

$$Z_3 = \overline{S}D_{03} + SD_{13}$$
$$Z_2 = \overline{S}D_{02} + SD_{12}$$
$$Z_1 = \overline{S}D_{01} + SD_{11}$$
$$Z_0 = \overline{S}D_{00} + SD_{10}$$

The 2 to 1 4 bit bus digital symbol is shown in Figure 4.34.

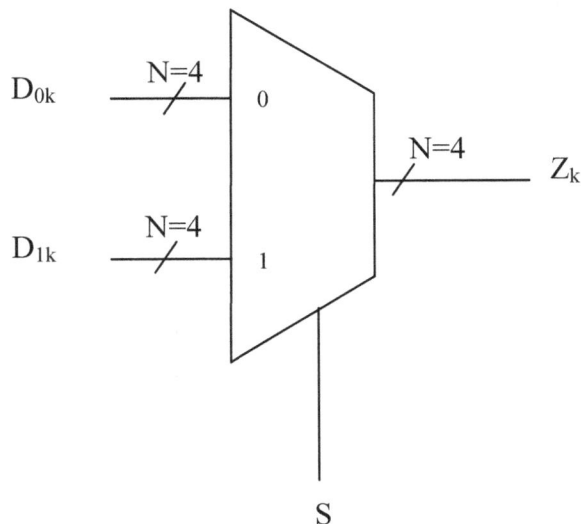

Figure 4.34. 2 to 1 4 bit bus MUX symbol.

4.11.2. Multiplexer Applications

Multiplexers are used in time division multiplexing (TDM) applications such as telephone line TDM as shown in Figure 4.35A. TDM is a technique of sending multiple channels over one transmission medium. Each channel is sent one at a time.

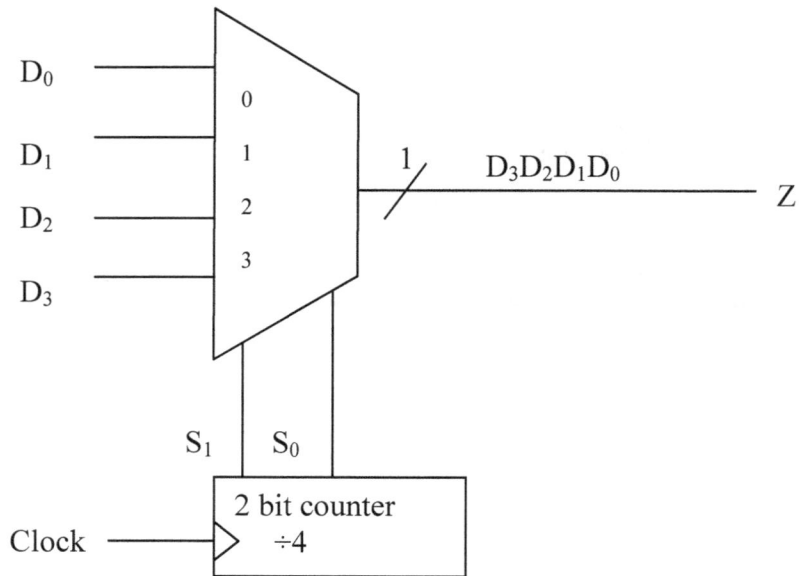

A.)

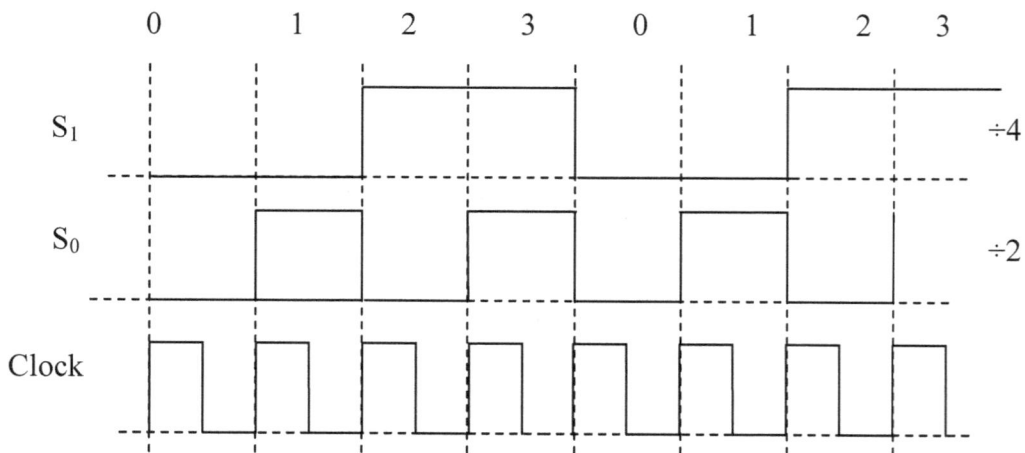

B.)

Figure 4.35. A.) Multiplexer application. B.) 2 bit counter timing diagram.

Figure 4.35A represents a four channel system using a 4x1 MUX to perform the multiplexing. $4=2^n$ implies that n must be 2 for a 4 channel MUX. A 2 bit counter is used to drive the select inputs to pass each data channel, $D_0, D_1, D_2,$ and D_3, one at a time to the output Z, in a serial fashion, $D_3D_2D_1D_0$, as shown in Figure 4.35A. D_0 is the first bit placed on Z as shown followed by $D_1, D_2,$ and D_3. Thus, a bit from each channel is placed on the output Z. This is called **bit multiplexing**. When a byte (8 bits) from each channel is sent, it is called **byte multiplexing**.

4.12. Binary Arithmetic Circuits

The binary addition truth Table for a **half adder** is given in Table 4.23.

B	A	B+A(Σ)	Carry Out (C_o)
0	0	0	0
0	1	1	0
1	0	1	0
1	1	0	1

Table 4.23. Half adder Truth Table.

In Table 4.23, the sum output (B+A) is represented by the summation symbol, Σ. The carry-out is represented by C_o. The sum column of Table 4.23 can be represented by the XOR function. The carry out column is represented by the AND function. Thus, the digital circuit for performing addition can be represented by a XOR and a AND gate as shown in Figure 4.36. This circuit is referred to as the **half adder**. It assumes no carry from a previous addition. Addition, the derivation of Σ and C_o can be obtained from the Truth Table in Table 4.23 as follows:

$$\Sigma = \overline{B}A + B\overline{A}$$
EQ. 4.9

and

$$C_o = BA$$
EQ. 4.10

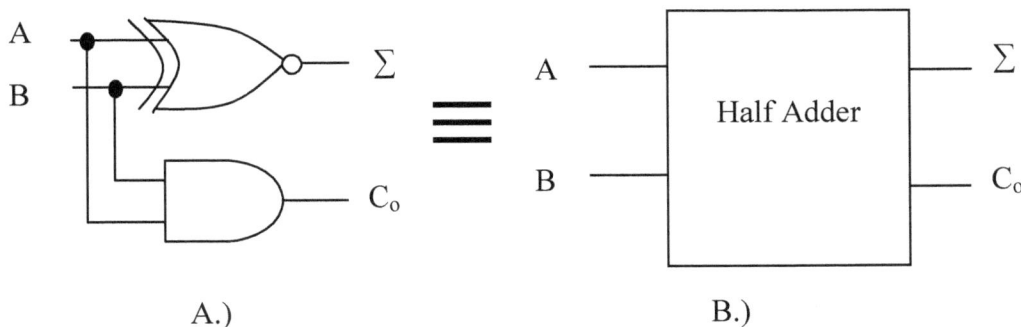

A.)   B.)

Figure 4.36. A.) The summing circuit. B.) Half adder symbol.

The limitation of the half adder is that it can add two binary 1s but not three 1s. The **full adder** provides this function. Thus, the full adder cover the case with a carry from a previous addition. It requires three inputs with all possible cases given in Table 4.24. The full adder can be built using half adders as illustrated in Figure 4.37A with symbol in Figure 4.37B. The VHDL code for a full adder is discussed in Section 7.2.3.1 on page 282.

C_{in}	B	A	$C_{in}+B+A(\sum)$	Carry Out (C_o)
0	0	0	0	0
0	0	1	1	0
0	1	0	1	0
0	1	1	0	1
1	0	0	1	0
1	0	1	0	1
1	1	0	0	1
1	1	1	1	1

Table 4.24. Full adder Truth Table

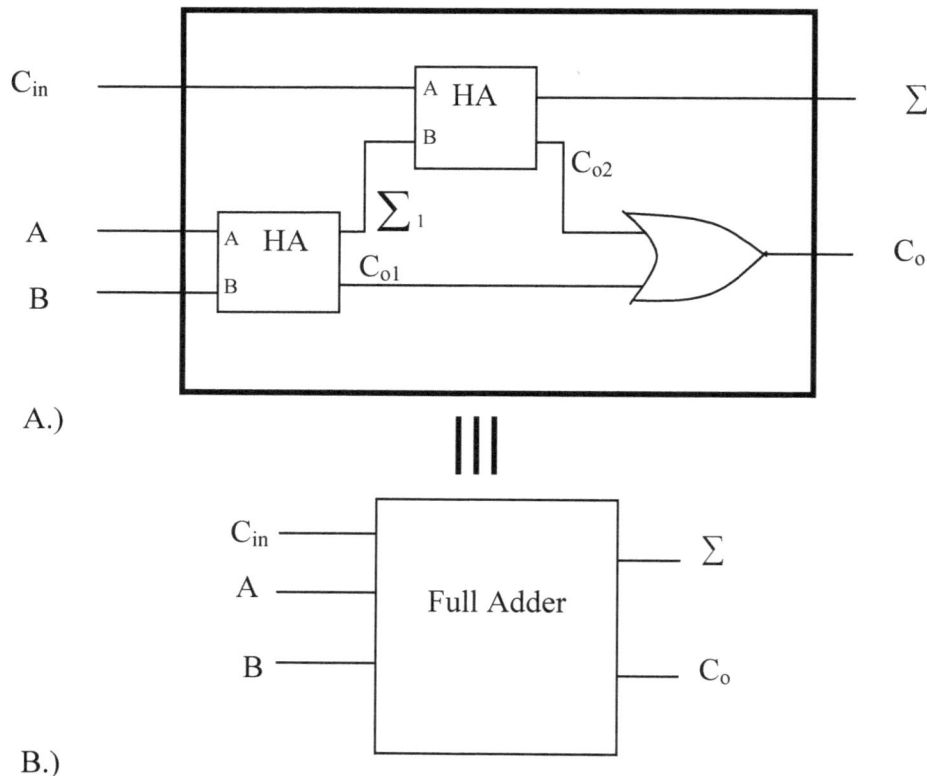

A.)

B.)

Figure 4.37. A.) Full adder summing circuit. B.) Full adder symbol.

PROBLEMS – Digital Circuits

1. A digital waveform is a series of ones and zeros with respect to time.
    A.) True   B.) False

2. A nonperiodic waveform repeats the same pattern of ones and zeros.
    A.) True   B.) False

s = seconds

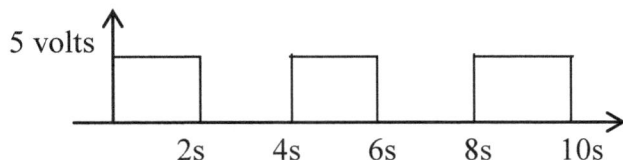

Figure 4.38.

3. The period of the digital waveform in Figure 1 is
    A.) 4s
    B.) 2s
    C.) 6s
    D.) 8s

4. If f is the frequency, the period of a digital waveform is calculated by which of the following
    A.) T=f
    B.) T=2f
    C.) T=T/2
    D.) T=1/f

5. The fall time of a pulse waveform is the time from 10% to 90% of the rising edge of a pulse.
    A.) True   B.) False

6. The correct VHDL expression for $Y = \overline{A} + \overline{BC}$ is:
    Y<=(not A) or ((not B) and (not C));
    A.) True   B.) False

7. The edge of a pulse is that part of the pulse that is logic HIGH.
    A.) True   B.) False

8. VHDL is a programming language that allows you to describe digital circuits without using a graphic editor.
    A.) True   B.) False

9. All logic gates and functions in Boolean Algebra can be created from three basic gates: AND, OR, and NOT.

     A.) True   B.) False

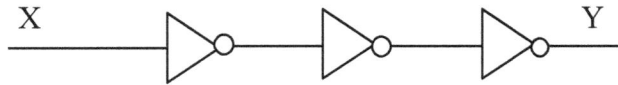

Figure 4.39.

10. The output of the circuit in Figure 4.39 is $Y = \overline{\overline{\overline{X}}}$.

     A.) True   B.) False

A	B	Y
0	0	0
0	1	1
1	0	1
1	1	1

A.)

A	B	Y
0	0	0
0	1	0
1	0	0
1	1	1

B.)

Figure 4.40.

11. The truth table in Figure 4.40A represents which logic function.

     A.) AND   B.) OR   C.) NOT   D.) XOR

12. The truth table in Figure 4.40B represents which logic function.

     A.) AND   B.) OR   C.) NOT   D.) XOR

13. A NAND gate is a logic function whose output Y=1 only when all the inputs are LOW.

     A.) True   B.) False

14. A NOR gate is a logic function whose output Y=0 when at least one input is HIGH.

     A.) True   B.) False

15. The bubble on the input or output of a logic gate means to complement or invert the Boolean signal.

     A.) True   B.) False

A	B	Y
0	0	0
0	1	1
1	0	1
1	1	0

A.)

A	B	Y
0	0	1
0	1	0
1	0	0
1	1	1

B.)

A	B	Y
0	0	0
0	1	1
1	0	1
1	1	1

C.)

A	B	Y
0	0	1
0	1	0
1	0	0
1	1	0

D.)

Figure 4.41.

16. The truth table in Figure 4.41A represents which of the following logic functions:
A.) OR B.) XOR C.) XNOR D.) AND

17. Which of the following Boolean expressions represent Y in the truth table in Figure 4.41A:

A.) $Y = AB + \overline{A}\,\overline{B}$ B.) $Y = \overline{AB + \overline{A}\,\overline{B}}$ C.) $Y = AB + \overline{AB}$ D.) $Y = \overline{A}B + A\overline{B}$

18. The truth table in Figure 4.41B represents which of the following logic functions:
A.) OR B.) XOR C.) XNOR D.) AND

19. The K-map for the truth table in Figure 4.41C is:

	$\overline{B}$	B
$\overline{A}$		1
A	1	1

A.)

	$\overline{B}$	B
$\overline{A}$	1	1
A	1	

B.)

	$\overline{B}$	B
$\overline{A}$	1	
A	1	1

C.)

	$\overline{B}$	B
$\overline{A}$	1	1
A		1

D.)

20. The simplified SOP Boolean expression for the truth table in Figure 4.41C is:

A.) $Y = AB + AB + AB$

B.) $Y = A + B$

C.) $Y = \overline{A}B + A\overline{B} + AB$

D.) $Y = (\overline{A} + B)(A + \overline{B})(A + B)$

21. The truth table in Figure 4.41D represents which of the following logic functions:
    A.) OR   B.) XOR   C.) NAND   D.) AND   E.) NOR

22. The XOR operator is represented by which of the following symbols:
    A.) +   B.) $\oplus$ C.) $\bullet$ D.) $\Leftarrow$

23. If Y=A+B+C then the application of DeMorgan's Theorem to Y gives
    A.) $Y = \overline{ABC}$
    B.) $Y = \overline{\overline{A}\,\overline{B}\,\overline{C}}$
    C.) $Y = \overline{A}\,\overline{B}+\overline{C}$
    D.) $Y = \overline{A}+\overline{B}+\overline{C}$

24. The application of DeMorgan's Theorem to $Y = \overline{AB+CD}$ gives
    A.) $Y = (A+B)(C+D)$
    B.) $Y = \overline{A}\,\overline{B}+\overline{C}\,\overline{D}$
    C.) $Y = \overline{AB}+\overline{CD}$
    D.) $Y = (\overline{A}+\overline{B})(\overline{C}+\overline{D})$

25. A logic gate is enabled if it allows a digital signal to pass from one of its inputs to its output.
    A.) True   B.) False

26. Inhibit means to allow a digital signal to pass from an input to an output.
    A.) True   B.) False

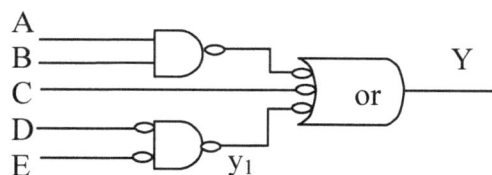

Figure 4.42.

27. The Boolean expression for signal y_1 in Figure 4.42 is
    A.) $\overline{\overline{D}+\overline{E}}$
    B.) $\overline{DE}$
    C.) $\overline{\overline{D}\,\overline{E}}$
    D.) $\overline{D+E}$

28. The Boolean expression for signal Y in Figure 4.42 is

    A.) $\overline{A}\,\overline{B}+\overline{C}+\overline{D}\,\overline{E}$

    B.) $\overline{AB}+\overline{C}+\overline{DE}$

    C.) $AB+\overline{C}+DE$

    D.) $AB+\overline{C}+\overline{D}\,\overline{E}$

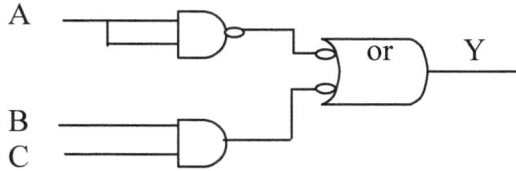

Figure 4.43.

29. The Boolean expression for signal Y in Figure 4.42 is

A.) $\overline{A}+\overline{B}+\overline{C}$

B.) $A+\overline{B}+C$

C.) $\overline{A+BC}$

D.) $A+\overline{BC}$

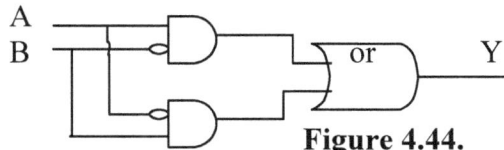

Figure 4.44.

30. The truth table of the logic circuit in Figure 4.44 is given by which of the following:

A	B	Y		A	B	Y		A	B	Y		A	B	Y
0	0	0		0	0	1		0	0	0		0	0	1
0	1	1		0	1	0		0	1	1		0	1	0
1	0	1		1	0	0		1	0	0		1	0	1
1	1	0		1	1	1		1	1	1		1	1	0

  A.)   B.)   C.)   D.)

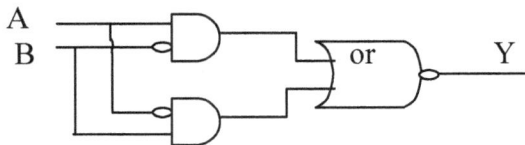

Figure 4.45.

31. The truth table of the logic circuit in Figure 4.45 is given by which of the following:

A	B	Y
0	0	0
0	1	1
1	0	0
1	1	1

A	B	Y
0	0	1
0	1	0
1	0	1
1	1	0

A	B	Y
0	0	0
0	1	1
1	0	1
1	1	0

A	B	Y
0	0	1
0	1	0
1	0	0
1	1	1

A.)   B.)   C.)   D.)

A	B	C	Y
0	0	0	0
0	0	1	1
0	1	0	1
0	1	1	0
1	0	0	0
1	0	1	1
1	1	0	1
1	1	1	1

Figure 4.46.

32. The POS Boolean expression for the truth table in Figure 4.46 is

A.) $Y = \overline{A}\,\overline{B}C + \overline{A}B\overline{C} + AB\overline{C} + ABC$

B.) $Y = (\overline{A}+\overline{B}+C)(\overline{A}+B+\overline{C})(A+B+\overline{C})(A+B+C)$

C.) $Y = (A+B+C)(A+\overline{B}+\overline{C})(\overline{A}+B+C)$

D.) $Y = \overline{A}\,\overline{B}\,\overline{C} + \overline{A}BC + \overline{A}B\overline{C} + A\overline{B}C$

33. Each term in an unsimplified SOP expression from the truth table is referred to as the
      A.) sum term
      B.) product term
      C.) None of the above

34. If A represents a Boolean variable then A+A=1.
      A.) True   B.) False

35. If A represents a Boolean variable then $A \cdot \overline{A} = A$.
      A.) True   B.) False

36. If A represents a Boolean variable then $A + 0 = A$.
    A.) True   B.) False

	$\overline{C}\,\overline{D}$	$\overline{C}D$	CD	$C\overline{D}$
$\overline{A}\,\overline{B}$		1	1	
$\overline{A}B$	1	1	1	1
AB	1	1	1	1
$A\overline{B}$		1	1	

Figure 4.47.

37. The simplified Boolean expression represented by the K-map in Figure 4.47 is
    A.) $Y = BD$
    B.) $Y = \overline{C}D$
    C.) $Y = \overline{D} + \overline{B}$
    D.) $Y = D + B$

	$\overline{C}\,\overline{D}$	$\overline{C}D$	CD	$C\overline{D}$
$\overline{A}\,\overline{B}$	1	1		
$\overline{A}B$	1	1		
AB	1	1		
$A\overline{B}$	1	1		1

Figure 4.48.

38. The simplified Boolean expression represented by the K-map in Figure 4.48 is
    A.) $Y = \overline{C} + \overline{B}\,\overline{D}$
    B.) $Y = \overline{C}\,\overline{D} + A\overline{B}C\overline{D}$
    C.) $Y = (\overline{C})(A\overline{B}D)$
    D.) $Y = \overline{C} + A\,\overline{B}\,\overline{D}$

39. Using Figure 4.49, determine the following:
A. truth table
B. Boolean expressions (for Y0, Y1, Y2)
C. digital circuit for the decimal to binary encoder

Keypad

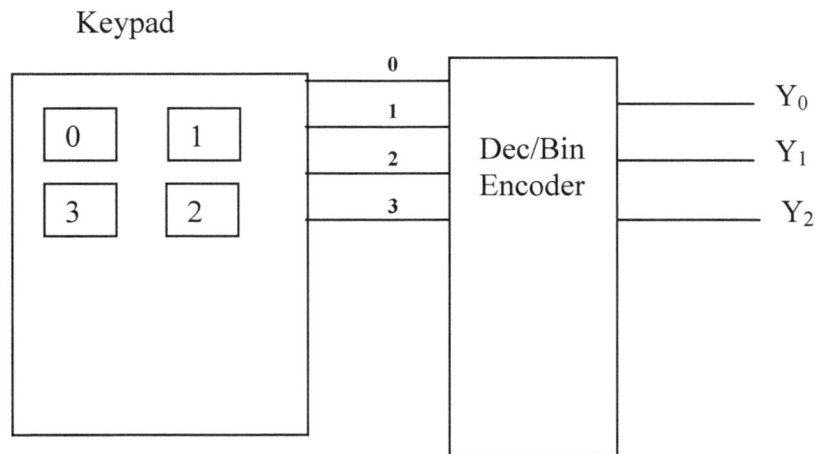

Figure 4.49. Combination logic circuit.

40. Using Figure 4.50, determine the following:
A. Truth Table for the binary to decimal decoder
B.   Boolean equations for the binary to decimal decoder
C. Digital circuit for the binary to decimal decoder

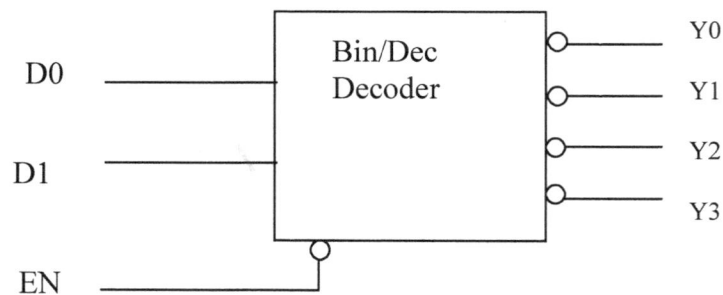

Figure 4.50. Combination logic circuit.

41. Using Figure 4.51, determine the fllowing:
A. Truth Table
B. Boolean equations
C. Digital circuit

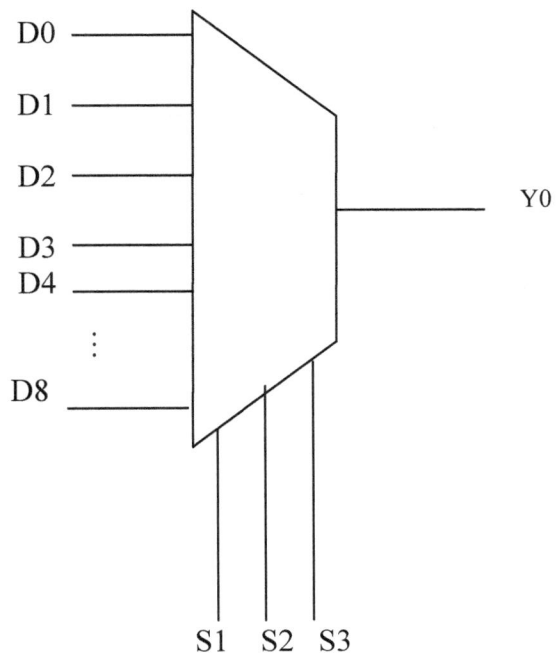

Figure 4.51. Combination logic circuit.

5. Sequential Logic Circuits

Sequential logic circuits are created from one digital element called the S-R latch. The S-R latch is the foundation of sequential logic circuits. As an example, a flip-flop is created from a S-R latch. SR latches are useful in various applications because of their ability to hold data temporarily. A Truth Table will be used to describe the input output relationship the latches and flip-flops in this Chapter. Remembering the truth tables will be very useful in the analysis of sequential circuits. The design or implementation of sequential circuits such as counters, shift registers, state machines, and memory devices shall also be discussed.

5.1. The S-R Latch

Figure 5.1 shows the basic Set-Reset Latches also referred to as S-R Latches. Two types of basic S-R latches are represented: A.) An active high input S-R latch and B.) An active low input S-R latch. Their differences will be explained in the following paragraphs.

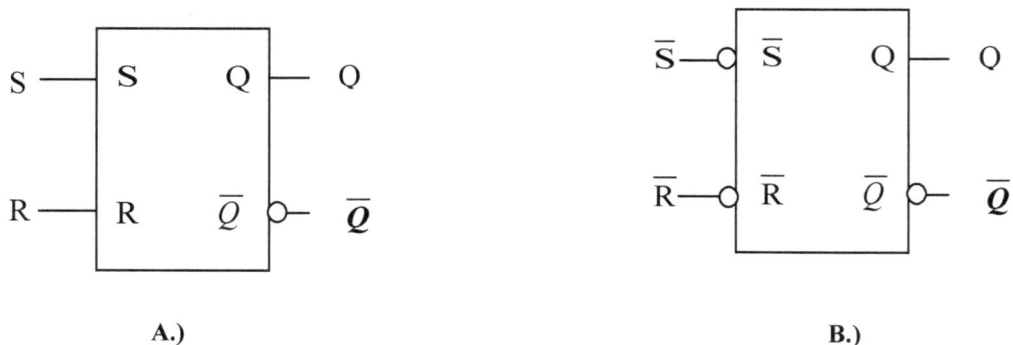

Figure 5.1. S-R Latches. A.) Active high input S-R latch. B.) Active low input S-R latch.

A S-R latch has two stable states. For this reason it is a bistable logic device. A S-R latch has two inputs and two outputs. The inputs are set (S) and reset (R). The outputs are Q and $\overline{Q}$. An active set means to set Q=1 and an active reset means to set Q=0. When Q=1 and $\overline{Q}$=0, the S-R latch is in the set state. When Q=0 and $\overline{Q}$=1, the S-R latch is in the reset state. When operating in the stable mode $\overline{Q}$ is always the opposite of logic value of Q. For example, if Q=0, then $\overline{Q}$=1. The inputs can be active low or active high depending on the gates being used (see Section 5.2 on NAND/NOR gates). The forbidden state is when both inputs have the same logic value. When S=R, and is the opposite state of their active inputs, i.e., active high or active low, the S-R latch is in the hold state. For example, if a S-R latch has active high inputs and S=R=0, the latch is in the hold state.

The truth tables of S-R latches are shown in Table 5.1 and Table 5.2. An active high input S-R latch has the truth table shown in Table 5.1. An active low input S-R latch has the truth table shown in Table 5.2.

S	R	Q	$\overline{Q}$	State
0	0	Q	$\overline{Q}$	Hold the previous values of Q and $\overline{Q}$
0	1	0	1	Reset State
1	0	1	0	Set State
1	1	0	0	Prohibited or Forbidden State

Table 5.1. Active high inputs S-R latch truth table.

$\overline{S}$	$\overline{R}$	Q	$\overline{Q}$	State
0	0	1	1	Prohibited or Forbidden State
0	1	1	0	Set State
1	0	0	1	Reset State
1	1	Q	$\overline{Q}$	Hold the previous values of Q and $\overline{Q}$

Table 5.2. Active low inputs S-R latch truth table.

EXAMPLE 5.1

Complete the timing diagram below for the given S-R latch in Figure 5.2.

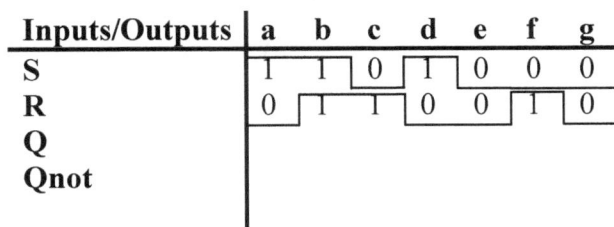

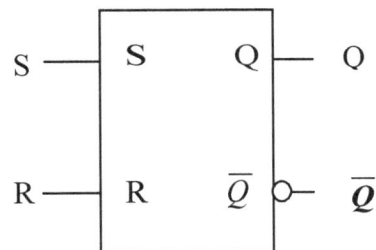

Figure 5.2. EXAMPLE 5.1 timing diagram and S-R latch.

Solution:

The timing diagram is as follows:

Inputs/Outputs	a	b	c	d	e	f	g
S	1	1	0	1	0	0	0
R	0	1	1	0	0	1	0
Q	1	0	0	1	1	0	0
$\overline{Q}$	0	0	1	0	0	1	1

Figure 5.3. EXAMPLE 5.2 timing diagram.

5.2. NAND/NOR Latches

Latches can be made out of NAND or NOR gates. A latch constructed from NOR gates is illustrated in Figure 5.4A with Truth Table given in Table 5.3. A latch constructed from NAND gates is illustrated in Figure 5.4B with Truth Table given in Table 5.4. In summary, an active low SR latch uses NAND gates while an active high SR latch uses NOR gates. The DeMorgan's theorem was applied to the NAND gates in Figure 5.4B which converted them to the inverted NOR gates in Figure 5.4B. We leave it as an exercise for you to analysis these circuits for the propagation of signals through the gates and feedback paths for hold, set, reset, and forbidden states.

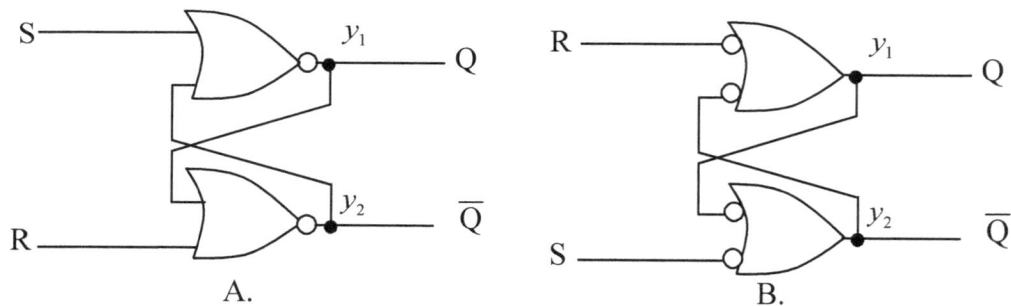

Figure 5.4. Latches. A.) NOR Latch. B.) NAND Latch.

S	R	Q	$\overline{Q}$	State
0	0	Q	$\overline{Q}$	Hold the previous values of Q and $\overline{Q}$
0	1	0	1	Reset State
1	0	1	0	Set State
1	1	0	0	Prohibited or Forbidden State

Table 5.3. NOR Latch - active high inputs S-R latch truth table.

S	R	Q	$\overline{Q}$	State
0	0	1	1	Prohibited or Forbidden State
0	1	1	0	Set State
1	0	0	1	Reset State
1	1	Q	$\overline{Q}$	Hold the previous values of Q and $\overline{Q}$

Table 5.4. NAND Latch - active low inputs R-S latch truth table.

It is better to use the latch primitives in the vendor supplied tools rather than draw your own gates because the vendor primitives have already been optimized and electrical characteristics such as propagation times, gate placements, etc., are consistent from latch to latch.

5.3. **Gated S-R Latch**

The gated S-R latch is the same as a S-R latch but has an added enable input as shown in Figure 5.5. The gated SR latch is a level triggered device. For example, when the enable is a clock source (see Figure 5.6) and the clock input is high, the data from the SR inputs are passed to the latch and the corresponding Q outputs. The truth table for the level triggered latch is shown in Table 5.5.

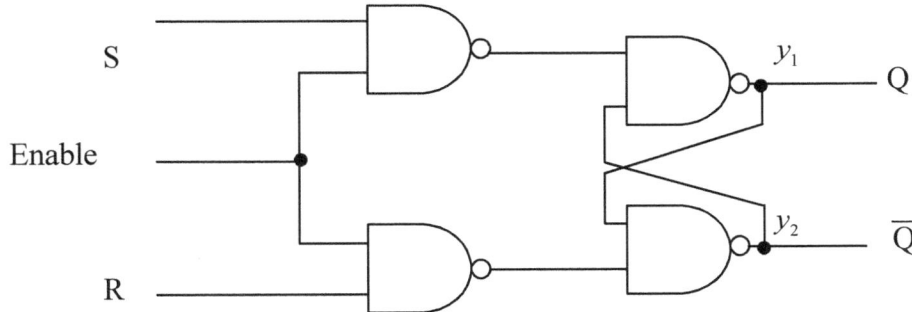

Figure 5.5. Gated NAND Latch with enable – active SR high inputs.

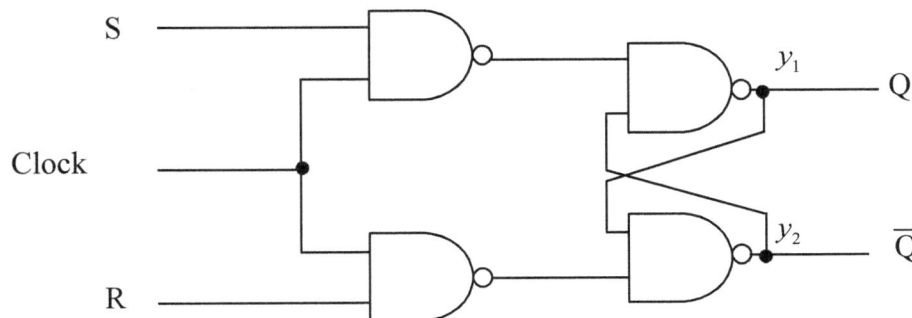

Figure 5.6. Gated NAND Latch with clock source – active SR high inputs.

EN	S	R	Q	$\overline{Q}$	State
...⊓	0	0	Q	$\overline{Q}$	Hold the previous values of Q and $\overline{Q}$
...⊓	0	1	1	0	Reset State
...⊓	1	0	0	1	Set State
...⊓	1	1	0	0	Prohibited or Forbidden State
⊐⊓	X	X	Q	$\overline{Q}$	Hold the previous values of Q and $\overline{Q}$

Table 5.5. Gated S-R latch truth table with level triggered clock source and active high inputs (S,R).

The **enable** (EN) input is used to determine when the Q and $\overline{Q}$ outputs are controlled by the inputs S and R. The enable input can be active high or active low. If the enable is an active high then the output will change when the enable input is high, otherwise, no change will occur. If the enable is an active low then the output will change when the enable input is low, otherwise, no change will occur. The enable feature can be used in two ways:

1.) Turning a device On or off
2.) Synchronizing signal such as synchronizing the response of a device to a clock

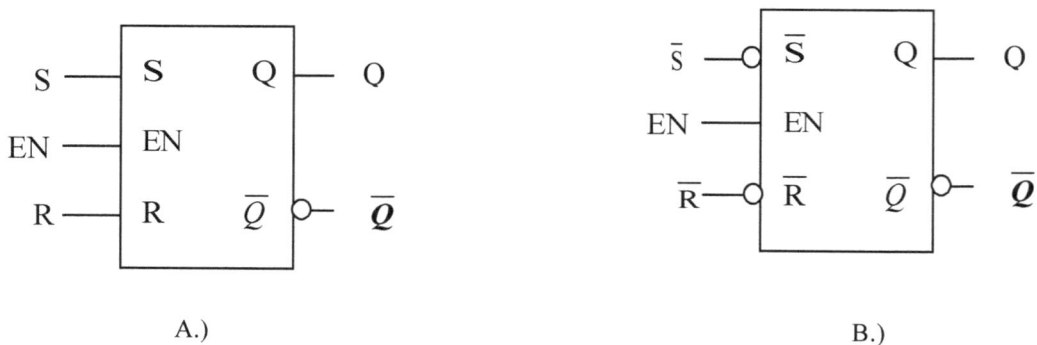

A.)   B.)

Figure 5.7. Gated S-R latches. A.) Gated S-R latch with active high inputs. B.) Gated S-R latch with active low inputs.

EN	S	R	Q	$\overline{Q}$	State
1	0	0	Q	$\overline{Q}$	Hold the previous values of Q and $\overline{Q}$
1	0	1	1	0	Reset State
1	1	0	0	1	Set State
1	1	1	0	0	Prohibited or Forbidden State
0	X	X	Q	$\overline{Q}$	Hold the previous values of Q and $\overline{Q}$

Table 5.6. Gated S-R latch truth table with active high enable (EN) and active high inputs (S,R).

EN	$\overline{S}$	$\overline{R}$	Q	$\overline{Q}$	State
1	0	0	1	1	Prohibited or Forbidden State
1	0	1	1	0	Set State
1	1	0	0	1	Reset State
1	1	1	Q	$\overline{Q}$	Hold the previous values of Q and $\overline{Q}$
0	X	X	Q	$\overline{Q}$	Hold the previous values of Q and $\overline{Q}$

Table 5.7. Gated S-R latch truth table with active high enable (EN) and active low inputs ($\overline{S}$, $\overline{R}$)

EXAMPLE 5.2

Complete the timing diagram in Figure 5.8.

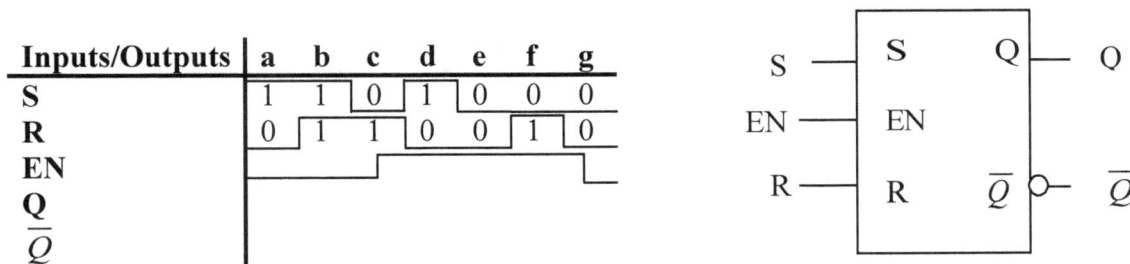

Figure 5.8. EXAMPLE 5.2 timing diagram and gated S-R latch.

Solution:

The completed timing diagram is as follows:

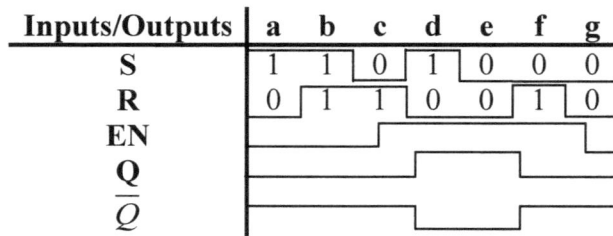

Inputs/Outputs	a	b	c	d	e	f	g
S	1	1	0	1	0	0	0
R	0	1	1	0	0	1	0
EN							
Q							
$\overline{Q}$							

Figure 5.9. EXAMPLE 5.2 solution.

The **clocked gated S-R latch** adds a synchronous feature to the gated S-R latch. The clock input enables the S-R latch when the clock goes high. When the clock goes high the data at the S-R inputs is transferred to the outputs. Table 5.8 shows the truth table of a clocked gated S-R latch. The clocked gated S-R latch adds a synchronizing feature to the gated S-R latch.

EN	S	R	Q	$\overline{Q}$	State
...⸬	0	0	Q	$\overline{Q}$	Hold the previous values of Q and $\overline{Q}$
...⸬	0	1	1	0	Reset State
...⸬	1	0	0	1	Set State
...⸬	1	1	0	0	Prohibited or Forbidden State
⌐	X	X	Q	$\overline{Q}$	Hold the previous values of Q and $\overline{Q}$

Table 5.8. Truth table of gated S-R latch with active high EN with CLK input as synchronizing signal.

5.4. **Clocked Device Basics**

Some basic definitions to know before going to the following sections are given next.
Clock waveform - A repeated waveform with period T has frequency $f = 1/T$.

Edge – A high to low or low to high transition of a waveform. Examples of positive and negative edges are shown in Figure 5.10 and Figure 5.11.

Figure 5.10. Low to high transition of clock – positive edge.

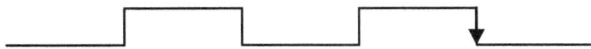

Figure 5.11. High to low transition of clock – negative edge.

Edge triggered – Enabled by the positive or negative edge.

Level sensitive – Enabled by a high or low level.

Edge detector – A circuit that detects the active edge (positive or negative) of a digital waveform. The sideways triangle, >, represents a positive edge device (Figure 5.12A) while the sideways triangle with a bubble represents a negative edge device (Figure 5.12B). Figure 5.13 and Figure 5.14 show the use of a positive edge detector circuit in the S-R flip-flop.

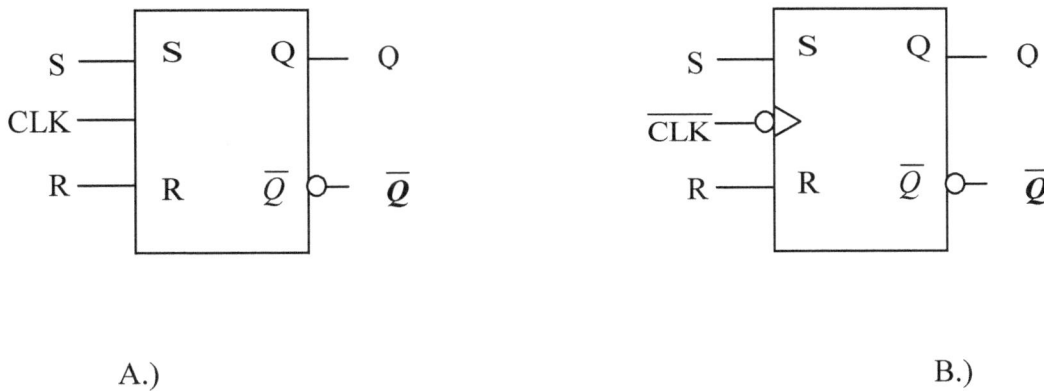

A.)   B.)

Figure 5.12. A.) This device has a positive edge detector in the CLK input. B.) This device has a negative edge detector in the $\overline{CLK}$ input.

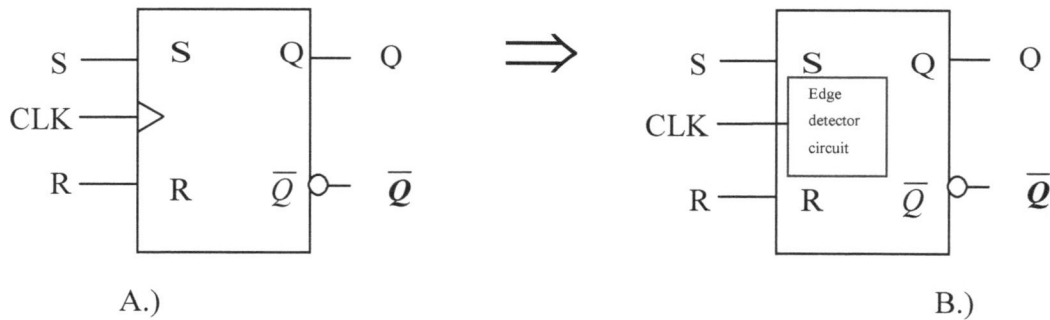

Figure 5.13. A.) This device has a positive edge detector with CLK as the input. B.) Same device with positive edge detector circuit.

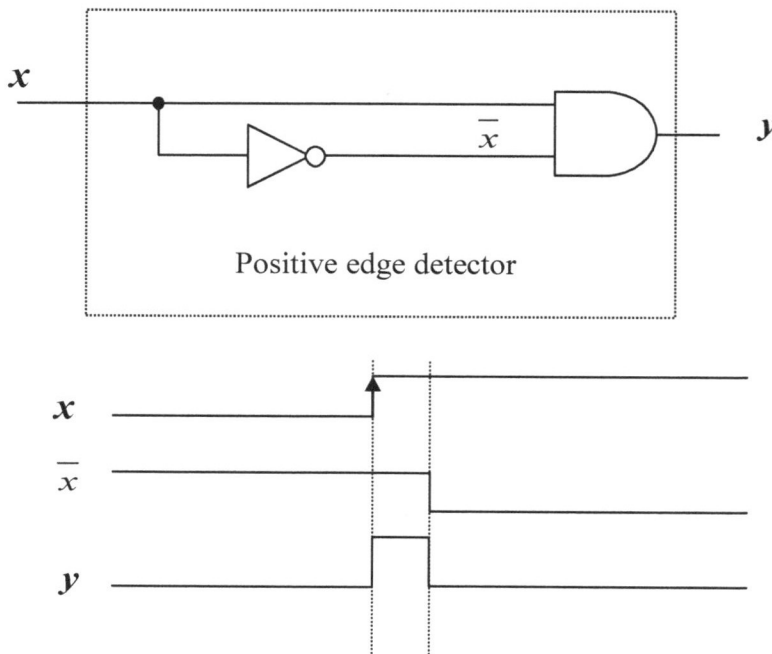

Figure 5.14. Digital positive edge detector circuit.

5.5. **Edge Triggered S-R Flip-Flops**

The basic RS latch does not operate in step with a clock. That is, when an input is applied to the set or reset input the output Q is immediately activated (just like combinational logic). This is referred to as **asynchronous logic**. On the other hand synchronous logic operates in step with a clock. A synchronous logic device is the synchronous SR latch referred to as the SR flip-flop. There are some widely accepted distinctions between latches and flip-flops. These distinctions are:

1.) Latches are level sensitive devices. Levels can be high or low. An example is the gated S-R latch.

2.) Flip-flops are edge triggered devices. Edges can be positive or negative. Examples are D-flip-flop, J-K flip-flops, S-R flip-flop, etc.

Examples of symbols for S-R flip-flops are shown in Figure 5.15. Figure 5.15 shows active high input S-R flip-flops with positive and negative edge triggered CLKs. Table 5.6 gives the truth table of an active high input S-R flip-flop with positive edge triggered CLK.

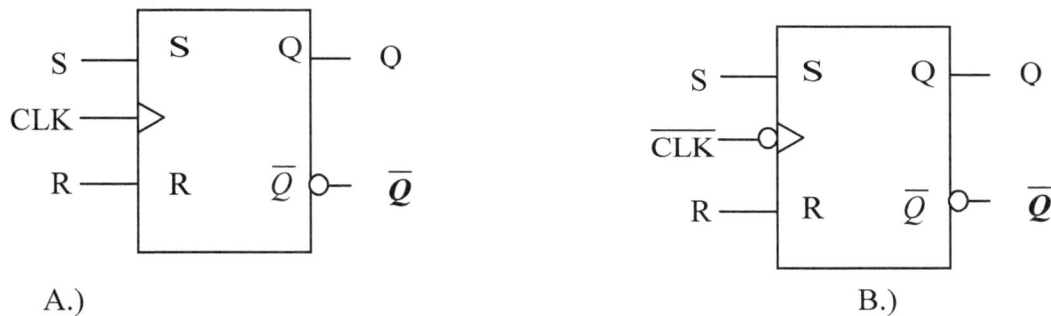

A.)   B.)

Figure 5.15. S-R Flip-flops. A.) Active high input S-R flip-flop with positive edge triggered clock (CLK). B.) Active high input S-R flip-flop with negative edge triggered clock ($\overline{CLK}$).

CLK	S	R	Q	$\overline{Q}$	State
⤒	0	0	Q	$\overline{Q}$	Hold the previous values of Q and $\overline{Q}$
⤒	0	1	1	0	Reset State
⤒	1	0	0	1	Set State
⤒	1	1	0	0	Prohibited or Forbidden State
⤓	X	X	Q	$\overline{Q}$	Hold the previous values of Q and $\overline{Q}$
⊓	X	X	Q	$\overline{Q}$	Hold the previous values of Q and $\overline{Q}$

Table 5.9. Truth table of active high input S-R flip-flop with positive edge triggered CLK.

5.6. Edged Triggered D Flip-Flops

The D-flip-flop has a data input and a clock input. The outputs are Q and $\overline{Q}$. Some examples of the D- flip-flops are shown in Figure 5.16.

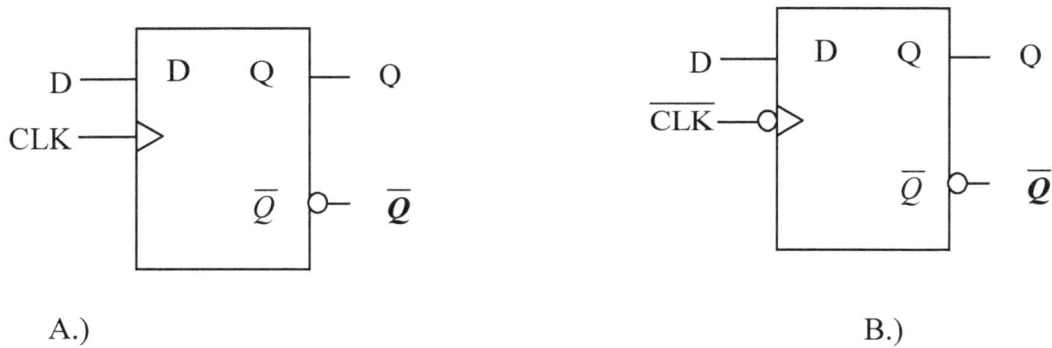

A.) B.)

Figure 5.16. D Flip-flops. A.) Active high input D flip-flop with positive edge triggered clock (CLK). B.) Active high input D flip-flop with negative edge triggered clock (CLK).

The D flip-flop is sometimes called a delay flip-flop. The data at the D input is delayed from getting to the output by one clock pulse.

CLK	D	Q	$\overline{Q}$	State
↗	0	0	1	Reset State
↗	1	1	0	Set State
↘	X	Q	$\overline{Q}$	Hold the previous values of Q and $\overline{Q}$
0	X	Q	$\overline{Q}$	Hold the previous values of Q and $\overline{Q}$
1	X	Q	$\overline{Q}$	Hold the previous values of Q and $\overline{Q}$

Table 5.10. Truth table of active high input D flip-flop with positive edge triggered CLK.

EXAMPLE 5.3

Complete the timing diagram in Figure 5.8.

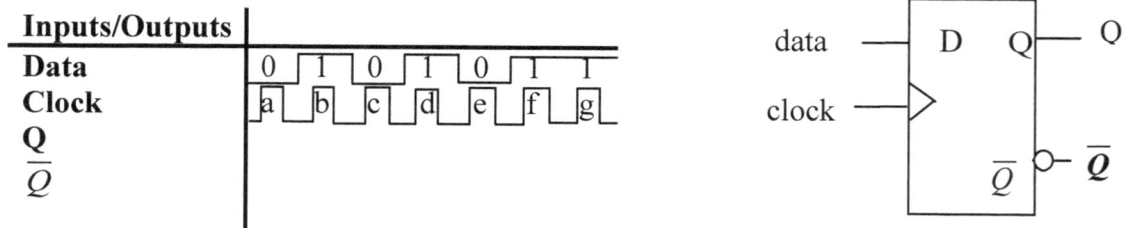

Figure 5.17. EXAMPLE 5.3 timing diagram and gated D flip-flop.

Solution:
The completed timing diagram is as follows:

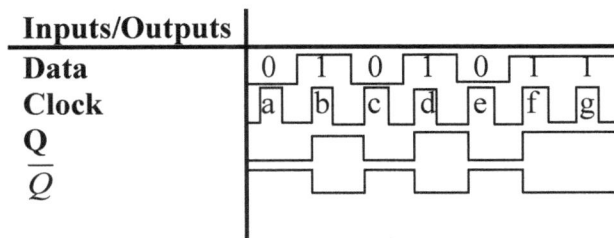

Figure 5.18. EXAMPLE 5.3 solution.

5.7. JK Flip-Flops

The **JK flip-flop** has the following main inputs: J, K, and clock. The outputs are Q and $\overline{Q}$. Some examples of JK flip-flops are shown in Figure 5.19A and Figure 5.19B. The corresponding truth tables are given in Table 5.11 and Table 5.12, respectively. In Figure 5.19A, the J can be thought of as the set input while the K can be thought of as the reset input.

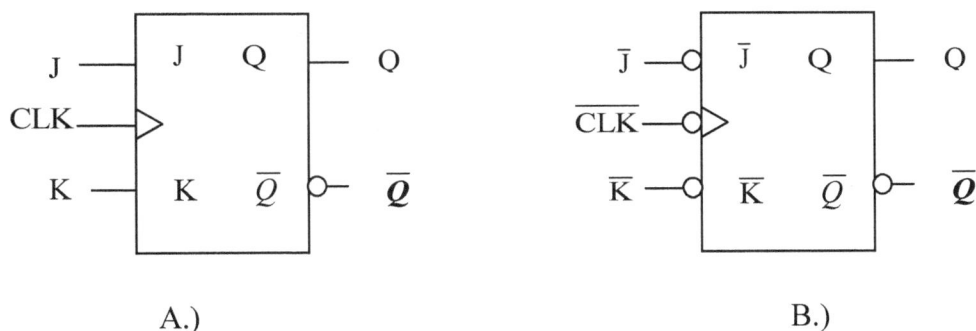

A.)   B.)

Figure 5.19. JK Flip-flops. A.) JK flip-flop with active high JK inputs and positive edge triggered clock (CLK). B.) JK flip-flop with active low JK inputs and negative edge triggered clock (CLK).

CLK	J	K	Q	$\overline{Q}$	State
↗	0	0	Q	$\overline{Q}$	Hold the previous values of Q and $\overline{Q}$
↗	0	1	0	1	Reset State
↗	1	0	1	0	Set State
↗	1	1	$\overline{Q}$	Q	Toggle state or change to opposite state
↘	X	X	Q	$\overline{Q}$	Hold the previous values of Q and $\overline{Q}$
0	X	X	Q	$\overline{Q}$	Hold the previous values of Q and $\overline{Q}$
1	X	X	Q	$\overline{Q}$	Hold the previous values of Q and $\overline{Q}$

Table 5.11. Truth table of JK flip-flop in Figure 5.19B.

CLK	J	K	Q	$\overline{Q}$	State
⎍↴	0	0	$\overline{Q}$	Q	Toggle state or change to opposite state
⎍↴	0	1	1	0	Set State
⎍↴	1	0	0	1	Reset State
⎍↴	1	1	Q	$\overline{Q}$	Hold the previous values of Q and $\overline{Q}$
⎍↑	X	X	Q	$\overline{Q}$	Hold the previous values of Q and $\overline{Q}$
0	X	X	Q	$\overline{Q}$	Hold the previous values of Q and $\overline{Q}$
1	X	X	Q	$\overline{Q}$	Hold the previous values of Q and $\overline{Q}$

Table 5.12. Truth table of JK flip-flop in Figure 5.19A.

In summary, when J=0, K=0, and the positive edge of the clock arrives then Q will hold the previous state, or Q=Q. When J=0, K=1, and the positive edge of the clock arrives then Q will change to 0, or Q=0. When J=1, K=0, and the positive edge of the clock arrives then Q will change to 1, or Q=1. When J=1, K=1, and the positive edge of the clock arrives then Q will change to the opposite state. This is referred to as the **toggle state**. or Q=1. If any other state of the clock occurs such as 0, 1, or its negative edge, then Q will hold the previous flip flop state, represented by Q=Q.

The JK flip-flop is sometimes referred to as the universal flip-flop because it can be configured as a D flip-flop, T flip-flop, etc. Figure 5.20 shows the JK flip flop configured as a D flip-flop while Figure 5.21 shows the JK flip-flop configured as a T flip-flop. T flip flops are useful in counter applications as will be seen Section 5.11 on page 139.

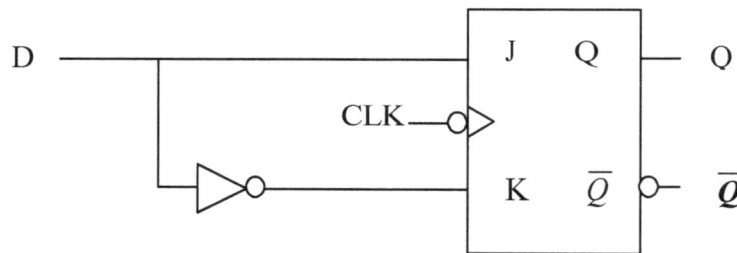

Figure 5.20. JK Flip-flop configured as a D flip-flop.

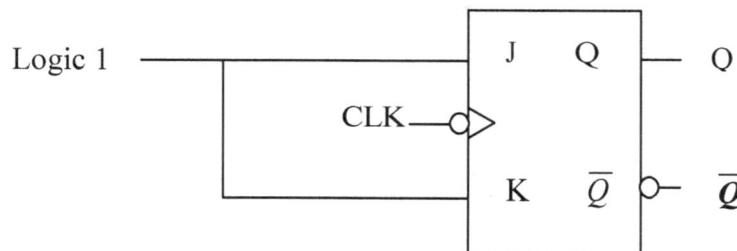

Figure 5.21. JK Flip-flop configured as a T flip-flop.

5.8. **Memory**

Memory is a device that can hold or store one or more bits. For example, an eight bit register can hold 8 bits. When a register is used to hold data it is referred to as a **memory register.**

Memory can be categorized into two main categories:
   A. Primary memory
   B. Secondary memory

Primary memory devices can further be broken down into one of the following:
   A. Read Only memory (ROM)
   B. Read/Write memory (RAM)

Examples of ROM memories are:
   A. Mask ROM (MOS, Bipolar)
   B. PROM (MOS, Bipolar)
   C. EPROM (MOS)
   D. EEPROM (MOS)

Examples of RAM devices are:
   A. Static RAM
   B. Dynamic RAM
   C. Non volatile RAM

Secondary memory devices are typically much cheaper than primary memory devices. Secondary memories can further be broken down into one of the following:
   A. Magnetic memories
   B. Optical memories

Examples of magnetic memories are:
   A. Floppy disk
   B. Hard disk
   C. Digital audio tape (DAT)
   D. magnetic bubble memories (MBM)

Example of optical memories are:
   A. Compact disk ROMs
   B. Compact disk RWMs

Although memories can be organized into primary and secondary memories they are classified as **volatile** and **non-volatile.** A volatile memory is memory that loses its data when power is turned off. A non-volatile memory is memory that does not lose it memory when power is turned off.

The data in RAM memory can be accessed (read) at any point in time. The locations in RAM can also be written to at any point in time. Thus, RAM memory can be read from or written to at any time using the appropriate control signals.

ROM memory is one in which its data can be accessed (read) at any point in time. The contents of this memory can only be read, not written to. Microcomputers must store permanent information is ROM. Examples of permanent data are: system programs, look-up tables, character generators, etc. ROM are used in high volume production applications because initial costs are high. Dedicated computers tend to use more ROM than RAM while general purpose computers tend to use more RAM than ROM.

Programmable Read Only Memories (PROMs) are used in low volume production applications. They can be programmed locally.

EPROMs are Erasable PROMs that can be erased and used again. Some variations are:

A. UV –Erasable PROMs are ultraviolet erasable PROMs using light through an open window in the chip for a length of time.

B. EEPROM or electrically erasable PROM which can be erased and programmed while the chip is in the circuit. Part of the chip is erased and programmed at a time.

C. Flash PROM which is the same as the EEPROM except the entire chip can be erased at a time and reprogrammed.

Storing data in memory is called a write operation while reading data from memory is called a read operation.

5.8.1. Memory Terms

Some common memory terms are:

b ≡bit
B≡Byte
K≡1024 = 2^10
M ≡Mega = 2^20
G ≡Giga = 2^30

5.8.2. Number of Address Lines

The number of address lines indicate how many memory locations are available within an IC. The number of memory locations, L, is related to the number of address lines, n, using:

$$L = 2^n$$

EQ. 5.1

Thus, a decoder of size $n \times 2^n$ is needed to select each memory location.

EXAMPLE 5.4

Determine the decoder size for the following memory sizes:
  A. 4
  B. 8
  C. 16
  D. 128
  E. 1024

Solution:
  A. $4 = 2^n \Rightarrow n = 2 \Rightarrow$ 2- line-to-4-line decoder

  B. $8 = 2^n \Rightarrow n = 3 \Rightarrow$ 3- line-to-8-line decoder

  C. $16 = 2^n \Rightarrow n = 4 \Rightarrow$ 4- line-to-16-line decoder

  D. $128 = 2^n \Rightarrow n = 7 \Rightarrow$ 128- line-to-7-line decoder

  E. $1024 = 2^n \Rightarrow n = 10 \Rightarrow$ 1024- line-to-10-line decoder

5.8.3. Memory Organization
Memory organization can be described by the following:
$$\text{number of locations} \times \text{bits/word} = L \times k$$
where the number of locations, L, is defined by EQ. 5.1 and the number of bits/word, k, can be 2, 4, 8, 16, 32, 64, 128, etc.

5.8.4. Memory Size
Total memory size is in bits. To get the total size multiply the number of words by the bits/per word. For example, Figure 5.23 shows a 4×8 bit memory. The total size is 32 bits. As another example, a 1K×4 memory has a size of 4096 bits.

5.8.5. Memory Access Time
The memory access time determines the speed of the chip. It is the time it takes to locate and read data from RAM, etc.

5.8.6. Memory Map
A memory map shows how data is organized in memory. It typically shows blocks of memory and their address range.

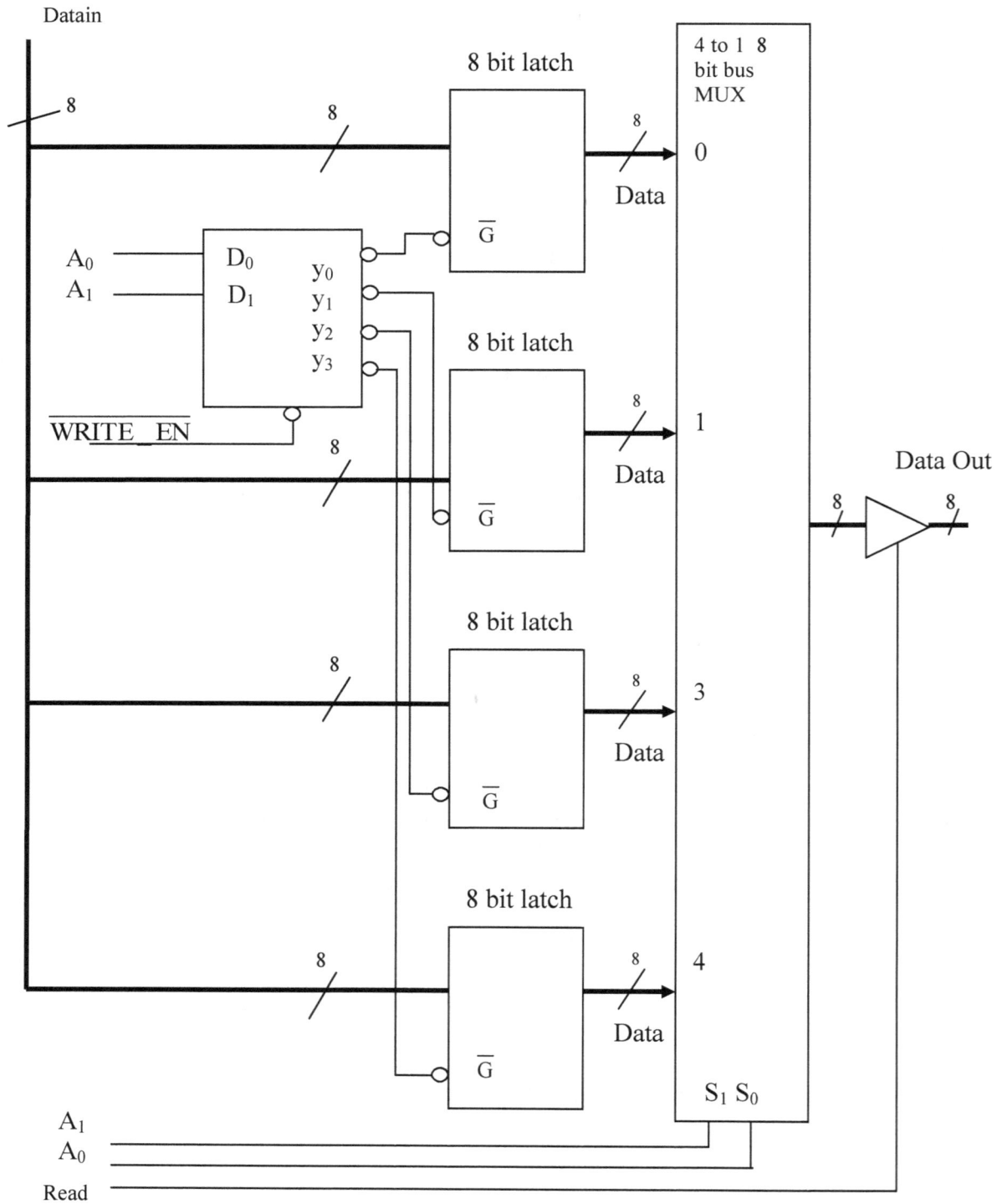

Figure 5.22. 4×8 bit memory.

5.8.7. Random Access Memory

A block diagram of a typical RAM memory unit is shown in Figure 5.23. The k data input lines provide the information to be stored in memory. The k data output lines is the path the data is coming out of memory. Typically, the number of data input lines equal the number of data output lines. The n address lines select a specific word in memory out of many. The other three inputs are control inputs. These are read, write, and select. The chip select input, $\overline{CS}$, selects the memory chip out of many. A write causes data to be transferred into memory while a read causes data to be transferred out of memory.

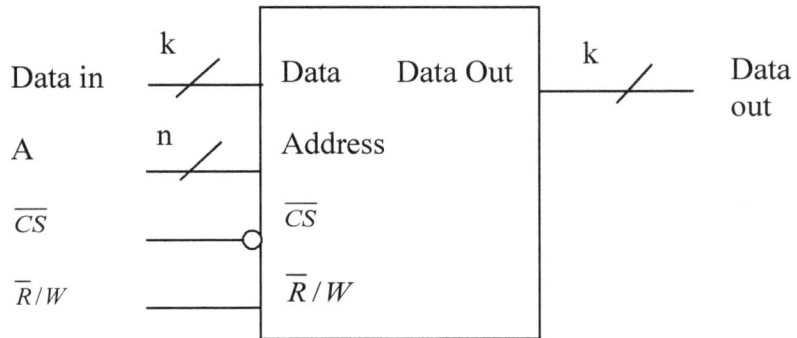

Figure 5.23. Representation of L × k bit memory.

5.8.8. Array of Memory Chips

A 1K × 8 bit memory chip is shown in Figure 5.24. The number of words can be doubled by adding another chip as shown in Figure 5.25. However, the word size will remain the same. Increasing the address bit by 1 increases the memory size by a factor of 2. To increase the memory size by a factor of 4, 2 more address bits are needed as shown in Figure 5.26. To increase the word size by 2 and keep the same number of words, e.g., to go from 1K×8 to 1K×16, connect the address lines, $\overline{CS}$, and $R/\overline{W}$ lines, leaving the data in and data out lines separately as shown in Figure 5.27.

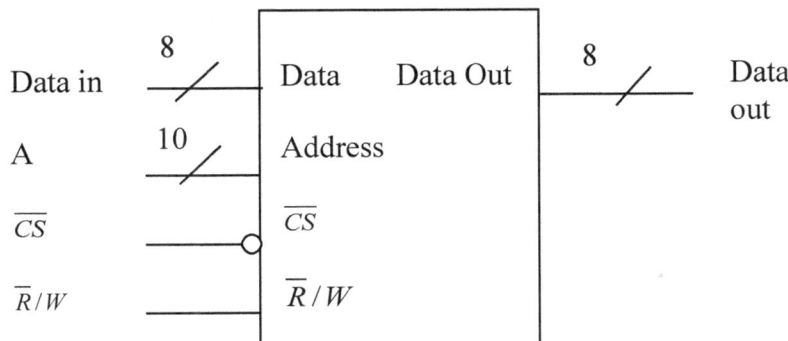

Figure 5.24. Representation of 1K × 8 bit memory.

To reduce the number of pins on a chip, the input and output data lines share the same pins. These are called **bidirectional**.

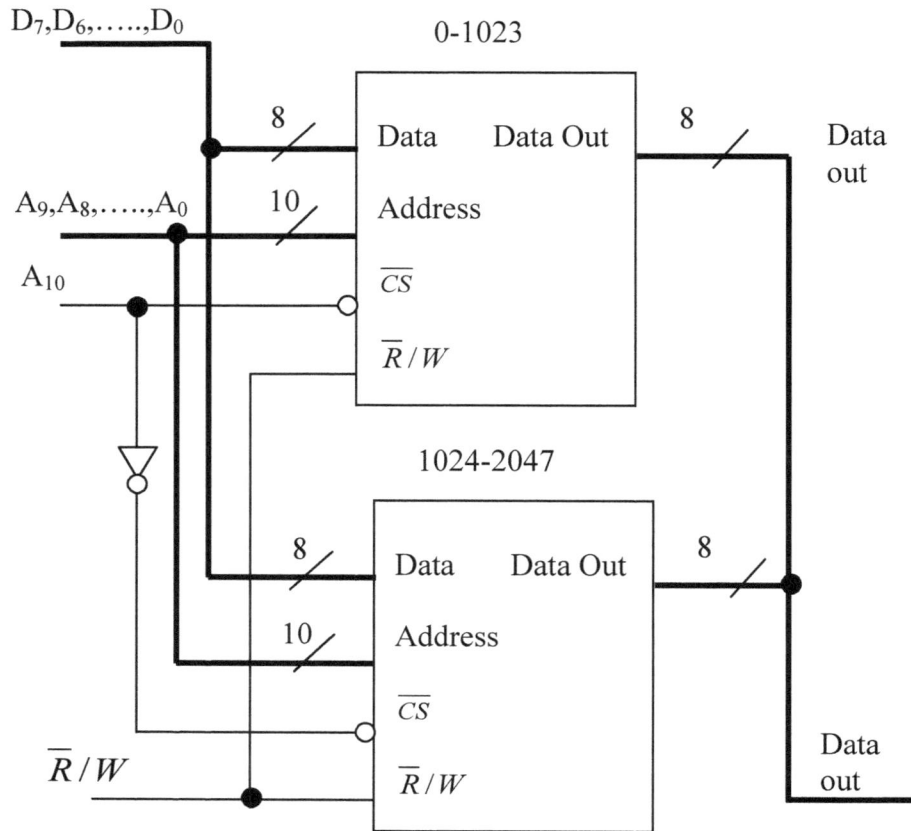

Figure 5.25. Representation of 2048 × 8 bit memory.

$D_7, D_6, \ldots, D_0$

0-1023

| 8 | Data | Data Out | 8 | Data out |

$A_9, A_8, \ldots, A_0$ 10 Address

$\overline{CS}$
$\overline{R}/W$

decoder 2x4

A_{10}
A_{11}

1024-2047

| 8 | Data | Data Out | 8 |

10 Address

$\overline{CS}$
$\overline{R}/W$

$\overline{R}/W$

Data out

2048-3071

| 8 | Data | Data Out | 8 | Data out |

10 Address

$\overline{CS}$
$\overline{R}/W$

3072-4095

| 8 | Data | Data Out | 8 |

10 Address

$\overline{CS}$
$\overline{R}/W$

$\overline{R}/W$

Data out

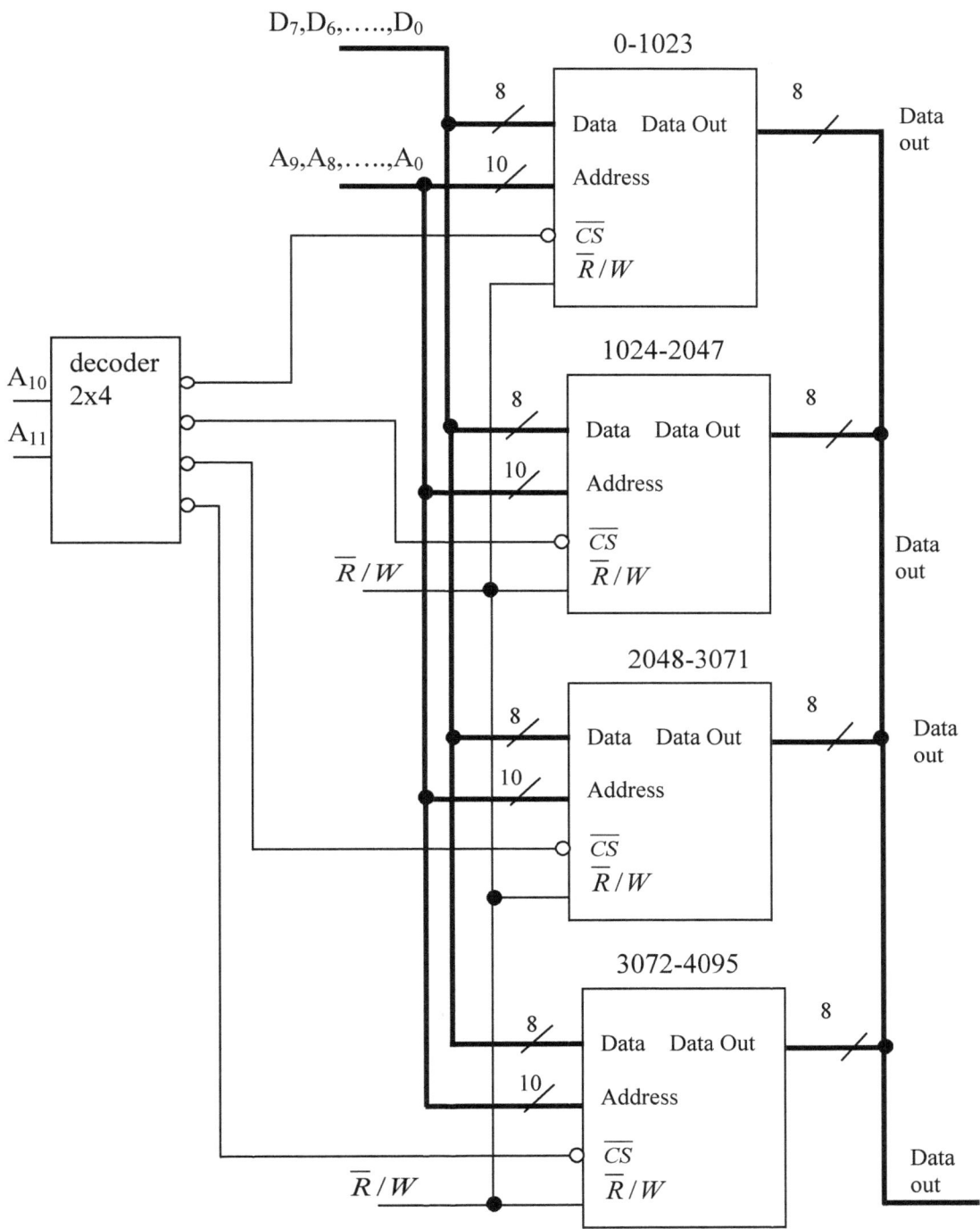

Figure 5.26. Representation of 4096×8 bit memory.

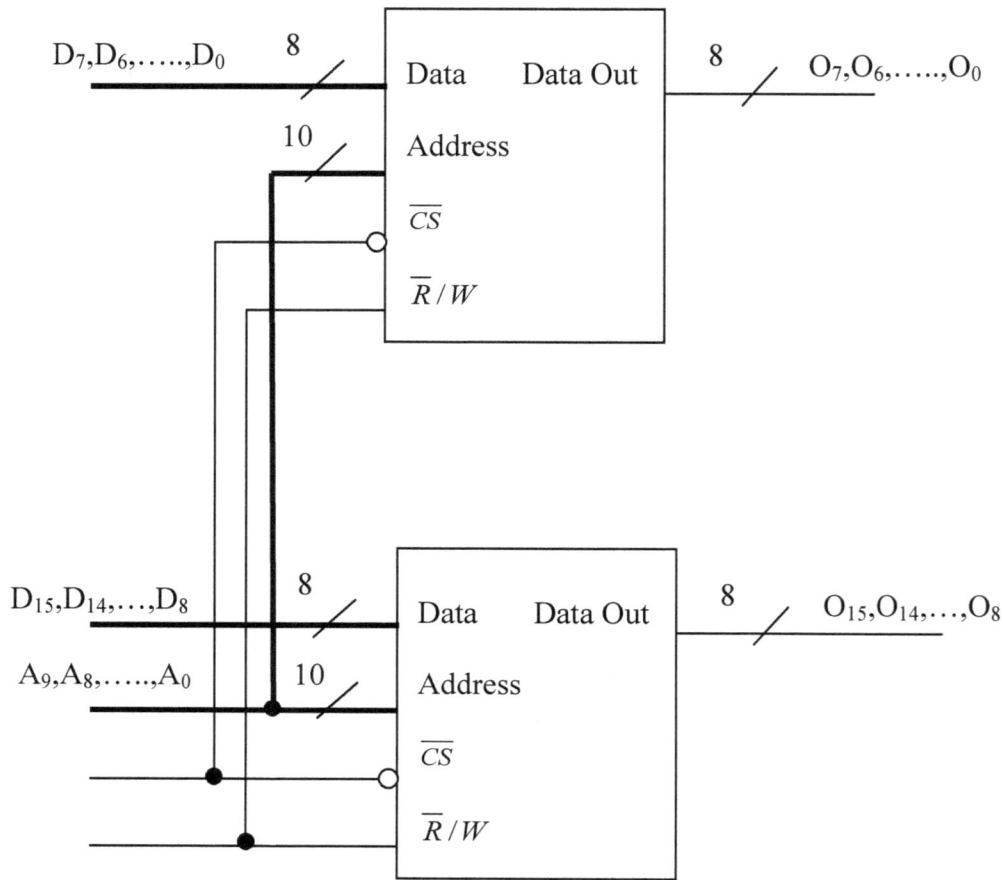

Figure 5.27. Representation of 1K × 16 bit memory.

5.9. **Counters**

A counter is a digital circuit that has binary outputs (states) that progress through a fixed sequence. The count sequence can be ascending (UP) or descending (DOWN). The count sequence is defined by the term **modulus**. The modulus is the number of states through which the counter progresses. For example, a 3 bit UP counter with a modulus of 8 counts from 000 to 111 in binary, or 0 to 7 in decimal. As another example, a 3 bit DOWN counter with a modulus of 8 counts from 111 to 00 in binary, or 7 to 0 in decimal. Both of these counters have 8 states and are modulus-8 or simply mod-8 counters. The maximum modulus of a n bit counter is defined by:

$$m_{max} = 2^n$$

EQ. 5.2

A **full sequence counter** counts to its maximum modulus. A **truncated sequence counter** is one whose modulus is less than the maximum modulus or $m < 2^n$.

The mod-8 UP counter count sequence Table is shown in Table 5.13. The mod-8 UP counter starts at 000 and wraps around to 000 at 111. This is a continuous cycle. The state diagram for the mod-8 UP counter is shown in Figure 5.28. The timing diagram for the mod-8 UP counter is shown in Figure 5.29.

count	Q_2	Q_1	Q_0
0	0	0	0
1	0	0	1
2	0	1	0
3	0	1	1
4	1	0	0
5	1	0	1
6	1	1	0
7	1	1	1

Table 5.13. Mod-8 count sequence Table.

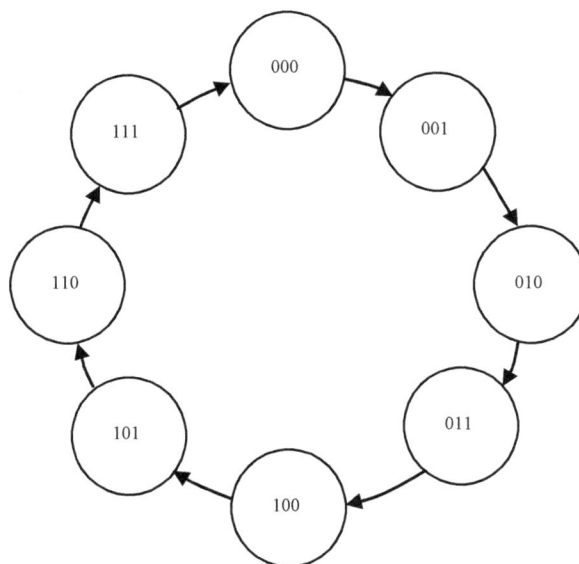

Figure 5.28. Mod-8 state diagram.

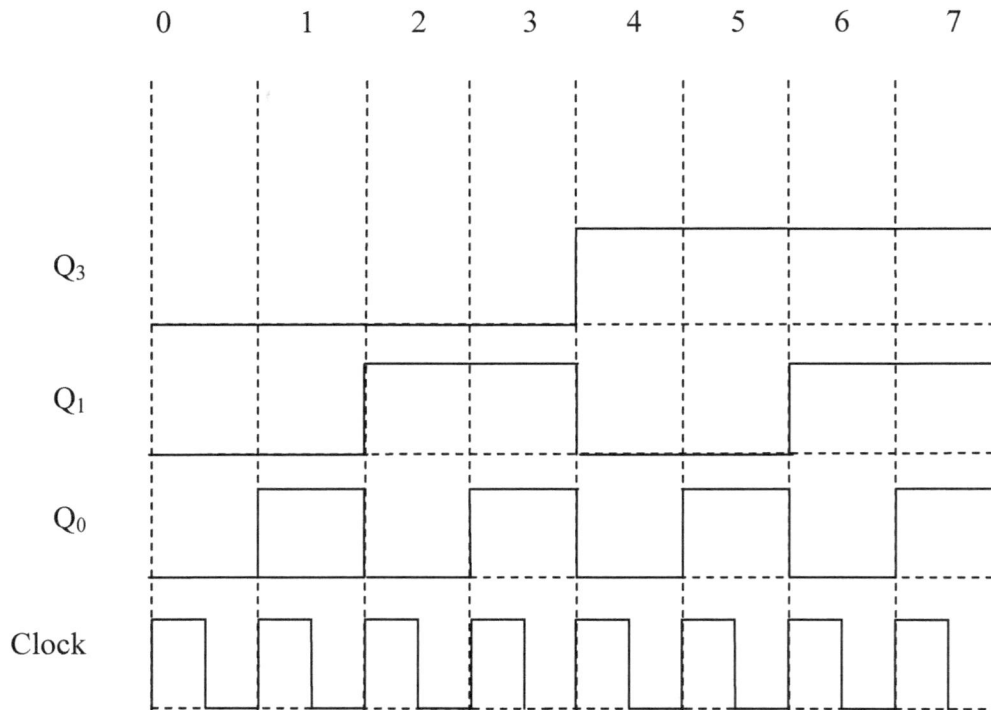

Figure 5.29. Mod-8 UP counter timing diagram.

In summary, **counters** can be used as follows:
   A. count the number of clock pulses in a given time
   B. divide frequency and store data
   C. sequential addressing as in some arithmetic circuits

Counters have the following important characteristics:
   A. maximum number of counts
   B. UP or DOWN count
   C. asynchronous (ripple) or synchronous (parallel) operation
   D. free running or self stopping

5.10. Counters in Cascade

Counters can be put in cascade to get a larger modulus. For example, to count from 0 to 99, use two mod-10 counters in cascade. In general, the overall modulus is equal to the product of the individual modulus as shown in Figure 5.30A, Figure 5.30B, and Figure 5.30C. This is done by connecting the last output of another counter to the clock input of the next counter as shown in Figure 5.30D.

Totol Modulus =16×16= 256

Input   Output

f_{in} ——→ | MOD-16 | ——→ | MOD-16 | ——→ $\dfrac{f_{in}}{256}$

A.)

Totol Modulus = 4×10×8=320

Input   Output

f_{in} ——→ | MOD-4 | →| MOD-10 | →| MOD-8 | ——→ $\dfrac{f_{in}}{320}$

B.)

Totol Modulus = 10×10=100

Output

Input

f_{in} ——→ | MOD-4 | →| MOD-10 | → $\dfrac{f_{in}}{100}$

C.)

CLK   A
                                 B
                                 C
                                 D   CLK   A
                                                              B
                                                              C
                                                              D

D.)

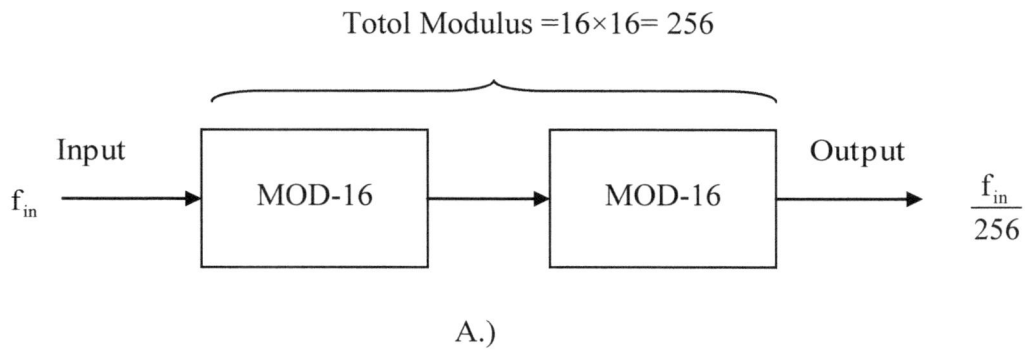

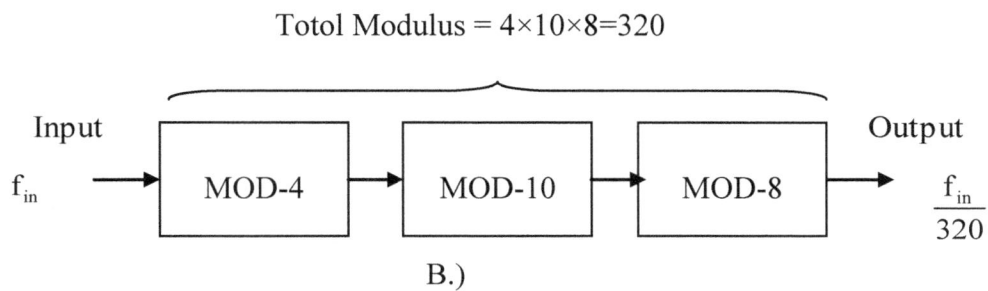

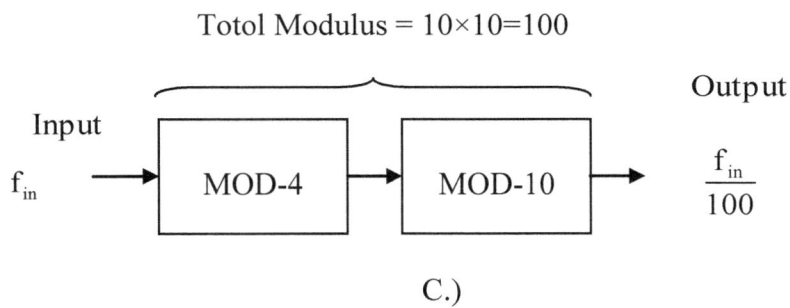

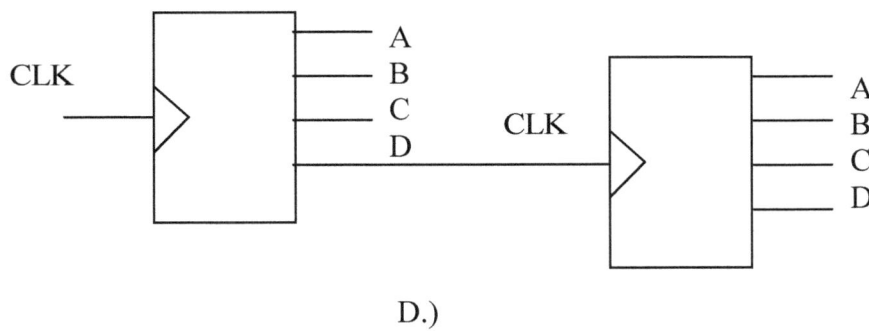

Figure 5.30. Examples of cascading counters.

5.11. Asynchronous Counters

A binary ripple counter consists of a series of connection of each flip-flop with the clock input of each flip-flop connected to the Q output of the previous flip-flop. These flip-flops are T or JK type. The flip-flops change one at a time in a ripple fashion through the flip-flops. Thus, ripple counters are also called asynchronous counters. With synchronous counters the clocked signal is connected to the input clock of all the flip-flops.

An example of a four bit ripple counter is shown Figure 5.31. This is an UP counter with count sequence given in Table 5.14.

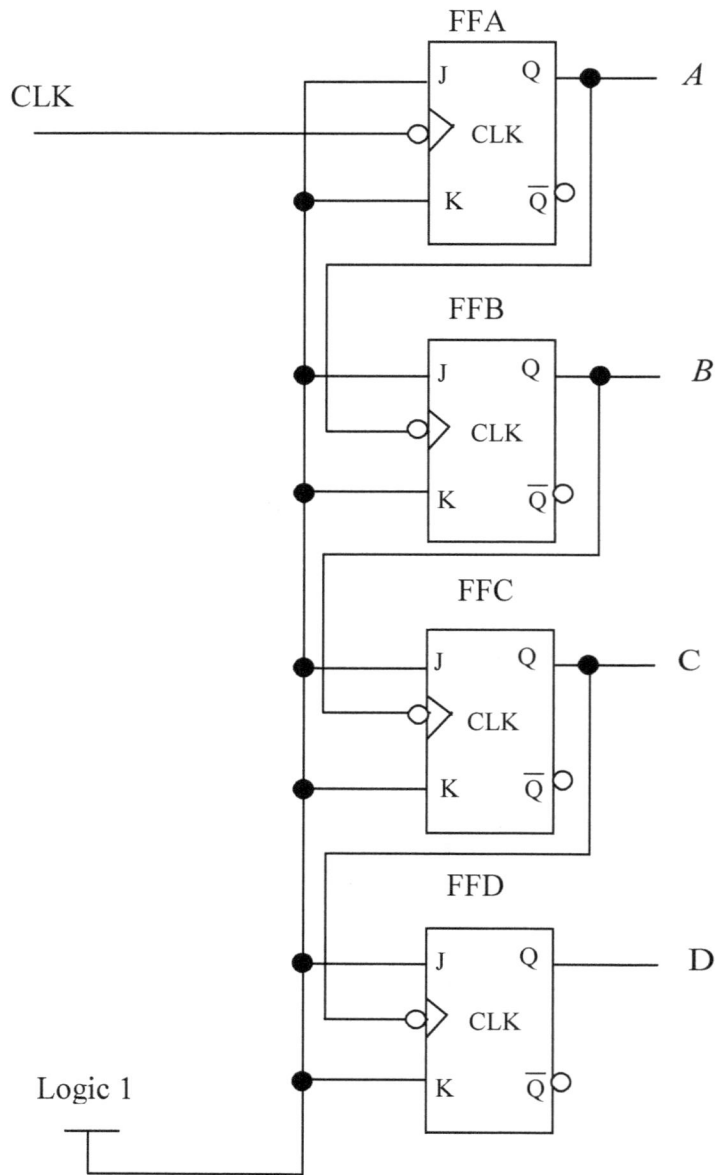

Figure 5.31. 4 bit binary ripple UP counter.

Decimal	D	C	B	A	D	C	B	A	Decimal
	UP Count				DOWN Count				
0	0	0	0	0	1	1	1	1	15
1	0	0	0	1	1	1	1	0	14
2	0	0	1	0	1	1	0	1	13
3	0	0	1	1	1	1	0	0	12
4	0	1	0	0	1	0	1	1	11
5	0	1	0	1	1	0	1	0	10
6	0	1	1	0	1	0	0	1	9
7	0	1	1	1	1	0	0	0	8
8	1	0	0	0	0	1	1	1	7
9	1	0	0	1	0	1	1	0	6
10	1	0	1	0	0	1	0	1	5
11	1	0	1	1	0	1	0	0	4
12	1	1	0	0	0	0	1	1	3
13	1	1	0	1	0	0	1	0	2
14	1	1	1	0	0	0	0	1	1
15	1	1	1	1	0	0	0	0	0

Table 5.14. UP count vs. DOWN count Truth Table.

An example of a four bit ripple counter is shown Figure 5.32. This is an DOWN counter with count sequence given in Table 5.14.

The count sequence of a **binary decimal** (BCD) counter is from 0 to 9, afterwards, it repeats. Thus, the BCD is a mod-10 counter. The state diagram of the BCD counter is shown in Figure 5.33A. The BCD count sequence is shown in Figure 5.33B. The key to recycling the mod-10 counter (as in any mod-N counter) is to look at the binary count immediately after the last count of the counter, which is 1010 for the mod-10 counter. Use a detection circuit to clear the JK flip-flops when this binary sequence occurs. This is shown in Figure 5.34.

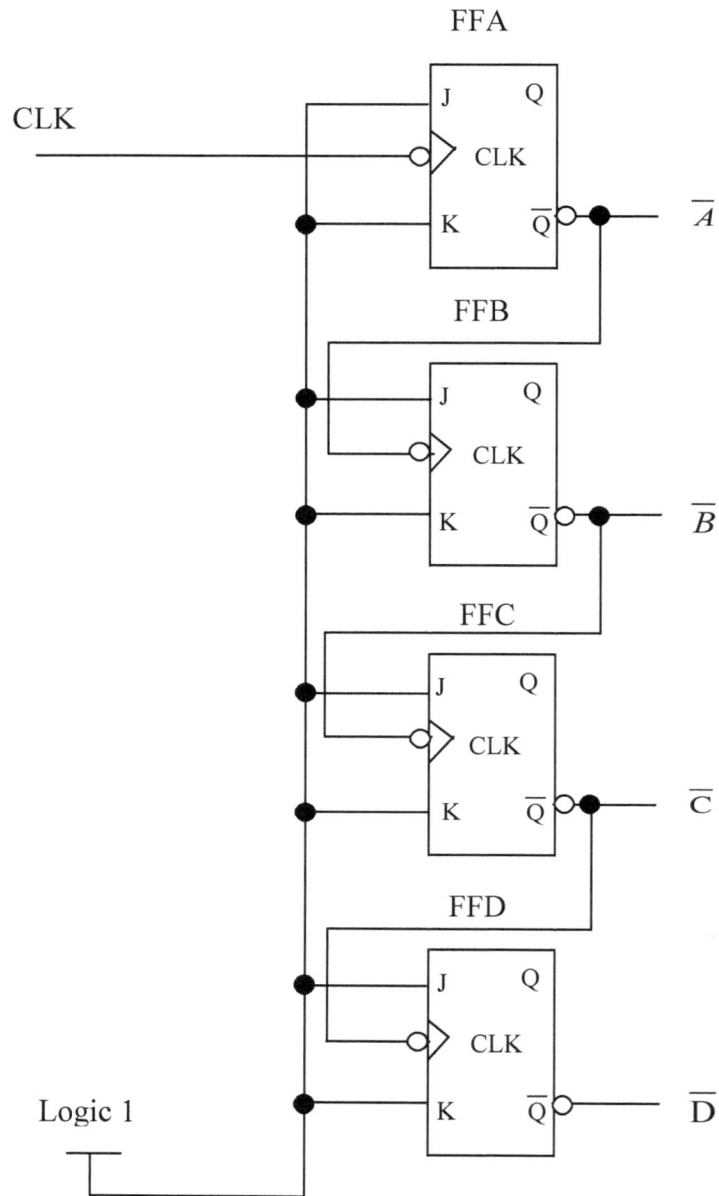

Figure 5.32. 4 bit binary ripple DOWN counter.

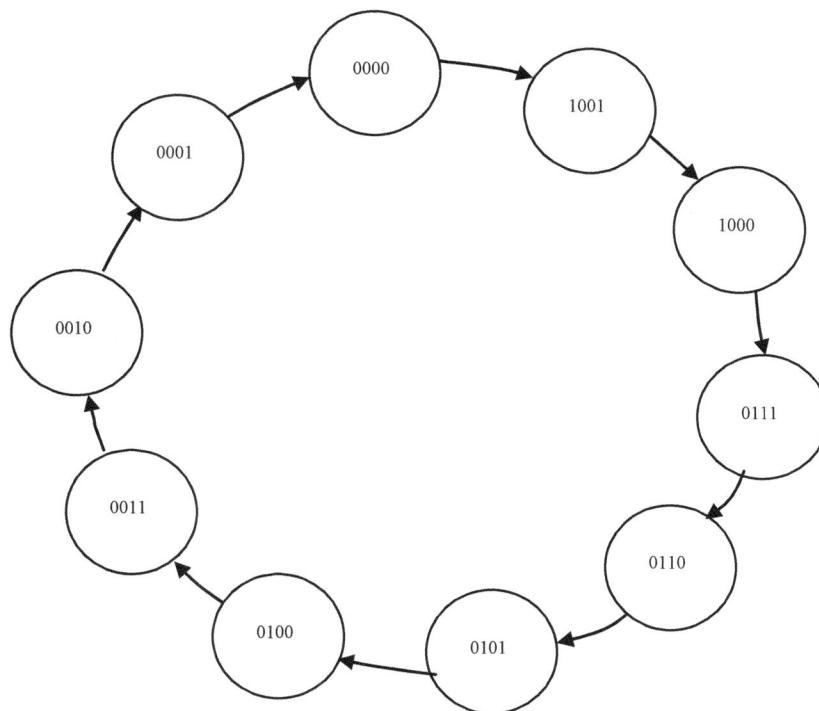

A.)

Decimal	D	C	B	A
0	0	0	0	0
1	0	0	0	1
2	0	0	1	0
3	0	0	1	1
4	0	1	0	0
5	0	1	0	1
6	0	1	1	0
7	0	1	1	1
8	1	0	0	0
9	1	0	0	1
10	1	0	1	0
11	1	0	1	1
12	1	1	0	0
13	1	1	0	1
14	1	1	1	0
15	1	1	1	1

B.)

Figure 5.33. A.) State diagram of mod-10 counter. B.) mod-10 count sequence.

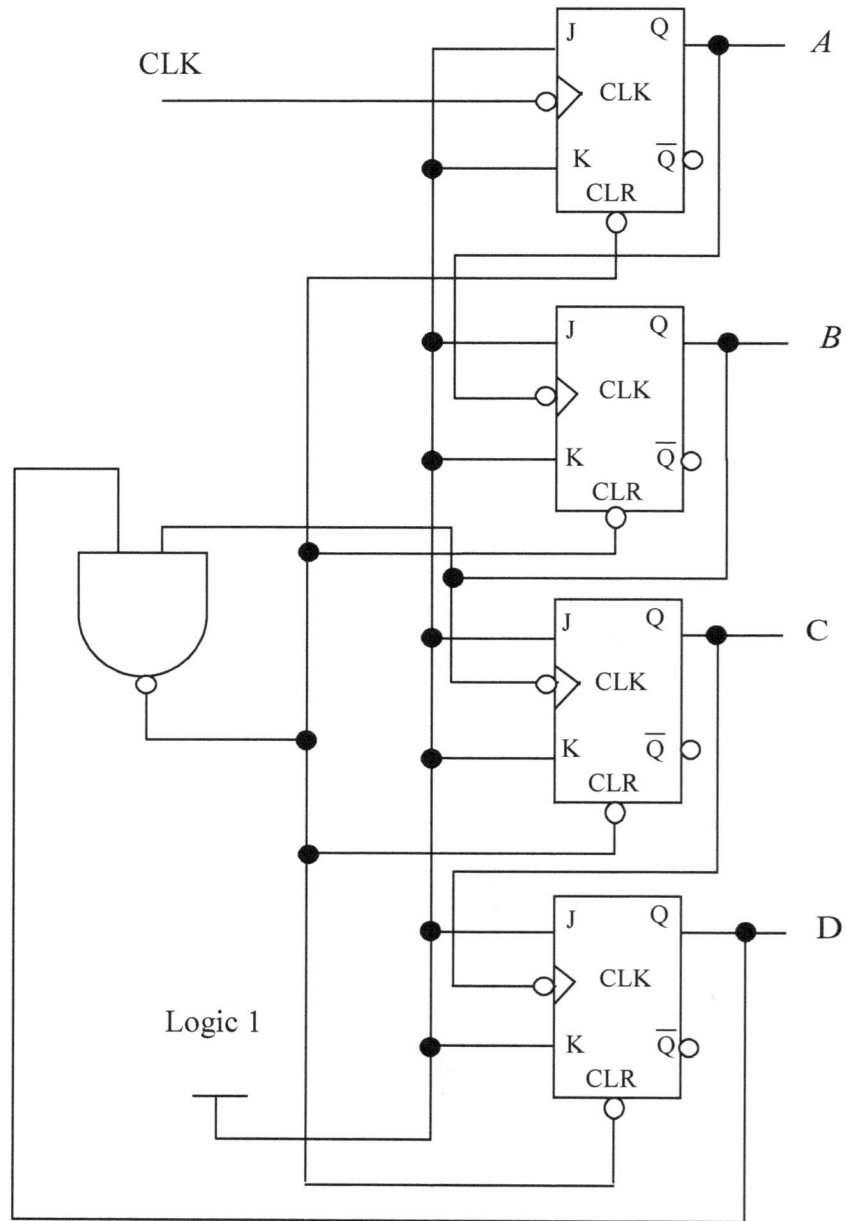

Figure 5.34. 4 bit binary ripple UP counter with detector circuit.

5.12.   From Sequential Circuits to the State Diagram

Sequential logic circuits can be analyzed using one of the following:
   A. State Table
   B. State Diagram

The **State Table** shows the time sequence of inputs, outputs, flip flop present states, and flip flop next states in Table form. The **state diagram** shows the time sequence of inputs and outputs, flip flop present states, and next states in a graphical form. The state diagram shows the transition between states using lines. Regardless of the representation, the outputs of combination logic circuits represent the present state. The outputs of flip flops, upon the next arrival of the clock pulse, represent the next state.

The State Table can be easily derived from a given digital circuit. The state diagram is then easily derived from the State Table. The only difference between the state table and the state diagram is the manner of representation.

EXAMPLE 5.5

Using D flip-flops, determine the State Table and state diagram of the clocked sequential circuit in Figure 5.35.

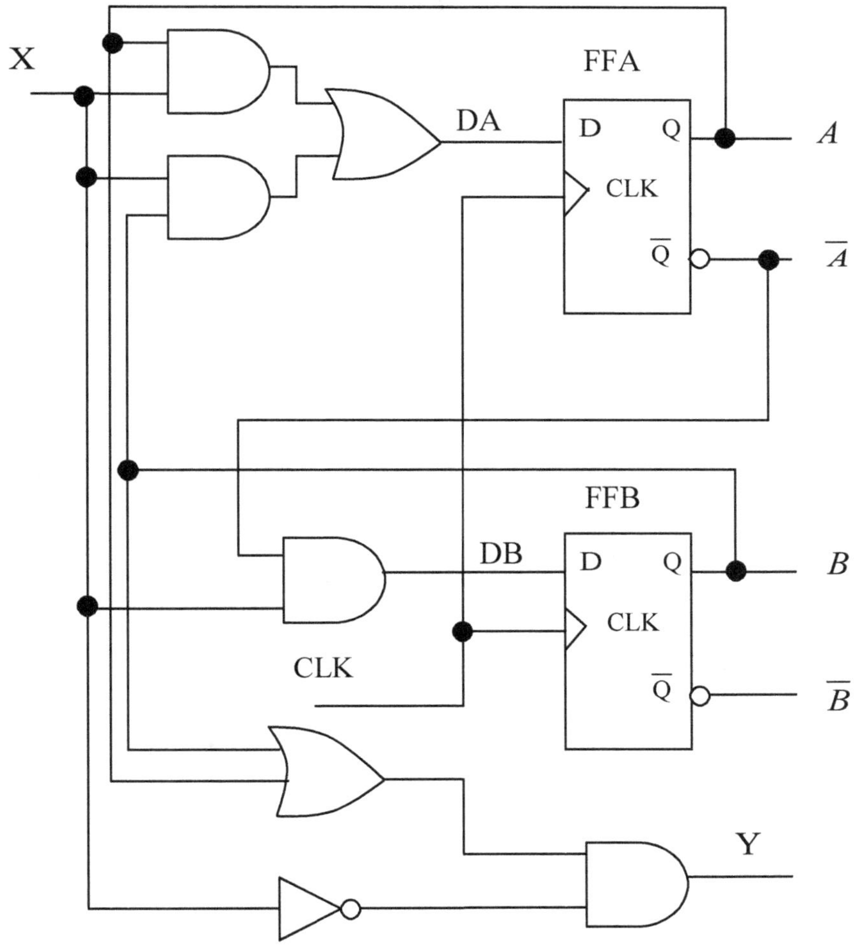

Figure 5.35. A D flip-flop sequential circuit.

Solution:

Figure 5.35 consists of two D flip flops named FFA and FFB. FFA has outputs A and $\overline{A}$ and while FFB has outputs B and $\overline{B}$. Figure 5.35 also consists of the input x and the output y. To determine the State Table:

1. Determine the Boolean equations of the D flip-flop inputs in terms of the present states, denoted by the variable n.
2. Determine the next state (n+1) Boolean expression for each flip-flop (FFA & FFB) output, in this case, A(n+1) and B(n+1), in terms of the present state inputs (X(n), A(n), and B(n)).
3. Determine the Boolean equations of the non-flip-flop circuit outputs, in this case, Y(n), in terms of the present state, denoted by the variable n.
4. Determine the State Table in term of all possible input combinations (in this case, X(n), A(n), and B(n)) for each output. This is **Form I** of the state Table.
5. Determine **Form II** of the State Table. Form II is the same as Form I but with the combinational inputs and outputs arranged a little differently in the State Table that makes it easier to derive the state diagram.

Using Figure 5.36, the Boolean equation for DA is

$$DA(n) = A(n)X(n) + B(n)X(n)$$

or using the short notation form gives

$$DA = AX + BX$$

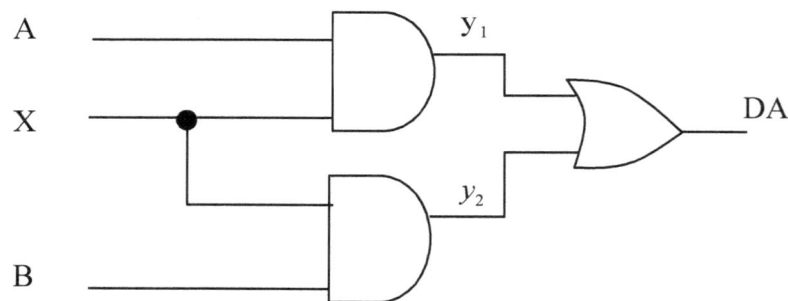

Figure 5.36. DA Circuit.

Therefore, the flip-flop FFA has next state equation

$$A(n+1) = A(n)X(n) + B(n)X(n)$$

or in short notation form gives

$$A_{n+1} = AX + BX$$

Similarly, flip-flop FFB has next state equation

$$B(n+1) = \overline{A(n)}X(n)$$

or in short notation form gives

$$B_{n+1} = \overline{A}X$$

The output y has Boolean equation

$$Y(n) = (A(n) + B(n))\overline{X(n)}$$

or in short notation form to give

$$Y = (A + B)\overline{X}$$

The State Table in terms of all possible input combinations for each output is given in Table 5.16. This form of the State Table will be referred to as **Form I** of the State Table because it gives the outputs for all possible input combinations similar to the standard Truth Table. This is where remembering the Truth Table for the D flip-flop comes in handy. It is given here in Table 5.15 for convenience.

CLK	D	Q	$\overline{Q}$	Next State
⬏	0	0	1	Reset State
⬏	1	1	0	Set State

Table 5.15. D flip-flop Truth Table.

Present States			Next States		Outputs
B	A	X	B_{n+1}	A_{n+1}	Y
0	0	0	0	0	0
0	0	1	1	0	0
0	1	0	0	0	1
0	1	1	0	1	0
1	0	0	0	0	1
1	0	1	1	1	0
1	1	0	0	0	1
1	1	1	0	1	0

Table 5.16. Form I of the State Table of Figure 5.35.

Form II of the State Table of Table 5.16 is shown Table 5.17.

Present State		Next State				Output	
		X=1		X=0		X=1	X=0
B	A	B_{n+1}	A_{n+1}	B_{n+1}	A_{n+1}	Y	Y
0	0	1	0	0	0	0	0
0	1	0	1	0	0	0	0
1	0	1	1	0	0	0	1
1	1	0	1	0	0	1	0

Table 5.17. Form II of the State Table of Figure 5.35.

The state diagram is taken from the reduced State Table in Table 5.17. Now, using Table 5.17, the state diagram of Figure 5.36 is provided in Figure 5.37. In Figure 5.37, the binary number inside each circle represents the state of the flip-flops. The lines between the states are the inputs in binary form separated by the outputs in binary form, e.g., X/Y. For example, using Figure 5.37, if the present state is BA=00 and the input X=1, then the output Y=0. Thus, at the next clock pulse the flip-flops transitions to state $A_{n+1}B_{n+1}=10$.

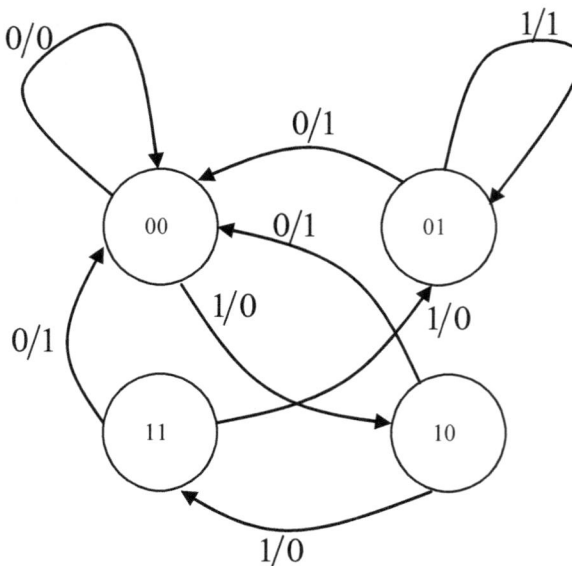

Figure 5.37. State diagram of Figure 5.36.

EXAMPLE 5.6

Using JK flip-flops, determine the State Table and state diagram of the clocked sequential circuit in Figure 5.38.

Solution:
Using Figure 5.38 for FFA, the Boolean equation for JA is
$$JA(n) = B(n)$$
or using the short notation form gives
$$JA = B$$

The Boolean equation for KA is
$$KA(n) = B(n)\overline{X(n)}$$
or using the short notation form gives

$$KA = B\overline{X}$$

Using Figure 5.38 for FFB, the Boolean equation for JB is

$$JB(n) = \overline{X(n)}$$

or using the short notation form gives

$$JB = \overline{X}$$

and the Boolean equation for KB is

$$KB(n) = A(n)\overline{X(n)} + \overline{A(n)}X(n)$$

or using the short notation form gives

$$KB = A\overline{X} + \overline{A}X$$

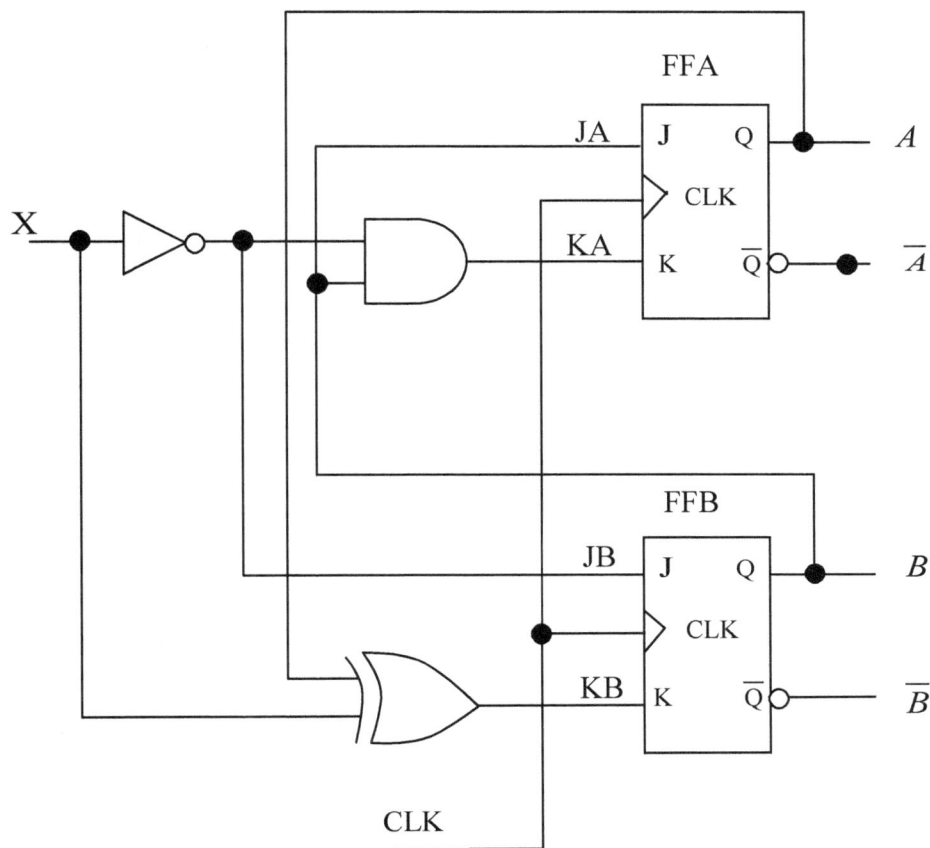

Figure 5.38. A JK flip-flop sequential circuit.

Recall that the Truth Table of the JK FF as indicated in Table 5.18. It will be useful to remember it for determining the next states in Form I or Form II of the State Table.

CLK	J	K	Q	$\overline{Q}$	Next State
⌐⌐	0	0	Q	$\overline{Q}$	Hold
⌐⌐	0	1	0	1	Reset State
⌐⌐	1	0	1	0	Set State
⌐⌐	1	1	$\overline{Q}$	Q	Toggle

Table 5.18. JK inputs present State Table.

The State Table in terms of all possible input combinations for each output is given in Table 5.19B. Table 5.19A shows the present state of the flip-flop JK inputs relative to all possible input combinations. Table 5.19A, representing the flip-flop inputs, is convenient for transitioning to Table 5.19B.

Present States			JK Present States			
B	A	X	JA	KA	JB	KB
0	0	0	0	0	1	0
0	0	1	0	0	0	1
0	1	0	0	0	1	1
0	1	1	0	0	0	0
1	0	0	1	1	1	0
1	0	1	1	0	0	1
1	1	0	1	1	1	1
1	1	1	1	0	0	0

Present States			Next States	
B	A	X	B_{n+1}	A_{n+1}
0	0	0	1	0
0	0	1	0	0
0	1	0	1	1
0	1	1	0	1
1	0	0	1	1
1	0	1	0	1
1	1	0	0	0
1	1	1	1	1

A.)   B.)

Table 5.19. A.) JK inputs Truth Table.   B.) Form I of the State Table of Figure 5.35.

Form II of the State Table is shown in Table 5.20 and the corresponding state diagram is given in Figure 5.39.

Present State		Next State			
		X=1		X=0	
B	A	B_{n+1}	A_{n+1}	B_{n+1}	A_{n+1}
0	0	0	0	1	0
0	1	0	1	1	1
1	0	0	1	1	1
1	1	1	1	0	0

Table 5.20. Form II of the State Table of Figure 5.35.

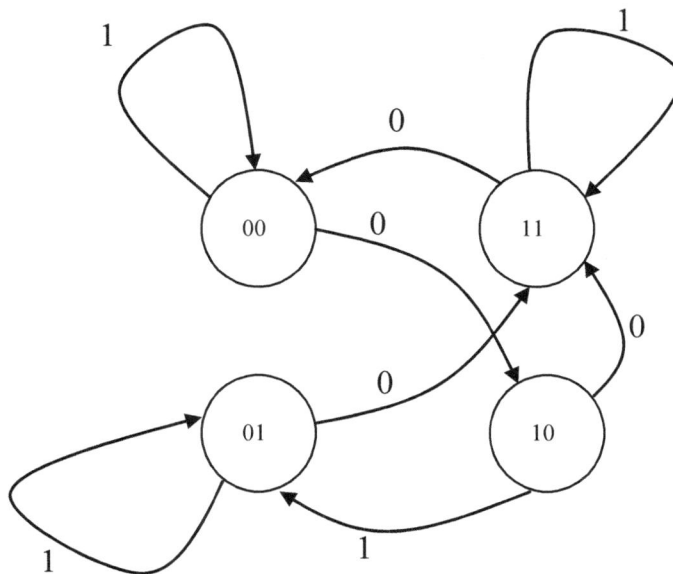

Figure 5.39. State diagram of Figure 5.38.

5.13. **From the State Diagram to Sequential Circuits**

The design procedure for deriving the sequential logic circuit from the state diagram is given as follows:

1. Obtain the state diagram or timing diagram from the word description or set of specifications.
2. Determine the number of flip-flops and assign letters to each.
3. Chose the type of flip-flop, JK, D, T, etc. The type of flip-flop to use may be included in the design specifications or depends on what flip-flops are available. Generally, it is recommended to use D flip-flops for data transfer applications(registers), T flip-flops for complementation applications (counters), and JK flip-flops for general applications.
4. Obtain Form I of the State Table.
5. Using the selected flip-flop excitation Table derive the combinational logic excitation Table. The excitation Table shows all possible input combinations for a flip-flop input that is required to go from the flip-flops's present state to its next state.
6. Derive the flip-flop input functions. Derive the combinational circuit's output functions.
7. Draw the digital circuit.

EXAMPLE 5.7. Design Example.

Using JK flip-flops, and the state diagram in Figure 5.40, determine the 1.) State Table and 2.) Boolean equations. The states within the bubbles represent the flip-flop next states in the following order, BA.

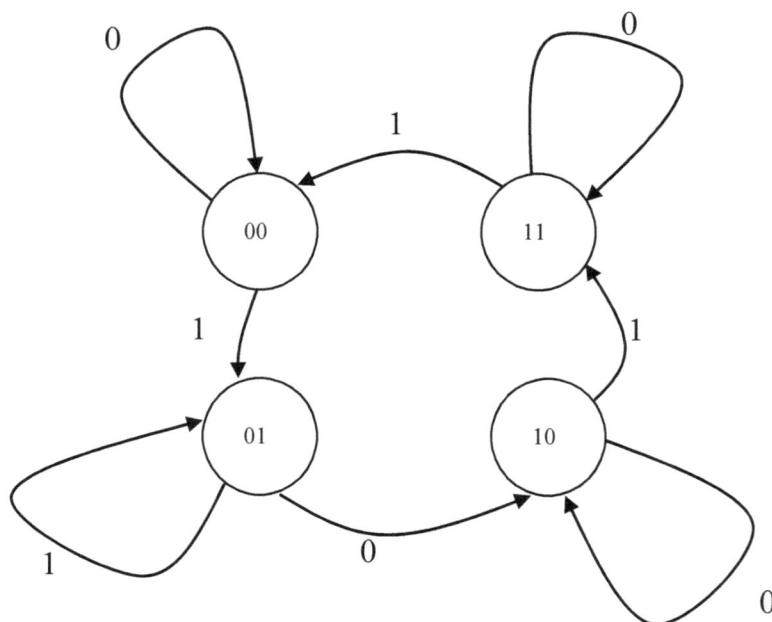

Figure 5.40. State diagram.

Solution:

1. Since the JK flip flop and state diagram has already been specified, we start by obtaining Form I of the State Table.

Present States			Next States	
B	A	X	B_{n+1}	A_{n+1}
0	0	0	0	0
0	0	1	0	1
0	1	0	1	0
0	1	1	0	1
1	0	0	1	0
1	0	1	1	1
1	1	0	1	1
1	1	1	0	0

Table 5.21. Form I of the State Table.

2. Obtain the JK flip-flop excitation Table as shown in Table 5.23 from Table 5.22. The last J and K column in Table 5.23 are derived by observing that in order to go from the present flip-flop state, $B_n=0$ (same as B=0), to the next state flip-flop state, $B_{n+1}=0$, JB and KB, must be a reset condition (JB=0, KB=0) or hold condition (JB=1, KB=0). Thus, both hold and reset condition have a combined result of JB=0 and KB=X, where X is a don't care condition, which means KB can be a 0 or 1. The same analysis is used to derive the JB, KB, JA, and KA columns in Table 5.24.

CLK	J	K	Q_{n+1}	Next State
⌐⌐	0	0	Q	Hold
⌐⌐	0	1	0	Reset State
⌐⌐	1	0	1	Set State
⌐⌐	1	1	$\overline{Q}$	Toggle

Table 5.22. JK flip-flop Truth Table. Note: where the subscript is not specified use present state n.

Q_n	Q_{n+1}	J	K		J	K		J	K
0	0	0	0		0	1		0	X
0	1	1	0	or	1	1	=	1	X
1	0	0	1		1	1		X	1
1	1	0	0		1	0		X	0

Table 5.23. JK flip-flop excitation Table. Note: where the subscript is not specified use present state n.

B	A	X	B_{n+1}	A_{n+1}	JB	KB	JA	KA
0	0	0	0	0	0	X	0	X
0	0	1	0	1	0	X	1	X
0	1	0	1	0	1	X	X	1
0	1	1	0	1	0	X	X	0
1	0	0	1	0	X	0	0	X
1	0	1	1	1	X	0	1	X
1	1	0	1	1	X	0	X	0
1	1	1	0	0	X	1	X	1

Table 5.24. Combinational logic circuit excitation Table. Note: where the subscript is not specified use present state n.

3. Now, derive the combinational logic equations for JA, KA, JB, and KB. Using Table 5.24 and the K-Map approach for JA gives:

	$\overline{X}$	X
$\overline{B}\,\overline{A}$		1
$\overline{B}A$	X	X
BA	X	X
$B\overline{A}$		1

$\Rightarrow$ $JA = X$

Using Table 5.24 and the K-Map approach to derive KA gives:

	$\overline{X}$	X
$\overline{B}\,\overline{A}$	X	X
$\overline{B}A$	1	
BA		1
$B\overline{A}$	X	X

$\Rightarrow$ $KA = \overline{B}\,\overline{X} + BX = \overline{B \oplus X}$

Using Table 5.24 and the K-Map approach for JB gives:

	$\overline{X}$	X
$\overline{B}\,\overline{A}$		
$\overline{B}A$	1	
BA	X	X
$B\overline{A}$	X	X

$\Rightarrow$ $JB = A\overline{X}$

Using Table 5.24 and the K-Map approach for KB gives:

	$\overline{X}$	X
$\overline{B}\,\overline{A}$	X	X
$\overline{B}A$	1	X
BA		1
$B\overline{A}$		

$\Rightarrow$ $KB = AX$

EXAMPLE 5.8. Design Example.

Using D flip-flops, and the state diagram in Figure 5.41:
1.) Determine the State Table
2.) Determine the Boolean equations
3.) Draw the sequential circuit

The states within the bubbles represent the flip-flop next states in the following order, BA.

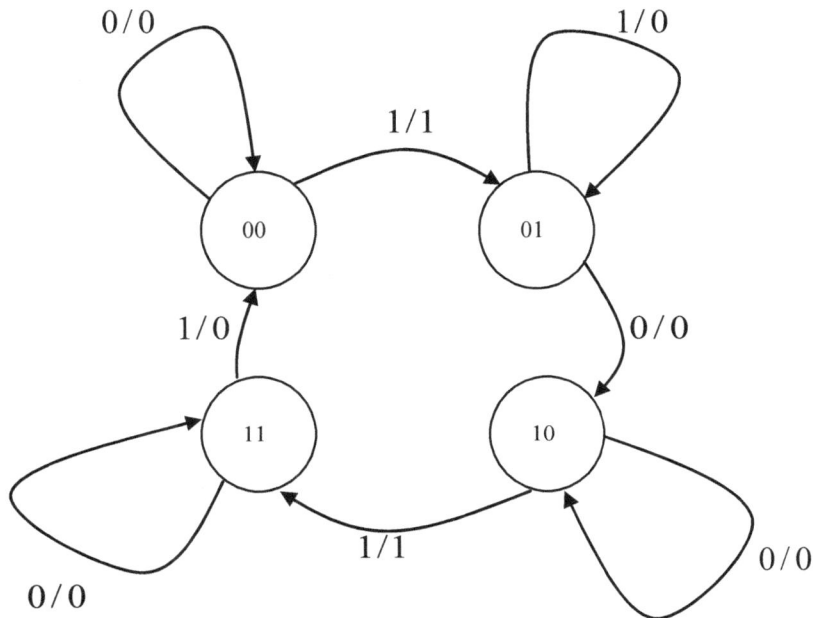

Figure 5.41. State diagram.

Solution:
1. Obtain Form I of the State Table.

Present States			Next States		Output
B	**A**	**X**	**B_{n+1}**	**A_{n+1}**	**Y**
0	0	0	0	0	0
0	0	1	0	1	1
0	1	0	1	0	0
0	1	1	0	1	0
1	0	0	1	0	0
1	0	1	1	1	1
1	1	0	1	1	0
1	1	1	0	0	0

Table 5.25. Form I of the State Table.

2. Obtain the D flip-flop **excitation Table** as shown in Table 5.27 from the D flip-flop Truth Table in Table 5.26. In Table 5.26, D=0 is required for the D flip-flop to go from a present state of $Q_n=0$ to a next state of $Q_{n+1}=0$. This same analysis is applied to the DB and DA columns in Table 5.28. Thus, the flip-flop excitation Table transforms the state diagram to a combinational circuit Truth Table in Table 5.28 needed to design the combinational logic circuit portion of the sequential circuit.

CLK	D_n	Q_{n+1}	Next State
⌐	0	0	Reset
⌐	1	1	Set

Table 5.26. D flip-flop Truth Table

Q_n	Q_{n+1}	D_n	Next State
0	1	1	Set
1	0	0	Reset

Table 5.27. D flip-flop excitation Table.

B	A	X	B_{n+1}	A_{n+1}	Y	DB	DA	DB Minterm	DA Minterm	Y Minterm
0	0	0	0	0	0	0	0			
0	0	1	0	1	1	0	0		$\bar{B}\bar{A}X$	$\bar{B}\bar{A}X$
0	1	0	1	0	0	0	1	$\bar{B}A\bar{X}$		
0	1	1	0	1	0	0	1		$\bar{B}AX$	
1	0	0	1	0	0	1	0	$B\bar{A}\bar{X}$		
1	0	1	1	1	1	1	0	$B\bar{A}X$	$B\bar{A}X$	$B\bar{A}X$
1	1	0	1	1	0	1	1	$BA\bar{X}$	$BA\bar{X}$	
1	1	1	0	0	0	1	1			

Table 5.28. Combinational circuit excitation Table and minterms (or maxterms) for DA, DB, and Y. Note: where the subscript is not specified use present state n.

3. Now, derive the combinational logic circuit Boolean equations for DA, DB, and Y. From Table 5.28, the unsimplified Boolean equations for the D flip-flop inputs, DA and DB, in the short form notation are:

$$DA = \overline{B}\overline{A}X + \overline{B}AX + B\overline{A}X + BA\overline{X}$$
$$DB = \overline{B}A\overline{X} + B\overline{A}\overline{X} + B\overline{A}X + BA\overline{X}$$
$$Y = B\overline{A}\overline{X} + B\overline{A}X$$

These Boolean equations can be reduced. Using the K-Map approach for DA gives:

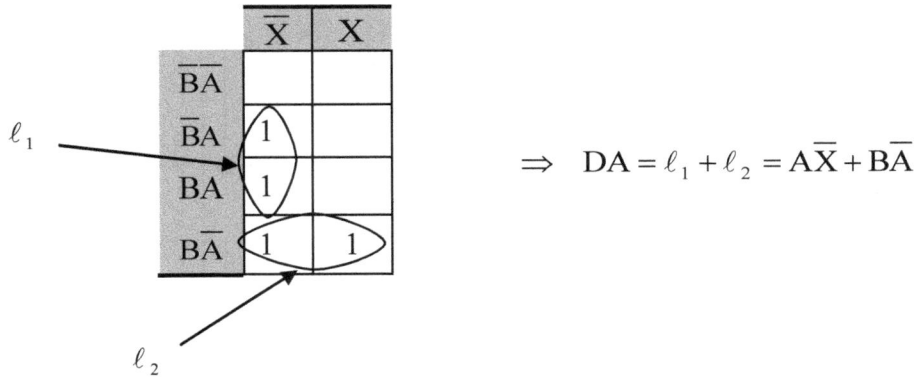

$$\Rightarrow \quad DA = \ell_1 + \ell_2 = A\overline{X} + B\overline{A}$$

Using the K-Map approach for DB gives:

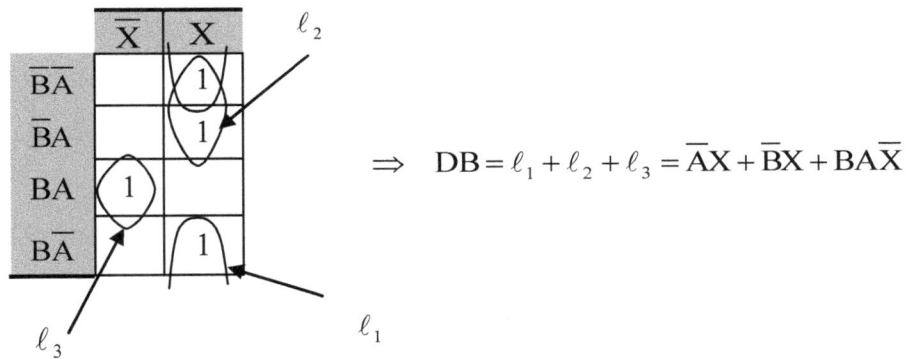

$$\Rightarrow \quad DB = \ell_1 + \ell_2 + \ell_3 = \overline{A}X + \overline{B}X + BA\overline{X}$$

Using the K-Map approach for Y gives:

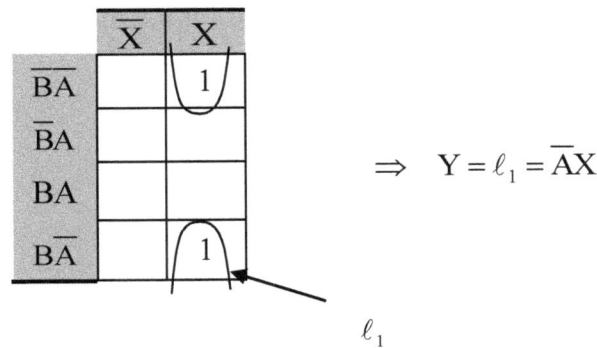

$$\Rightarrow \quad Y = \ell_1 = \overline{A}X$$

4. Finally, the D flip-flop sequential logic circuit is drawn in Figure 5.42.

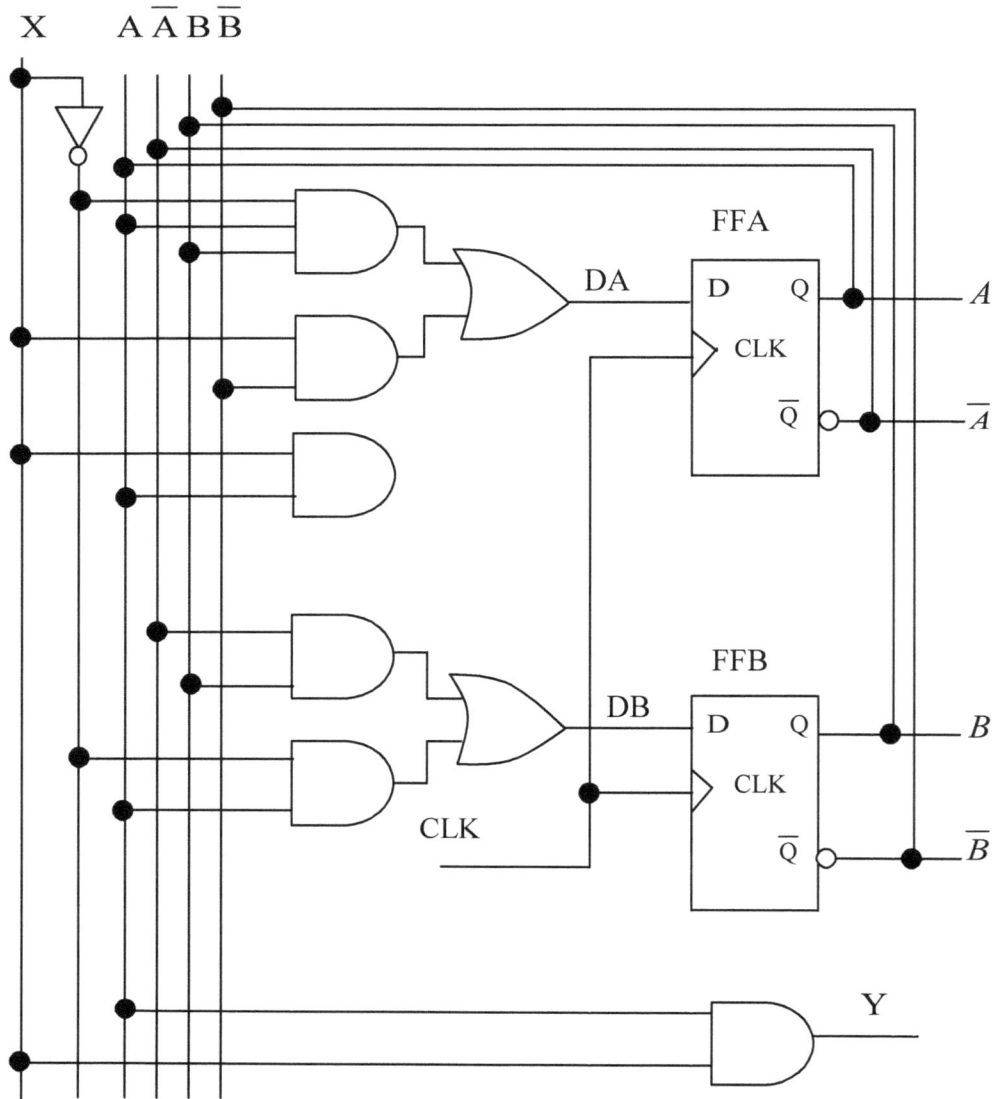

Figure 5.42. A D flip-flop sequential circuit for EXAMPLE 5.8.

5.14. Shift Registers

Shift registers are sequential logic circuits which are constructed of flip-flops. They are used for moving data and data storage. They are also used for changing serial data to parallel data and vice versa, etc. Registers can be classified into one of the following categories:

A. Serial in serial out (SISO)
B. Serial in parallel out (SIPO)
C. Parallel in serial out (PISO)
D. Parallel in parallel out (PIPO)
E. Rotate right
F. Rotate left

5.14.1. Serial In Serial Out (SISO)

A 4 bit SISO register is shown in Figure 5.43. On each clock pulse the shift register shifts data to the right one bit at a time. This is called serial data loading. An illustration of how data, 1010, is shifted through the 4 bit register is shown in Table 5.29. The first bit (left most) of data is sent to the D input of flip-flop A (FFA), followed by the second bit of data to D input of flip-flop A, and so on, until the register has all the data in and then out.

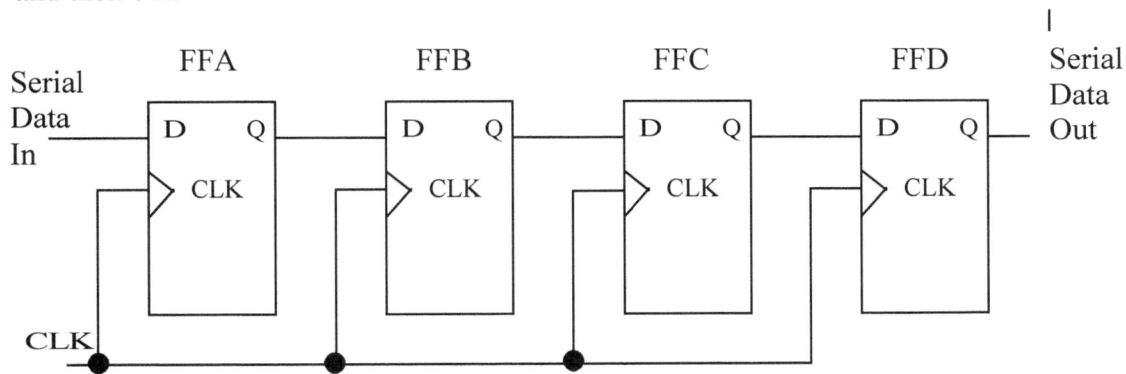

Figure 5.43. A 4 bit SISO register.

Clock Count	D	C	B	A	Serial Data In
0	0	0	0	0	Register is cleared
1	0	0	0	(1	1
2	0	0	(1	0	0
3	0	(1	0	1	1
4	(1	0	1	0)	0 (register contains all data)
5	0	1	0)	0	0
6	1	0)	0	0	0
7	0)	0	0	0	0
8)	0	0	0	0	0 (Register is clear)

Table 5.29. 4 bit SISO Truth Table.

The VHDL code for the 4 bit SISO register is given as follows:

```
library IEEE
use IEEE.std_logic_1164.all;

entity siso4_reg is
port (    serial_in, clk: in std_logic;
          q:buffer std_logic_vector(3 downto 0));
end siso4_reg;

architecture right_shift of siso4_reg is
signal d: std_logic_vector(3 downto 0);
begin

process(clk)
begin
     if CLK'event and CLK='1' then
          q=d;
     end if;
end process;

d<= serial_in & q(3 downto 1);

end right_shift;
```

```
d(3) <= serial_in;
d(2) <= q(3);
d(1) <=q(2);
d(0) <=q(1);
```

Another way to perform a shift right

5.14.2. Serial In Parallel Out (SIPO)

A 4 bit SIPO register is shown in Figure 5.44. An illustration of how data, 1001, is shifted through the 4 bit register is shown in Table 5.30. The first bit (MSB) of data is sent to the D input of flip-flop A (FFA), followed by the second bit of data to D input of flip-flop A, and so on until the register has all the data. The timing diagram in Figure 5.45 shows how serial data 000 and 100 are shifted into the 4 bit SIPO register.

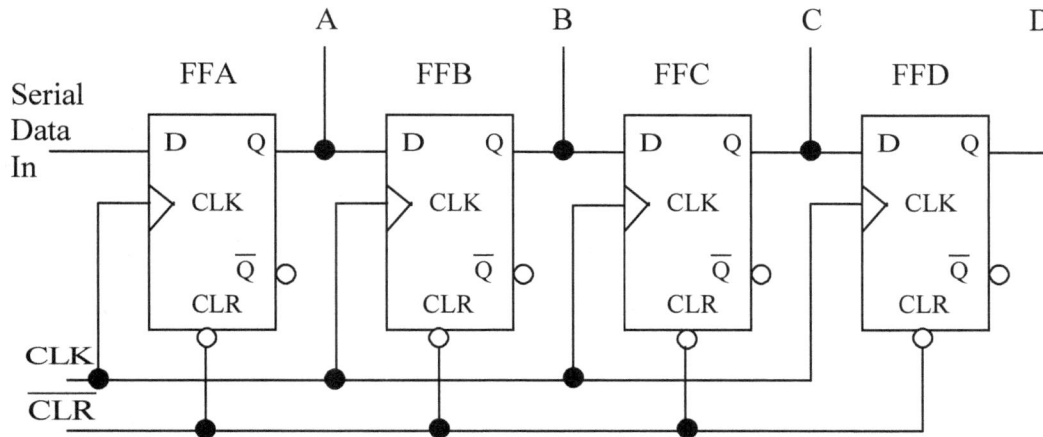

Figure 5.44. A 4 bit SIPO register.

D	C	B	A	Serial Data In
(0	0	0	0)	Register is cleared
0	0	0)	(1	1
0	0)	(1	0	0
0)	(1	0	0	0
(1	0	0	1)	1

Register is full, now data is ready to be read in, etc.

Table 5.30. 4 bit SIPO Truth Table.

		1	2	3	4	5	6	7	8	9	10	11	12	13	14	15
CLK																
Data In		1	0	0	0	0	1	0	0	1						
		CLR														
A	0	1	0	0	0	0	1	0	0	1						
B	0	0	1	0	0	0	0	1	0	0						
C	0	0	0	1	0	0	0	0	1	0						
D	0	0	0	0	1	0	0	0	0	1						

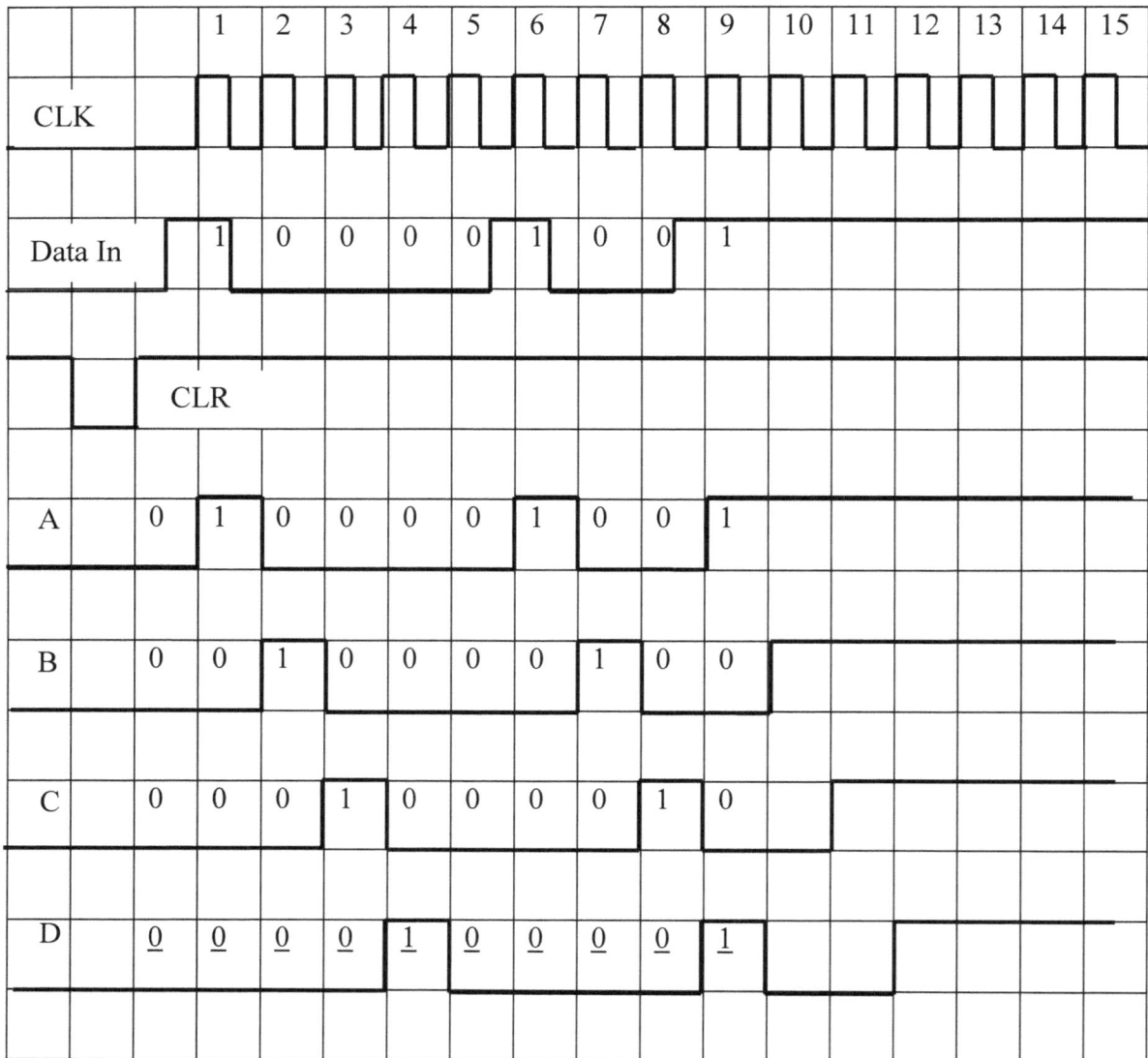

Figure 5.45. Timing diagram of A 4 bit SIPO register using data 0001 and 1001, right most bit goes in first, followed by remaining bits.

5.14.3. Parallel In Serial Out (PISO)

A 4 bit **PISO register** is shown in Figure 5.46. Four bit data is loaded into the register when shift/$\overline{\text{load}}$ is active low. The data is then shifted out serially when shift/ is active high.

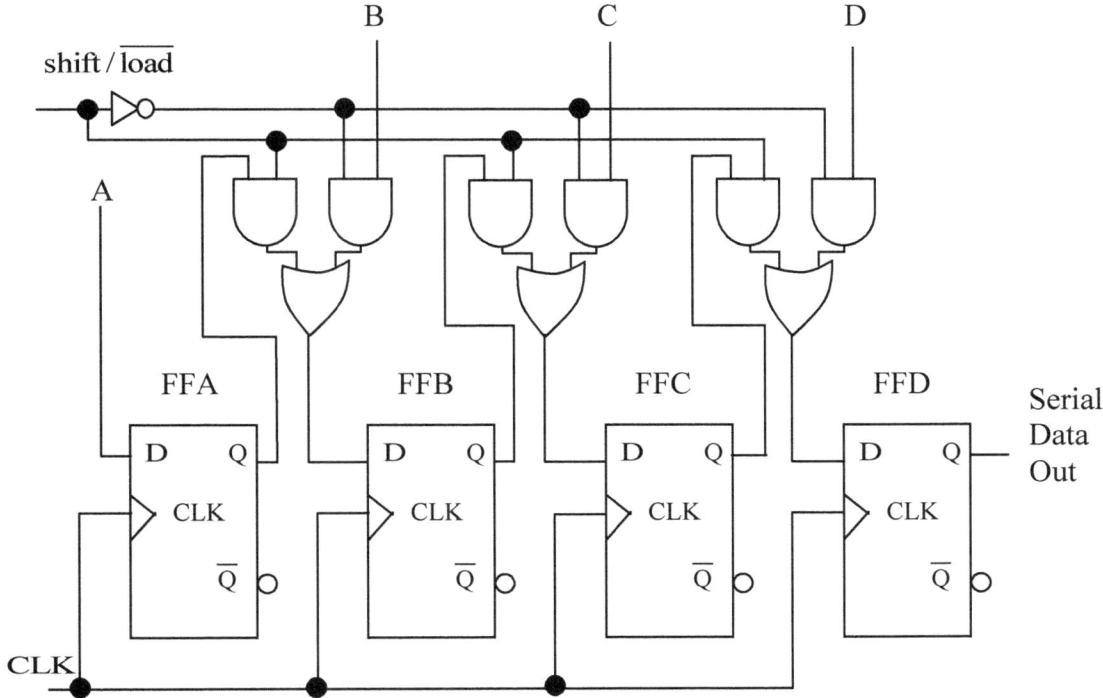

Figure 5.46. A 4 bit PISO register.

5.14.4. Parallel In Parallel Out (PIPO)

A 4 bit **PIPO register** is shown in Figure 5.47. Data is shifted in or out immediately.

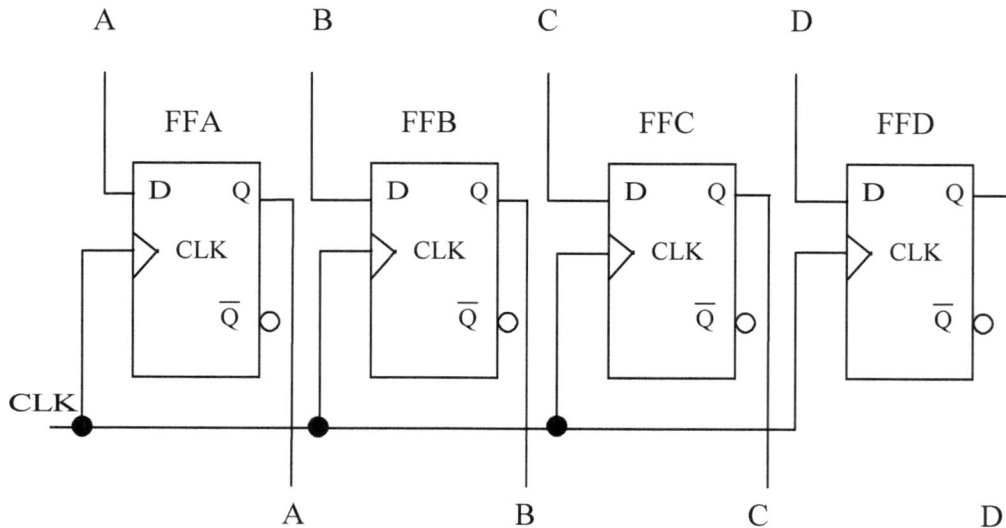

Figure 5.47. A 4 bit PIPO register.

5.14.5. Other Registers - Circulating Data

Up until now we have considered registers that shift data out one end . This results in losing the data. Two registers that circulate data without losing it are:

1. Ring register counter
2. Johnson register counter

Both the ring register counter and the Johnson counter register are SISO registers with their outputs tied back to their inputs as shown in Figure 5.48. However, the Johnson counter has an inverter tied between SO and SI which is the only difference from the ring counter.

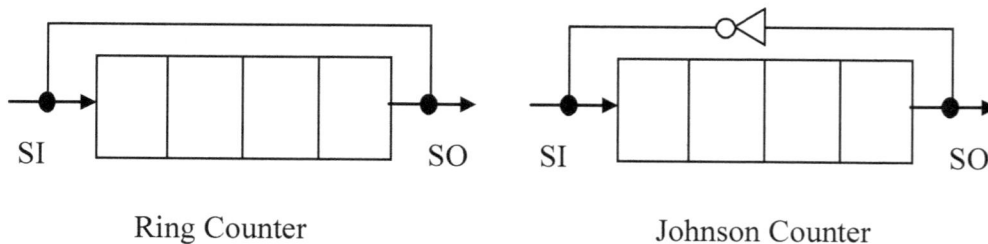

Ring Counter   Johnson Counter

Figure 5.48. Ring and Johnson counter register.

The ring counter register is shown in Figure 5.49. The corresponding 4 bit ring counter Truth Table is given in Table 5.31. Initially, a one is preset in the first flip-flop, A (FFA). The one is shifted around the "ring" for each clock pulse. This pattern is rotated continuously. The ring counter counts the number of clock pulses from 0 to n-1. For example, a 4 bit ring counter counts from 0 to 3. n is the number of flip-flop or bits.

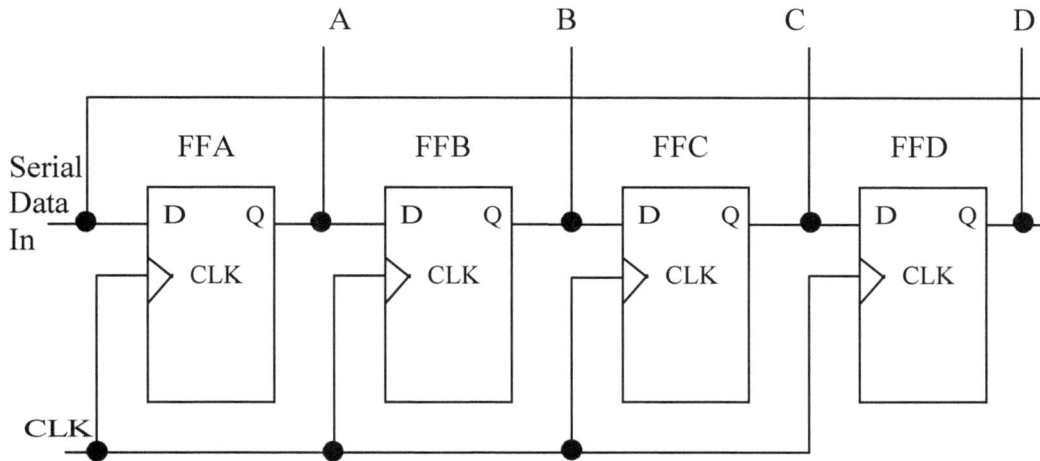

Figure 5.49. A 4 bit ring counter register.

Clock Count	D	C	B	A
0	(0	0	0	1)
1	0	0	1)	(0
2	0	1)	(0	0
3	1)	0(	0	0
4	(0	0	0	1)
5	0	0	1)	(0
6	0	1)	(0	0
7	1)	(0	0	0
8	(0	0	0	1)

Table 5.31. 4 bit ring counter register Truth Table.

Figure 5.50 shows a 4 bit Johnson counter register. The corresponding Truth Table is given in Table 5.32. A Johnson counter counts the number of clock pulses from 0 to 2n-1. For example, a 4 bit Johnson counter counts from 0 to 7. n is the number of flip-flops or bits. In Table 5.32 there are eight different states with the initial state being 0000. In general, a Johnson counter with n flip-flops will produce 2n states or a mod-2n counter.

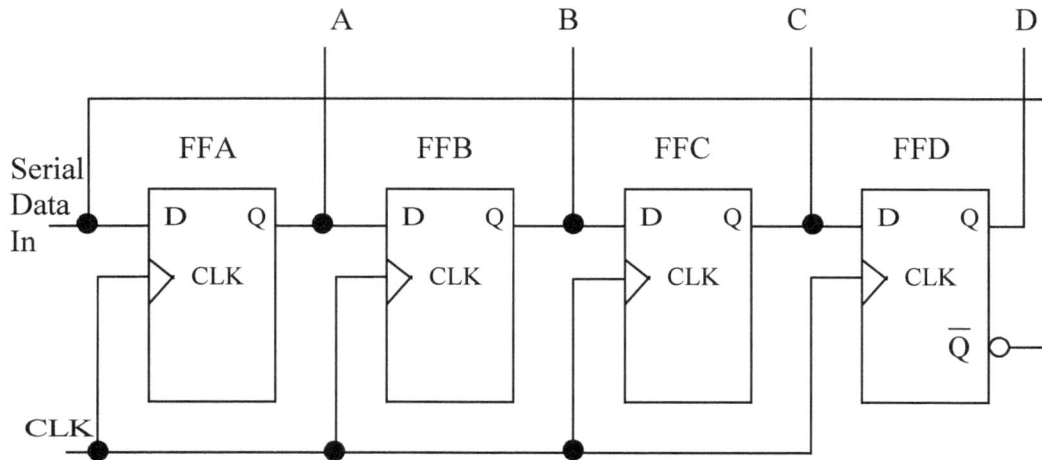

Figure 5.50. A 4 bit Johnson counter register.

Clock Count	D	C	B	A
0	(0	0	0	0)
1	0	0	0)	(1
2	0	0)	(1	1
3	0)	(1	1	1
4	(1	1	1	1)
5	1	1	1)	(0
6	1	1)	(0	0
7	1)	(0	0	0
8	(0	0	0	0)

Table 5.32. 4 bit Johnson counter register Truth Table.

5.15. State Machines

Another word for a sequential logic circuit is a state machine. Counters, registers, etc., are state machines. There are two types of state machines:

A. Moore machine – outputs depend only on present states as shown in Figure 5.51
B. Mealy machine – outputs depend on both present states and controlled inputs as shown in Figure 5.52.

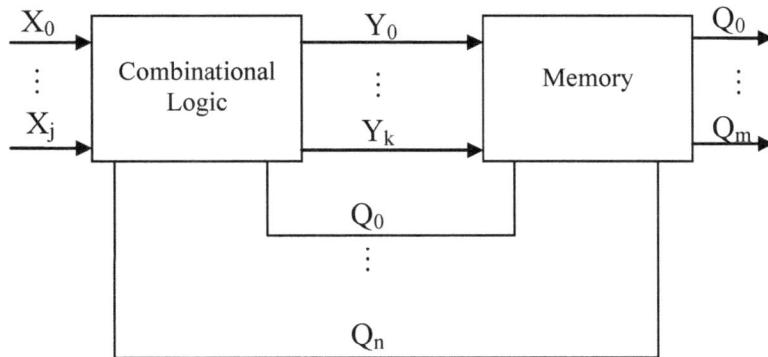

Figure 5.51. Moore machine – outputs depend on present states.

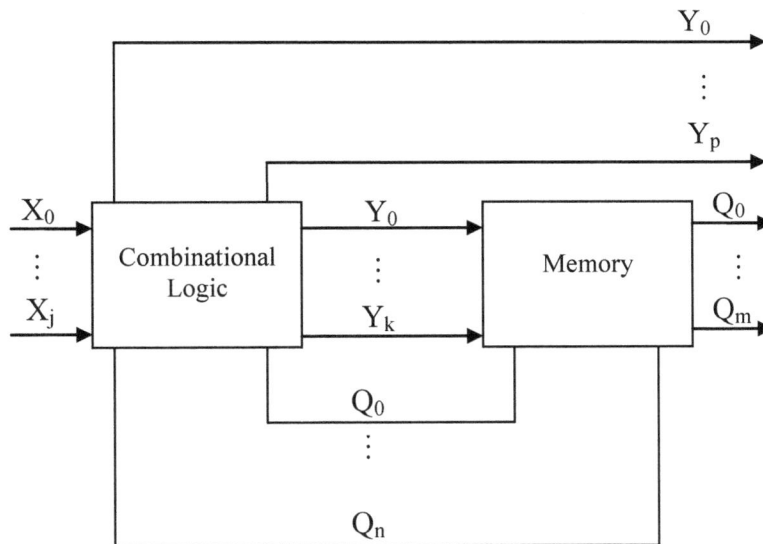

Figure 5.52. Mealy machine – outputs depend on present states and inputs.

A controlled input is an input that tells the state machine to go to the next state or not as shown in Figure 5.53. A state machine can have control inputs or no control inputs as shown in Figure 5.54 and Figure 5.55, respectively.

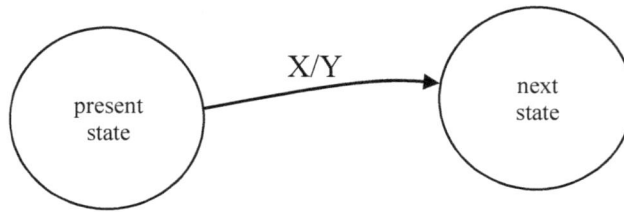

Figure 5.53. State machine transition from present state to next state with control input X.

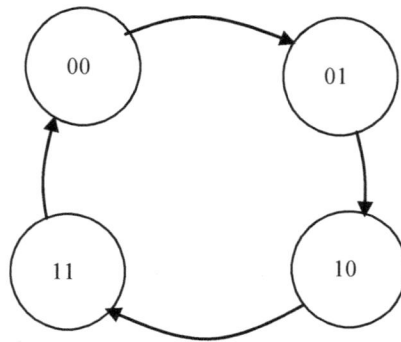

Figure 5.54. State machine with no control inputs.

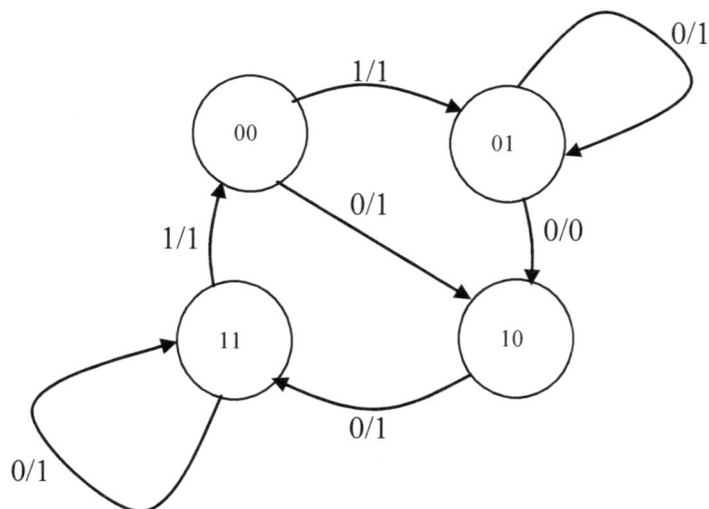

Figure 5.55. State machine with 1 control input.

PROBLEMS I – Flip Flops, Counters

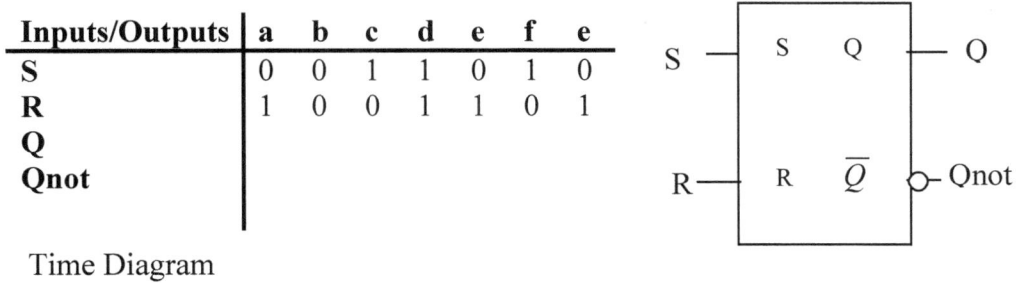

Inputs/Outputs	a	b	c	d	e	f	e
S	0	0	1	1	0	1	0
R	1	0	0	1	1	0	1
Q							
Qnot							

Time Diagram

Figure 5.56.

Use the Time Diagram in Figure 5.56 to answer questions 1-10 below.

	Set	*Reset*	*Hold*	*Forbidden*
1.) At time a, Q is	A.)	B.)	C.)	D.)
2.) At time b, Q is	A.)	B.)	C.)	D.)
3.) At time c, Q is	A.)	B.)	C.)	D.)
4.) At time d, Q is	A.)	B.)	C.)	D.)
5.) At time e, Q is	A.)	B.)	C.)	D.)

6. The input S in Figure 5.56 is which of the following:
A.) Active low   B.) Active high   C.) Active high and low   D.) None

7. The input R in Figure 5.56 is which of the following:
A.) Active low   B.) Active high   C.) Active high and low   D.) None

8. The output Q in Figure 5.56 is which of the following:
A.) Active low   B.) Active high   C.) Active high and low   D.) None

9. The output Qnot in Figure 5.56 is which of the following:
A.) Active low   B.) Active high   C.) Active high and low   D.) None

10. The input S in Figure 5.56 stands for:
A.) reset   B.) set   C.) Synchronous   D.) None

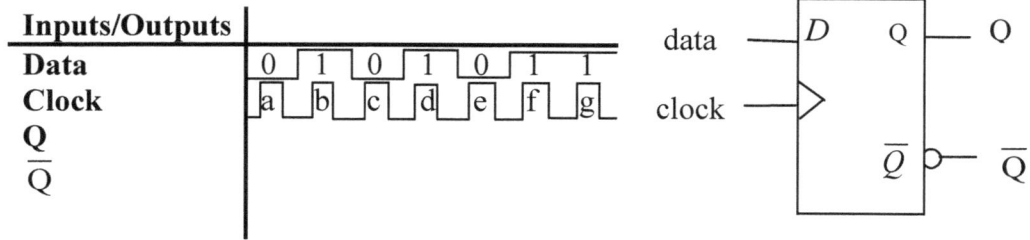

Inputs/Outputs

Data	0	1	0	1	0	1	1
Clock	a	b	c	d	e	f	g
Q							
Q̄							

data — D   Q — Q

clock — ▷

Q̄ ○— Q̄

Time Diagram

Figure 5.57.

Use the Time Diagram and Figure 5.57 to answer questions 11-20 below.

	Set	*Reset*	*Inhibit*	*Forbidden*
11.) On clock pulse a, Q is	A.)	B.)	C.)	D.)
12.) On clock pulse b, Q is	A.)	B.)	C.)	D.)
13.) On clock pulse c, Q is	A.)	B.)	C.)	D.)
14.) On clock pulse d, Q is	A.)	B.)	C.)	D.)
15.) On clock pulse e, Q is	A.)	B.)	C.)	D.)

16. The D flip flop in Figure 5.57 is in the inhibit mode when clock is on the negative edge.
A.) True   B.) False

17. The D flip flop in Figure 5.57 is in the forbidden mode when clock is 0.
A.) True   B.) False

18. For the D flip flop shown in Figure 5.57, Qnot is always the opposite of Q.
A.) True   B.) False

19.) The D flip flop can be constructed by connecting a not gate between the S and R input of SR latch.
A.) True   B.) False

20. The D flip flop in Figure 5.57 can be drawn to represent a gated D latch by doing which of the following:

A.) replacing D with an enable input (EN) only

B.) adding an enable input (EN) only

C.) replacing the clk input (CLK) with an enable input (EN) and removing the angle looking symbol

D.) doing nothing

E.) removing the clk input (CLK) and the angle looking symbol only

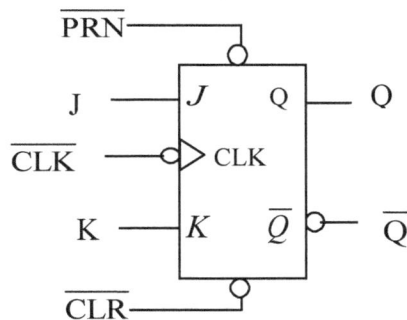

Figure 5.58. JK Flip-Flop.

Use Figure 5.58 to answer questions 21-24 below:

21.) Which one of the following is an asynchronous input:

A.) PRN   B.) Clock   C.) J   D.) K

22.) The JK flip-flop is active on which of the following:

A.) on the rising edge of $\overline{CLK}$

B.) on the falling edge of $\overline{CLK}$

C.) when clock is 1

D.) none of the above

23.) Which one of the following conditions must be true for the JK flip-flop to toggle:

    A.) J=0, K=1

    B.) CLR=0, PRN=0

    C.) CLR=1, PRN=1

    D.) J=1, K=1

24.) Which one of the following conditions puts the JK flip-flop in the forbidden state:

    A.) CLR=0, PRN=0

    B.) CLR=1, PRN=1

    C.) J=1, K=1

    D.) None of the above

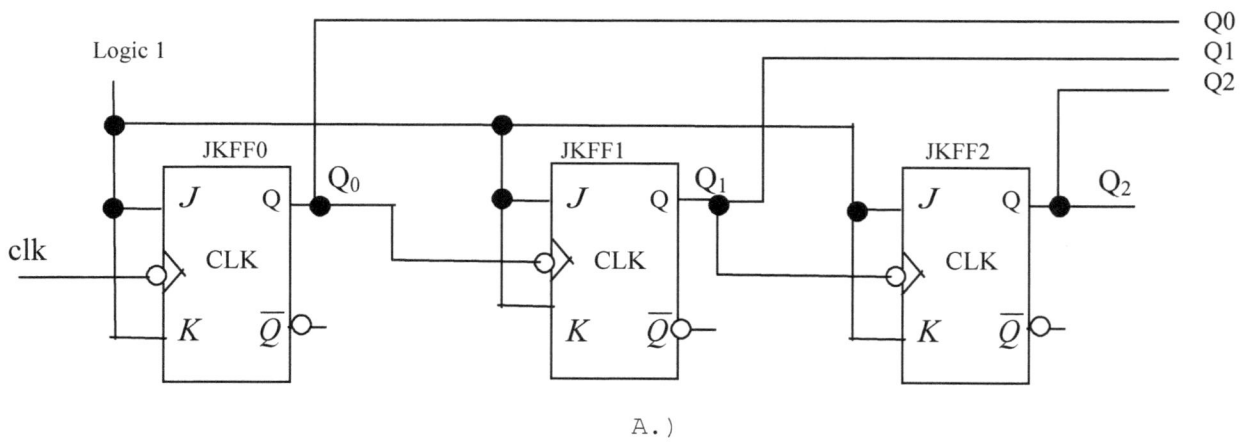

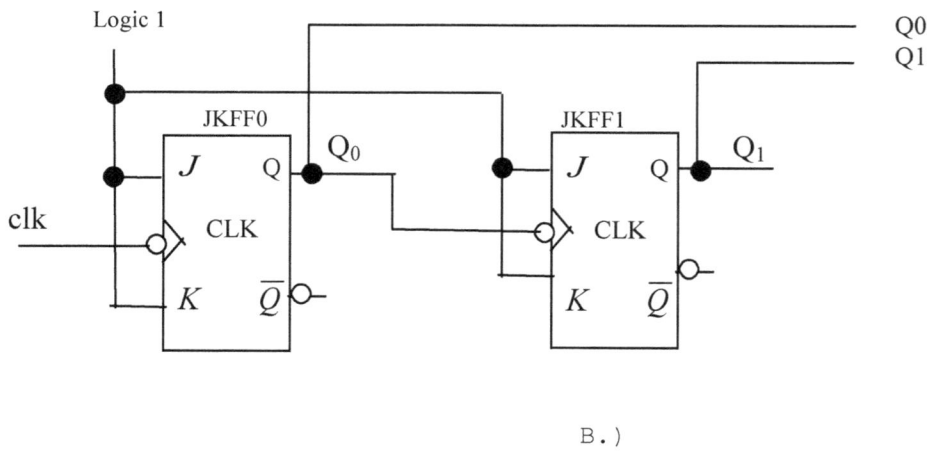

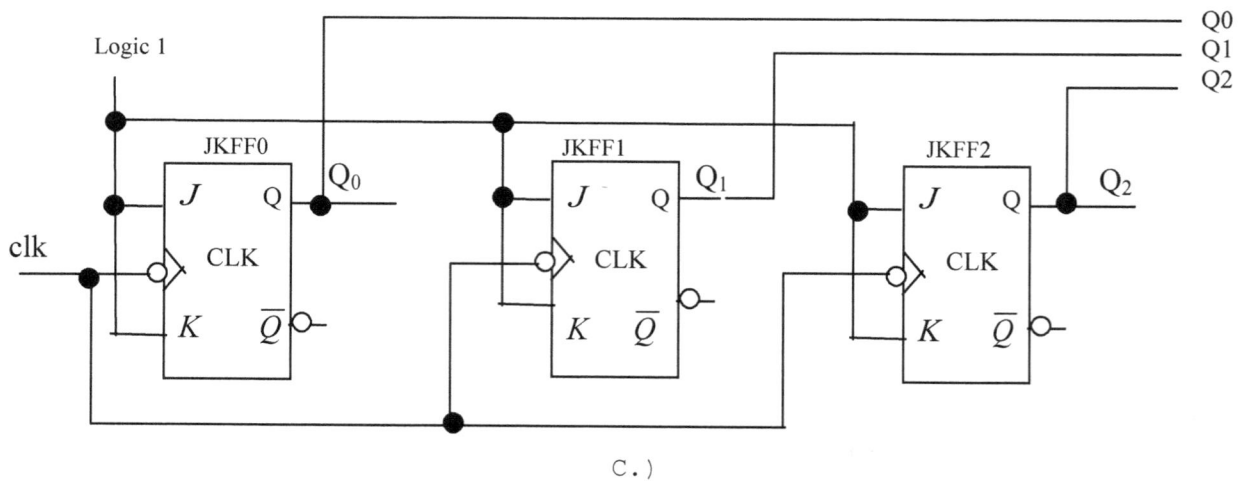

Figure 5.59. Sequential Circuit.

25. Which one of the circuits in Figure 5.59 is a synchronous circuit.
A.) Figure 5.59a
B.) Figure 5.59b
C.) Figure 5.59c
D.) Figure 5.59a and b
E.) None

26. The modulus of the circuit in Figure 5.59b is:
A.) 2   B.) 4   C.) 8   D.) 16

27. The maximum count of the circuit in Figure 5.59b is:
A.) 0   B.) 1   C.) 2   D.) 3   E.) 4

28. If the initial state of the counter in Figure 5.59b is $Q_1Q_0=01$, on the next clock pulse the counter will go to which of the following states:
A.) $Q_1Q_0=01$ B.) $Q_1Q_0=00$ C.) $Q_1Q_0=11$ D.) $Q_1Q_0=10$

29. If the initial state of the counter in Figure 5.59b is $Q_1Q_0=11$, on the next clock pulse the counter will go to which of the following states:
A.) $Q_1Q_0=01$ B.) $Q_1Q_0=00$ C.) $Q_1Q_0=11$ D.) $Q_1Q_0=10$

30. If the initial state of the counter in Figure 5.59b is $Q_1Q_0=00$, on the next clock pulse the counter will go to which of the following states:
A.) $Q_1Q_0=01$ B.) $Q_1Q_0=00$ C.) $Q_1Q_0=11$ D.) $Q_1Q_0=10$

31. The modulus of the circuit in Figure 5.59a is:
A.) 2   B.) 4   C.) 8   D.) 16   E.) 32

32. The maximum count of the circuit in Figure 5.59a is:
A.) 4   B.) 6   C.) 7   D.) 8   E.) 9

33. The counter in Figure 5.59a counts:
A.) up   B.) down   C.) None

34. The counter in Figure 5.59b counts:
A.) up   B.) down   C.) None

35. The counter in Figure 5.59a has how many possible output states:
A.) 2   B.) 4   C.) 8   D.) 16   E.) 32

36. The terminal count of the counter in Figure 5.59a is 0000_2.
A.) True   B.) False

37. The terminal count of the counter in Figure 5.59b is 11_2.
A.) True   B.) False

38. To make the counter in Figure 5.59a count in the reverse direction you should connect the previous Qnot to the clock input of the next FF and leave the clk as negative edge triggered.
A.) True   B.) False

39. To make the counter in Figure 5.59b count in the reverse direction you should connect the previous Qnot to the clock input of the next FF and make the clk positive edge triggered.
A.) True   B.) False

40.) The terminal count of a counter is the initial state of the counter.
A.) True   B.) False

PROBLEMS II – Counters, Memory, Devices

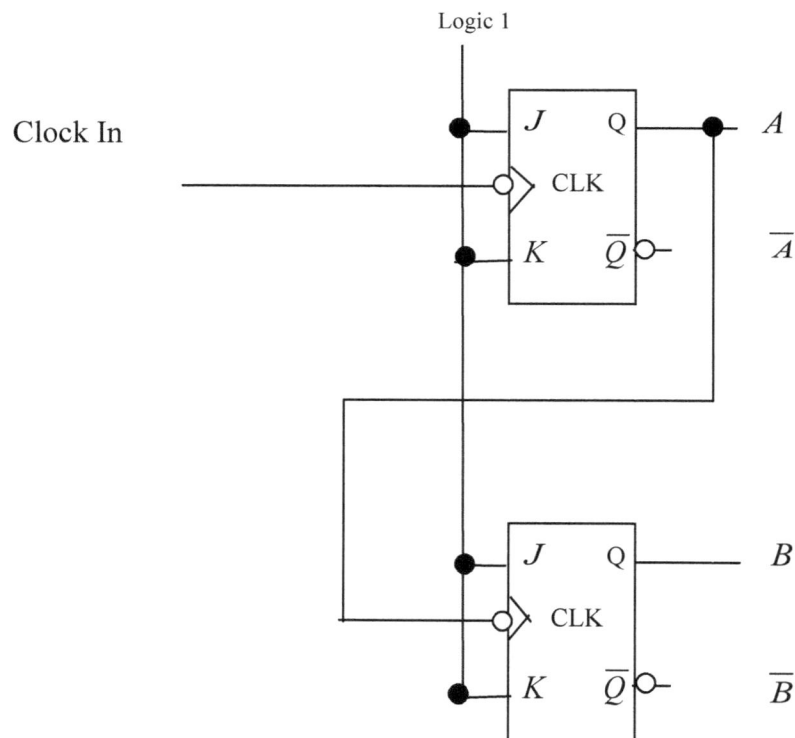

Figure 5.60. A sequential circuit.

1. Consider the sequential circuit in Figure 5.60. The state diagram has how many bubbles:
A. 1   B. 2   C. 3   D. 4   E. 5

2. Consider the sequential circuit in Figure 5.60. If **clock in** has a frequency of 1000 Hz the frequency of the output at A is:
A. 500 Hz
B. 1000 Hz
C. 2000 Hz
D. 4000 Hz

3. The circuit in Figure 5.60
A. counts from 0 to 4
B. counts from 0 to 3
C. counts from 0 to 7
D. counts from 7 to 0
E. counts from 3 to 0

4. The sequential circuit in Figure 5.60 is:
A. A synchronous circuit
B. An asynchronous circuit
C. A serial in serial out register
D. A Johnson register

5. A mod12 counter has how many states:
A. 4
B. 8
C. 12
D. 16

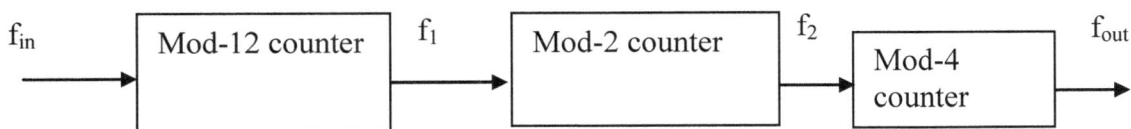

Figure 5.61.

6. In Figure 5.61, if f_{in} is 41,472 Hz, f_1 is:
A. 12 Hz
B. 288 Hz
C. 41472 Hz
D. 3456 Hz
E. 123.4 Hz

7. In Figure 5.61, if f_{in} is 1728 Hz, f_2 is:
A. 14 Hz
B. 288 Hz
C. 41472 Hz
D. 1728 Hz
E. 72

8. The total modulus of the counters in Figure 5.61 is:
A. 12
B. 18
C. 36
D. 96
E. 192

9. A state machine is not a sequential circuit.
A. True
B. False

10. When referring to A/D converters a R-2R Ladder resistor network has more of the same resistors than a resistive network.
A. True
B. False

11. The Thevenin equivalent circuit between terminals A and B of a circuit is obtained by:
A. Finding the current across terminals A and B using and circuit analysis method, then finding the Thevenin resistance after opening all independent current sources and shorting all independent voltage sources
B. Shorting all independent current sources and opening all independent voltage sources and determining the Thevenin voltage across terminals A and B, then killing all sources and finding the Thevenin resistance across terminals A and B
C. Shorting all sources and determining the voltage across terminal A and B
D. Finding the Thevenin voltage across terminals A and B using any circuit analysis method, then shorting all independent voltage sources and opening all independent current sources and determining the Thevenin resistance across terminals A and B

12. A Flash analog to digital converter has which of the following main components:
A. SAR, comparator, digital to analog converter
B. Comparators, priority encoder
C. Priority encoder, DAC, SAR
D. Comparators, ADC, SAR

13. The sampling frequency of the ADC should be equal to the maximum frequency of the analog input signal.
A. True
B. False

14. RAM is a memory which can only be read.
A. True
B. False

15. One difference between the PROM and ROM is that the ROM can be burned locally.
A. True
B. False

16. A Flash PROM can only be burned off the circuit board.
A. True
B. False

17. Volatile memory is one that retains its data when power is turned off.
A. True
B. False

18. When referring to memory capacity the letter K stands for
A. 1,000
B. 2,000
C. 1,024
D. 2,024

19. When referring to memory capacity the letter M stands for
A. 14,800
B. 148,000
C. 1,000,000
D. 1,048,576

20. Memory can be divided into which of the following main categories discussed in class:
A. RAM and ROM
B. Primary and main
C. Primary and secondary
D. Magnetic, Optic, RAM

21. An example of magnetic memory is
A. RAM   B. ROM   C. CDROM   D. Hard drive

22. An IDT6116 IC is which of the following:
A. ROM   B. RAM   C. Counter   D. multivibrater

Consider the RAM organization with the contents indicated by the Table 5.33 below for the next several questions (23 to 27).

Location	D	C	B	A
0	1	1	0	0
1	1	1	1	1
2	1	1	1	1
3	1	1	0	0
4	0	1	0	0
5	1	1	0	0
6	0	1	1	1
7	1	1	1	1
8	0	1	0	0
9	1	0	1	1
10	0	1	0	1
11	1	0	1	1
12	1	1	0	1
13	0	0	1	1
14	0	1	0	1
15	0	0	0	0

Table 5.33.

23. The capacity of the RAM indicated by Table 5.33 is:
A. 16 x 4 bits B. 64K bits   C. 4 x 16 bits D. 64 bits

24. The organization of the RAM indicated by Table 5.33 is:
A. 64 bits   B. 4 x 16 bits C 16 x 4 bits   D. 64 x 1 bits E. 1 x 64 bits

25. How many address inputs are required for the memory represented in Table 5.33:
A. 1   B. 2   C. 4   D. 8   E. 16

26. How many data lines are required for the memory represented in Table 5.33:
A. 1   B. 2   C. 4   D. 8   E. 16

27. What are the contents of address 00001101 of the RAM in Table 5.33:
A. 1011   B. 1101   C. 0011   D. 0101

28. Fan-out is
A. The maximum number of outputs connected to an input
B. The maximum number of inputs connected to another input
C. The maximum number of inputs connected to an output
D. The maximum number of outputs connected to another output

29. Interfacing is the method of finding the voltage in a circuit.
A. True   B. False

30. TTL stands for transistor totem logic.
A. True   B. False

31. A 74ALS04 IC is fabricated using which one of the following technologies:
A. Advanced Schottky TTL logic
B. Standard TTL logic
C. Schottky TTL logic
D. Low power TTL logic
E. Advanced low power Schottky logic

32. VLSI stands for very large skew integration.
A. True   B. False

33 A MOS IC cannot be directly connected to a TTL IC.
A. True   B. False

34. A device with a tristated output can be connected to only one other device with a tristated output where the tristated outputs are directly connected.
A. True   B. False

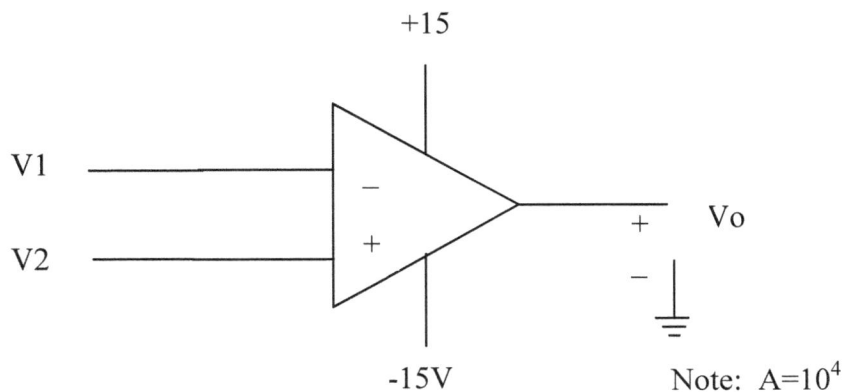

Figure 5.62.

35. Given Figure 3, if V1=-10 volts and V2=-10 volts then the output voltage, Vo, is:
A. 0 volts   B. +15 volts   C. -15 volts   D. infinite

36. Given Figure 3, if V1=-10 volts and V2=+10 volts then the output voltage, Vo, is:
A. +20 volts   B. +15 volts   C. -15 volts   D. +200000 volts

37. Given Figure 3, if V1=+10 volts and V2=-10 volts then the output voltage, Vo, is:

A. +200000 volts   B. +15 volts   C. -15 volts   D. –200000 volts

38. Given Figure 3, if V1=+10 volts and V2=+10 volts then the output voltage, Vo, is:
A. 0 volts   B. +15 volts   C. -15 volts   D. infinite

Use Figure 5.63 to answer questions 39 through 43.

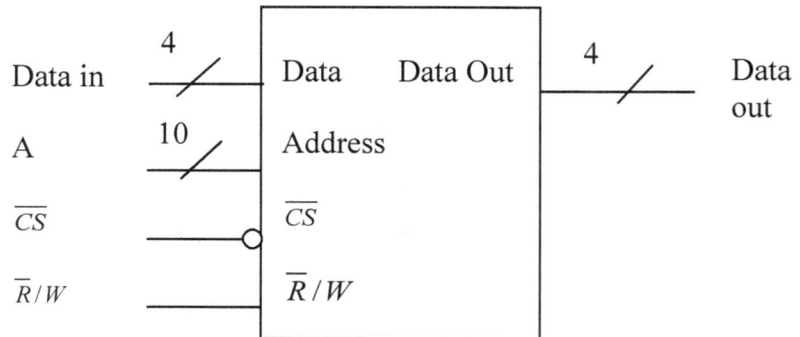

Figure 5.63. Representation of 1K x 4 bit memory.

A company proposes a new product that requires an 8K x 4 bit memory unit. They must use several of the memory chips indicated in Figure 5.63 to fulfill this requirement. Therefore:

39. How many of the chips are needed for the required memory unit:
A.) 1   B.) 3   C.) 4   D.) 8   E.) 12

40. What is the total number of address lines needed for the required memory unit:
A.) 8   B.) 11 C.) 12 D.) 13 E.) 14

41. What is the number of bits per word needed for the required memory unit:
A.) 2   B.) 3   C.) 4   D.) 8   E.) 12

42. What is the memory capacity needed for the required memory unit:
A.) 1024 bits B.) 4096 bits   C.) 8192 bits D.) 32768 bits E.) 131072 bits

43. What is the size of the decoder needed for the required memory unit:
A.) 2 by 4   B.) 3 by 8   C.) 4 by 16   D.) 1 by 8   E.) None

44. Draw the inner circuit of Figure 5.64. Use D flip-flops. Label all inputs, outputs, and Flip Flops.

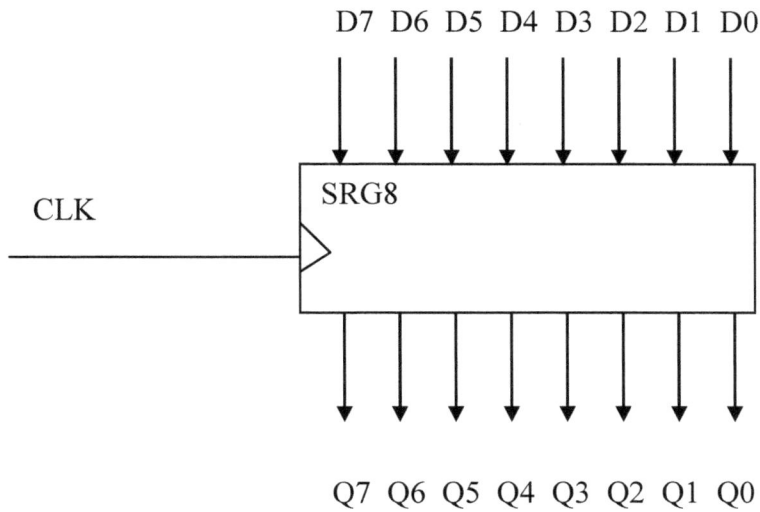

Figure 5.64. Eight bit shift register.

45. Draw the inner circuit of Figure 5.65. Use D flip-flops. Label all inputs, outputs, and Flip Flops.

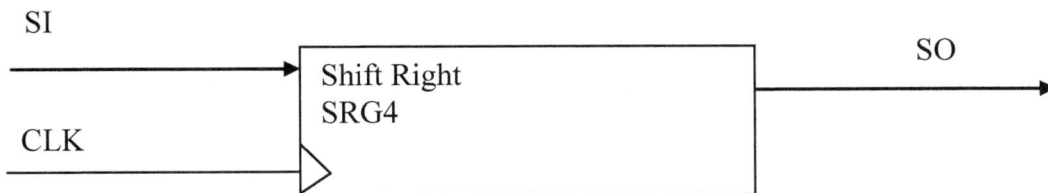

Figure 5.65. Four bit shift register.

46. Draw the inner circuit of Figure 5.66. Use D flip-flops. Label all inputs, outputs, and Flip Flops.

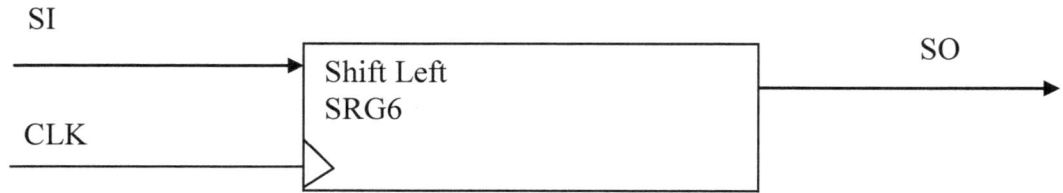

SI

Shift Left
SRG6

SO

CLK

Figure 5.66. Six bit shift register.

47. Draw the inner circuit of Figure 5.67. Use D flip-flops. Label all inputs, outputs, and Flip Flops.

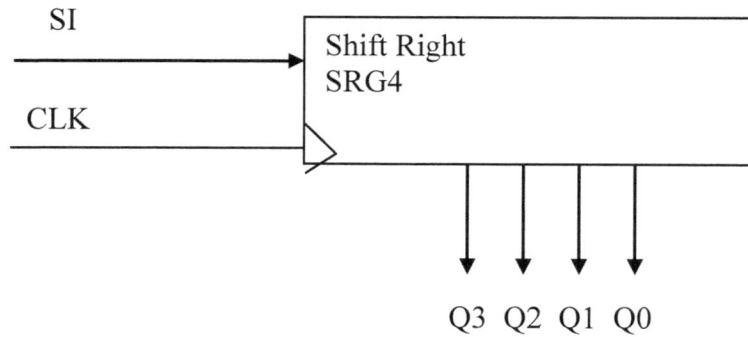

SI

Shift Right
SRG4

CLK

Q3 Q2 Q1 Q0

Figure 5.67. Four bit shift register.

48. Draw the inner circuit of Figure 5.68. Use D flip-flops. Label all inputs, outputs, and Flip Flops.

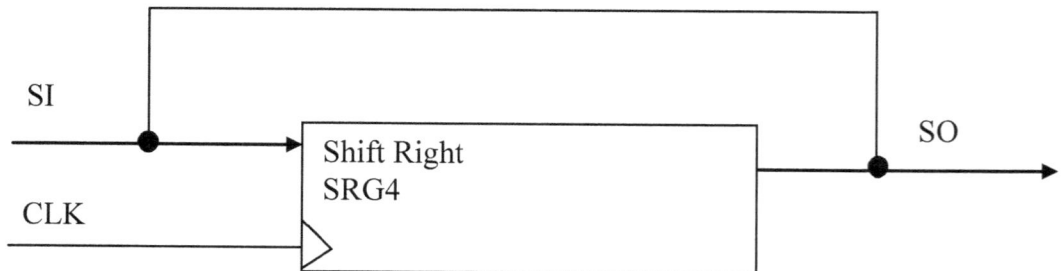

SI

Shift Right
SRG4

SO

CLK

Figure 5.68. 8 bit ring counter register.

49. Draw the inner circuit of Figure 5.69. Use D flip-flops. Label all inputs, outputs, and Flip Flops.

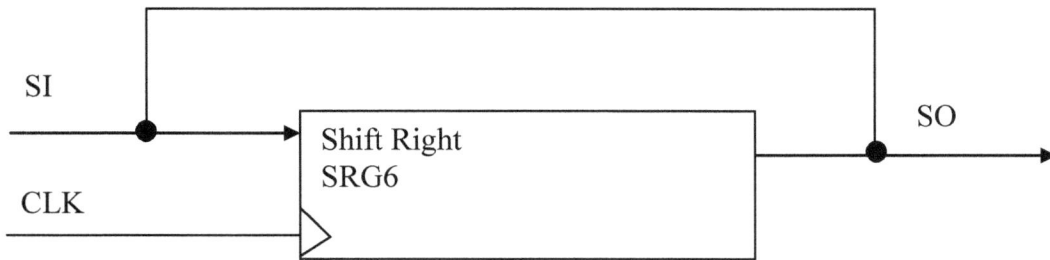

Figure 5.69. Six bit Johnson counter register.

50. Complete the timing diagram for a 6 bit SISO (shift right) register which is positive edge triggered. Use the following serial data: 011011_2. The MSB is shifted in first. Assume the initial state of all flip-flops are zero.

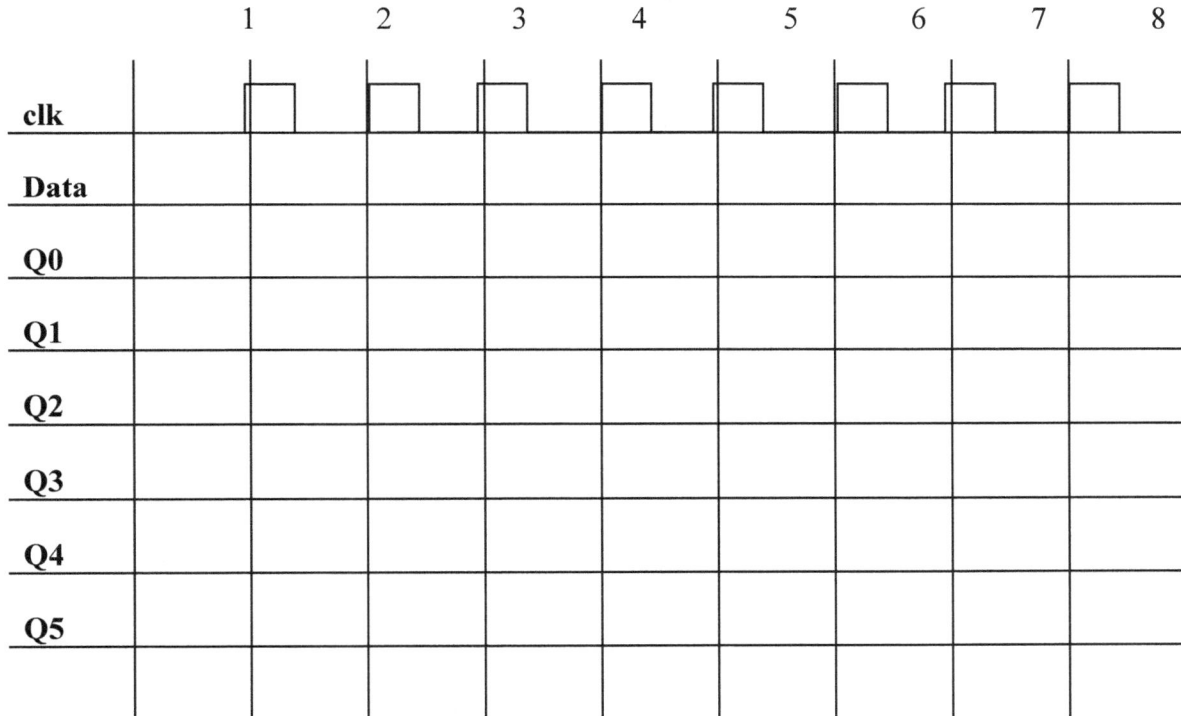

Figure 5.70. Timing diagram.

51. If the Johnson counter in problem 6 is initialized to 1000_2 on clock pulse number 0, what is the clock number when the pattern repeats?

52. The state diagram for a synchronous counter with initial state of 11_2 is given in Figure 5.71. Determine the following using positive edge JK flip flops:

a. The circuit excitation table.
b. The Boolean equations.
c. Draw the sequential circuit. Correctly label all flip flop inputs and outputs, etc.
d. The timing diagram.

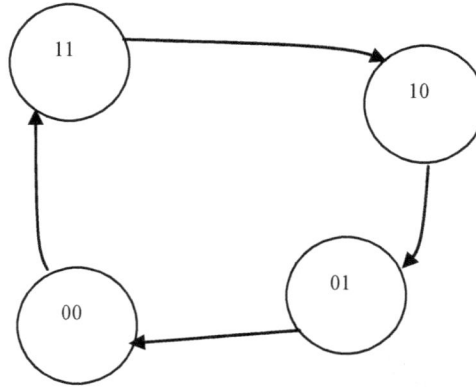

Figure 5.71. State diagram.

53. Given the synchronous sequential circuit in Figure 5.72 determine the following:

a. The Boolean equations.
b. The circuit state table.
c. Draw the state diagram. Correctly label all states, etc.
d. The timing diagram.

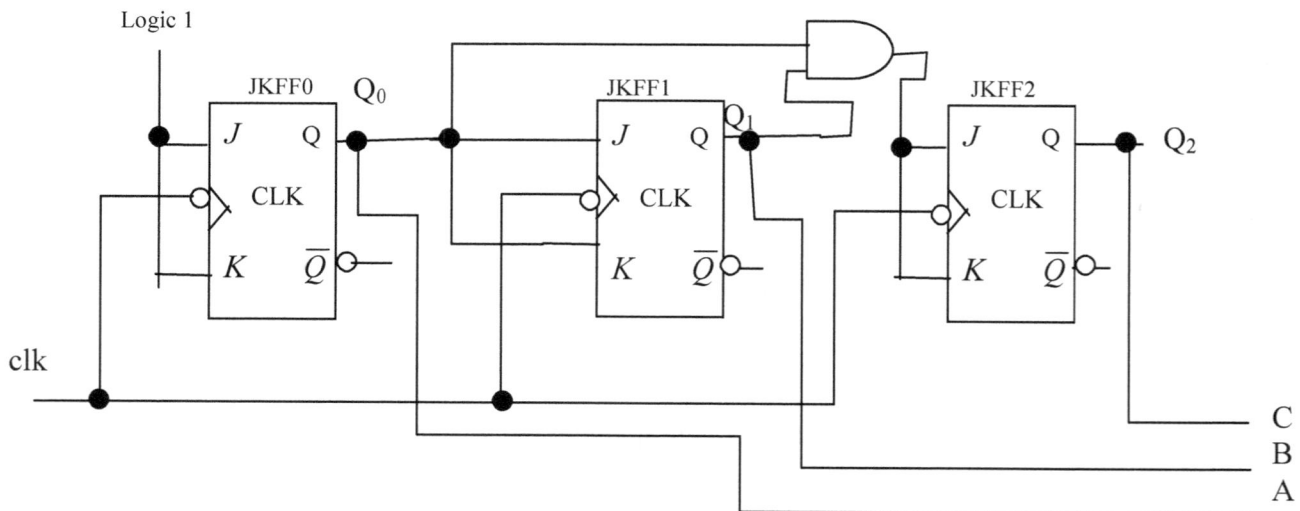

Figure 5.72. Sequential circuit.

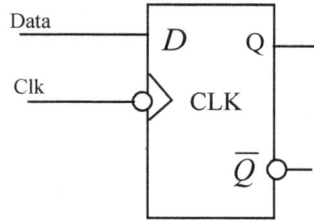

Figure 5.73. Flip-flop.

54. Write the VHDL code for the device in Figure 5.73. The name of the device is DFF.

55. Determine the **state table** and **state diagram** of the sequential logic circuit in Figure 5.74.

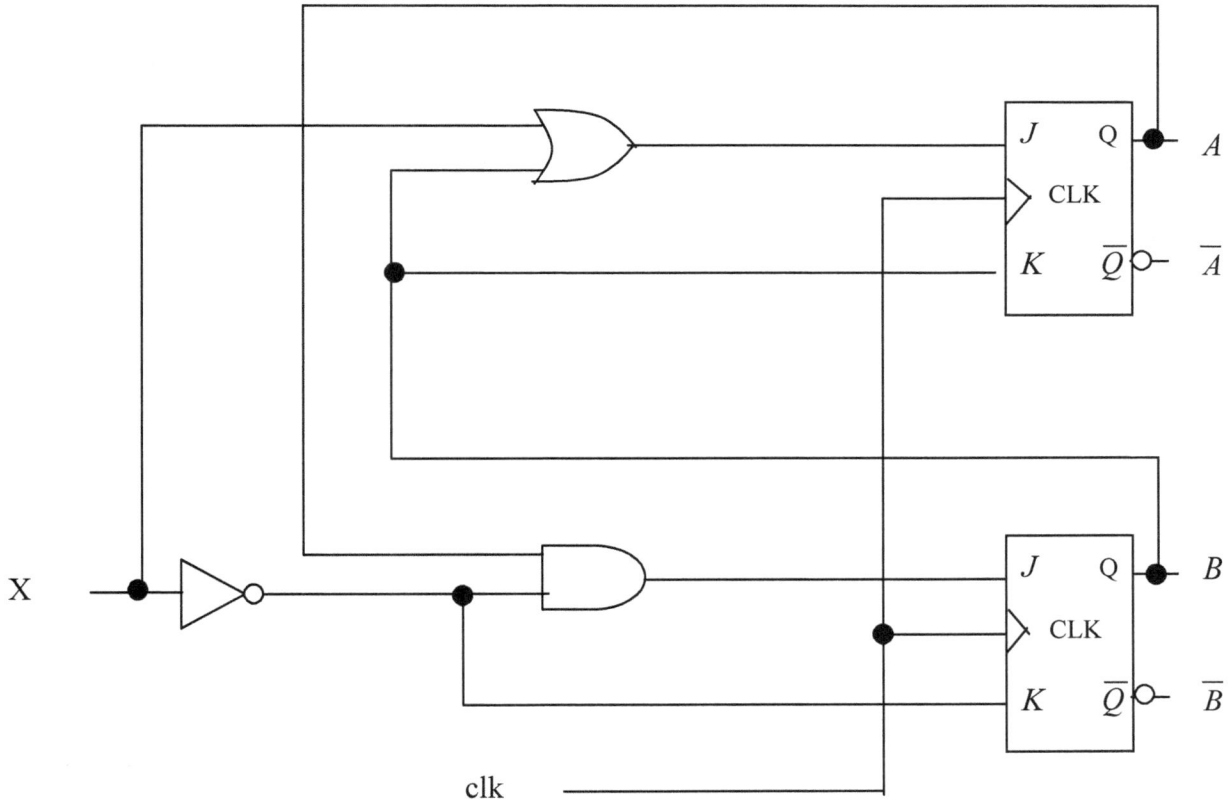

Figure 5.74. Sequential circuit.

6. Communications Systems

T his chapter shall discuss the two types of communication systems: analog and digital. The analog or digital communication system consists of the following main parts: information source, transmitter, transmission medium, receiver, and information destination. The prerequisite to understanding the various communication concepts is to have a basic understanding of mathematical concepts dealing with properties of signals, frequency domain, time domain, and mixing. These concepts shall be discussed in Section 6.1. A general discussion of basic communication systems is provided in Section 6.2. Section 6.3 summarizes the various transmission modes. Section 6.4 provides a basic description of analog and digital signals in communication systems. A very important topic in communication systems is oscillators that generate the high frequency carrier signals. Section 6.5 explains the basic concepts of oscillators. Analog communication systems in Section 6.6 discuss amplitude modulation, frequency modulation, and phase modulation. Digital communication systems in Section 6.7 covers amplitude shift keying, frequency shift keying, phase shift keying, and quadrature phase shift keying.

6.1. Basic Communication Mathematics and Concepts

This section will cover some basic concepts necessary for understanding communication systems.

6.1.1. Frequency

Frequency is the number of time an event occurs in some given time. Frequency is measured in cycles per second. Common prefixes are:

1. K   Kilo   103
2. M   Mega   106
3. G   Giga   109
4. T   Tera   1012

For example, 1KHz = 1000 cycles per second.

6.1.2. Wavelength

Wavelength is related to frequency by using velocity, distance, and time formula given by:

$$v = \frac{d}{t}$$ **EQ. 6.1**

where
v is velocity in meters/second (m/s)
d is the distance in meters (m)
t is the time in seconds (s)

Since wavelength, λ, is the distance represented by one cycle of a waveform such as the sinusoidal waveform shown in Figure 6.1, then EQ. 6.1 becomes

$$\lambda = \frac{c}{f}$$ **EQ. 6.2**

where c is the speed of light, 3×10^8 meters per second (m/s) and f is the frequency of the signal in Hz.

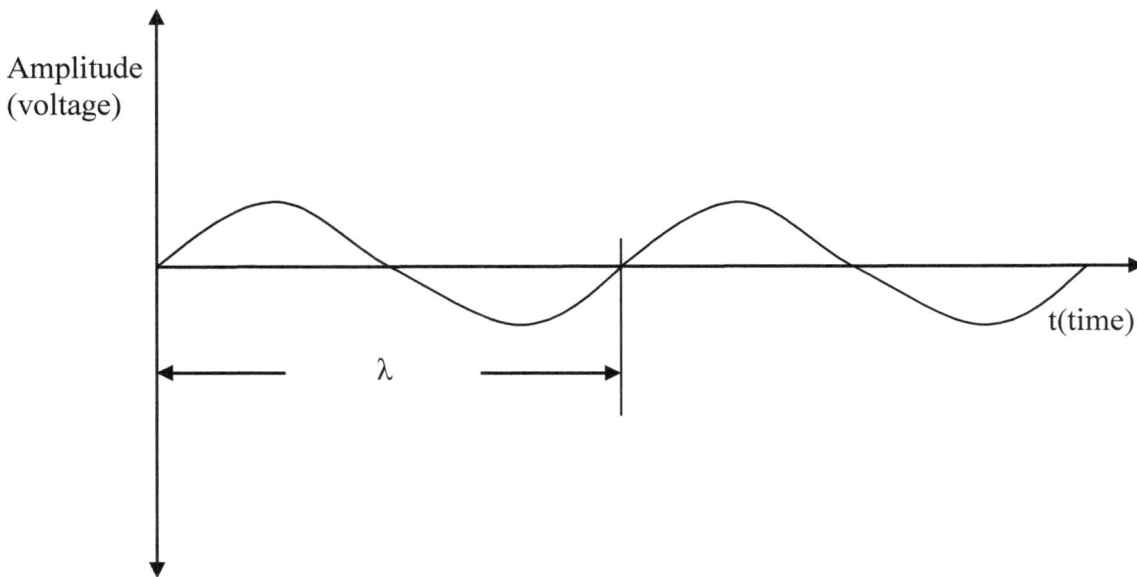

Figure 6.1. Sinusoidal Waveform.

6.1.3. Signals

A signal typically represents a physical quantity relative to time which is transmitted from one point to another through a media such as free space, wire, optical, etc. Signals can be represented in two main ways:
1. Time domain representation
2. Frequency domain representation

There are both complex and simple signals. Most signals can be broken down into simpler signals call sinusoidals. Sinusoidals are the simplest signals which cannot be broken down any further. An example of a complex waveform is the square wave, sawtooth, etc., which can be broken down into an infinite number of sinusoidals.

6.1.4. Sinusoidal Signal

A sinusoidal signal is represented by one of the following mathematical expressions:

$$y(t) = A\sin(\omega t + \theta)$$

EQ. 6.3

$$y(t) = A\cos(\omega t + \theta)$$

EQ. 6.4

where
$\omega = 2\pi f$ in radians per second
t=time in seconds
θ=phase angle radians
A= amplitude in voltage, current, etc
$y(t)$ = instantaneous value of the voltage or current sinusoidal at time t

6.1.5. Time Domain Representation of Signals

The time domain representation of a signal shows how its amplitude varies with time. The time domain representation of a sinusoidal signal is shown in Figure 6.2. The sinusoidal is periodic with period T.

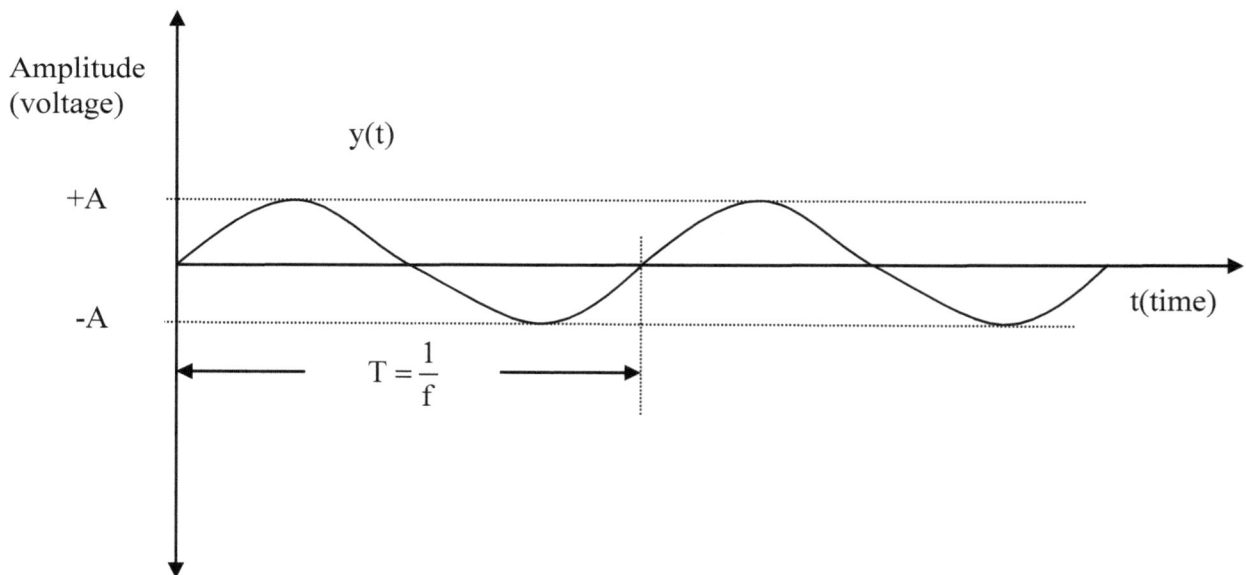

Figure 6.2. Time domain representation of a sinusoidal waveform.

6.1.6. Frequency Domain Representation of Signals

The frequency domain representation of a signal shows how its amplitude varies with frequency. The frequency domain representation of a sinusoidal signal is shown in Figure 6.3. The sinusoidal is represented by a single spike in the frequency domain.

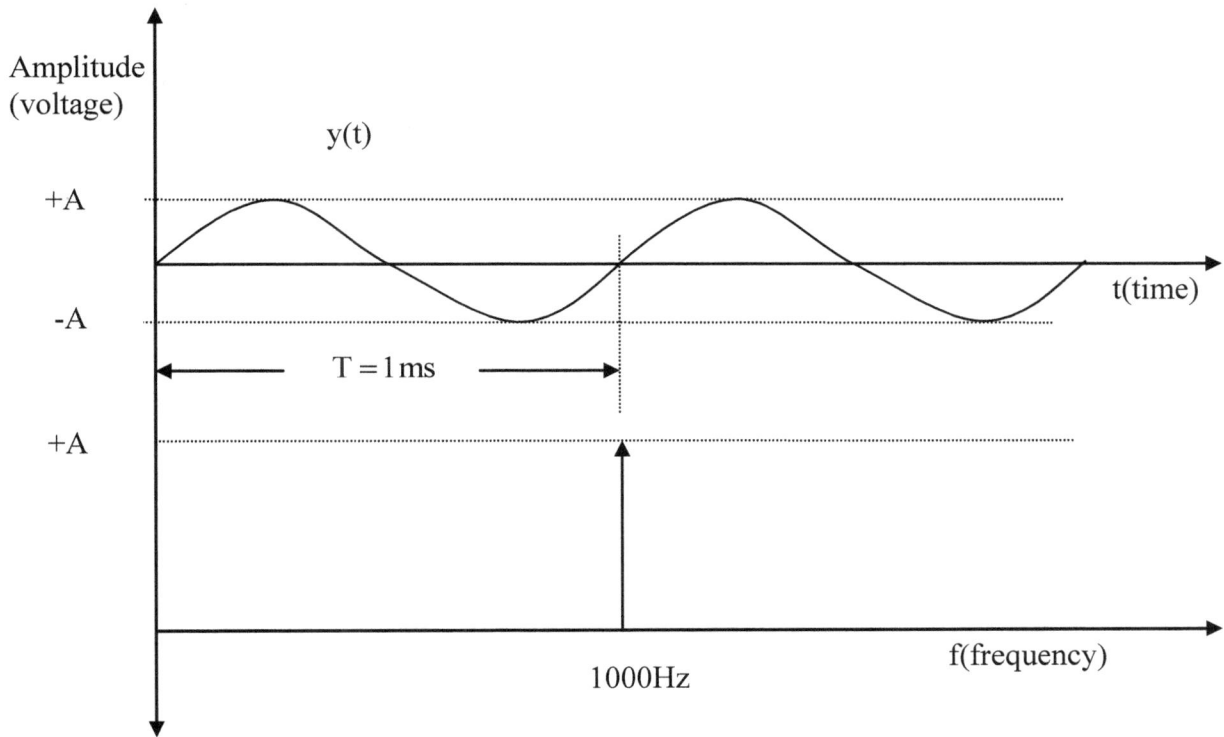

Figure 6.3. Time domain (top waveform) and Frequency domain (bottom waveform) representation of a sinusoidal waveform.

6.1.7. Properties of Signals

Wave symmetry describes the symmetry of a signal in the time domain. When a waveform is even about the origin it referred to as an even signal which meets the following condition:

$$y(t) = y(-t)$$

EQ. 6.5

Thus, the cosine function in EQ. 6.4 represents an even signal. On the other hand, when a waveform is odd about the origin it referred to as an odd signal. An odd signal meets the following condition:

$$y(t) = -y(-t)$$

EQ. 6.6

The sine function in EQ. 6.3 is representative of an odd signal. If one half of a periodic signal is equal to the other half mirrored across the x-axis it is referred to as half wave symmetric. A half wave symmetric signal meets the following condition :

$$y(t) = -y\left(\frac{T}{2} + t\right)$$

EQ. 6.7

A half wave symmetric signal is shown in `Figure 6.4.`

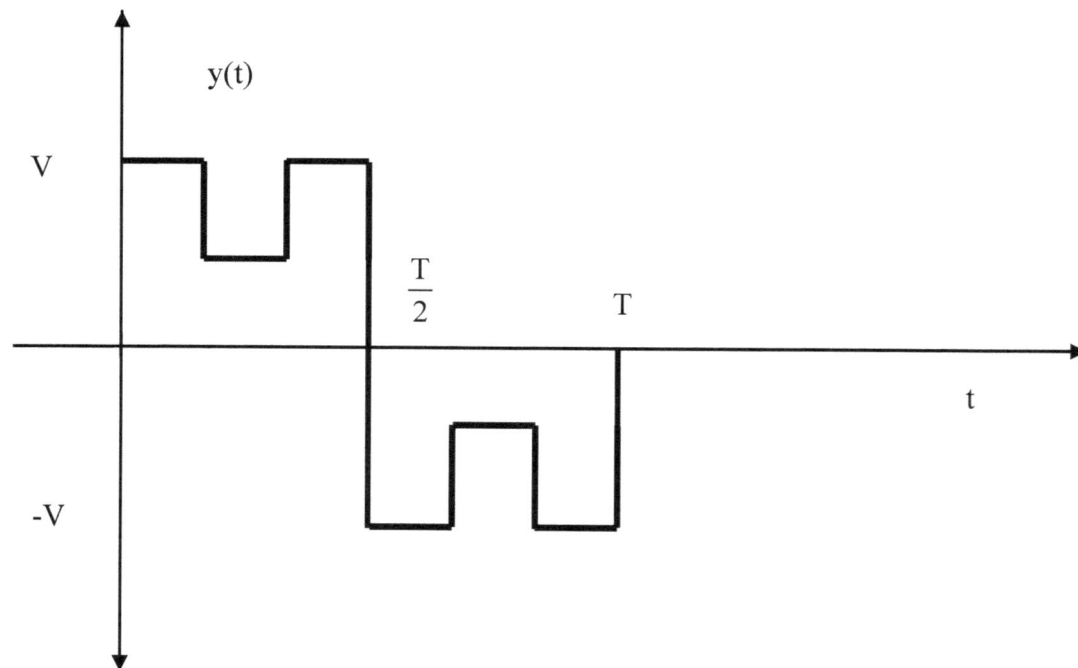

Figure 6.4. Time domain representation of the half wave symmetric signal.

6.1.8. Complex Waveforms

Complex waveforms are created by adding many sinusoidals. The periodic square waveform in Figure 6.5 is a complex waveform. It is created by the addition of many sinusoidals as shown in Figure 6.6. These sinusoidals are obtained by a mathematical process known as the Fourier series. The Fourier series of a periodic signal is represented by:

$$y(t) = A_0 + A_1 \cos \omega_1 t + A_2 \cos 2\omega t + \cdots + A_n \cos n\omega t$$
$$+ A_0 + B_1 \sin \omega t + B_2 \sin 2\omega t + \cdots + B_n \sin n\omega t$$

EQ. 6.8

where

$$\omega = 2\pi f_0$$

EQ. 6.9

and

$$f_0 = \frac{1}{T_0}$$

EQ. 6.10

where T_0 is the fundamental period and f_0 is the corresponding fundamental frequency. Thus, a periodic waveform consists of a dc component and a series of harmonically related sine and cosine waves. The fundamental is the first harmonic and is equal to the frequency of the waveform. The fundamental is the minimum frequency required to represent the waveform. The second multiple of the fundamental is called the second harmonic, the third multiple of the fundamental is called the third harmonic, and so on. The nth harmonic frequency of a periodic waveform is determined by:

$$f_n = nf_0 \quad n=1,2,3, \cdots$$

EQ. 6.11

The coefficients A_0, A_n, and B_n are determined by:

$$A_0 = \frac{1}{T} \int_0^T y(t) dt$$

EQ. 6.12

$$A_n = \frac{2}{T} \int_0^T y(t) \cos(n\omega t) dt$$

EQ. 6.13

$$B_n = \frac{1}{T} \int_0^T y(t) \sin(n\omega t) dt$$

EQ. 6.14

The solution to EQ. 6.12, EQ. 6.13, and EQ. 6.14 requires integral calculus which will not be done here. Therefore, Fourier series solutions will be provided where applicable in this book.

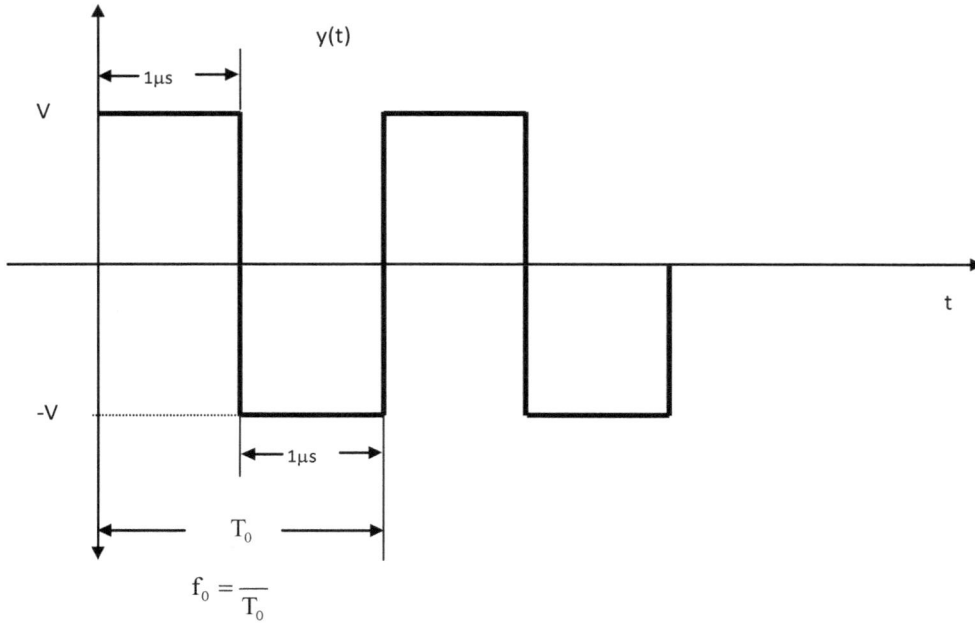

Figure 6.5. Time domain representation of the periodic square waveform with fundamental period of 1000000μs and fundamental frequency of 1MHz.

The sinusoidal series of the square wave in Figure 6.5 is illustrated in Figure 6.6 and is mathematically represented by:

DC	1st	2nd	nth
Term	harmonic	harmonic	harmonic

$$y(t) = V_0 + V_1 \sin \omega t + V_2 \sin 2\omega t + ... + V_n \sin n\omega t \qquad \text{EQ. 6.15}$$

where $V_0 = \sin(2\pi \cdot 0) = 0$ since at DC f=0, $\omega = 2\pi f_0$ and

$$V_n = \frac{4V}{n\pi} \quad n = 1,3,5,7...... \qquad \text{EQ. 6.16}$$

The condensed version of the Fourier series of the square wave is

$$y(t) = \sum_{n=odd}^{\infty} \frac{4V}{n\pi} \sin(n\omega t) \qquad \text{EQ. 6.17}$$

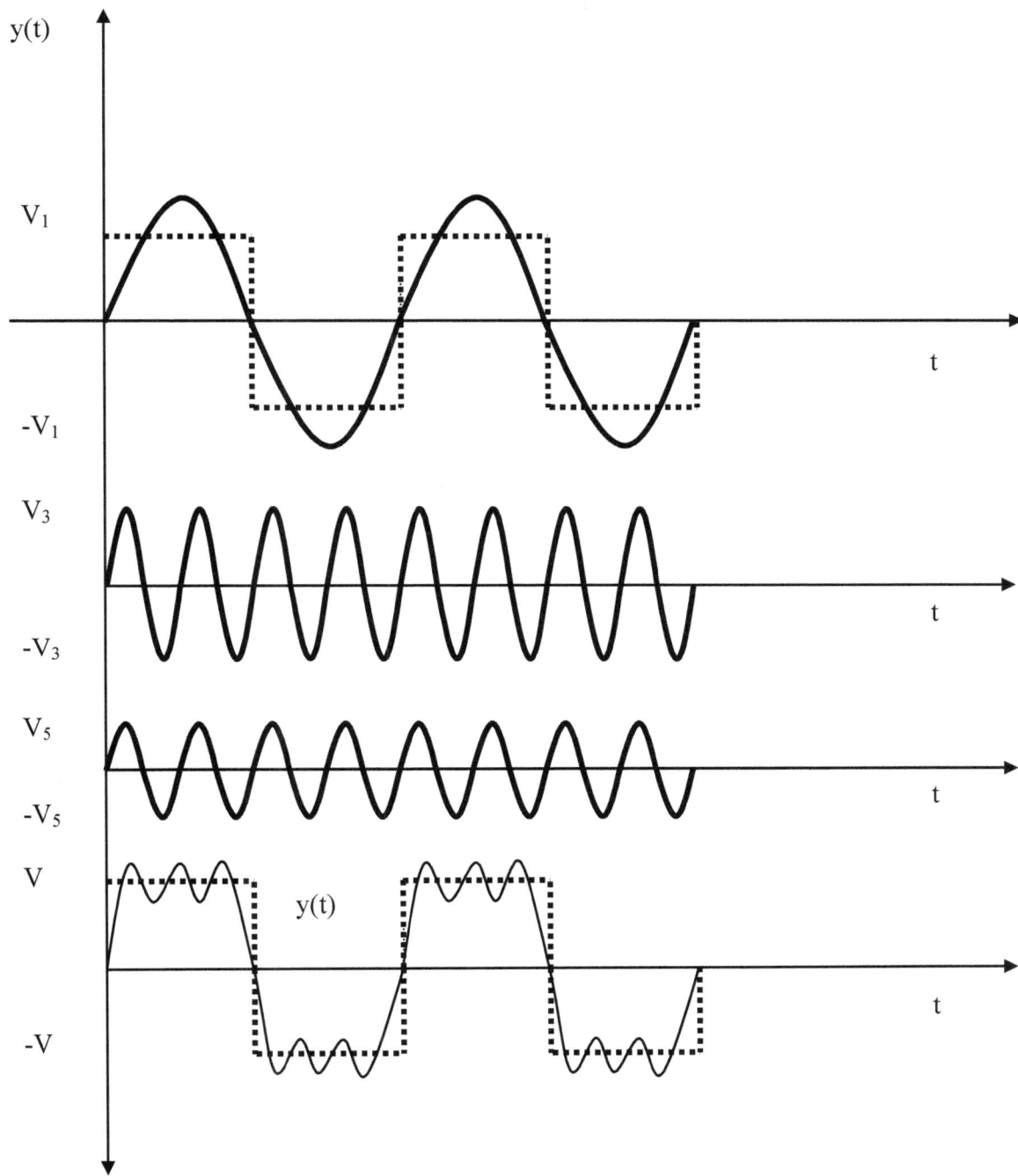

Figure 6.6. Fourier Series representation of the periodic square waveform using the first three harmonics.

6.1.9. Mixing

Mixing is a type of multiplication. It is represented by the symbol × enclosed in the circle shown in Figure 6.7. Thus, from Figure 6.7 y(t) is determined by the multiplication of $v_m(t)$ and $v_c(t)$ as follows:

$$y(t) = V_c \sin(\omega_c t) V_m \sin(\omega_m t)$$

$$= \frac{V_c V_m}{2} \left[\cos((\omega_c - \omega_m)t) - \cos((\omega_c + \omega_m)t) \right]$$

$$= \frac{V_c V_m}{2} \left[\cos(2\pi(f_c - f_m)t) - \cos(2\pi(f_c + f_m)t) \right]$$

EQ. 6.18

where the following trig identity was made useful:

$$\sin\alpha\sin\beta = \frac{1}{2}\left[\cos(\alpha-\beta) - \cos(\alpha+\beta)\right]$$

A nonlinear device such as a diode can be used to perform the mixing operation. The diode has a nonlinear current to voltage relationship as shown Figure 6.8.

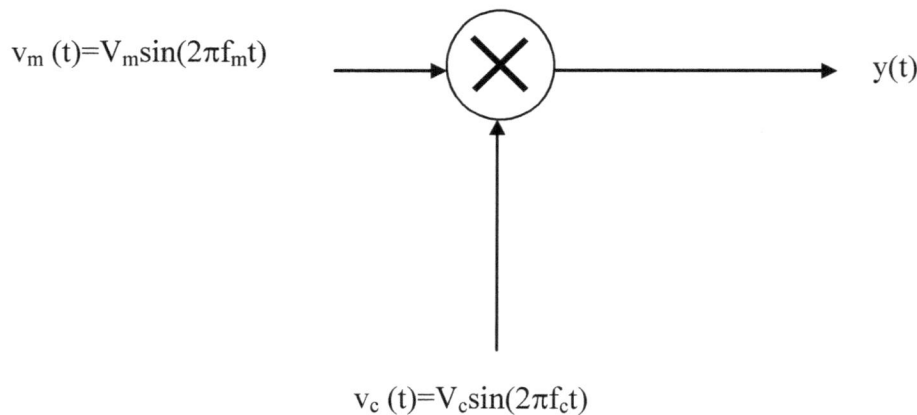

Figure 6.7. Mixing of two sinusoidal signals: $v_m(t)$ and $v_c(t)$.

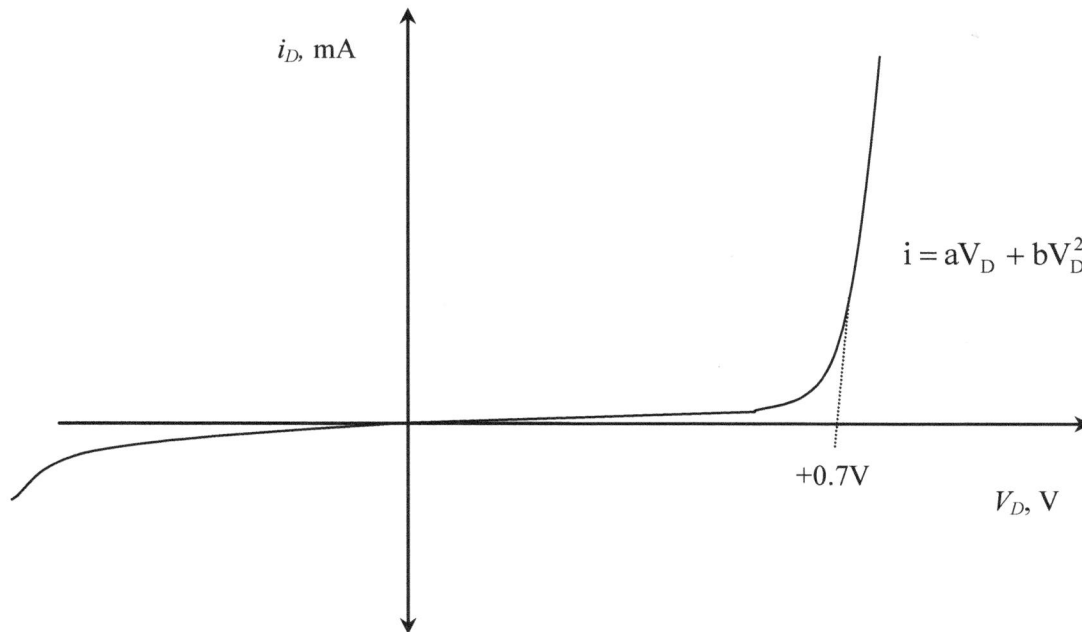

Figure 6.8. Typical diode electrical characteristics.

If the voltage across the diode is the sum of two sinusoidals : $v_m(t) = \sin(\omega_m t)$ and $v_c(t) = \sin(\omega_c t)$ as shown in Figure 6.9. Let $v_m = v_m(t)$ and $v_c = v_c(t)$. Then:

$$i = a v_D + b v_D^2 \qquad\qquad\qquad \textbf{EQ. 6.19}$$

$$= a(v_c + v_m) + b(v_c + v_m)^2$$

$$= a v_c + a v_m + b v_c^2 + b v_c v_m + b v_m^2$$

$$= a\sin(\omega_c t) + a\sin(\omega_m t) + b\sin^2(\omega_c t) + b\sin(\omega_c t)\sin(\omega_m t) + b\sin^2(\omega_m t)$$

$$= a\sin(\omega_c t) + a\sin(\omega_m t) + \frac{b}{2} - b\cos(2\omega_c t) + \frac{b}{2}\left[\cos((\omega_c - \omega_m)t) - \cos((\omega_c + \omega_m)t)\right]$$

$$+ \frac{b}{2} - b\cos(2\omega_m t)$$

If a bandpass filter is applied to the output of the diode circuit as shown in Figure 6.10 the following output results in the removal of the extra harmonics, DC components b, f_m, $2f_c$, $2f_m$ to give the following:

$$i = a\sin(\omega_c t) + \frac{b}{2}\left[\cos((\omega_c - \omega_m)t) - \cos((\omega_c + \omega_m)t)\right] \qquad \textbf{EQ. 6.20}$$

This is an AM like signal. AM will be discussed in more details in 6.6.1.

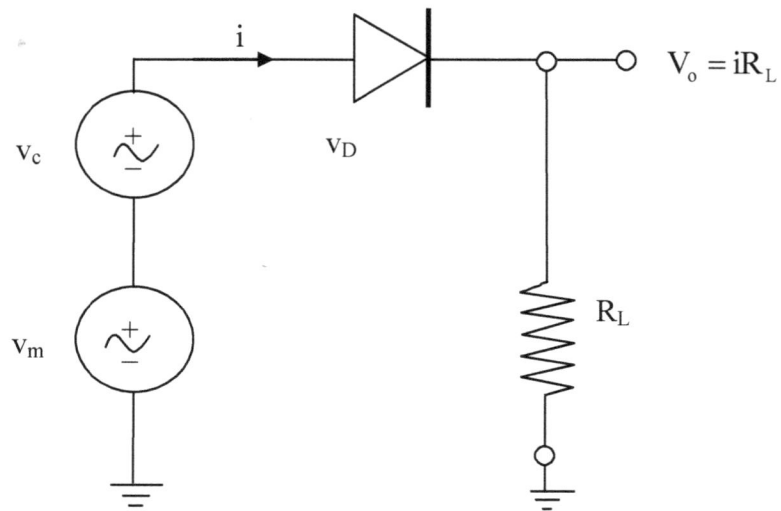

Figure 6.9. Nonlinear circuit using diode.

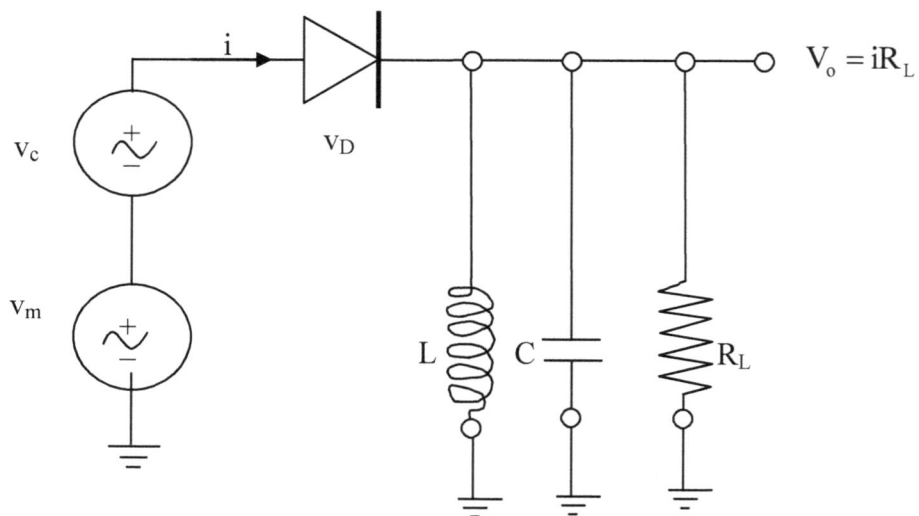

Figure 6.10. Bandpass filtering on output of nonlinear diode circuit.

6.2. **Basic Communication Systems**

A communication system is a connection of individual subsystems working together to transmit information from one point to another with minimum distortion such as by noise. An example of a basic communication system is shown in Figure 6.11. Information is sent via the transmitter to a receiver over a transmission medium. The transmitter prepares the electronic message for transmission over the communication channel. This is typically done using an electronic message to modulate a high frequency carrier. The modulated carrier is typical amplified and then transmitted over the communication channel. At the receiver it recovered as the information destination. Figure 6.12 gives more details of the same system.

There are two types of digital communication systems: Digital and Analog. A digital communication system is one in which the information source is in a digital form, e.g., a series of zeros and ones, etc. An analog communication system is one in which the information source is in analog form, e.g., continuous varying sine wave, etc. For either type, the transmission medium can be wire (coaxial, fiber, etc) or wireless (air, water, etc). The communication channel is the medium over which the electronic message is sent over from the transmitter. In summary, the medium could be any of the following:

1. Electrical conductor, e.g., coaxial, cable
2. Free space, e.g., air
3. Optical, e.g., optical fiber
4. Sound conductor, e.g., water
5. Sound conductor, e.g., earth

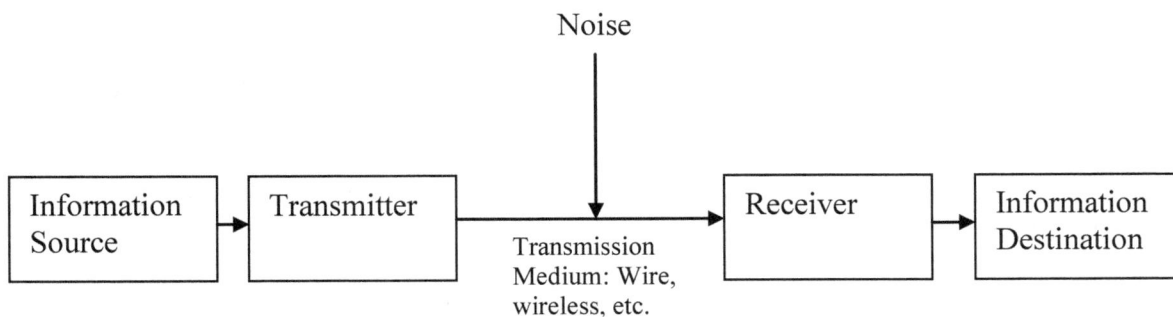

Figure 6.11. Basic communication system.

C: RF Oscillator

G: Local Oscillator

| B: Modulator/Up Conversion | D: Amplifier | | E: Amplifier and Detector | F:Down Conversion and/or Demodulator |

Transmission Medium: Wire, wireless, etc.

A: Modulation Signal: Low Frequency Information Source

H: Demodulated Signal: Low Frequency Information Source

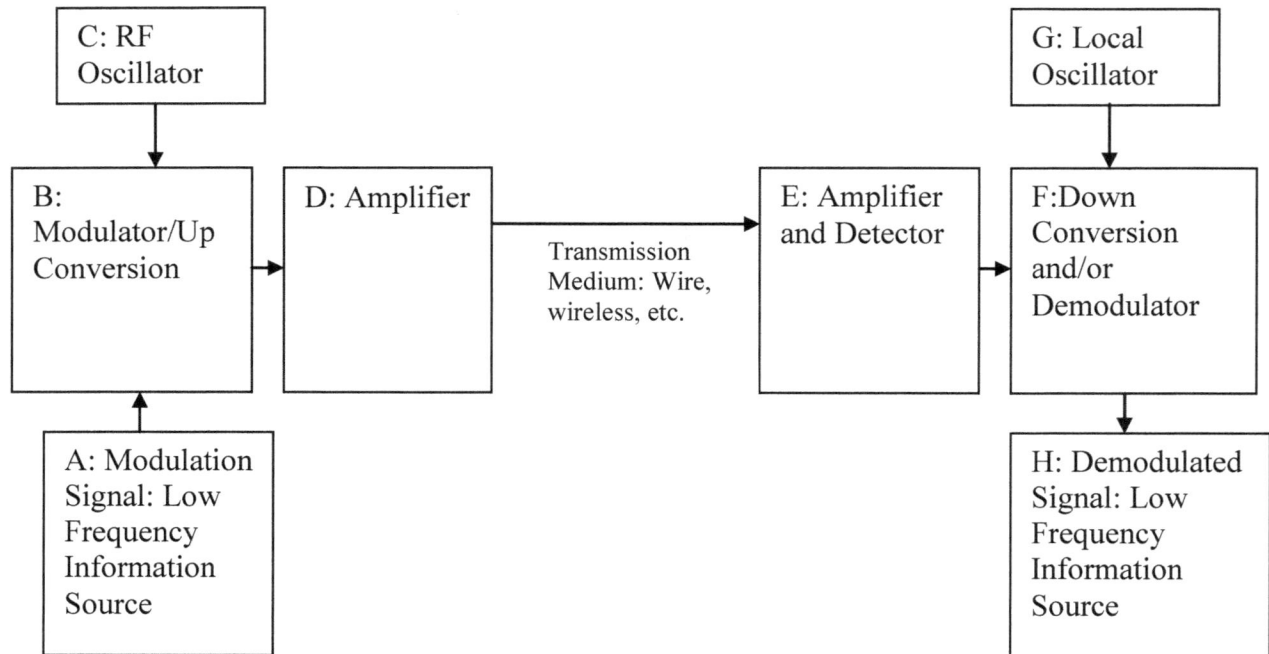

Figure 6.12. Basic communication system details.

Figure 6.12 gives a more detailed representation of the basic communication system. In block A, the **modulating signal,** which will be represented as $v_m(t)$ throughout this chapter, is the low frequency message signal containing the information. Before a message can be transmitted it must be converted to an electrical form which can then be processed by an electronic circuit that prepares it for transmission over the communication channel. For voice messages a microphone is used to convert the audio signal to an electrical signal. For the TV set, a camera is used to convert the light energy to an electrical energy for each scene. For a computer system, a keyboard is used to convert input messages into binary code which can be processed by the computer system and then transmitted elsewhere. **Modulation** is the process by which some characteristic of the carrier signal is change relative to a modulating signal. This is performed by the modulator circuit in block B. In this stage, the information obtained from the modulating signal stage is up-converted to the information in a high frequency carrier signal. This is typically done by mixing the modulating signal with a radio frequency (RF) oscillator signal (carrier). Thus, depending on the modulation scheme used, the carrier's amplitude, phase, frequency (or a combination of), changes. The **carrier** signal is represented by block C. An example of a **carrier** in communication systems is a high frequency sinusoidal waveform. The carrier is created by an oscillator circuit, etc. In block D, the **amplifier** filters and amplifies the resulting modulated signal for transmission across the transmission medium. In block E, the amplifier on the receive end amplifies and detects the modulated signal. Block F represents the **demodulation** process which may include a down-converter. For example, a direct receiver directly demodulates the signal without the down-conversion process which typically involves mixing the received signal with a **local oscillator** (block G). Thus, **demodulation** is the process of recovering the message signal from the modulated signal. The final signal is the original message signal in block H.

Modulation is required for systems using radio (radiation of electromagnetic waves) for the following reasons:

  A. Assume the low frequency information source is a digital message with a spectrum from near DC to 200KHz. The range of transmission by directly radiating the low frequency information source is extremely low due to the inefficiency of practical antennas in this frequency range.
  B. There will be no means of eliminating interference from multiple users since all will be operating on the same band of frequencies.

By modulation the information signal onto a much higher radio frequency signal such as 300 MHz, the total bandwidth as a percentage of the operating frequency is now very small and antenna bandwidth issues are of no concern. In addition, very efficient antenna can be constructed for 300 MHz with very small size (a ½ wavelength antenna is about 18 inches).

The carrier signal allows low frequency information to be transmitted over higher frequency channels supporting various frequencies. In the U.S., these are regulated by the Federal Communication Commission (FCC). The FCC controls commercial AM/FM and Satellite broadcasting, 2 way radio, pagers, amateur radio, cell phones, and various other services in the frequency range from 535KHz to 25HHz. There are also frequency bands where unlicensed, low power transmissions are allowed, such as in the 300MHz range. television (channels 2-13) occupies frequencies from 54 MHz to 216 MHz.

6.3. Transmission Modes
Electronic communication systems can be classified according to how they transmit and receive information:

  A. simplex
  B. half duplex
  C. full duplex

A simplex communication system is the simplest. The information is transmitted in one direction. Some examples are radio and TV broadcasting, beeper, etc. In half duplex systems the information is transmitted in one direction at a time. For example, citizen band (CB) used by the police, military, etc., are half duplex. In full duplex systems, the information is allowed to travel in both directions at the same time. Some examples are the standard telephone, cell phones, etc.

Most modern communication systems require two way communication. This requires each side to receive or transmit. Thus, a transmitter and receiver circuit are needed for each side in order to achieve full duplex communication. Collectively, a transmitter and receiver circuit together are referred to as a transceiver.

6.4. **Analog and Digital Signals**

Time domain signal refers to the representation of signals that vary with time. A time domain signal is represented as a function of amplitude vs. time generally denoted by $x(t)$ where t is the independent variable representing time. There are two types of signals:

a.) Analog - Continuous amplitudes
b.) Digital - Discrete amplitudes

These shall be briefly discussed in the following sections.

6.4.1. **Analog Signals**

Continuous time signals, $x(t)$, can take on any amplitude between the two points a and b. That is, $x(t)$ is continuous in the interval (a,b) where **a** may be $-\infty$ and **b** may be $+\infty$. This type of signal is referred to as an analog signal. Figure 6.13 shows an example of an analog signal.

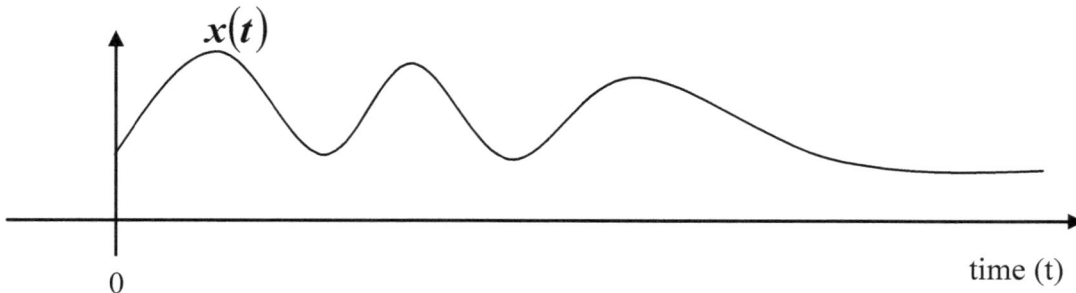

Figure 6.13. Example of analog signal.

6.4.2. **Digital Signals**

A digital signal is represented by $x[n]$ where n are discrete points in time. $x[n]$ are the amplitudes of the signal at each instance in time represented by n(Ifeachor, et al., 2002) (Vegte, 2002). $x[n]$ are obtained by sampling a continuous time signal $x(t)$, i.e., $x[n] = x(nT_s)$ where T_s is the sampling interval. An example of $x[n]$ is shown in Figure 6.14. Each sample $x[n]$ is then converted into a digital code (8 bits, 16 bits, 32 bits, etc.) that can be understood and processed by a computer. A device that performs this digitization process is called an analog to digital converter (ADC). A digital to analog converter (DAC) is used to convert the digital signal back to continuous or analog form.

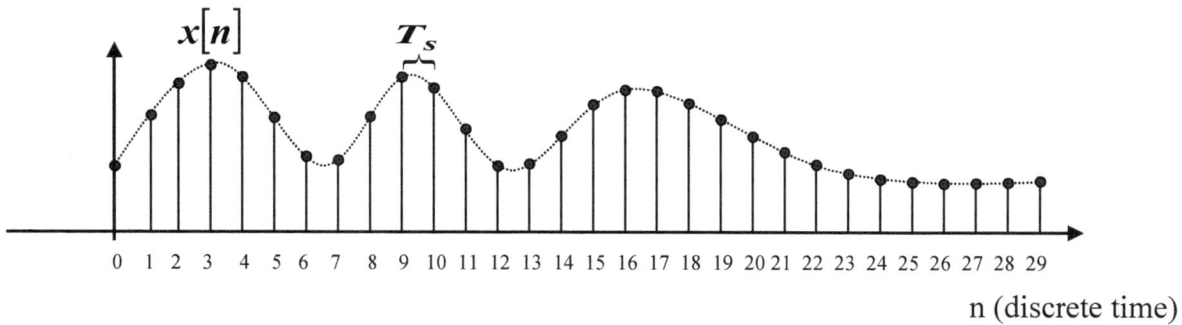

Figure 6.14. Example of digitizing an analog signal to get a digital signal (also called discrete signal).

6.5. Oscillators

Electronic circuits in communication systems require one or more frequencies. This section discusses the basic concepts of circuits used to generate these frequencies. Such circuits are:

    A. oscillators
    B. phase lock loops
    C. frequency synthesizers

Figure 6.15 illustrates the basic oscillator circuit. It consists of the following main parts: 1.) open loop gain circuit with gain A, 2.) and feedback circuit with gain β.

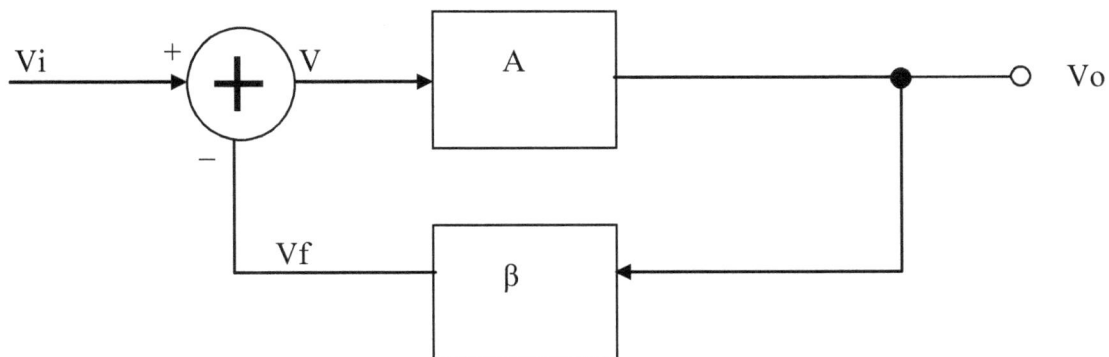

Figure 6.15. Block diagram of basic oscillator circuit.

Using Figure 6.15, the gain of the circuit is derived as follows:

$$(1) \quad V = V_i - V_f$$

$$(2) \quad V_f = \beta V_o$$

$$(3) \quad V_o = AV$$

$$= A(V_i - V_f)$$

$$= A(V_i - \beta V_o)$$

$$= AV_i - A\beta V_o$$

$$\Rightarrow V_o(1 + A\beta) = AV_i$$

$$\Rightarrow V_o = \frac{AV_i}{1 + A\beta}$$

$$\Rightarrow \frac{V_o}{V_i} = \frac{A}{1 + A\beta}$$

EQ. 6.21

Thus, EQ. 6.21 shows how the gain is affected by the factor $1+A\beta$. The condition $A\beta = 1$ is referred to as **Barkhausen criterion** for oscillation. Under such condition the circuit will be able to sustain self oscillations. In practice, $A\beta$ is made greater than 1 knowing that saturation effects in the circuit will cause an average $A\beta$ of 1.

6.5.1. Tuned Oscillator

In general, a tuned oscillator circuit takes the form given by the block diagram in Figure 6.16. It has an amplifier with gain A and three reactive circuits. Table 6.1 lists two types of configurations of this tuned circuit: A.) Collpitts Oscillator, B.) and Hartley Oscillator.

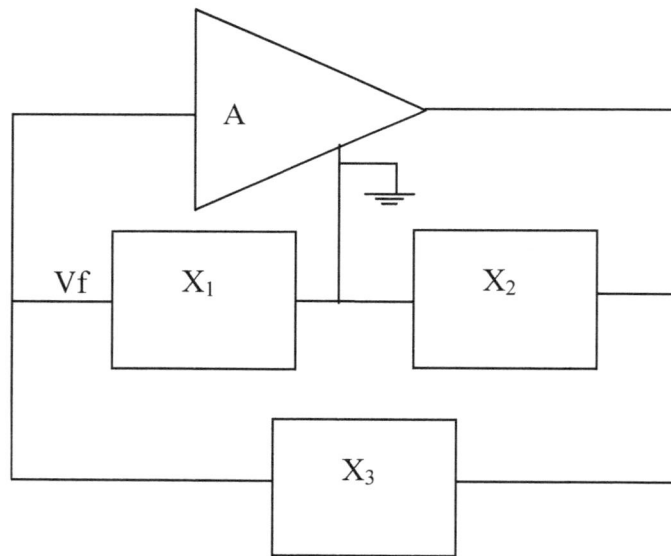

Figure 6.16. Block diagram of basic tuned oscillator circuit.

Oscillator Type	X_1	X_2	X_3
Colpitts	C	C	L
Hartley	L	L	C

Table 6.1. Various configurations of the tuned oscillator circuit shown in Figure 6.16. L is an inductor and C is a capacitor.

The Collpits oscillator can be created in several ways:
  A. Bipolar junction transistor (BJT) oscillator circuit
  B. Field effect transistor (FET) oscillator circuit
  C. Operational amplifier oscillator circuit

A BJT Colpitts oscillator is shown in Figure 6.17. For self sustain oscillations the following condition must be met

$$A\beta = -1$$

EQ. 6.22

In EQ. 6.22, the phase angle is 180 degrees. This occurs at resonance when

$$f_o = \frac{1}{2\pi\sqrt{L \cdot C_{eq}}}$$

EQ. 6.23

where f_o is the oscillator frequency, L is the inductance, and C_{eq} is the equivalent capacitance seen looking into the feedback capacitive ircuit. Additionally, the gain must meet the following condition given by

$$A = \frac{1}{\beta} = \frac{C_1}{C_2}$$

EQ. 6.24

Practically, select

$$A \geq \frac{C_1}{C_2}$$

EQ. 6.25

to sustain oscillations so that AB is a little greater than 1.

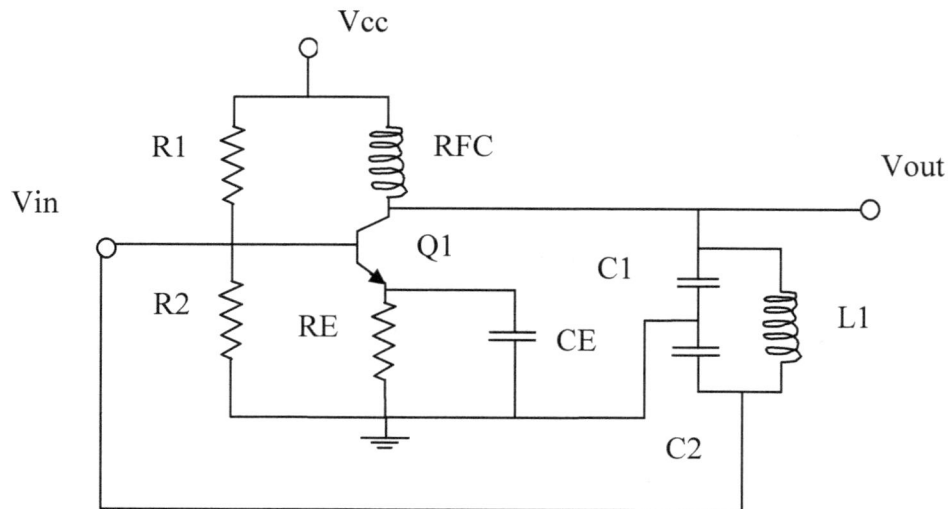

Figure 6.17. Colpitts oscillator circuit.

EXAMPLE 6.1

Using Figure 6.15, if A = 20, determine β for oscillation.

$$\beta = \frac{1}{A} = \frac{1}{20} = 0.05$$

EXAMPLE 6.2

Using Figure 6.15, if A = 120, β = 0.01 will the circuit oscillate?

$$A\beta = 120(0.01) = 1.2$$
Yes, the circuit will oscillate.

6.6. Analog Communications

The various modulation techniques can be summarized using the following carrier signal:

$$v(t) = V_c \sin(2\pi f_c t + \theta)$$
EQ. 6.27

where v(t) is a time varying sine wave of the voltage, V_c is the peak carrier voltage in volts, f_c is the frequency in Hertz, and θ is the phase shift in radians. Typically, for a carrier signal the frequency f_c is a high frequency value. If the information signal is analog and the amplitude V of the carrier is varied in proportion to the information signal, then the resulting signal is amplitude modulation or AM. If the information signal is analog and the frequency f_c of the carrier is varied in proportion to the information signal then the resulting signal is frequency modulation or FM. If the information signal is analog and the phase θ of the carrier is varied in proportion to the information signal then the resulting signal is phase modulation or PM. These will be discussed in more details in the following Sections.

6.6.1. Amplitude Modulation Transmission

There are several types of amplitude modulation schemes: double side band (DSB) modulation, standard amplitude modulation (AM), single sideband (SSB) modulation, and vestigial sideband (VSB) modulation. We shall only be concerned with standard AM in this section. An example of a standard AM signal is shown in Figure 6.18 where

$$V_m = \frac{(V\max - V\min)}{2}$$
EQ. 6.28

and

$$V_c = \frac{(V\max + V\min)}{2}$$
EQ. 6.29

Therefore, the modulation index, also referred as modulation factor, represents the amount of modulation and is given by where

$$m = \frac{V_m}{V_c} = \frac{(V\max - V\min)}{(V\max + V\min)}$$
EQ. 6.30

An undistorted AM signal has values for m between 0 and 1. The ideal condition for AM is m=1 which gives 100% modulation. This results in the greatest power output at the transmitter and received signal at the receiver and with no distortion.

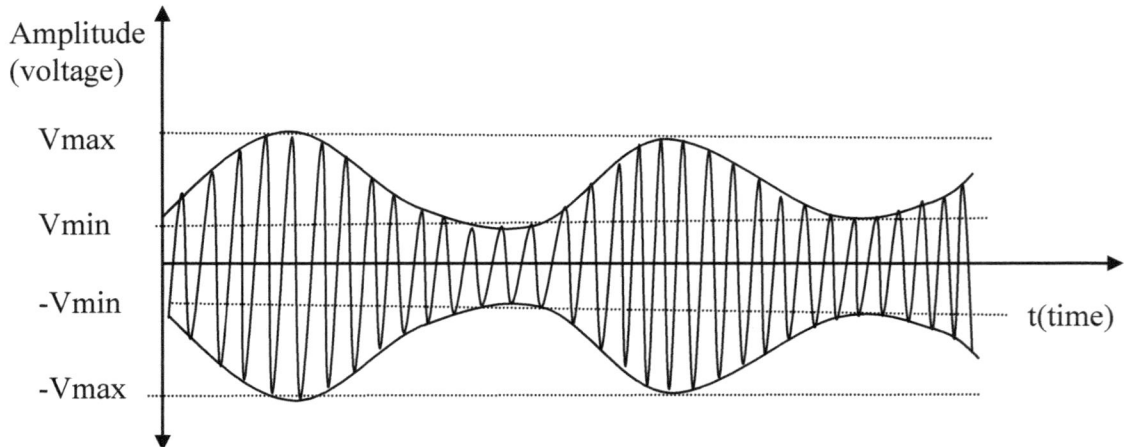

Figure 6.18. Example of time domain representation of a standard AM signal.

The standard AM signal is created by adding a DC component, V_c, to the modulating signal, $v_m(t)$, as shown in Figure 6.19, resulting in the following standard AM expression:

$$v_{AM}(t) = (V_m \sin(\omega_m t) + V_c) \sin(\omega_c t)$$

EQ. 6.31

$$= V_c \sin(\omega_c t) + \frac{V_m}{2} [\cos((\omega_c - \omega_m)t) - \cos((\omega_c + \omega_m)t)]$$

$$= V_c \sin(2\pi f_c t) + \frac{V_m}{2} [\cos(2\pi(f_c - f_m)t) - \cos(2\pi(f_c + f_m)t)]$$

The corresponding frequency domain representation of the AM signal represented by EQ. 6.31 is shown in Figure 6.20.

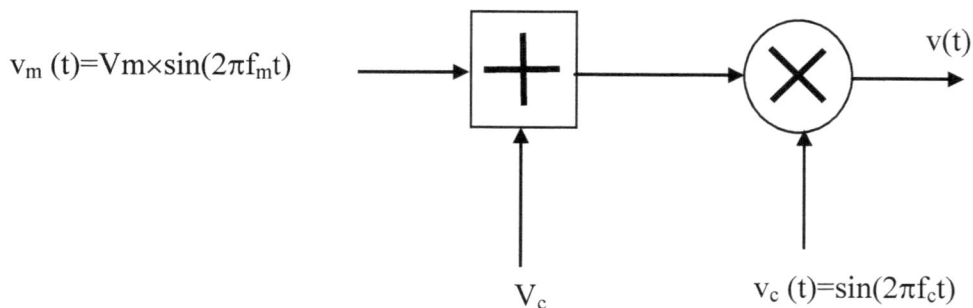

Figure 6.19. Standard model to generate an AM signal using a single sinusoidal, $v_m(t)$, as the modulating signal and carrier, $v_c(t)$.

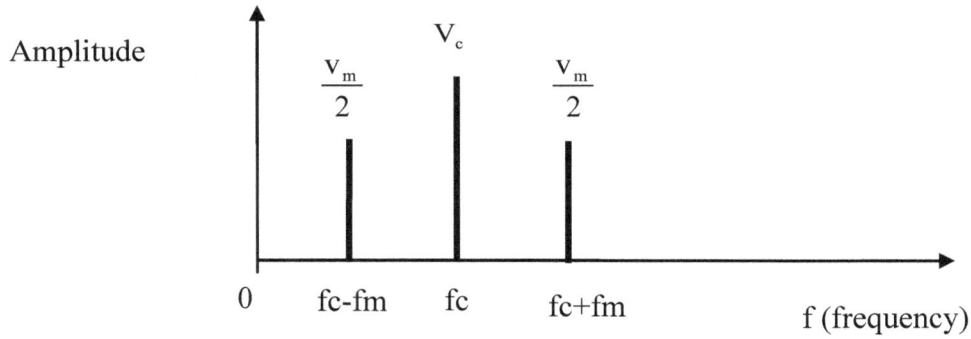

Figure 6.20. Frequency domain representation of an AM signal.

Using the modulation index m given by EQ. 6.30 and rearranging to $V_m = mV_c$ results in a modified version of EQ. 6.31 as follows:

$$v_{AM}(t) = V_c \sin(2\pi f_c t) + \frac{mV_c}{2}\left[\cos(2\pi(f_c - f_m)t) - \cos(2\pi(f_c + f_m)t)\right]$$ EQ. 6.32

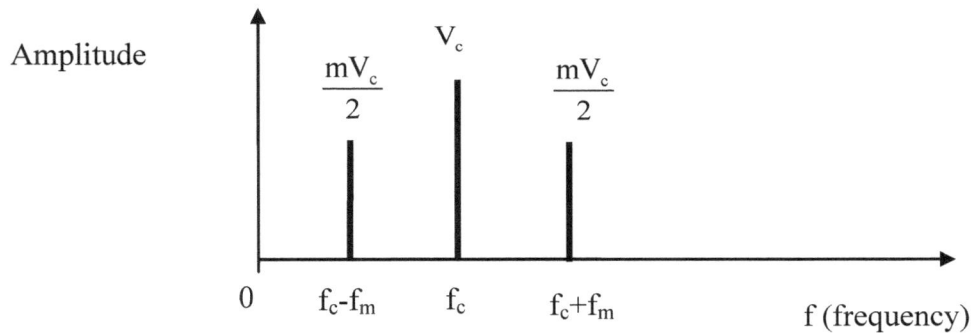

Figure 6.21. Frequency domain representation of an AM signal in terms of the modulation index m.

6.6.2. Amplitude Modulation Bandwidth

The frequency domain of an AM signal can be obtained by an instrument called a spectrum analyzer which shows the signal amplitude vs. frequency.

In general, the lower sideband frequency of an AM signal is determined by

$$f_{lsb} = f_c - f_m$$ EQ. 6.33

while the upper sideband frequency is determined by

$$f_{usb} = f_c + f_m$$ EQ. 6.34

Therefore, the bandwidth is

$$BW = f_{usb} - f_{lsb}$$ EQ. 6.35

Figure 6.22 shows the Bandwidth of an audio AM signal. In general, the BW of an audio signal has voice frequencies from 300Hz to 3000Hz. Additionally, as illustrated in Figure 6.22 for an AM signal, the upper and lower sidebands are f+3000Hz and f-3000Hz, respectively.

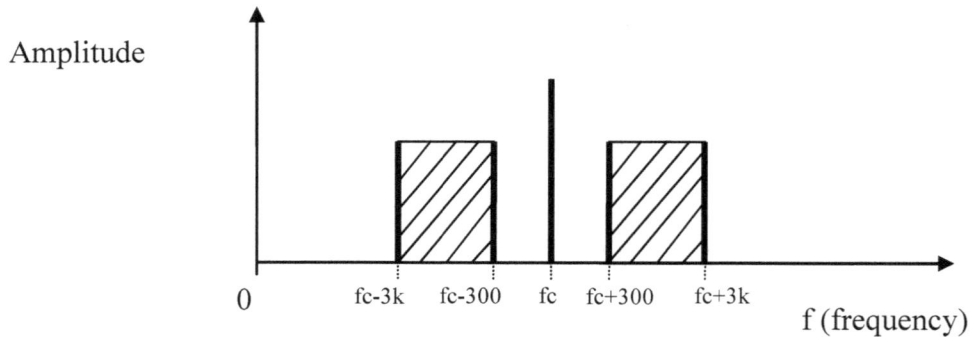

Figure 6.22. Bandwidth representation of an audio AM signal.

6.6.3. Amplitude Modulation Power

The AM power in an AM signal is produced by the carrier and its two sidebands:

    A. f_{usb}
    B. f_{lsb}

Therefore, the total power in a an AM signal is determined by

- $$P_T = P_c + P_{lsb} + P_{usb}$$
- EQ. 6.36

- Power is typically indicated in RMS given by

$$P_{RMS} = \frac{V^2}{R} = \frac{(RMS\,Voltage)^2}{load\,impedance}$$ EQ. 6.37

So, going back to the standard AM equation and converting to RMS by dividing by $\sqrt{2}$ (or multiply by 0.707) gives

$$v_{AMRMS}(t) = V_c \sin(2\pi f_c t) + \frac{mV_c}{2}\left[\cos(2\pi(f_c - f_m)t) - \cos(2\pi(f_c + f_m)t)\right]$$ EQ. 6.38

$$= \frac{V_c}{\sqrt{2}} \sin(2\pi f_c t) + \frac{mV_c}{2\sqrt{2}}\left[\cos(2\pi(f_c - f_m)t) - \cos(2\pi(f_c + f_m)t)\right]$$

Therefore, the total power in RMS using the modulation index, m, is given by

$$P_T = \frac{\left(V_c/\sqrt{2}\right)^2}{R} + \frac{\left(V_m/2\sqrt{2}\right)^2}{R} + \frac{\left(V_m/2\sqrt{2}\right)^2}{R}$$

$$= \frac{V_c^2}{2R} + \frac{V_m^2}{8R} + \frac{V_m^2}{8R}$$

$$= \frac{V_c^2}{2R} + \frac{m^2 V_c^2}{8R} + \frac{m^2 V_c^2}{8R}$$

$$= \frac{V_c^2}{2R}\left(1 + \frac{m^2}{4R} + \frac{m^2}{4R}\right)$$

$$= P_c\left(1 + \frac{m^2}{2}\right) \qquad \text{EQ. 6.39}$$

EXAMPLE 6.3

An AM transmitter has a carrier power of 1000W and is modulated 100%. Determine the total AM signal power.

$$P_T = P_c\left(1 + \frac{m^2}{2}\right) = 1000\left(1 + \frac{1^2}{2}\right) = 1000(1.5) = 1500\text{W}$$

It is more practical or easier to measure current in the load than voltage. Therefore, it is common to see a RF ammeter in series with an antenna to measure current as shown in Figure 6.23. When the antenna impedance is known the total power using current is

$$P_T = \left(I_T\right)^2 R \qquad \text{EQ. 6.40}$$

where

$$I_T = I_c\sqrt{\left(1 + \frac{m^2}{2}\right)} \qquad \text{EQ. 6.41}$$

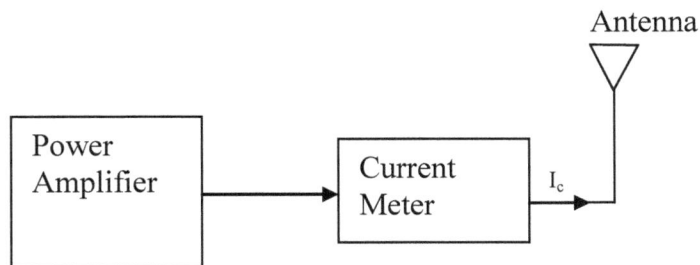

Figure 6.23. Measurement of current for calculating total transmitted power.

6.6.4. Amplitude Modulation Circuits

The circuit shown in Figure 6.24 is a diode modulator circuit. Two signals, $V_m(t)$ and $V_c(t)$, are linearly mixed using R1, R2, and R3. The output of the resistive circuit is then input into a nonlinear device, a diode, which performs rectification. The output is filtered to remove the undesired harmonics to get an AM waveform. See Figure 6.9 and associated discussion.

An improved version of the diode modulator is the transistor modulator in Figure 6.25 because it has gain. A transistor has gain, a diode does not. Rectification occurs due to the base emitter junction. The amplified signal is then sent through a tuned circuit which produces the missing half cycle to give an AM signal.

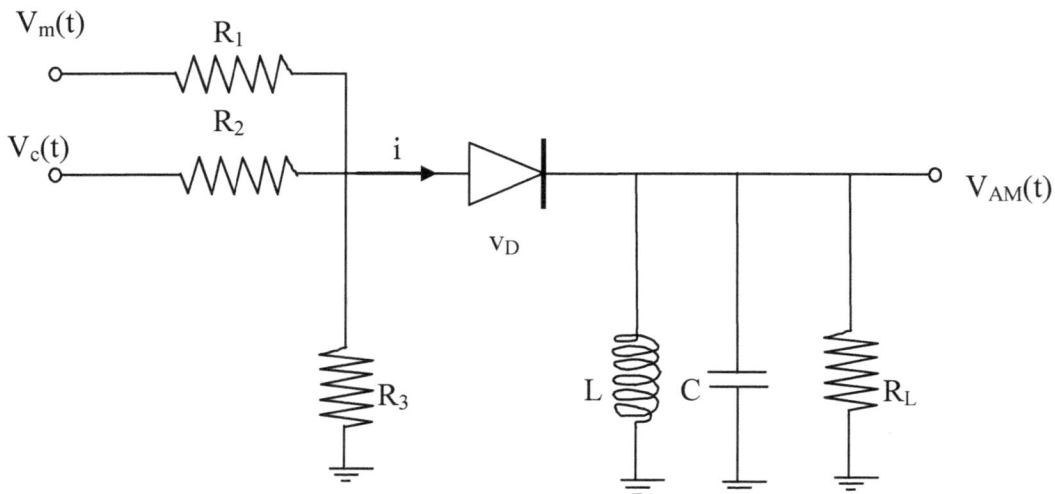

Figure 6.24. Diode modulator circuit.

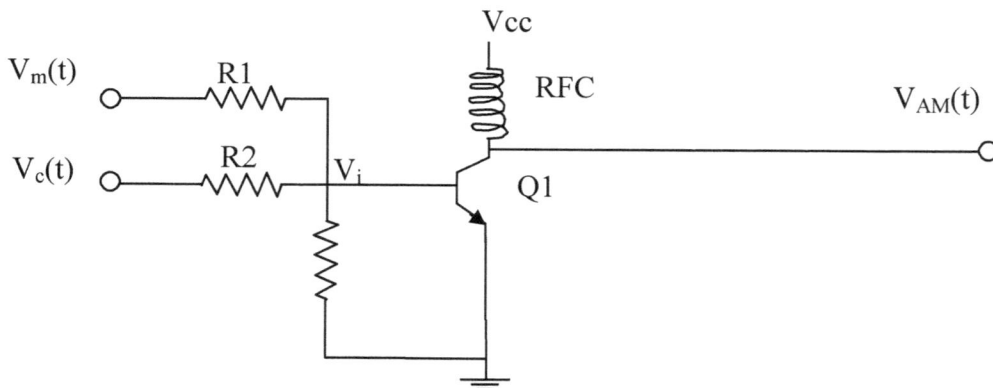

Figure 6.25. Transistor modulator circuit.

6.6.5. AM Reception

Figure 6.30 shows the block diagram of an AM transmitter. An AM transmitter transmits an AM signal. It is an electronic system that performs amplitude modulation on the information signal. To do this the AM transmitter either uses an internal or external oscillator that generates the carrier frequency. The AM transmitter has to meet the following general requirements:

A. It must generate a carrier signal of the desired frequency.

B. It must provide AM modulation on the carrier

C. It must provide sufficient amplification to insure that the signal travels the desired distance. This includes impedance matching of the power amplifier to the antenna. Antennas are discussed in Chapter 12.

Figure 6.31 and Figure 6.32 shows the block diagram of an AM tuned radio frequency receiver and superheterodyne receiver, respectively. The AM receiver accepts the AM signal of a specific carrier frequency. Two types of AM receivers are:

A. Tuned radio frequency receiver

B. Superheterodyne receiver

The tuned radio frequency receiver is the simplest. It consists of the following stages:

A. Selector and RF amplifier combination.

B. Demodulator

C. Audio Amplifier

The selector and RF amplifier select the desired AM signal and rejects all others. The selected signal is then amplified. The demodulator recovers the original information. A basic detector is a diode in series with a lowpass filter as shown Figure 6.26. The audio amplifier amplifies the audio signal from the detector for preparation for the speaker, etc. Figure 6.27, Figure 6.28, and Figure 6.29 shows the time domain representation of the diode detector at various points in its circuit.

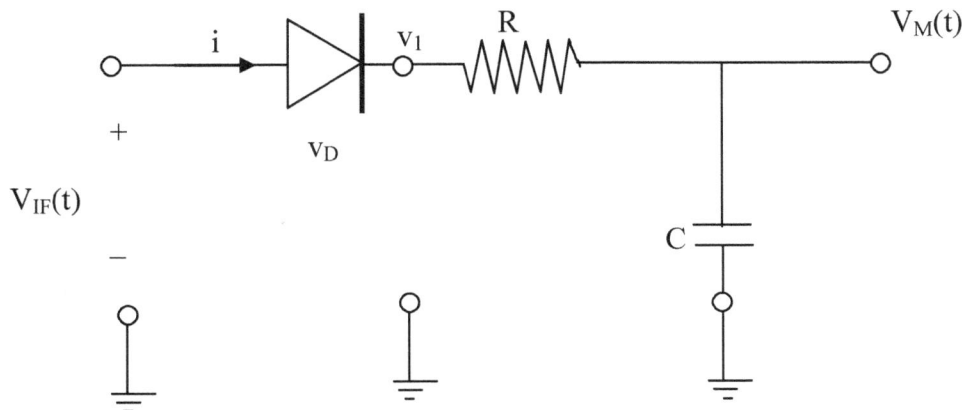

Figure 6.26. Detector circuit.

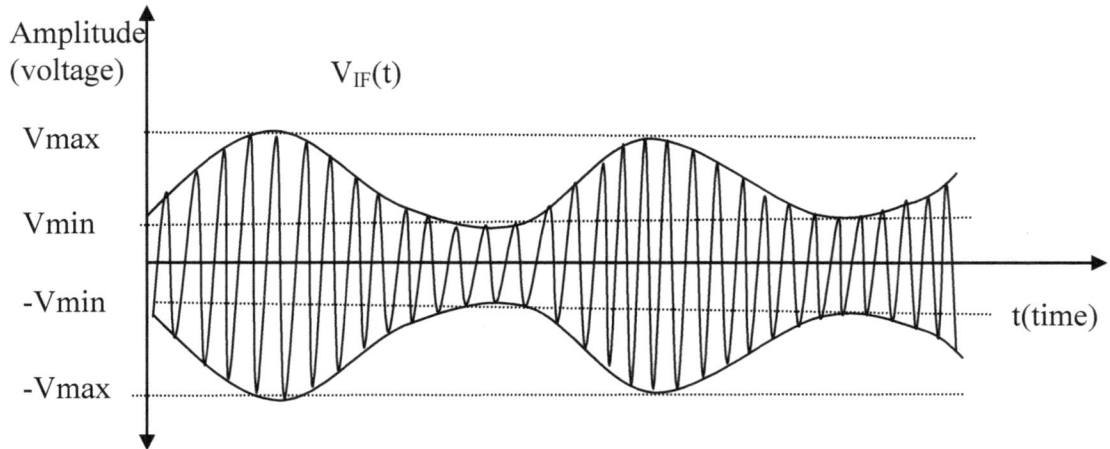

Figure 6.27. Time domain representation of $v_{IF}(t)$.

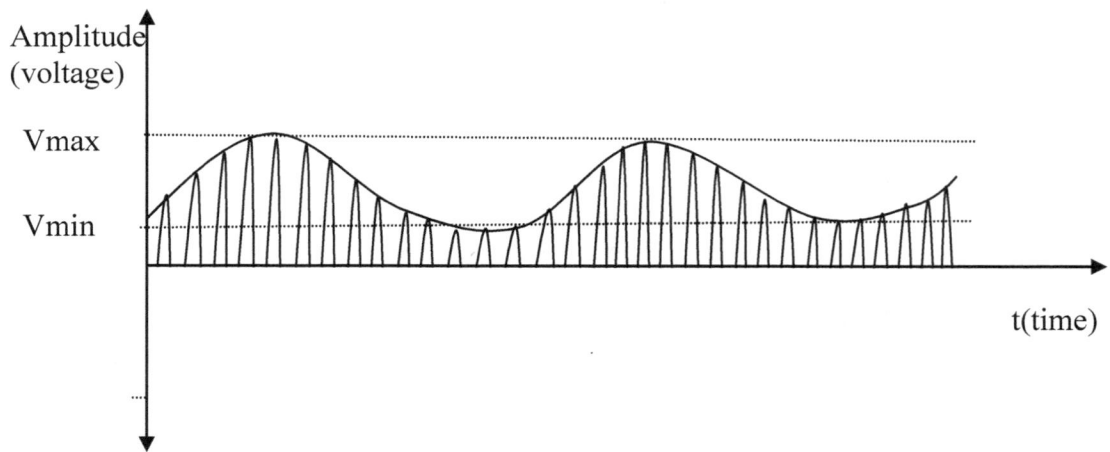

Figure 6.28. Time domain representation of $v_1(t)$.

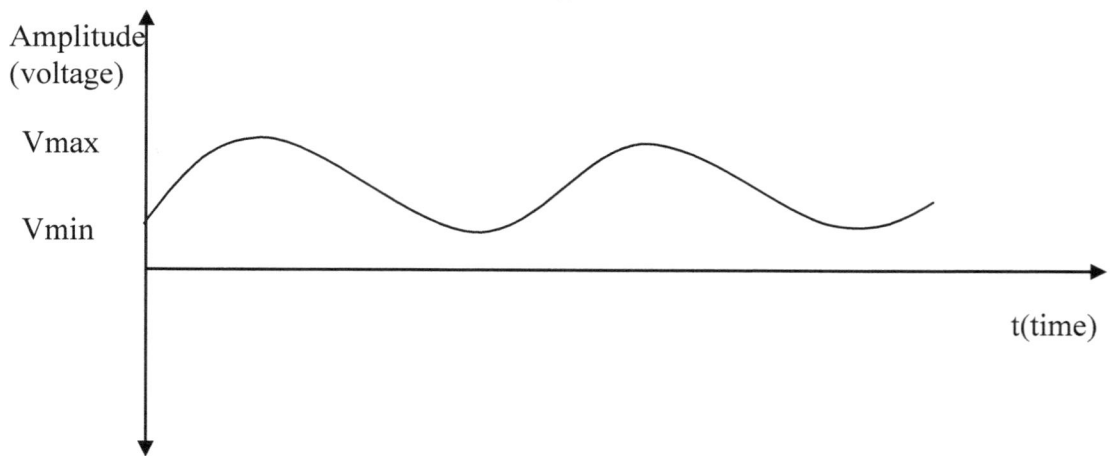

Figure 6.29. Time domain representation of $v_M(t)$.

The superheterodyne receiver consists of the following stages:
  A. RF Amplifier
  B. modulator – mixer
  C. local oscillator
  D. intermediate amplifier
  E. detector
  F. audio amplifier

The RF amplifier selects the desired input and amplifies it. The mixer mixes the local oscillator signal, f_{LO}, with the incoming AM signal. For high-side injection the local oscillator frequency, f_{LO}, is the carrier frequency plus the intermediate frequency:

$$f_{LO} = f_c + f_{IF}$$

EQ. 6.44

For low-side injection the local oscillator frequency, f_{LO}, is the carrier frequency minus the intermediate frequency:

$$f_{LO} = f_c - f_{IF}$$

EQ. 6.45

The intermediate frequency amplifier is tuned to the IF frequency. It selects the IF and its sum and difference frequencies. The AM demodulator or simply detector restores the audio signal using an envelope detector circuit such as the one just discussed in Figure 6.26. The audio amplifier amplifies the audio signal from the detector.

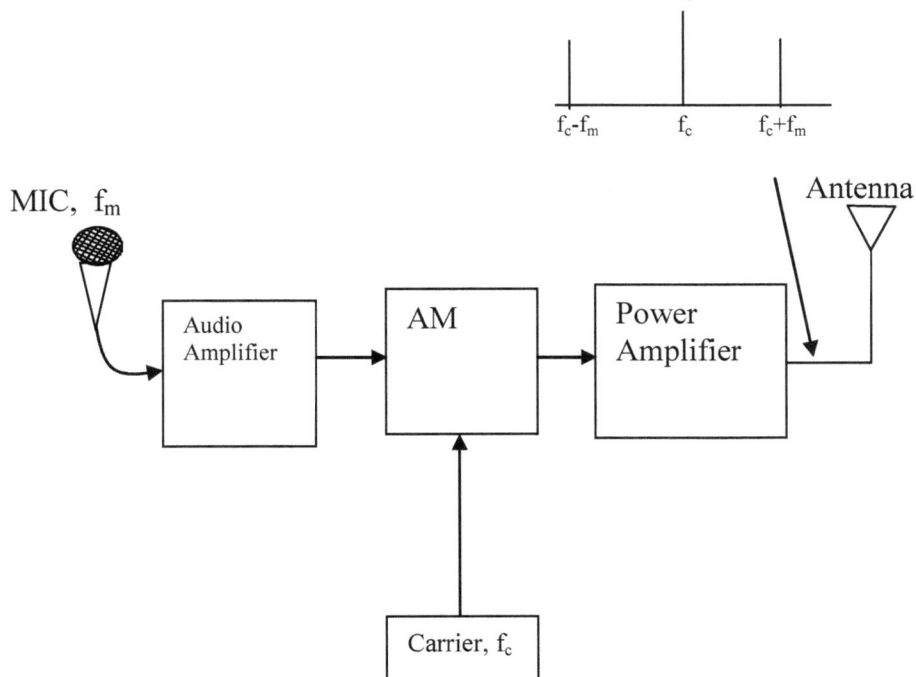

Figure 6.30. AM Transmitter.

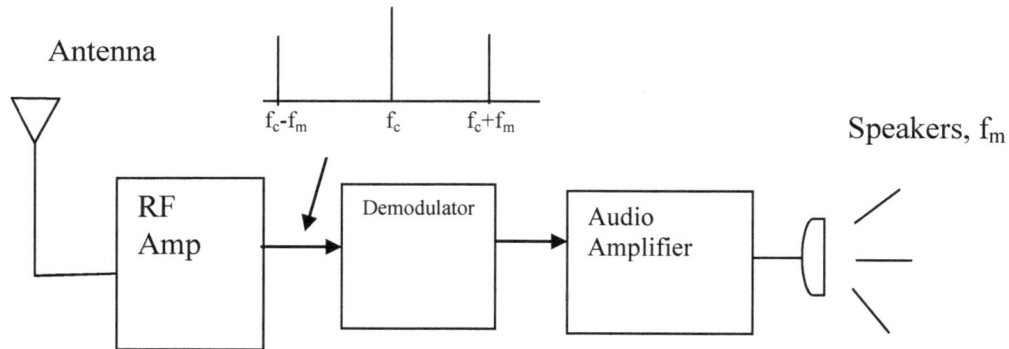

Figure 6.31. Tuned radio frequency receiver.

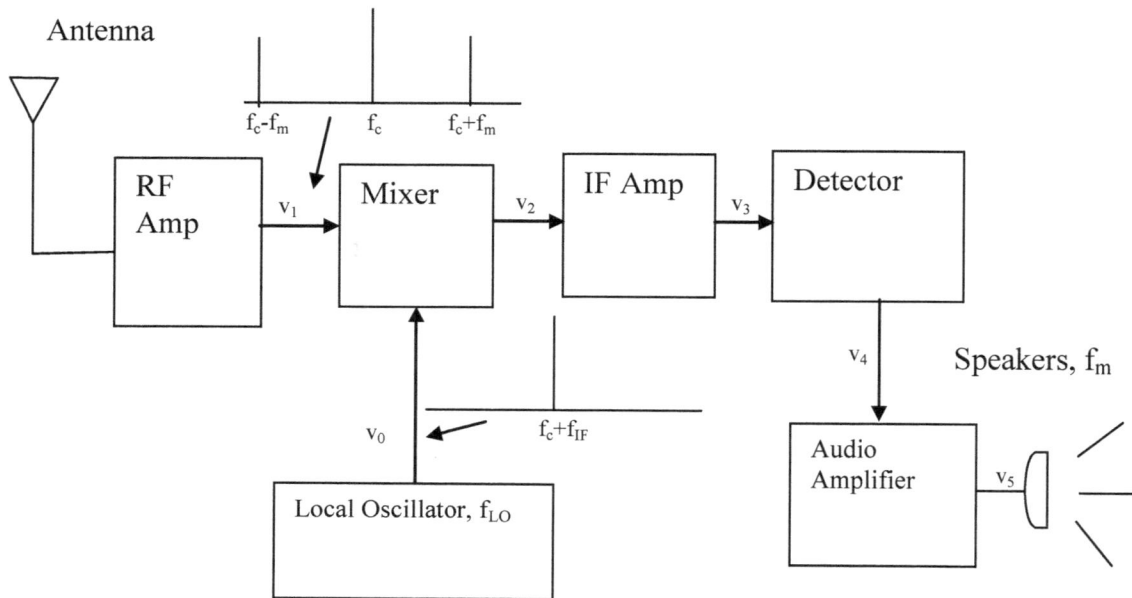

Figure 6.32. AM Superheterodyne Receiver.

Let's consider a more detail analysis of the AM superheterodyne receiver. At the AM transmitter, the input information signal is a pure sinusoidal, $V_m(t) = V_m \sin(\omega_m t)$ and the carrier is $V_c(t) = V_c \sin(\omega_c t)$, with antenna output

$$v_{AM}(t) = V_c \sin 2\pi f_c t + \frac{mV_c}{2} \left[\cos(2\pi(f_c - f_m)t) - \cos(2\pi(f_c + f_m)t) \right]$$ **EQ. 6.46**

In the frequency domain the AM frequencies f_c-f_m, f_c, and f_c+f_m are transmitted and radiated from the antenna. These frequencies are selected by the RF detector and amplifier of the superheterodyne receiver as shown in Figure 6.32. The received version of the AM signal can be represented by:

$$v_1(t) = V_c{}' \sin 2\pi f_c t + \left(\frac{mV_c}{2} \right)' \left[\cos(2\pi(f_c - f_m)t) - \cos(2\pi(f_c + f_m)t) \right]$$ **EQ. 6.47**

where $V_c{}'$ and $mV_c{}'/2$ are the received voltage levels. The superheterodyne receiver uses high side injection with the local oscillator frequency f_{LO}=f_c+f_{IF} as indicated by EQ. 6.44. The time domain representation of f_{LO} is:

$$v_{LO}(t) = V_{LO} \sin(2\pi(f_c + f_{IF})t)$$ **EQ. 6.48**

Thus, the time domain equation for the mixer output, $v_2(t)$, is given by

$$v_2(t) = (V_{LO} \sin(2\pi(f_c + f_{IF})t)) \times$$ **EQ. 6.49**

$$\left(V_c{}' \sin 2\pi f_c t + \left(\frac{mV_c}{2} \right)' \left[\cos(2\pi(f_c - f_m)t) - \cos(2\pi(f_c + f_m)t) \right] \right)$$

So, this can get kind of messy. Since we are more interested in the received frequencies rather than the received voltage amplitudes, a more convenient method is to work in the frequency domain for each product of v_1 and v_0 and take their sum and difference frequencies. There are three frequency products as indicated by EQ. 6.49. These frequency products are:
    A. f_c+f_{IF} and f_c
    B. f_c+f_{IF} and f_c-f_m
    C. f_c+f_{IF} and f_c+f_m

For A the sum and difference frequencies are respectively:

$$\text{sum} \Rightarrow f_c + f_{IF} + f_c = 2f_c + f_{IF}$$
$$\text{difference} \Rightarrow f_c + f_{IF} - f_c = f_{IF}$$

For B the sum and difference frequencies are respectively:

$$\text{sum} \Rightarrow f_c + f_{IF} + f_c - f_m = 2f_c - f_m + f_{IF} = (2f_c + f_{IF}) - f_m$$

$$\text{difference} \Rightarrow f_c + f_{IF} - f_c + f_m = f_{IF} + f_m$$

For C the sum and difference frequencies are respectively:

$$\text{sum} \Rightarrow f_c + f_{IF} + f_c + f_m = 2f_c + f_m + f_{IF} = (2f_c + f_{IF}) + f_m$$

$$\text{difference} \Rightarrow f_c + f_{IF} - f_c - f_m = f_{IF} - f_m$$

Therefore, sum and difference frequencies of $v_2(f)$ when observed in the frequency domain are shown in Figure 6.33.

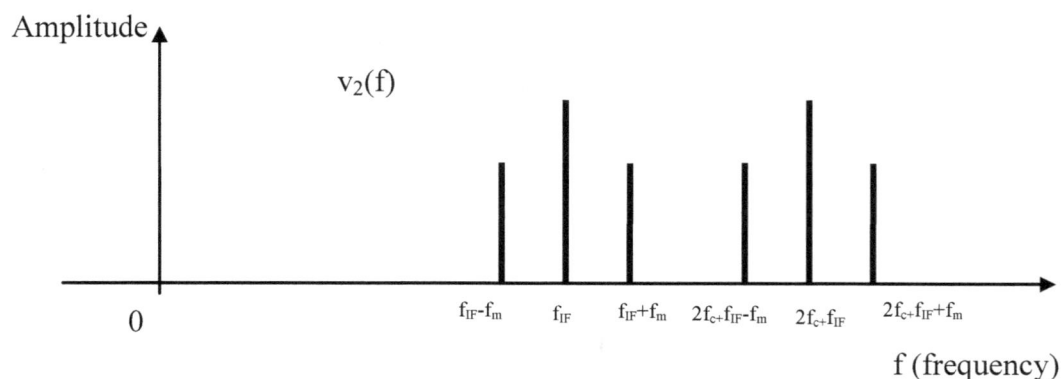

Figure 6.33. Frequency domain representation $v_2(t)$.

The frequency domain representation of $v_3(t)$ is shown in Figure 6.34. The IF amplifier selects the desired frequencies around the IF frequency using a bandpass filter centered at f_{IF} as shown in Figure 6.34. The corresponding time domain representation of $v_3(t)$ is shown in Figure 6.35. The output of the detector is illustrated in Figure 6.36.

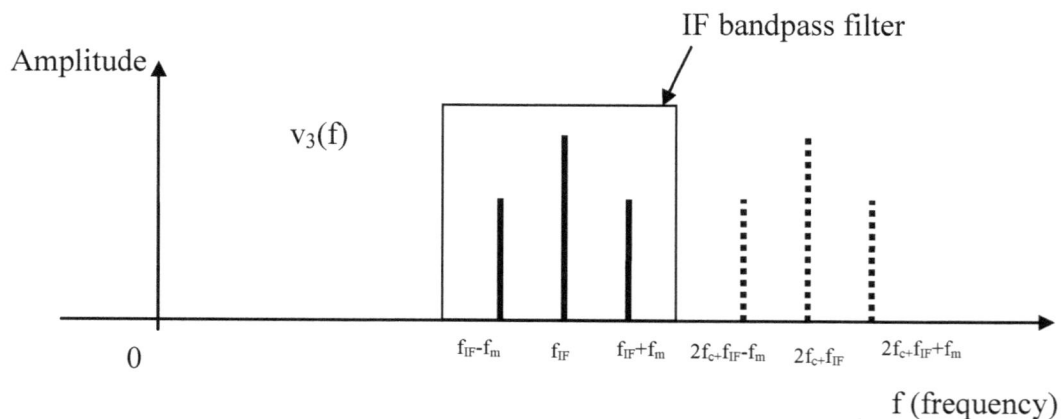

Figure 6.34. Frequency domain representation of $v_3(t)$ showing the elimination of the undesired frequencies using the IF bandpass filter.

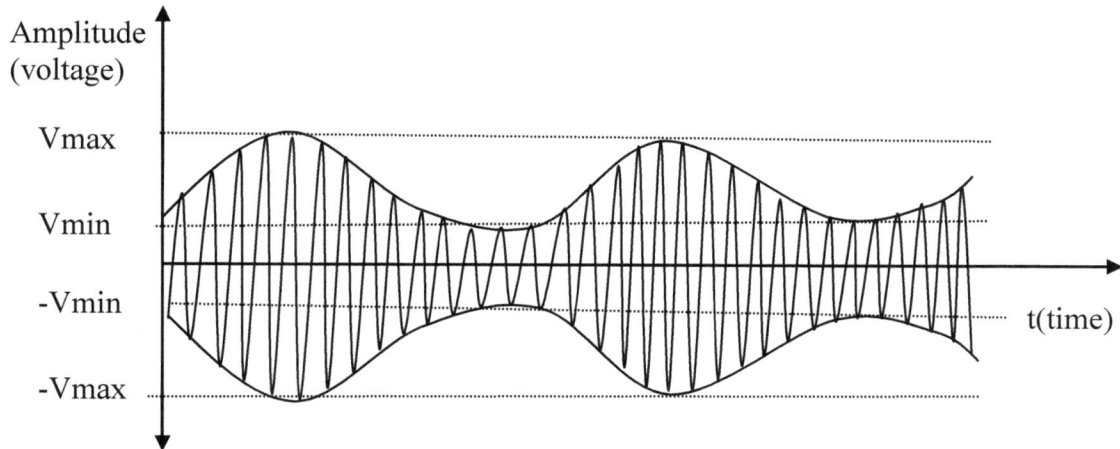

Figure 6.35. Time domain representation of $v_3(t)$.

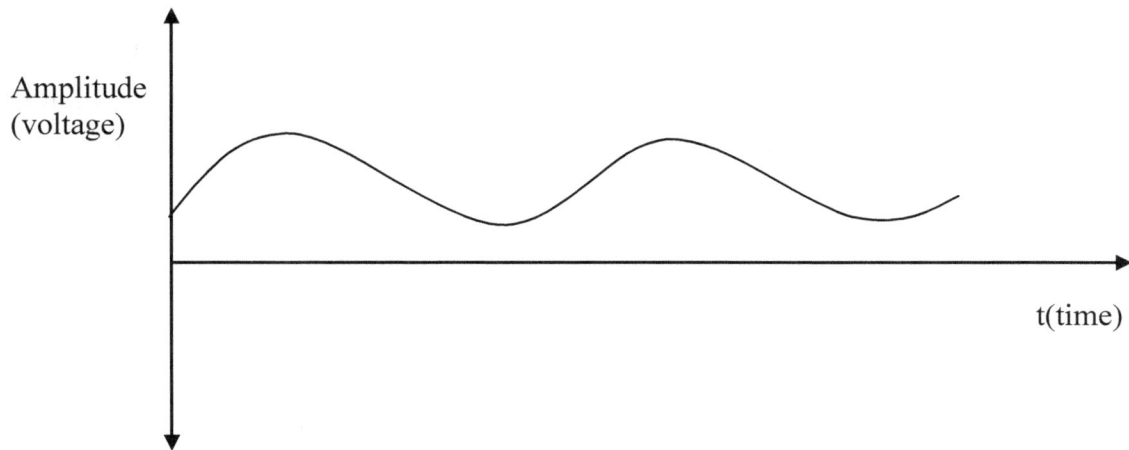

Figure 6.36. Time domain representation of $v_4(t)$.

In summary there are two types of radio receivers:

    A. **Coherent** – the receiver depends on the carrier transmitted. Thus, the receiver must have some means of recovering the carrier and synchronizing to it. Such receivers typically perform a product of the received signal with a recovered carrier. An example of a coherent receiver is the superheterodyne receiver shown in Figure 6.32.

    B. **Non-coherent** - the receiver is completely independent of the carrier transmitted. In this type of receiver the information is recovered by detecting the envelope of the signal and thus not dependent on a carrier. An example of the non-coherent receiver is the tuned radio frequency receiver shown in Figure 6.31.

6.6.6. Angle Modulation Concepts

Angle modulation encompasses frequency modulation (FM) and phase modulation (PM). In angle modulation, the phase angle of the sinusoidal wave varies in proportional to the modulating signal. An angle modulated signal is represented by

$$v_c(t) = V_c \sin(\omega_c t + \theta(t))$$

EQ. 6.50

Some text use

$$v_c(t) = V_c \cos(\omega_c t + \theta(t))$$

EQ. 6.51

as an angle modulated signal. The only difference between EQ. 6.50 and EQ. 6.51 is a phase difference of $+90°$. For the purposes of this text we will use EQ. 6.50. Using EQ. 6.50 and letting

$$v_c(t) = V_c \sin(\phi(t))$$

EQ. 6.52

where

$$\phi(t) = \omega_c t + \theta(t)$$

EQ. 6.53

The **instantaneous radian frequency**, ω_i, is obtained by taking the derivative as follows

$$\omega_i = \frac{d\phi(t)}{dt}$$

EQ. 6.54

$$= \frac{d(\omega_c t)}{dt} + \frac{d\theta(t)}{dt}$$

$$= \omega_c + \frac{d\theta(t)}{dt}$$

When $\theta(t)$ is constant $\omega_i = \omega_c$. Otherwise, $\theta(t)$ represents the instantaneous phase deviation and it's derivative, $d\theta(t)/dt$ represents the instantaneous frequency deviation of the angle modulated waveform.

In phase modulation (PM), the instantaneous phase deviation, $\theta(t)$, of the carrier, $v_c(t)$, is proportional to the modulating signal, $v_m(t)$, represented by

$$\theta(t) = k_p v_m(t)$$

EQ. 6.55

where k_p is the phase deviation constant in radians per unit of $v_m(t)$.

In frequency modulation (FM), the instantaneous frequency deviation of the carrier, $v_c(t)$, is proportional the modulating signal, $v_m(t)$, represented by

$$\frac{d\theta(t)}{dt} = k_f v_m(t)$$

EQ. 6.56

where k_f is the frequency deviation constant in radians per second per unit of $v_m(t)$. Integrating EQ. 6.56 gives the phase angle as a function of time as

$$\int_{t_0}^{t} \frac{d\theta(t)}{dt} = \int_{t_0}^{t} k_f v_m(t) \qquad \text{EQ. 6.57}$$

$$\theta(t) = k_f \int_{t_0}^{t} v_m(\lambda) d\lambda + \theta(t_0)$$

where $\theta(t_0)$ is the initial phase angle at $t=t_0$. Typically, $t_0=-\infty$ and $\theta(-\infty)=0$. Therefore, the phase modulated carrier is represented by

$$v_{PM}(t) = V_c \sin(\omega_c t + k_p v_m(t)) \qquad \text{EQ. 6.58}$$

and the frequency modulated carrier is represented by

$$v_{FM}(t) = V_c \sin\left(\omega_c t + k_f \int_{-\infty}^{t} v_m(\lambda) d\lambda\right) \qquad \text{EQ. 6.59}$$

For the case of a single sinusoidal, $v_m(t) = V_m \cos(\omega_m t)$, then EQ. 6.86 becomes

$$v_{FM}(t) = V_c \sin\left(\omega_c t + \frac{k_f V_m}{\omega_m} \sin(\omega_m t)\right) \qquad \text{EQ. 6.60}$$

where the modulation index for a sinusoidal FM signal is

$$m_f = \frac{k_f V_m}{\omega_m} = \frac{\Delta f}{\omega_m} \qquad \text{EQ. 6.61}$$

where $\Delta f = k_f V_m$ is the peak frequency deviation. For phase modulation, using the case of a single sinusoidal and for consistency to FM, let $v_m(t) = V_m \sin(\omega_m t)$, then EQ. 6.58 becomes

$$v_{PM}(t) = V_c \sin(\omega_c t + k_f V_m \sin(\omega_m t)) \qquad \text{EQ. 6.62}$$

where the modulation index for a sinusoidal PM signal is

$$m_p = k_p V_m \qquad \text{EQ. 6.63}$$

The modulation index m_f and m_p is the maximum phase deviation for a FM and PM signal, respectively. Thus, for FM or PM modulation, given the modulation index, the number of amplitudes of the significant sidebands can be determined by solving the general equation of an angle modulated signal, using EQ. 6.87 and EQ. 6.62 :

$$v_c(t) = V_c \sin(\omega_c t + m \sin(\omega_m t))$$

EQ. 6.64

where m is the modulation index of the FM or PM signal, $v_c(t)$. Using the Fourier series on EQ. 6.62 the result is represented in terms of the Bessel functions, J_0, J_1, J_2, ... as follows

$$v_c(t) = V_c \{ J_0 [\sin \omega_c t]$$
$$+ J_1 [\sin(\omega_c + \omega_m) - \sin(\omega_c - \omega_m)]$$
$$+ J_2 [\sin(\omega_c + 2\omega_m) + \sin(\omega_c - 2\omega_m)]$$
$$+ J_3 [\sin(\omega_c + 3\omega_m) - \sin(\omega_c - 3\omega_m)]$$
$$+ J_4 [\sin(\omega_c + 4\omega_m) + \sin(\omega_c - 4\omega_m)]$$
$$+ ....\}$$

EQ. 6.65

This can be written in a more compact form as follows:

$$v_c(t) = V_c \sum_{n=-\infty}^{n=+\infty} J_n(m) \sin(\omega_c \pm n\omega_m) t$$

EQ. 6.66

where $\omega_c = 2\pi f_c$ is the carrier angular frequency, $\omega_m = 2\pi f_m$ is the modulating signal angular frequency, and V_c is the peak voltage of the unmodulated carrier. Thus, the FM signal is a composite of sine waves of different frequencies and amplitudes that when added together produces a FM signal. The Bessel Table is provided in Table 6.2.

	J_0	J_1	J_2	J_3	J_4	J_5	J_6	J_7	J_8	J_9	J_{10}	J_{11}	J_{12}	J_{13}	J_{14}	J_{15}	J_{16}
m	Carrier	1st	2nd	3rd	4th	5th	6th	7th	8th	9th	10th	11th	12th	13th	14th	15th	16th
0.00	1.00	--	--	--	--	--	--	--	--	--	--	--	--	--	--	--	--
0.25	0.98	0.12	--	--	--	--	--	--	--	--	--	--	--	--	--	--	--
0.5	0.94	0.24	0.03	--	--	--	--	--	--	--	--	--	--	--	--	--	--
1.0	0.77	0.44	0.11	0.02	--	--	--	--	--	--	--	--	--	--	--	--	--
1.5	0.51	0.56	0.23	0.06	0.01	--	--	--	--	--	--	--	--	--	--	--	--
2.0	0.22	0.58	0.35	0.13	0.03	--	--	--	--	--	--	--	--	--	--	--	--
2.5	-0.05	0.50	0.45	0.22	0.07	0.02	--	--	--	--	--	--	--	--	--	--	--
3.0	-0.26	0.34	0.49	0.31	0.13	0.04	0.01	--	--	--	--	--	--	--	--	--	--
4.0	-0.40	-0.07	0.36	0.43	0.28	0.13	0.05	0.02	--	--	--	--	--	--	--	--	--
5.0	-0.18	-0.33	0.05	0.36	0.39	0.26	0.13	0.05	0.02	--	--	--	--	--	--	--	--
6.0	0.15	-0.28	-0.24	0.11	0.36	0.36	0.25	0.13	0.06	0.02	--	--	--	--	--	--	--
7.0	0.30	0.00	-0.30	-0.17	0.16	0.35	0.34	0.23	0.13	0.06	0.02	--	--	--	--	--	--
8.0	0.17	0.23	-0.11	-0.29	-0.10	0.19	0.34	0.32	0.22	0.13	0.06	0.03	--	--	--	--	--
9.0	-0.09	0.24	0.14	-0.18	-0.27	-0.06	0.20	0.33	0.30	0.21	0.12	0.06	0.03	0.01	--	--	--
10.0	-0.25	0.04	0.25	-0.06	-0.22	-0.23	-0.01	09.22	0.31	0.29	0.20	0.12	0.06	0.03	0.01	--	--
12.0	-0.05	-0.22	-0.08	0.20	0.18	-0.07	-0.24	-0.17	0.05	0.23	0.30	0.27	0.20	0.12	0.07	0.03	0.01
15.0	-0.01	0.21	-0.04	0.19	-0.12	0.13	0.21	0.03	-0.17	-0.22	-0.09	0.10	0.24	0.28	0.25	0.18	0.12

Table 6.2. Carrier, J_0, and the significant sideband amplitudes, 1st, 2nd, 3rd, 4th, ..., for the different modulation indexes, m, of an FM or PM signal using the Bessel functions, J_0, J_1, J_2, ..., J_n.

From EQ. 6.92 and Table 6.2 we have the following summarized points:
A. The spectrum consists of a carrier frequency component, n=0, plus an infinite number of sideband components at frequencies $f_c \pm n f_m$ where n=1, 2, 3,.......
B. The amplitudes of the spectral lines depend on the value of $J_n(m)$ as the values of $J_n(m)$ becomes very small for large n.

C. The number of significant sidebands is a function of m. For m<<1, only J_0 and J_1 are significant. For m>>1 many sidebands will be present in the spectrum.

The total bandwidth of an FM signal can be determined using its modulation index, Thus, from Table 6.2 the bandwidth of the significant sidebands for an FM signal is

$$BW_{FM} = 2f_m n$$

EQ. 6.67

where n is the number of significant sidebands in the signal. The FM signals for various modulation indexes are illustrated in Figure 6.37. The FM signal with modulation index $m_f = 0.25$ resembles that of an AM signal. This type of FM signal that occupies the same bandwidth of an AM signal or less is referred to as a narrow-band FM signal or NBFM. By definition, a FM signal is a NBFM signal if it meets the following condition:

$$m_f < \frac{\pi}{2} = 1.57$$

EQ. 6.68

However, values of true NBFM has a modulation index that ranges from 0.2 to 0.25.

It can be shown that 98% of the normalized total power resides within the bandwidth of an angle modulated signal which is calculated by

$$BW \approx 2(m+1)f_m$$

EQ. 6.69

where BW is in Hertz. Another method to determine the bandwidth is known as **Carson's rule** given by:

$$BW_{FM} \approx 2(D+1)f_m = 2(\Delta f + f_m)$$

EQ. 6.70

where

$$D = \frac{max\,imum\ frequency\ deviation}{bandwidth\ of\ v_m(t)} = \frac{\Delta f}{f_m}$$

EQ. 6.71

is defined as the **deviation ratio**. The deviation ratio can be thought of as the worse case modulation index where you choose the maximum deviation of the FM signal, Δf, and the maximum modulating signal frequency, f_m.

EXAMPLE 6.4

Obtain the Fourier spectra of EQ. 6.51.

Solution:
EQ. 6.51 can be represented in exponential notation and taking "the real part of", denoted by Re, as follows:

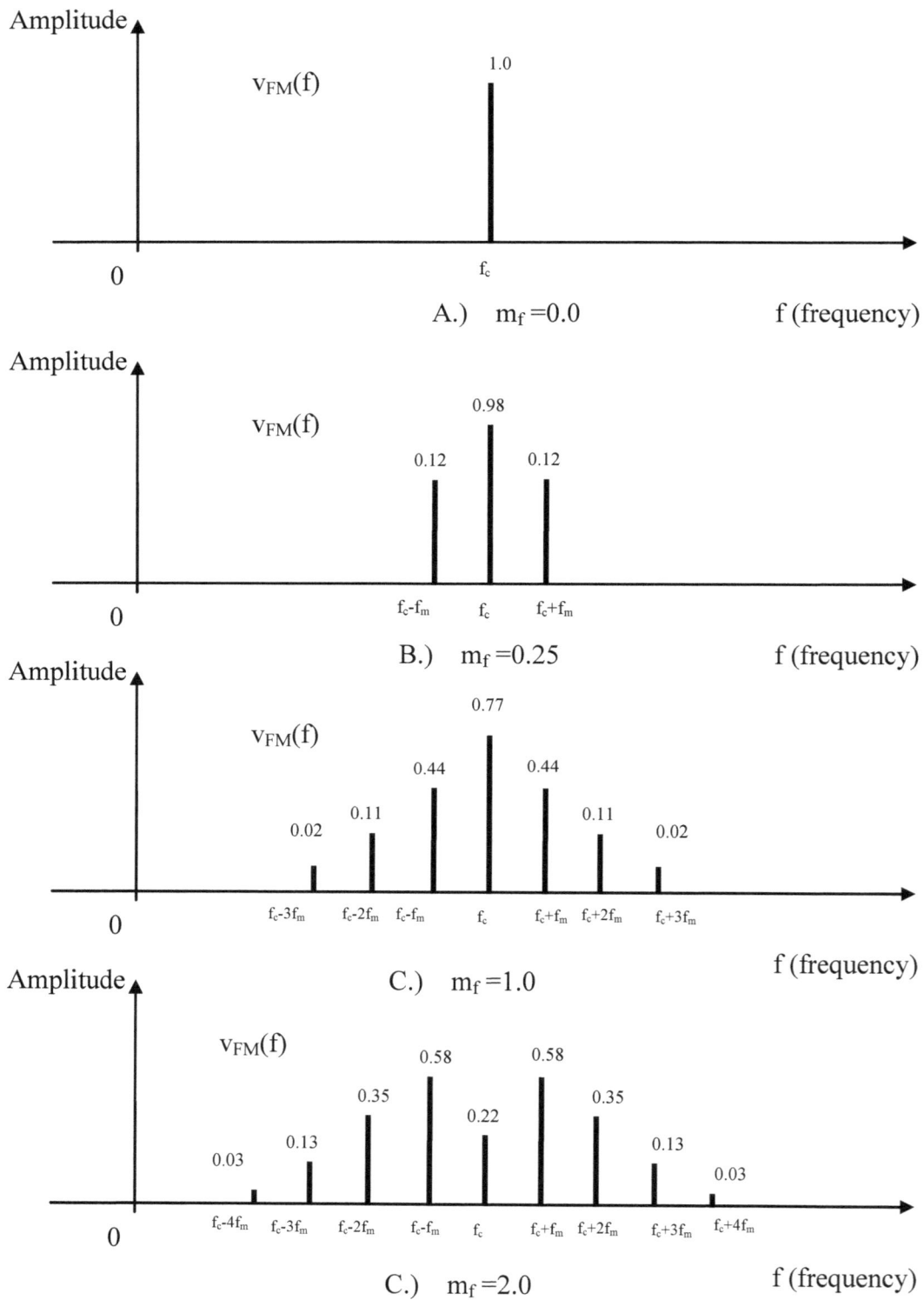

Figure 6.37. FM Signals with various modulation indexes as indicated with f_m fixed.

$$v_c(t) = V_c \cos(\omega_c t + \theta(t)) = \text{Re}\left(V_c e^{j(\omega_c + \theta(t))}\right) = \text{Re}\left(V_c e^{j\omega_c} e^{j\theta(t)}\right) \qquad \text{EQ. 6.72}$$

$$= \text{Re}\left\{ V_c e^{j\omega_c}\left(1 + j\theta - \frac{\theta^2}{2!} - \dots + j^n \frac{\theta^n}{n!} + \dots\right)\right\}$$

$$= V_c\left(\cos\omega_c t - \theta\sin\omega_c t - \frac{\theta^2}{2!}\cos\omega_c t + \frac{\theta^3}{2!}\sin\omega_c t + \dots\right)$$

where $\theta = \theta(t)$.

EXAMPLE 6.5

A. Find the reduced version of EQ. 6.72 by letting $\theta \ll 1$. Remove second order terms and higher for narrowband angle modulated signals. B. Determine the narrowband FM signal. C. Determine the narrowband PM signal. D. Then draw the narrowband FM block diagram.

Solution:

A.

$$v_c(t) \approx V_c(\cos\omega_c t - \theta\sin\omega_c t) \qquad \text{EQ. 6.73}$$

$$\approx V_c \cos\omega_c t - V_c\theta\sin\omega_c t$$

B. EQ. 6.73 is referred to as a narrowband angle modulated signal. Thus, narrowband FM or NBFM, using EQ. 6.51, is represented by

$$v_{\text{NBFM}}(t) \approx V_c \cos\omega_c t - V_c\theta\sin\omega_c t \qquad \text{EQ. 6.74}$$

$$\approx V_c \cos\omega_c t - V_c\left\{k_f \int_{-\infty}^{t} v_m(\lambda)d\lambda\right\}\sin\omega_c t$$

C. Thus, narrowband PM or NBPM, using EQ. 6.51, is represented by

$$v_{\text{NBPM}}(t) \approx V_c \cos\omega_c t - V_c k_p v_m(t)\sin\omega_c t \qquad \text{EQ. 6.75}$$

D. The narrowband FM block diagram is drawn in Figure 6.38.

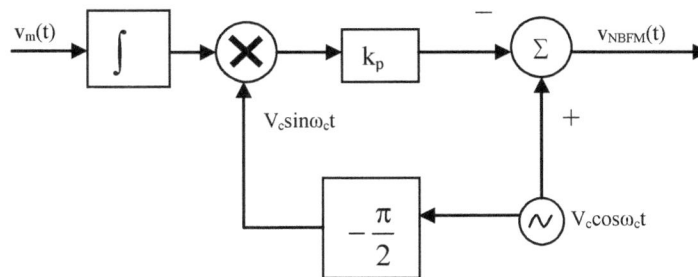

Figure 6.38. Narrowband FM block diagram.

You've seen the generation of a narrowband angle modulated signal in EXAMPLE 6.5. There are also wideband (WB) angle modulated signals. There are two methods to generate WB angle modulated signals:

  A. Direct method
  B. Indirect method

For FM, the direct angle modulated method involves the direct control of a carrier frequency. This is typically done using the modulating signal, $v_m(t)$. The voltage of $v_m(t)$ directly varies the inductance or capacitance of a tuned oscillator. This type of circuit is called **voltage controlled oscillator** (VCO). The advantage of direct FM is the large frequency deviations while the major disadvantage is that the carrier frequency tends to drift and therefore requires extra circuitry for frequency stabilization.

Indirect FM is used to produce WBFM from a narrowband FM signal by a multiplier circuit as shown Figure 6.39. For example, if the input is EQ. 6.50, then the output y(t) is

$$y(t) = V_c \sin(n\omega_c t + n\theta(t))$$

EQ. 6.76

Multiplication tends to increase the carrier to an impractical frequency and typically requires a frequency shift circuit following the multiplier.

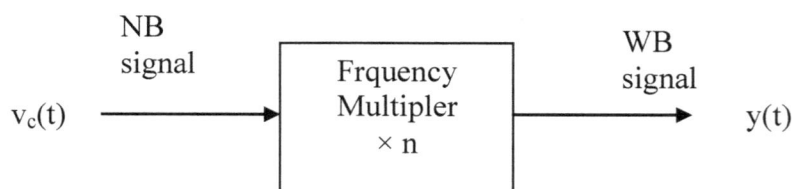

Figure 6.39. Wideband FM/PM block diagram.

Demodulation of FM signal is typically done using a **frequency discriminator**. A frequency discriminator gives an output that is proportional to the frequency deviation of the input signal as shown in Figure 6.40A. For example, if the input to the frequency discriminator is EQ. 6.50 (given here again for convenience):

$$v_c(t) = V_c \sin(\omega_c t + \theta(t))$$

EQ. 6.77

then the output is

$$y(t) = k_d \frac{d\theta}{dt}$$

EQ. 6.78

$$\theta(t) = k_d \frac{d}{dt} \left(k_f \int_{t_0}^{t} v_m(\lambda) d\lambda + \theta(t_0) \right)$$

$$= k_d k_f v_m(t)$$

where k_d is the discriminator sensitivity. As shown here, a scaled ($k_d k_f$) version of the modulating signal, $v_m(t)$, has been recovered. A practical frequency discriminator is shown in Figure 6.40B.

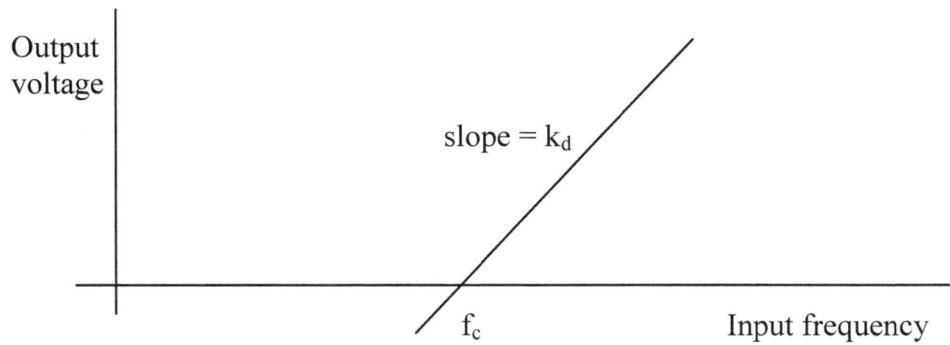

A.)

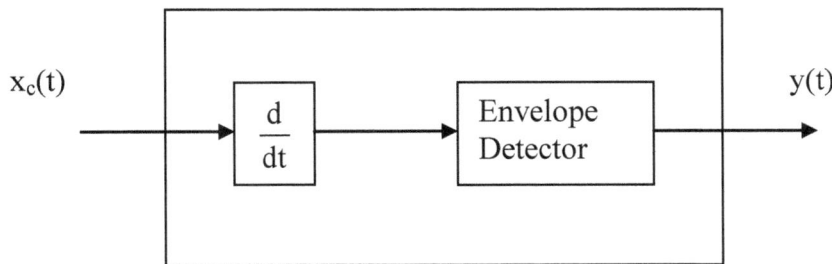

B.)

Figure 6.40. A.) Ideal frequency discriminator. B.) Block diagram of a typical frequency discriminator.

6.6.7. Frequency Modulation Transmission

As shown in the previous Section frequency modulation is a type of angular modulation. Frequency modulation is the process of changing carrier frequency relative to the to the amplitude of the modulating signal. One method to accomplish this is to use direct FM modulation using a LC oscillator. Recall that an oscillator produces its own signal with oscillating frequency according to

$$f_o = \frac{1}{2\pi\sqrt{LC}}$$ EQ. 6.79

In this case, the oscillator produces a high frequency signal. As shown in Figure 6.41 the capacitance can be controlled by a microphone that uses air pressure to vary capacitance, and thus the frequency of oscillation. Pre-emphasis, as shown in Figure 6.42, consists of amplifying the high frequency information more than the low frequency information. In de-emphasis (the reverse of pre-emphasis) the gain is removed from the high frequency information. This reduces noise in the signal at the front end of the transmitter and when the signal is in transmission. For example, a large amplitude audio signal causes a large frequency of oscillation while a small amplitude audio signal causes a small frequency of oscillation as shown in Figure 6.43.

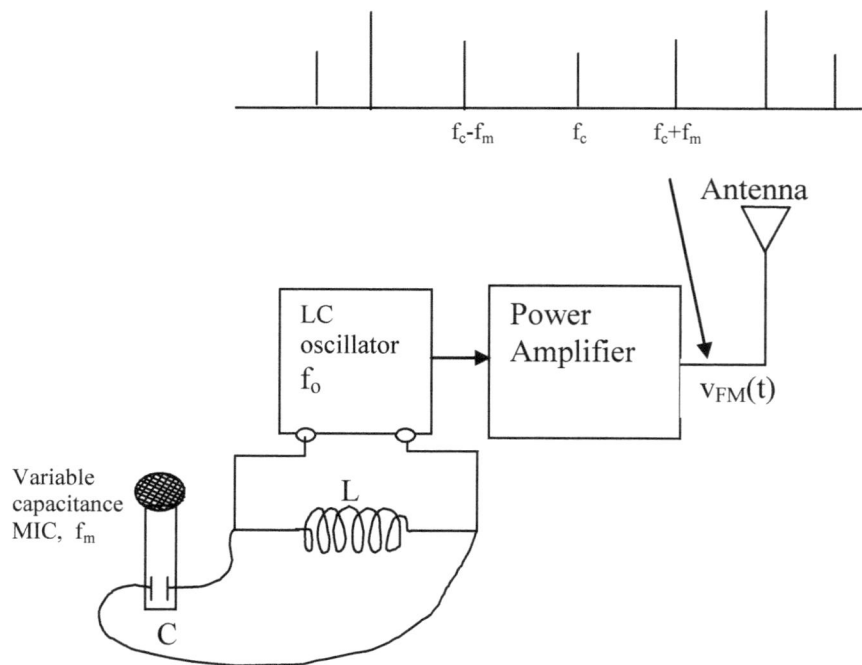

Figure 6.41. FM Transmitter using LC oscillator also called VCO.

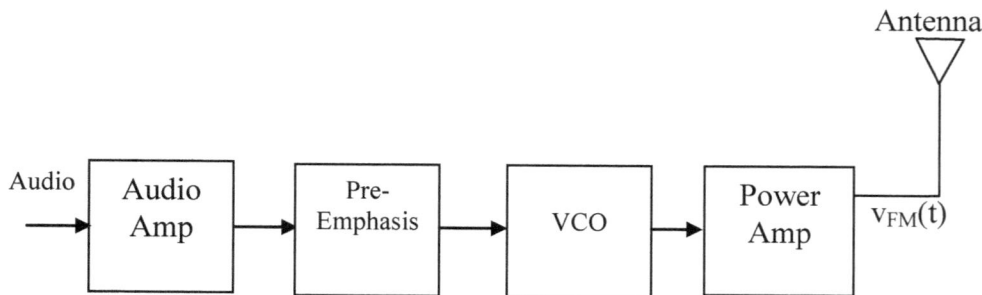

Figure 6.42. FM Transmitter using VCO.

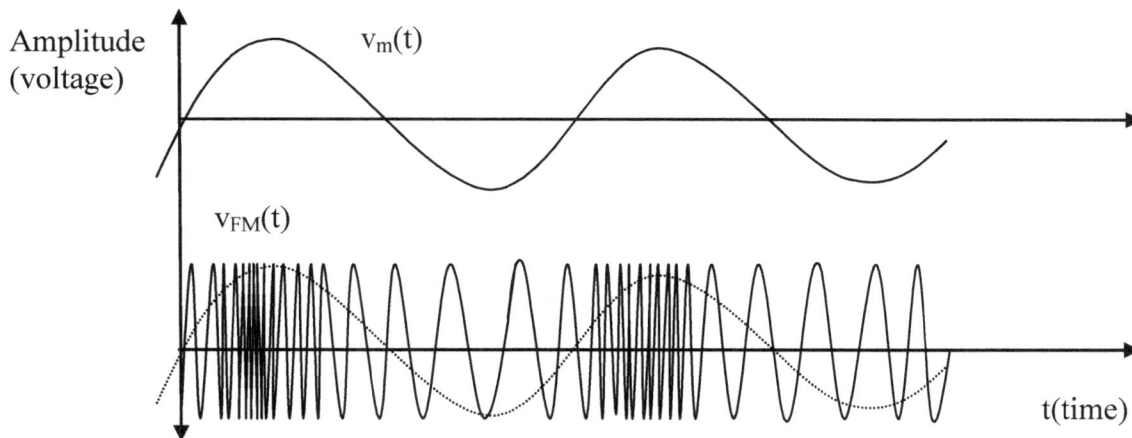

Figure 6.43. Frequency modulated carrier using LC oscillator. Increasing input signal amplitude, $v_m(t)$ causes increasing oscillator frequency, $v_{FM}(t)$.

The carrier frequency changes with the amplitude of the modulating signal as shown in Figure 6.44. The frequency of deviation, Δf, is the maximum change in frequency that the carrier undergoes. The carrier swing is the total variation from the highest to the lowest frequency and is determined by

$$f_{CS} = 2\Delta f$$ EQ. 6.80

The modulation index of an FM signal is given by

$$m_f = \frac{\Delta f}{f_m}$$ EQ. 6.81

The percent modulation is given by

$$m_f = \frac{\Delta f_{actual}}{\Delta f_{max}} \times 100$$ EQ. 6.82

EXAMPLE 6.6

An FM signal with carrier frequency of 107.6 MHz is modulated by a 7KHz sine wave with frequency of deviation of 50KHz. Determine:
- A. f_{cs}
- B. highest frequency and lowest frequency attained
- C. m_f

Solution:
- A. $f_{cs} = 2\Delta f = 2 \times 50KHz = 100KHz$
- B. $f_H = f_c + \Delta f = 107.6MHz + 50KHz = 107.65MHz$; $f_L = f_c - \Delta f = 107.6MHz - 50KHz = 107.55MHz$

C. $m_f = \dfrac{\Delta f}{f_m} = \dfrac{50KHz}{7KHz} = 7.143$

EXAMPLE 6.7

An FM signal has frequency of deviation of 20KHz. Determine the percent modulation in the 88MHz to 108MHz band.

Solution:
By the FCC the maximum frequency deviation in the 88MHz to 108MHz band is 75KHz, so

$$m_f = \dfrac{\Delta f_{actual}}{\Delta f_{max}} \times 100 = \dfrac{20KHz}{75KHz} \times 100 = 26.6\%$$

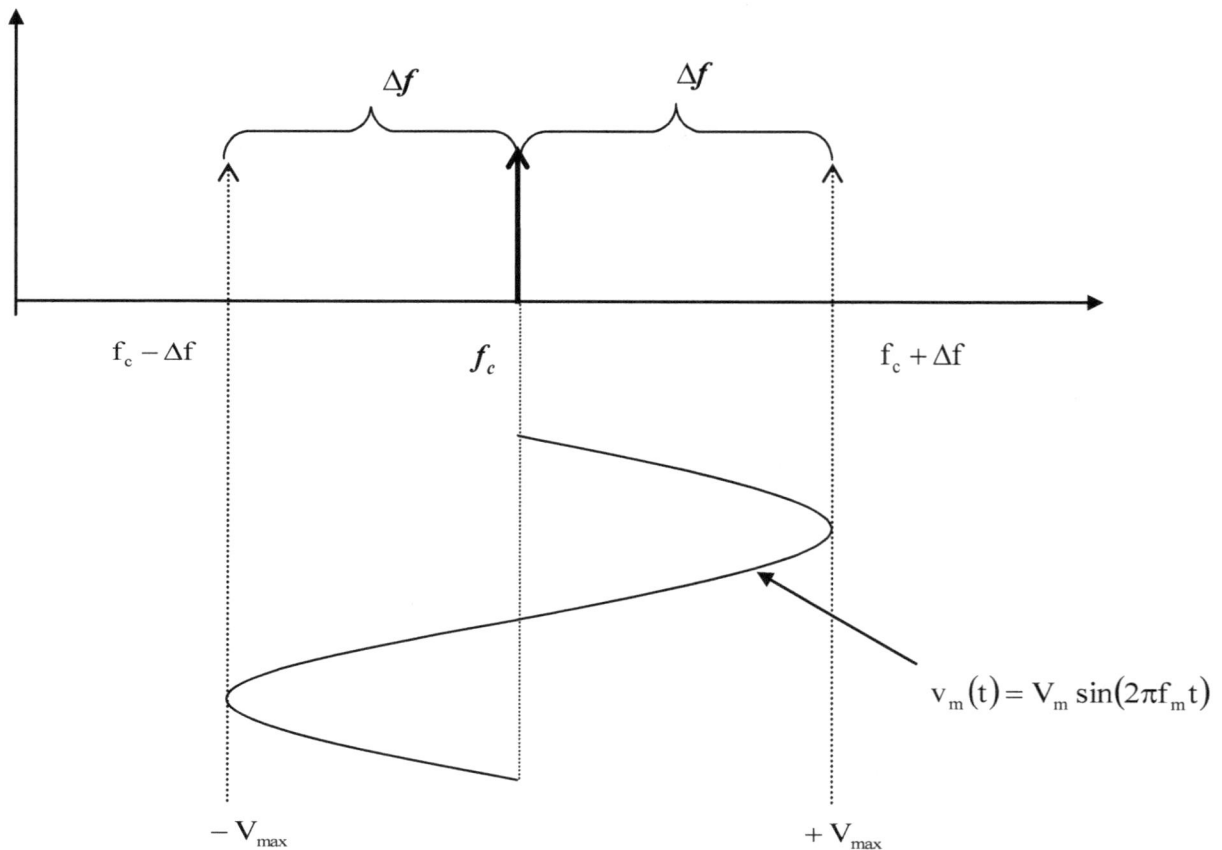

Figure 6.44. FM signal in frequency domain, showing frequency deviation, Δf (Hz), from the center frequency f_c, relative to time domain signal $v_m(t)$.

6.6.8. FM Reception

The FM receiver is the same as the AM receiver up to the IF amplifier as indicated in Figure 6.45 and Figure 6.46, respectively. The only difference is the intermediate frequency between the two:

A. $f_{IF} = 455kHz$ for standard AM
B. $f_{IF} = 10.7MHz$ for standard FM

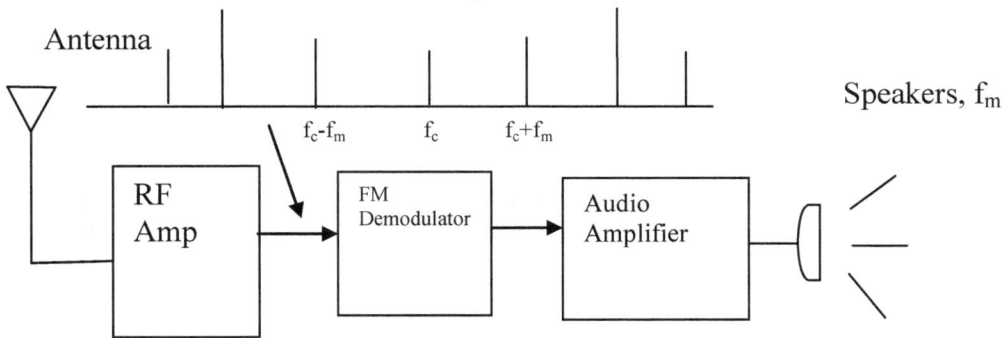

Figure 6.45. Tuned radio frequency receiver.

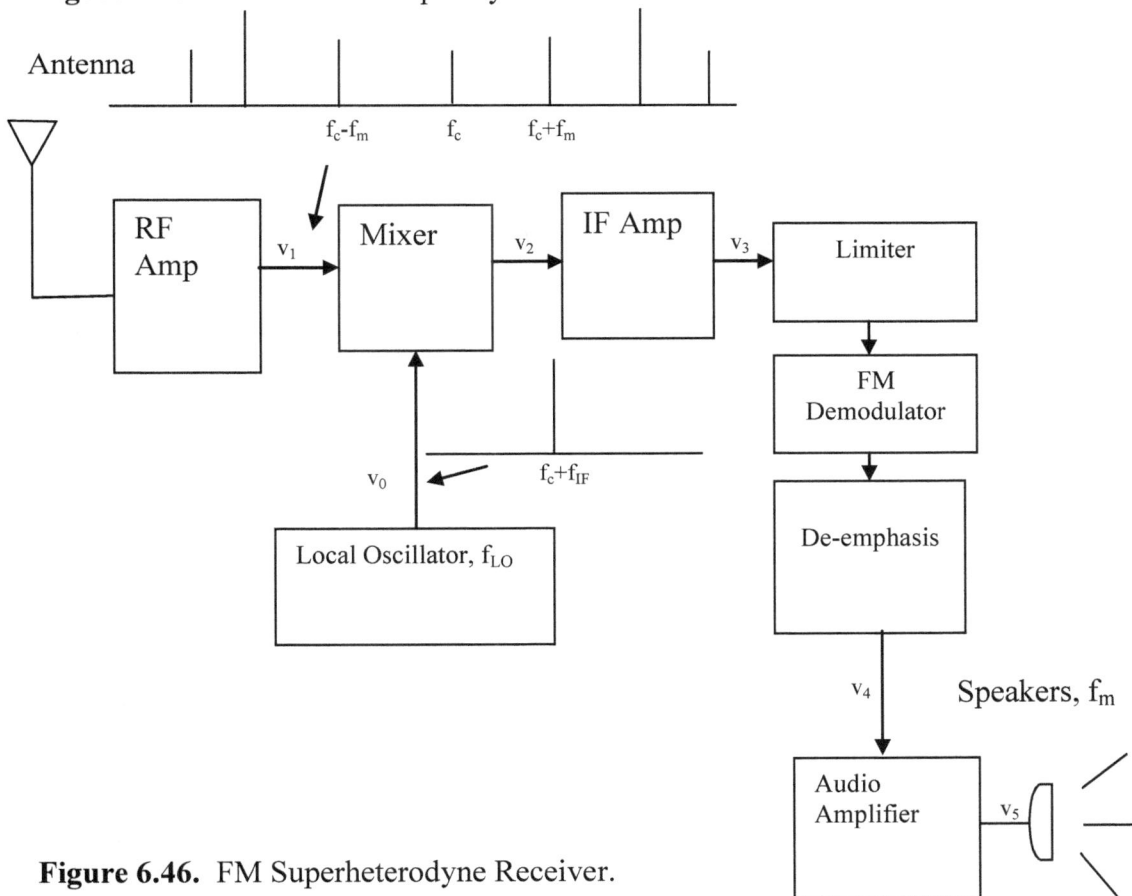

Figure 6.46. FM Superheterodyne Receiver.

The major difference is in the demodulation section of the FM receiver. The demodulation section uses circuits that output a voltage which is proportional to the input frequency, that is, a frequency discriminator, as illustrated in Figure 6.40. Techniques are also used to reduce distortion such as emphasis (at the transmitter) and de-emphasis(at the receiver) and limiters (at the receiver). The types of FM demodulation circuits are:

A. Slope detector
B. Balance slope detector
C. Foster –Seely discriminator
D. Ratio detector
E. Phase locked loop (PLL)
F. Quadrature detector

A few of the FM demodulators will be discussed here – slope detector and quadrature detector. The slope detector is one of the simplest FM demodulator circuits. Figure 6.47 shows the slope detector. It consists of a LC circuit(L_2 and C_1), a diode (D_1), and a lowpass filter (R_1 and C_2) to recover the modulating signal. The LC circuit has the frequency response characteristics as shown in Figure 6.48.

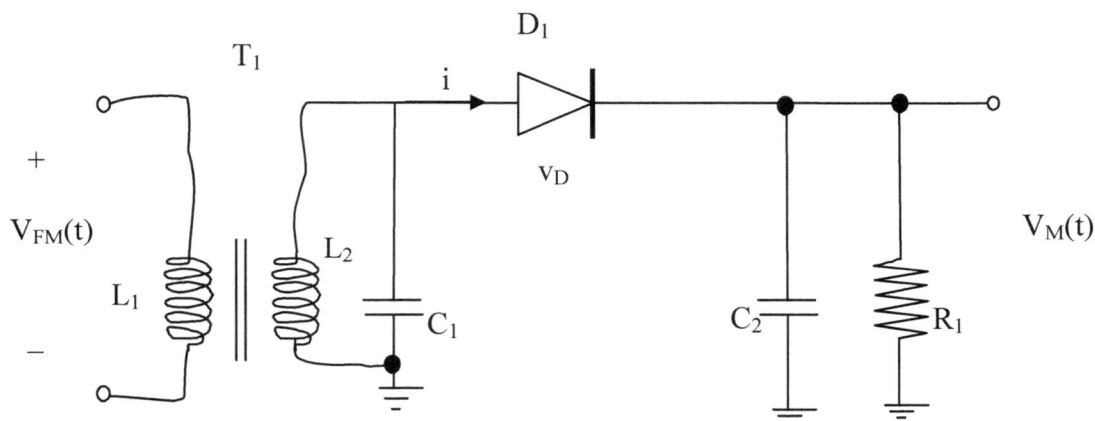

Figure 6.47. FM slope detector circuit. T_1 is a transformer and D_1 is a diode.

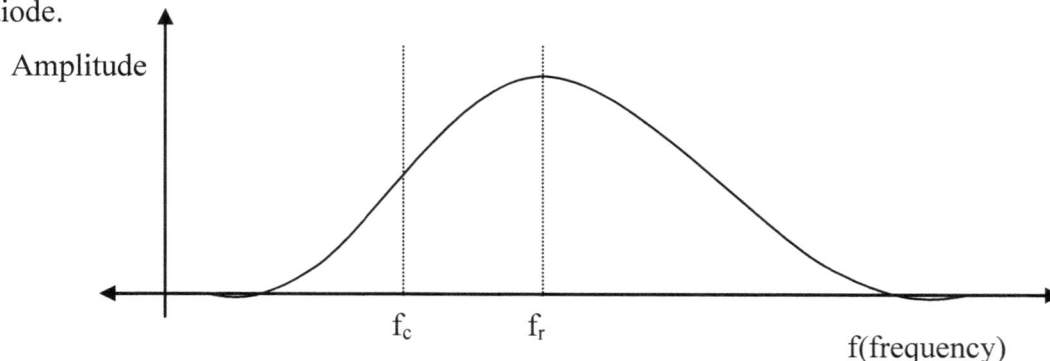

Figure 6.48. FM slope detector LC circuit (L_2 and C_1) frequency response characteristics.

In Figure 6.48, f_r is the resonant frequency of the LC circuit consisting of L_2 and C_1. The amplitude peaks at the resonant frequency f_r. The LC circuit is designed such that the carrier frequency is centered on the rising (or falling) edge of the frequency response curve. If the input frequency is lower than the carrier frequency the output voltage is smaller otherwise if the input frequency is greater than the carrier frequency the output voltage is larger. This voltage is then rectified and filtered using a lowpass filter.

Some disadvantages of the slope detector is:
A. The process of centering the carrier on the rising (or falling) edge of the frequency response is difficult
B. The rising (or falling) edge is not perfectly linear which introduces distortion. It is linear over a small range.

The Quadrature detector is very popular and is shown in Figure 6.49. It consists of a capacitor C_1 which provides a 90 degrees phase shift when the carrier is not modulated. C_1 has a very high impedance than the tuned RLC circuit at resonance. The RLC circuit is centered at the f_{IF} frequency. It produces an additional phase shift, θ, which is proportional to the frequency deviation. At the f_{IF} frequency the RLC circuit is at resonance and therefore purely resistive. Otherwise, the frequency variations will produce an additional positive or negative phase shift.

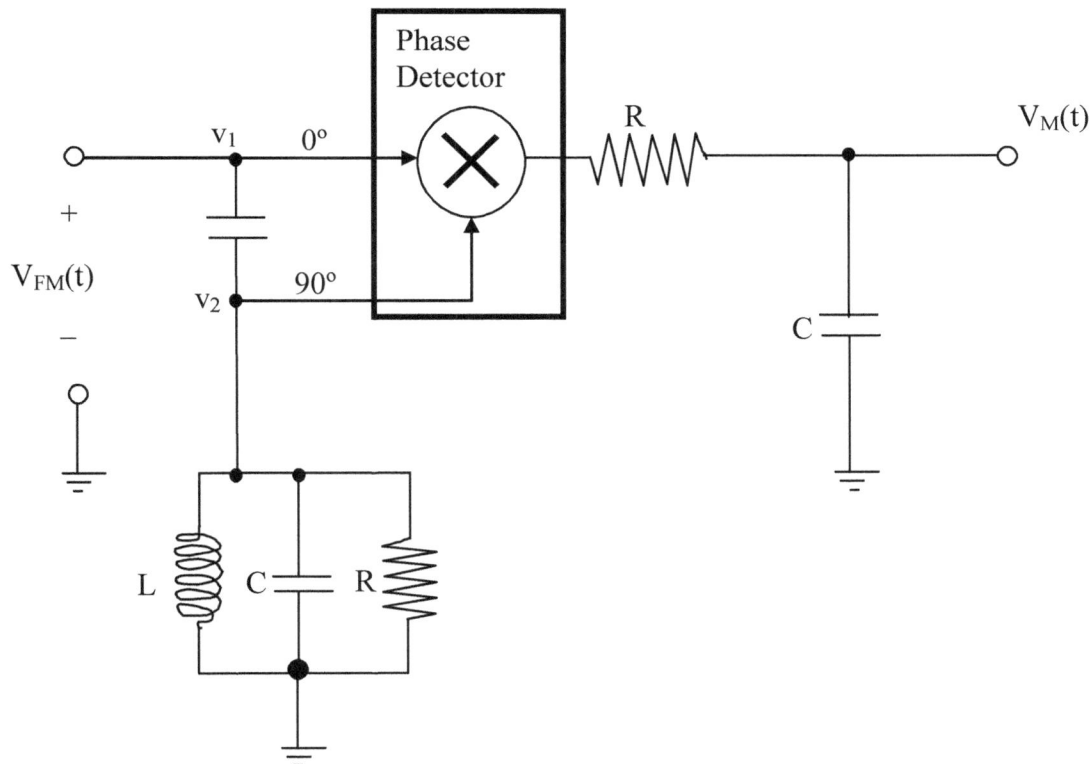

Figure 6.49. Quadrature detector circuit.

The output of the Quadrature circuit is represented by the multiplication of the IF signal, $v_1(t)$, and the Quadrature signal, $v_2(t)$, as follows

$$v_M(t) = v_1(t) \times v_2(t)$$

$$= V_1 \sin(\omega t + \theta) V_2 \sin(\omega t + \theta + 90)$$

$$= V_1 \sin(\omega t + \theta) V_2 \cos(\omega t)$$

$$= \frac{V_1 V_2}{2} \left[\sin(2\omega t + \theta) + \sin(\theta) \right]$$

EQ. 6.85

After the lowpass filter the above becomes

$$v_M(t) = \frac{V_1 V_2}{2} \left[\sin(\theta) \right]$$

EQ. 6.86

The following ICs can be used as FM receivers:
   A. NE/SA 614 A IC by Signetics corporation up to 25MHz
   B. NE/SA616 by Signetics corporation up to 150MHz
   C. TDA7000

6.6.9. Frequency Division Multiplexing

Multiplexing is a technique where several signals can be combined into a composite signal for transmission over one communication channel. Thus, the signals must be kept apart, i.e., no interference from one another, such that they can be easily demodulated at the receiver. Two basic multiplexing techniques are frequency division multiplexing (FDM) and time division multiplexing (TDM). In FDM signals are separated in frequency whereas in TDM signals are separated in time. Code division multiplexing (CDM) is another technique that involves the spread-spectrum technology. A code is used to distinguish between the users instead of a time slot or frequency slot used in TDM and FDM, respectively.

The FDM technique is illustrated in Figure 6.50. Three simultaneous signals, $v_1(t)$, $v_2(t)$, and $v_3(t)$ are individually frequency shifted using double side band modulation (DSB) and then all summed to form one composite signal. Any type of modulation can be used as long as there is sufficient spacing to avoid overlap in the frequency domain. However, the most popular modulation in FDM is single sideband modulation (SSB) to allow for more bandwidth. The frequency domain of each signal $V_1(f)$, $V_2(f)$, and $V_3(f)$ are illustrated along with the frequency domain of the summed signal.

Commercial AM radio stations use carrier frequencies spaced at 10kHz apart in the frequency range 540kHz to 1600kHz. This separation is not sufficient for a high fidelity AM signal with frequency range 50Hz to 15kHz. So AM stations on adjacent carriers have to be placed apart to avoid overlapping. Commercial FM use carrier frequencies spaced 200kHz apart.

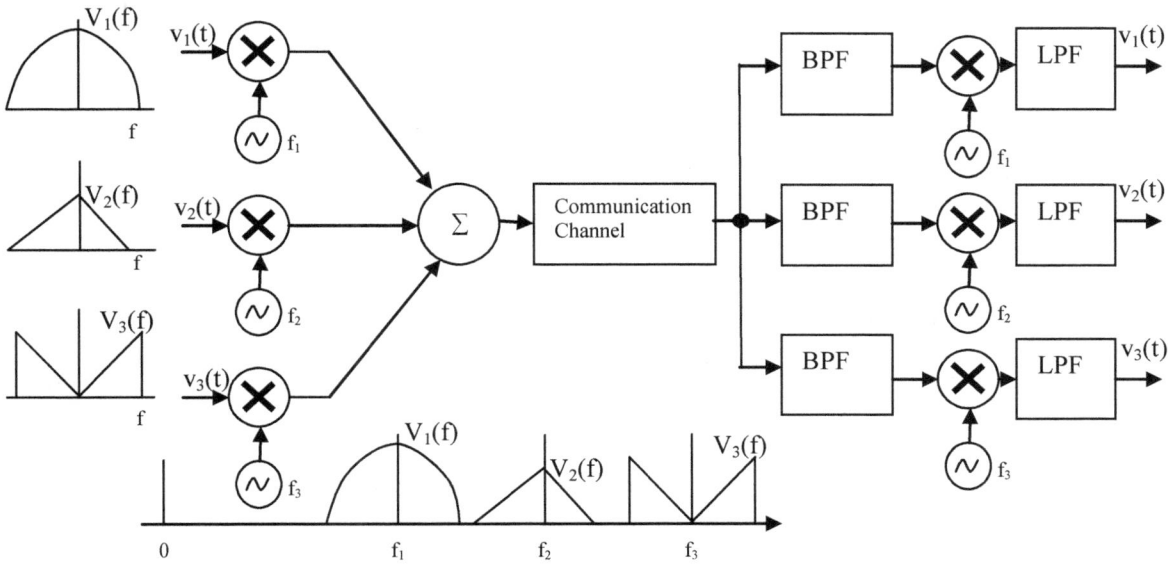

Figure 6.50. Frequency Division Multiplexing.

6.7. Digital Communications

A practical representation of a digital (or analog) communication system is shown in Figure 6.52. The details are explained therein as indicated in A through X. Figure 6.51 shows the details of the format block (A, B, and C) shown in Figure 6.52.

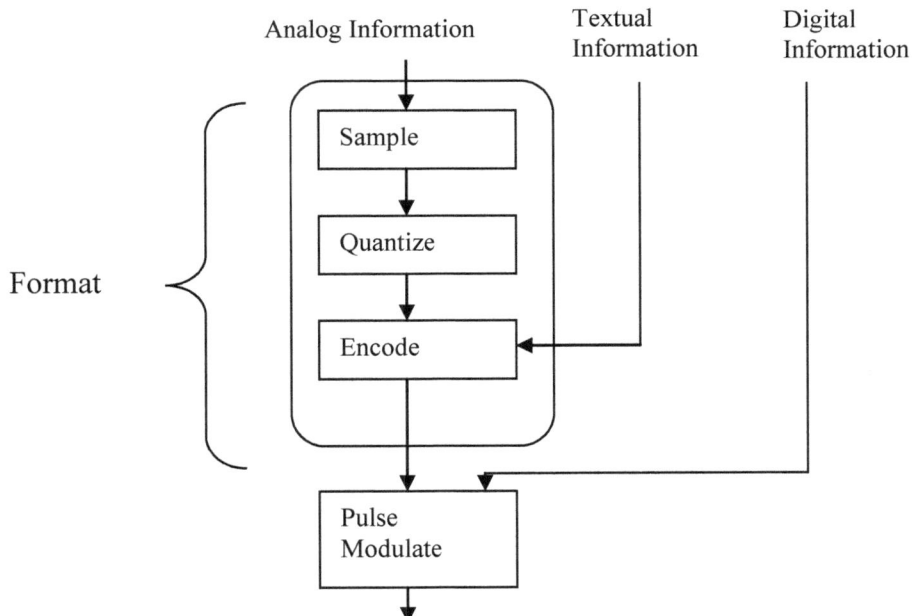

Figure 6.51. Details of formatting in Figure 6.52.

x(t)

| Format | A. A/D conversion is performed if input signal source, x(t), is analog.
B. If signal source is textual no A/D is required, bits are passed.
C. Digital data is passed. |

| Source Encode | D. Reduce data redundancy: Examples are predictive coding, block encoding, lossless compression, lossy compression, variable length coding. This is optional. |

| Encrypt | E. Make data private. This is optional. |

| Channel Encode | F. Reduce probability of error for a given data rate at the expense of bandwidth, for example, Trellis code modulation, etc. This is optional. |

| Multiplexer | G. Combine signals that originate from different sources to share same channel, for example, FDM, TDM, CDM, etc. This is optional. |

| Pulse Modulator | H. Conversion from binary to baseband. When pulse modulation is applied to binary signals the resulting waveform is referred to as pulse code modulation (PCM), for example, nonreturn-to-zero (NRZ), return to zero (RZ), Manchester, etc.
I. Conversion from nonbinary to baseband. When pulse modulation is applied to nonbinary signals the resulting waveform is called pulse modulation waveform, e.g., pulse amplitude modulator (PAM), pulse position modulation (PPM), etc. |

| Bandpass Modulator | J. Can be coherent or noncoherent. Bandpass modulation is required for RF applications. It is required when the transmiossion channel does not support pulse like signals. The coherent technique uses phase and amplitude while the noncoherent technique uses only amplitude. |

$s_1(t)$

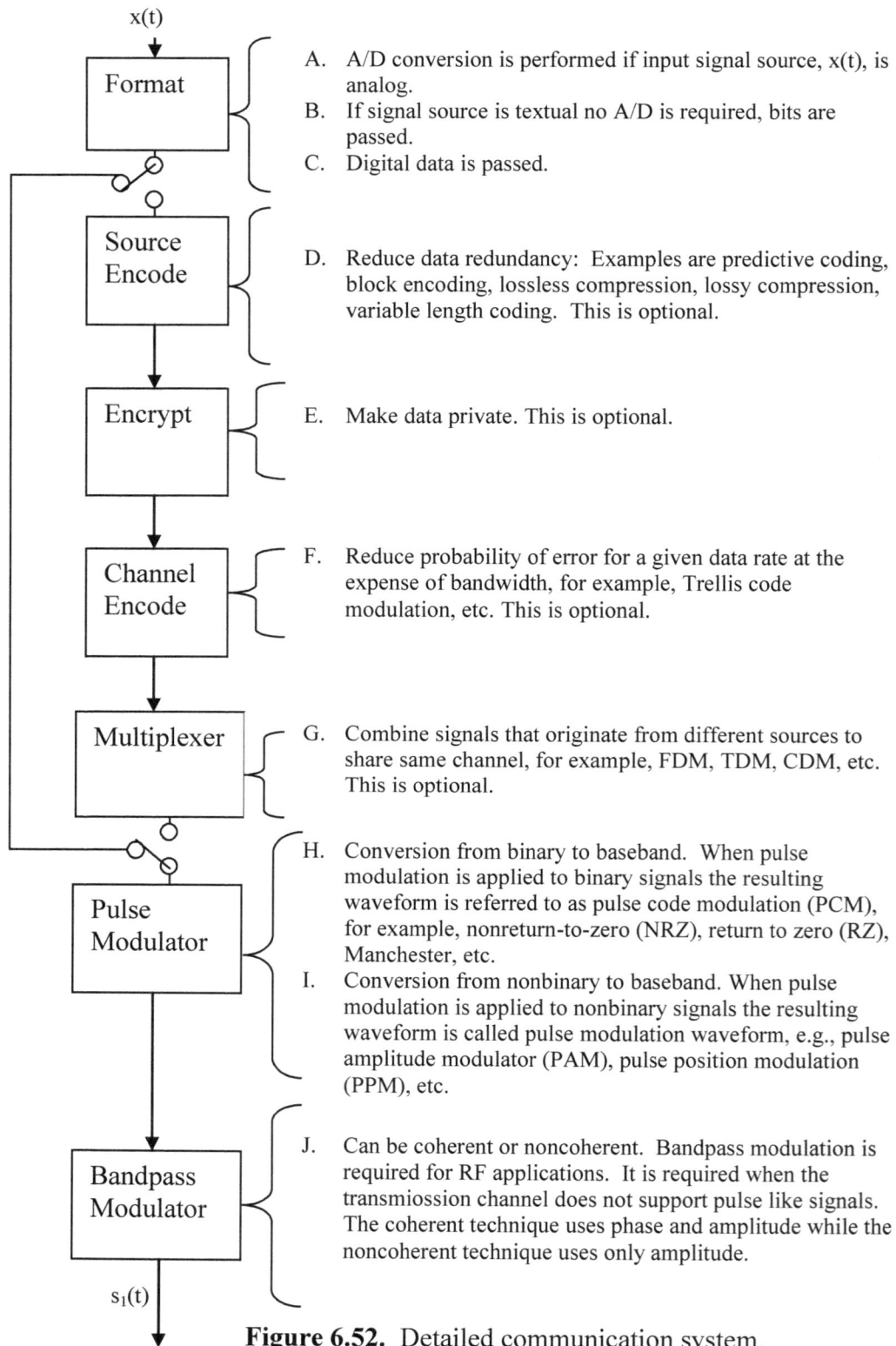

Figure 6.52. Detailed communication system.

$s_1(t)$

Frequency Spread

Multiple Access

Transmitter

Cable

Receiver

$s_2(t)$

K. Make signal tolerable to interefence from other sources. This is optional.

L. Combine signals from different sources or of different characteristics for transmission over the same channel, e.g., frequency division [frequency division multiplexing (FDM), frequency division Multiple Access (FDMA)], time division [time division multiplexing (TDM), time division multiple access (TDMA)], code division [code division multiplexing (CDM), code division multiple access (CDMA)], etc. This is optional.

M. For wireless applications the transmitter consists of a frequency up conversion stage to RF, a high power amplifier, and an antenna.

N. The transmission channel, $h_c(t)$, can be a cable or wireless. In transmission the signal can be distorted by noise, $n(t)$, etc.

O. The received signal is $r(t)=s(t)*h_c(t) + n(t)$ where $r(t)$ is the received signal, $s(t)$ is the transmitted signal, $h_c(t)$ is the transmission channel, and $n(t)$ is the channel noise.

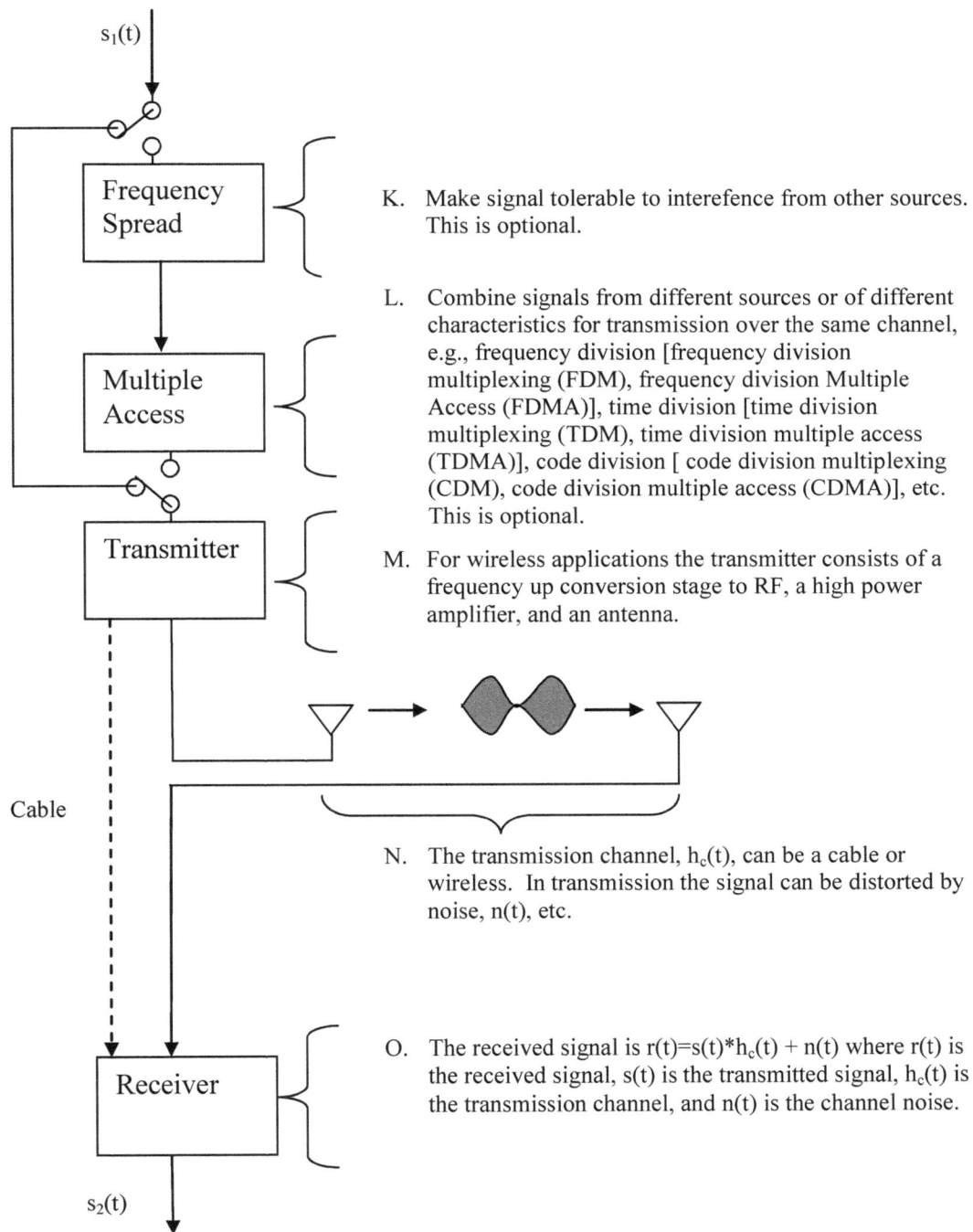

Figure 6.52. Continued.

$s_2(t)$

Block	Description
Multiple Access	P. Perform the reverse of the transmitter multiple access. This is optional.
Frequency Despread	Q. Perform the reverse of the transmitter frequency spread. This is optional.
Demodulate	R. Perform the reverse of the transmitter modulation.
Detect	S. Recover signals.
Demultiplex	T. Perform the reverse of the transmitter multiplexing. This is optional.
Channel Decode	U. Perform the reverse of the transmitter coding. This is optional.
Decrypt	V. Perform the reverse of the transmitter encrypting. This is optional.
Source Decode	W. Perform the reverse of the transmitter source coding. This is optional.
Format	X. Perform the reverse of the transmitter formating

Output Information

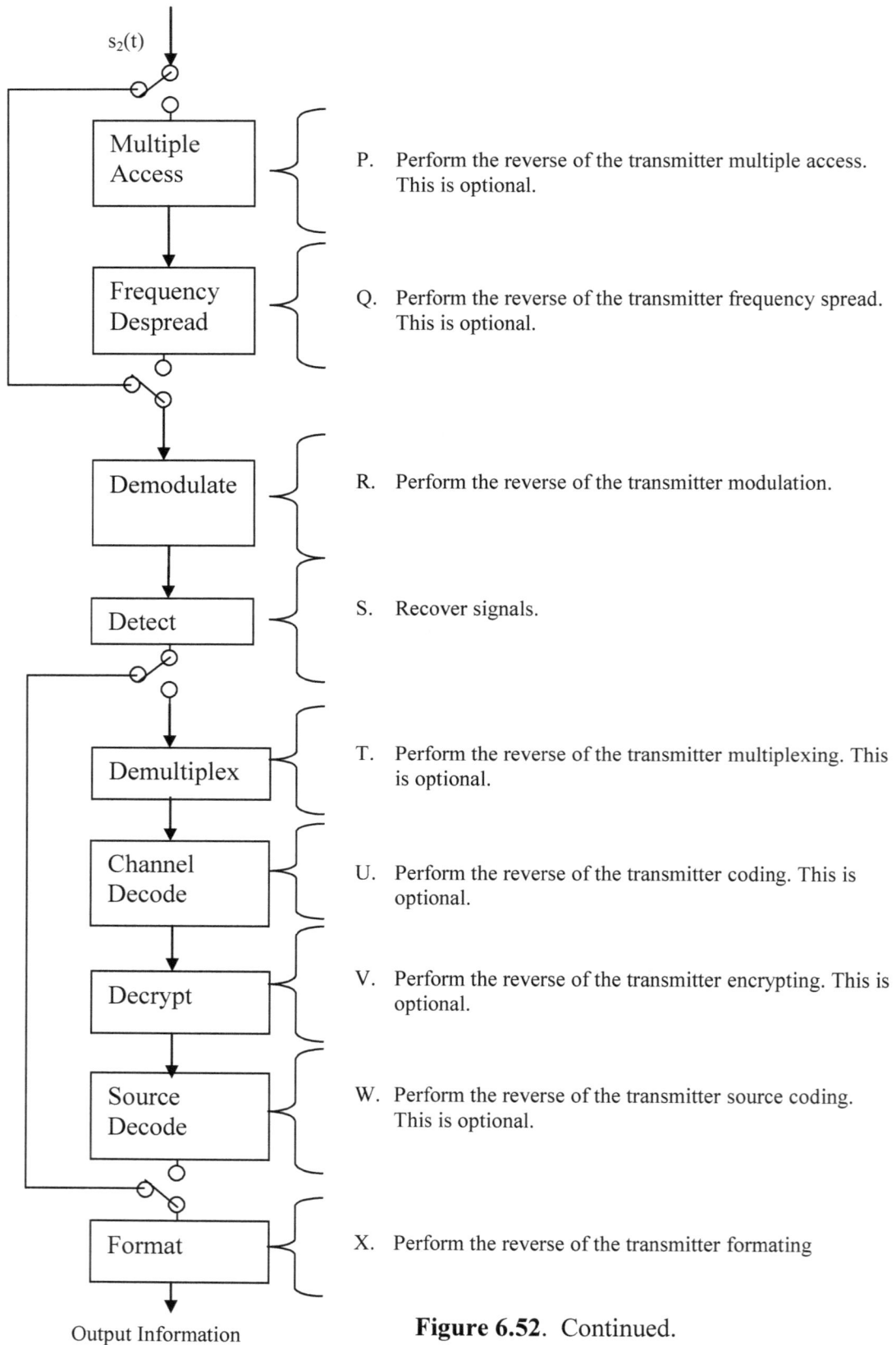

Figure 6.52. Continued.

The rest of this section shall explain in more details some of the digital modulation techniques indicated in Figure 6.52.

Digital data can be modulated using one of the following modulation techniques:
   a.) AM (Amplitude modulation)
   b.) FM (Frequency modulation)
   c.) PM (Phase modulation)
If amplitude modulation is used and the modulating signal is a logic 1 or 0 then this type of modulation is referred to as amplitude shift keying. If frequency modulation is used and the modulating signal is a logic 0 or 1 then this type of modulation is referred to as frequency shift keying. If phase modulation is used and the modulating signal is a logic 1 or 0 then this type of signal is referred to as phase shift keying. The digital data consists of a series of zeros and ones, 1101000, as shown in Figure 6.53. Here, the most significant bit is the most left bit (1) while the least significant bit is the bit farthest to the right (0).

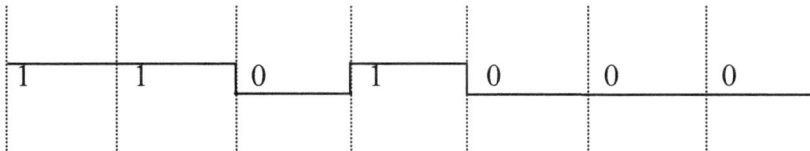

Figure 6.53. Digital serial data.

6.7.1. Bit Rate

Consider the serial data represented in Figure 6.54, t_b (seconds) is called the bit interval. The bit interval is the duration of 1 bit, 0 or 1. The bit rate is determined by

$$f_b = \frac{1}{t_b}$$ EQ. 6.87

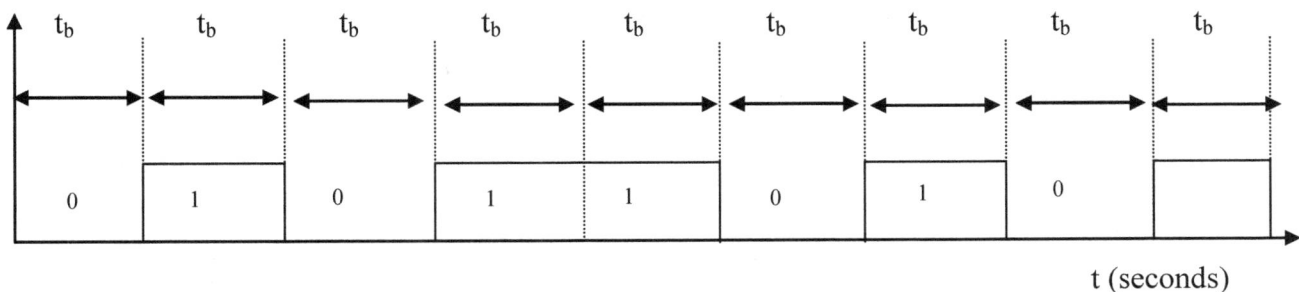

t (seconds)

Figure 6.54. Digital serial data and bit rate.

6.7.2. Amplitude Shift Keying

 In analog amplitude modulation systems analog data is the modulating signal. However, in digital amplitude modulation systems digital data (binary data) is the modulating signal. This is called amplitude shift keying or ASK. Since one bit is modulated at a time this is often referred to as binary ASK or simply BASK. This is represented by:

$$v_{ask}(t) = [1 + v_m(t)]\left(\frac{A}{2}\cos(2\pi f_c t)\right)$$

EQ. 6.88

where A is the carrier amplitude (volts), f_c is the carrier frequency (Hz), and $V_m(t)$ is the modulating signal (volts). The carrier is either ON or OFF depending on whether the input is a logic 1 or logic 0. If the binary signal in Figure 6.53 is the modulating signal then the AM signal would look like that shown in Figure 6.55. Such an amplitude modulated signal is sometimes called an ON-OFF Keying (OOK) signal.

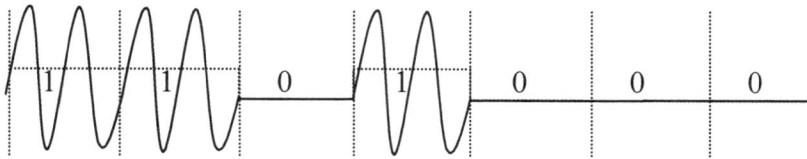

Figure 6.55. Digital amplitude modulation using the serial data as the modulating signal given in Figure 6.53.

6.7.3. Frequency Shift Keying

Digital frequency modulation is when waveform when the modulating signal is binary data. This is also called frequency shift keying or FSK. Since one bit is modulated at a time this is often referred to as binary FSK or simply BFSK. This type of frequency modulated signal is represented by:

$$v_{fsk}(t) = V_c \cos(2\pi f_c t + 2\pi V_m(t)\Delta f t)$$

EQ. 6.89

where V_c is the carrier amplitude (volts), f_c is the carrier frequency (Hz), $V_m(t)$ is the modulating signal (volts), and Δf is the frequency deviation (Hz). The peak frequency deviation is given by:

$$\Delta f = \frac{|f_{mark} - f_{space}|}{2}$$

EQ. 6.90

where Δf is the frequency deviation, f_{mark} is the mark frequency (Hz), and f_{space} is the space frequency (Hz). Here, a mark is when the input bit is a logic 1, otherwise, it is referred to as a space for a logic 0. This can be visa –versa. In our text, we represent a mark as a logic 1. We take the magnitude because sometimes the mark frequency is greater than the space frequency or visa-versa. A FSK signal where $f_{mark} > f_{space}$ is shown in Figure 6.56.

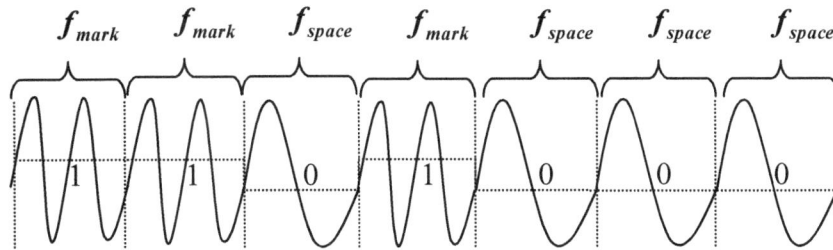

Figure 6.56. Frequency shift keying using the serial data as the modulating signal in given in Figure 6.53. f_{mark} is the frequency that occurs when the input to the FSK modulator is a logic 1. f_{space} is the frequency that occurs when the input to the FSK modulator is a logic 0.

The frequency spectrum of a FSK signal consists of the spectrum shown in Figure 6.57.

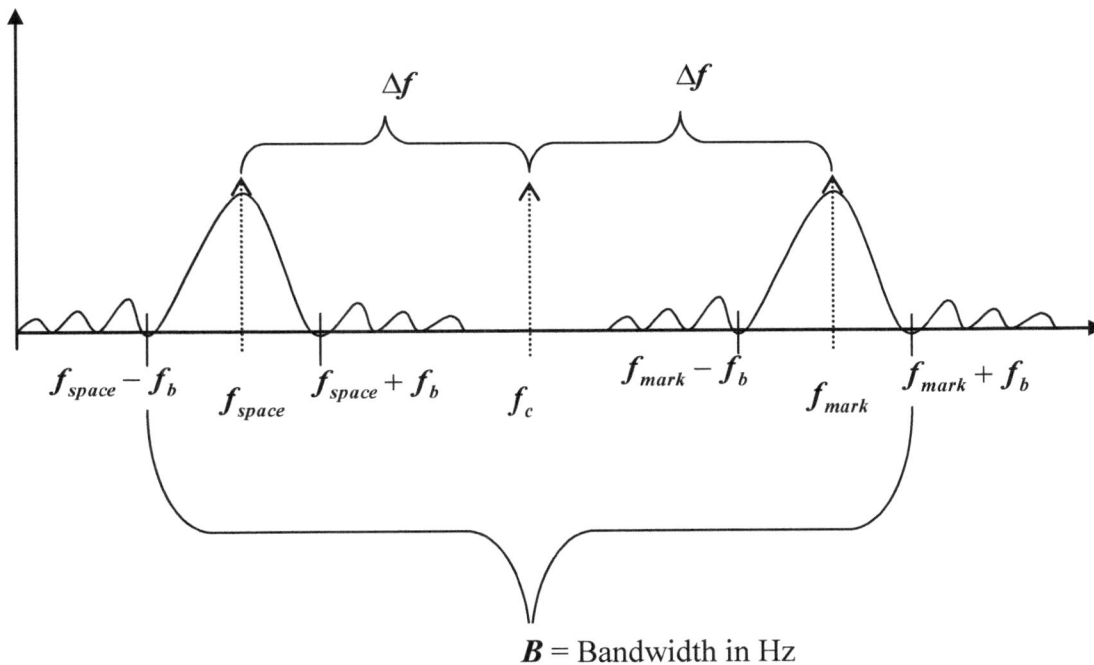

B = Bandwidth in Hz

Figure 6.57. FSK Spectrum showing bandwidth, B. Δf (Hz) is defined as the mark or space frequency deviation from the center frequency f_c.

In Figure 6.57 the FSK signal minimum bandwidth B is calculated by taking the difference of the two extreme frequencies. Δf is the frequency deviation of the mark or space frequencies from the carrier frequency. Thus, we have:

$$B = \left|\left(f_{mark} + f_b\right) - \left(f_{space} - f_b\right)\right|$$

EQ. 6.91

$$= \left|f_{mark} - f_{space} + 2f_b\right|$$

$$= \left|f_{mark} - f_{space}\right| + 2f_b$$

$$= 2\Delta f + 2f_b$$

A FSK transmitter is shown in Figure 6.58. The voltage control oscillator (VCO) receives the binary data, most significant bit first. The power amplifier amplifies the FSK signal from the VCO before it is sent to the antenna.

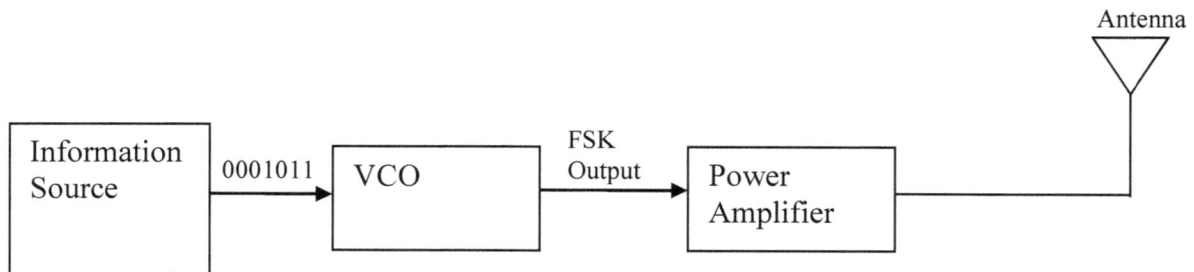

Figure 6.58. FSK transmitter using a voltage control oscillator (VCO).

A FSK receiver is shown in Figure 6.59. It uses a **noncoherent** demodulation to recover the information. Noncoherent detection involves no frequencies in the demodulation process. **Coherent** detection involves the use of the carrier frequency, etc., to recover the information as shown in Figure 6.60. The coherent receiver is not practical and is hardly used. For the noncoherent FSK receiver in Figure 6.59 the power splitter accepts the FSK input and outputs two copies without loading down the band pass filters (BPFs). The space BPF has a center frequency equal to the space frequency and the mark BPF has a center frequency equal to the mark frequency. The slope detector consists of a rectifier and a low pass filter (LPF) for the mark and space BPF outputs. The f_{space} slope detector has LPF with a cutoff frequency at f_{space} while the f_{mark} slope detector has a LPF with a cutoff frequency at f_{mark}. The comparator compares its two inputs from f_{space} and f_{mark} slope detector. When the output of the f_{mark} slope detector is greater than the output of the f_{space} slope detector a binary one (logic 1) is produced at the output of the comparator, otherwise a binary zero (logic 0) is produced.

Another method for FSK demodulation is the phase locked loop (PLL) based receiver. The PLL is the most common method for demodulating FSK signals. The input into the PLL has two possible frequencies, f_{space} and f_{mark} which cause the outputs of the PLL to shift between two output levels, logic 0 and logic 1. This results in the original binary signal. The PLL FSK modulator is shown in Figure 6.61.

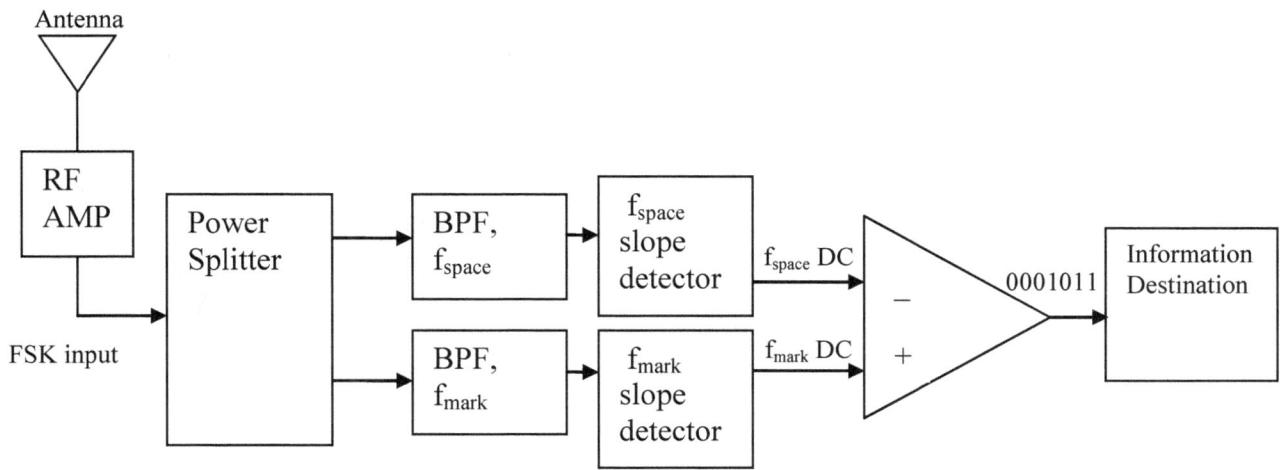

Figure 6.59. Noncoherent FSK receiver.

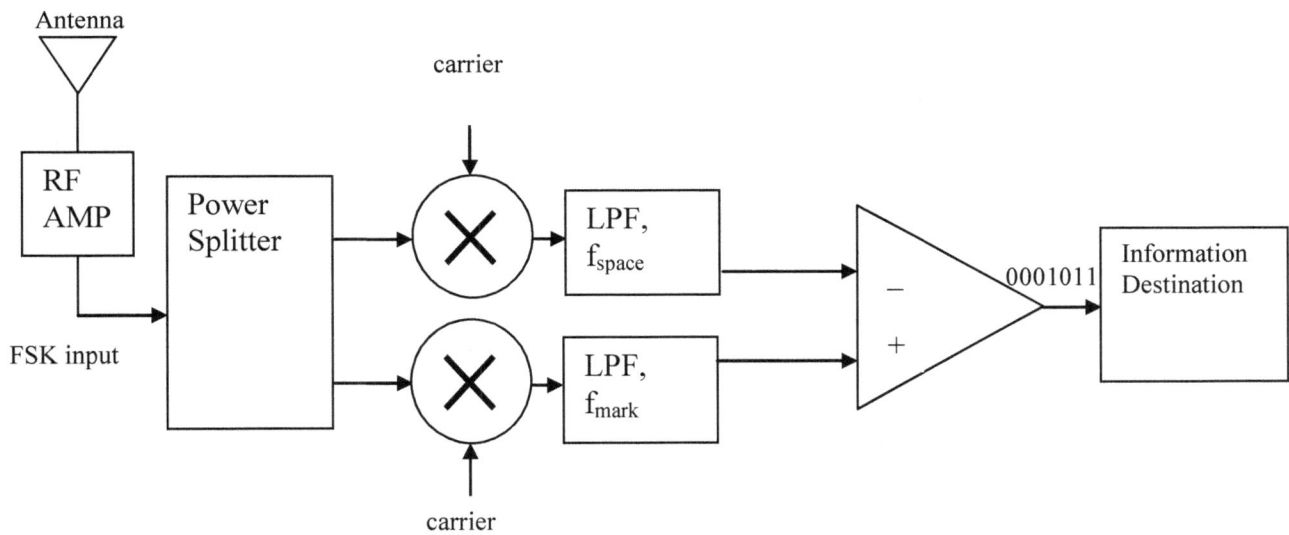

Figure 6.60. Coherent FSK receiver.

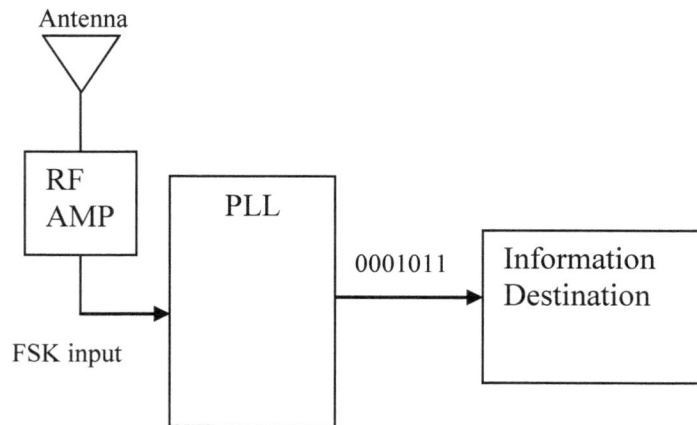

Figure 6.61. Phase lock loop (PLL) FSK receiver.

6.7.4. Phase Shift Keying

Digital phase modulation is when waveform when the modulating signal is binary data. This is also called **phase shift keying** or PSK. Since one bit is modulated at a time this is often referred to as binary PSK or simply BPSK. Simply stated the phase shift of the carrier changes depending on whether the input modulating signal is a logic 1 or logic 0. For example, a logic zero can cause 180 degrees phase shift and a logic 1 can cause 0 degrees phase shift. Such a PSK signal is shown in Figure 6.62.

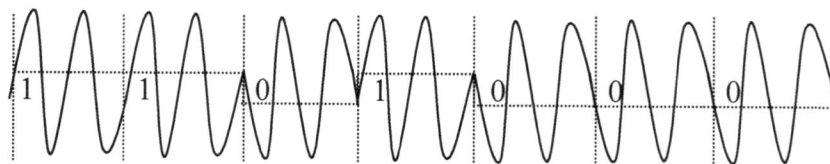

Figure 6.62. Phase shift keying using the serial data as the modulating signal given in Figure 6.53. Logic 0 => 180 degrees phase shift, logic 1 => 0 degrees phase shift.

6.7.5. M-ary Encoding

The digital modulation techniques used so far use binary data as the modulation data. Binary data has two possible states: 0 or 1. As seen with BASK, BFSK, and BPSK, shown earlier, this results in two possible modulated output states. Multilevel modulation or multiphase modulation is often referred to as **M-ary** encoding. Thus, for 2-ary (binary) each output represents one of two possible states. Examples of 2-ary systems are BASK, FSK, BPSK, etc. For M-ary coding M is the number of possible states that each modulated output represents. The number of output states, M, for N bits is given by:

$$M = 2^N \qquad\qquad \text{EQ. 6.92}$$

or

$$N = \log_2 M \qquad\qquad \text{EQ. 6.93}$$

N (number of bit encoded)	M (number of possible output states)
1	2
2	4
3	8
4	16
5	32

Table 6.3. M and N values for M-ary system.

The minimum bandwidth, B, required for an M-ary system is given by

$$B = \frac{f_b}{\log_2 M} = \frac{f_b}{N} \qquad\qquad \text{EQ. 6.94}$$

where f_b is the input bit rate, N is the number of input bits, and M is the number of output states.

6.7.6. Baud Rate

As seen earlier, bit rate, f_b is the rate of change of raw digital data. Baud rate, F_b is the rate of change on a transmission line medium after encoding and modulation have taken place. The Baud rate, F_b, is calculated by:

$$Baud = F_b = \frac{1}{T_b} \qquad\qquad \text{EQ. 6.95}$$

Where baud is the symbol rate in Bauds per second, T_b is the duration of one signaling element in seconds. A signaling element is also called a symbol. Often, F_b is referred to as the symbol rate in symbols per second. So Baud rate is equivalent to symbol rate. In Figure 6.63 we use BASK, BFSK, and BPSK waveforms to illustrate a signaling element and it's duration. One signaling element, also called symbol, is represented by one or more bits that have been modulated, and has a duration given by T_b. For binay systems such as BASK, BFSK, and BPSK, 1 bit is modulated at a time to get a signaling element. Thus, for BASK, BFSK, and BPSK, the baud rate and bit rate are equivalent.

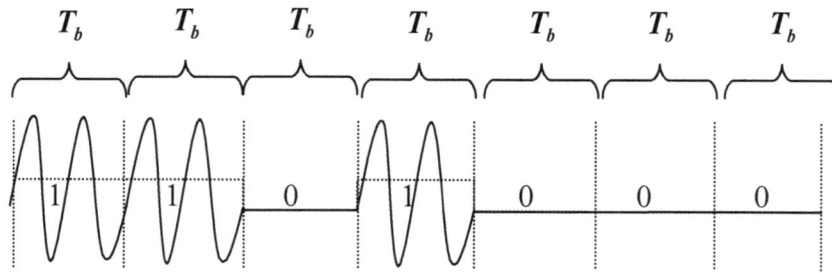

A.)

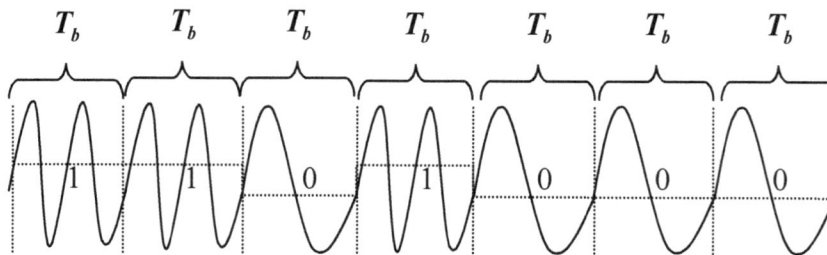

B.)

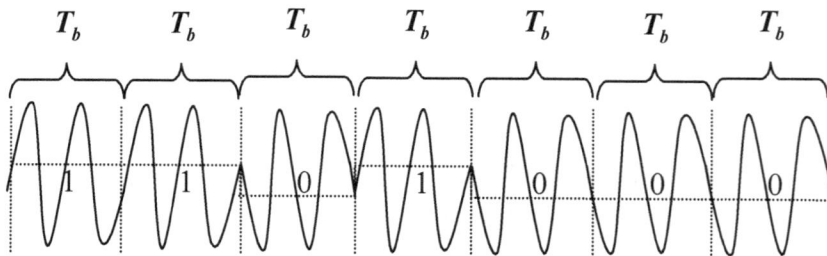

C.)

Figure 6.63. Duration of a signaling element T_b. A signaling element is also called a symbol. A.) BASK, B.) BFSK, C.) BPSK.

6.7.7. Information Capacity
The information capacity (maximum bits per second) of a channel is given by:

$$f_b = 3.32 B \log_{10}\left(1 + \frac{S}{N}\right)$$ **EQ. 6.96**

where f_b is the information capacity or maximum bit rate in bits per second, B is the bandwidth in Hertz, and S/N is the signal to noise ratio (unitless). The information capacity can be increased by using more bits to represent a symbol. This is done with M-ary systems such as quadrature phase shift keying (QPSK), quadrature amplitude modulation (QAM), 8 QAM, 16 QAM, etc.

6.7.8.   Quadrature Phase Shift Keying

For 4 Phase Shift Keying, referred to as quadrature phase shift keying (QPSK) there are 4 possible output symbols for each bit pair. Thus, QPSK is a 4-ary system where M=4 and N=2. Each pair of successive bits are assigned a unique phase shift. A block diagram of a QPSK modulator is shown Figure 6.64. The serial data is clocked into a serial input – parallel output (SIPO) shift register, most significant bit first, at the bit rate f_b. The parallel input – parallel output (PIPO) shift register accepts the bits from the SIPO register at a rate of $f_b/2$. So a 4MHz clock would result in f1=2MHz. Of the two bits in the PIPO, the most significant bit goes to the Q channel and the least significant to the I channel at the data rate $f_b/2$. The unipolar non-return-to-zero (UNRZ) line code is applied to the I and Q bit after it leaves the PIPO. Some popular line codes are discussed in Section 11.5 on page 422. So, logic 0 = -1, and logic 1 = +1. The UNRZ line code I bit is mixed, or multiplied by the zero degree phase shifted carrier, $\sin(\omega t + 0) = \sin(\omega t)$. Thus, this bit is called the in-phase (I) bit. Simultaneously, the UNRZ line coded Q bit is mixed, or multiplied by the +90 degree phase shifted carrier, $\sin(\omega t + 90)$. Thus, this bit is called the +90 degree out of phase (Q) bit. Their outputs are bandpass filtered to removed any extra harmonics as a result of the mixing. Finally, their outputs (Q and I) are summed before going further on downstream in the transmitter. This summed output is the QPSK modulated signal.

The block diagram of a QPSK demodulator is shown in Figure 6.65. The RF amplifier amplifies the received signal enough for a good reception. The power splitter takes this single input and copies it to three outputs: two mixers and a carrier recovery circuit. The carrier recovery circuit recovers the carrier to be in be in synchroization with the transmitted carrier. The top mixer multiples the incoming QPSK signal with a zero degrees carrier while the bottom mixer multiplies the incoming QPSK signal with a 90 degrees phase shift carrier. Their mixed outputs are lowpass filtered and shaped to recover the original transmitted data.

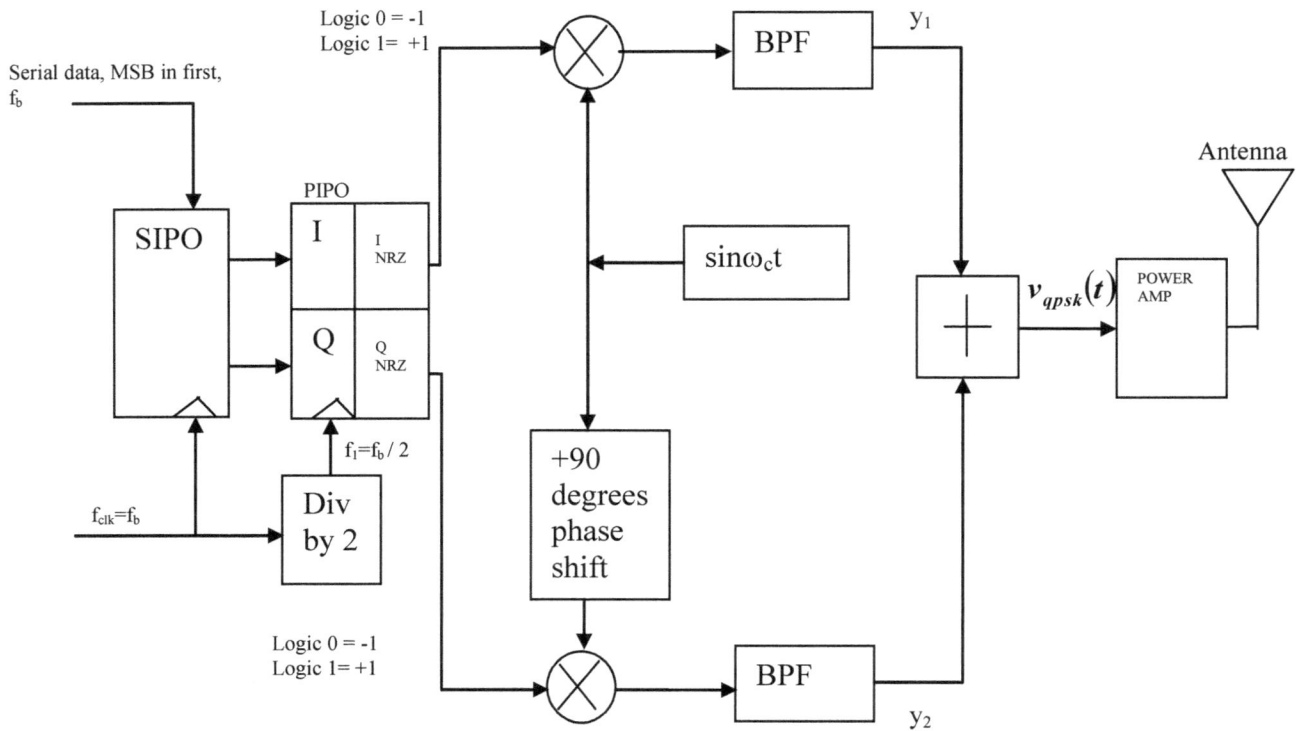

Figure 6.64. Block diagram of a QPSK modulator.

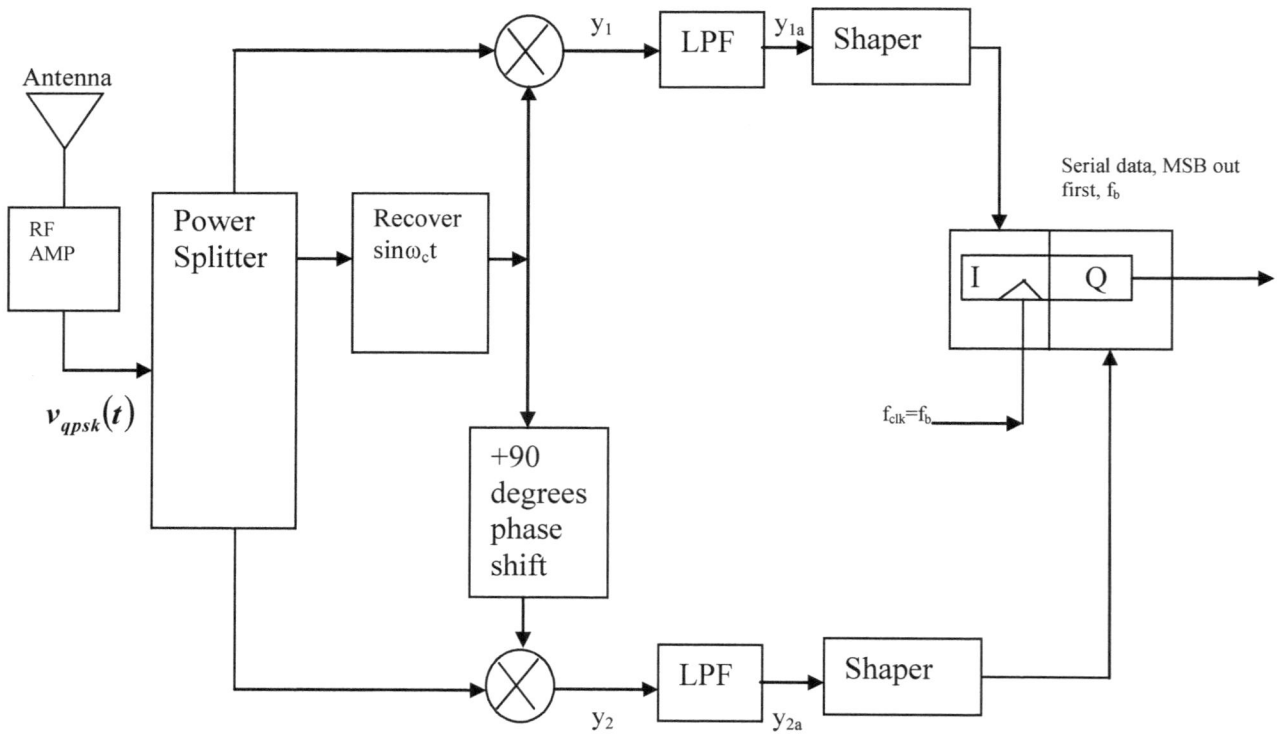

Figure 6.65. Block diagram of a QPSK demodulator.

The input output relationship of a QPSK modulator can be represented by a truth Table, constellation diagram, and phasors diagram. The input I and Q bit pair verses output phase shift values for Figure 6.64 is given in Truth Table 6.4.

Bit Pair		Phase Shift
Q	I	
0	0	-135
0	1	-45
1	0	+135
1	1	+45

Table 6.4. The Truth Table for QPSK modulator in Figure 6.64.

The phasor diagram of the QPSK system represented by Table 6.4 is illustrated in Figure 6.66. The corresponding constellation diagram is shown in Figure 6.67.

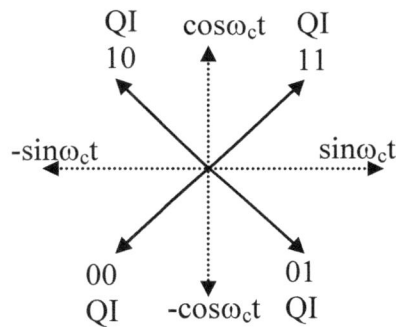

Figure 6.66. Phasor diagram of QPSK modulator in Figure 6.64.

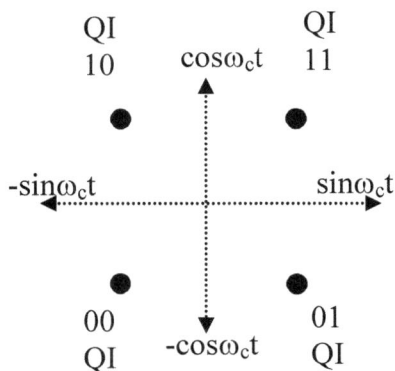

Figure 6.67. Constellation diagram of QPSK modulator in Figure 6.64.

EXAMPLE 6.8

Using Figure 6.64, let the input QI=10, find the following:
   a.) y_1 and y_2.
   b.) Time domain expression of $v_{qpsk}(t)$
   c.) Phasor expression of $v_{qpsk}(t)$

Solution:
Use the following diagram in Figure 6.68 to assist with basic trig identities and relating polar coordinates to rectangular coordinates:

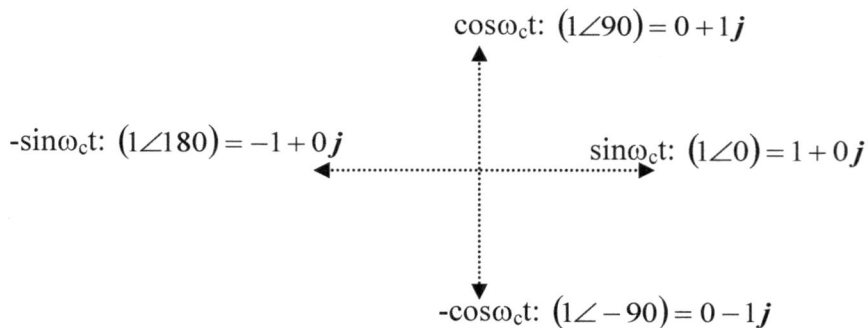

$$\cos\omega_c t: \ (1\angle 90) = 0 + 1j$$

$$-\sin\omega_c t: \ (1\angle 180) = -1 + 0j \qquad\qquad \sin\omega_c t: \ (1\angle 0) = 1 + 0j$$

$$-\cos\omega_c t: \ (1\angle -90) = 0 - 1j$$

Figure 6.68. Phasor diagram with reference sin(ωt).

a.)

For Q=1,

$$
\begin{aligned}
y_1 &= Q \times \sin(\omega t + 90°) \\
&= 1 \times \sin(\omega t + 90°) \\
&= \sin(\omega t + 90°) \\
&= \cos(\omega t)
\end{aligned}
$$

For I=0,

$$
\begin{aligned}
y_2 &= I \times \sin(\omega t) \\
&= -1 \times \sin(\omega t) \\
&= -\sin(\omega t)
\end{aligned}
$$

b.)

$$v_{qpsk}(t) = y_1 + y_2$$
$$= \cos(\omega t) - \sin(\omega t)$$

c.)

$$v_{qpsk}(t) = \cos(\omega t) - \sin(\omega t)$$
$$\therefore \left| v_{qpsk} \right| \angle\theta = 0 + 1j - (1 + 0j)$$
$$= -1 + j$$
$$= \sqrt{2}\angle 135°$$

EXAMPLE 6.9

Using Figure 6.65 and letting

$$v_{qpsk}(t) = -\cos(\omega t) + \sin(\omega t)$$

find the following:

    a.) y_1
    b.) y_2
    c.) I and Q

Solution:
The following trig identities are helpful:

$$\sin^2(\theta) = \frac{1}{2}[1 - \cos(2\theta)] \qquad \text{EQ. 6.97}$$

$$\sin\alpha \sin\beta = \frac{1}{2}[\cos(\alpha - \beta) - \cos(\alpha + \beta)] \qquad \text{EQ. 6.98}$$

$$\sin\alpha \cos\beta = \frac{1}{2}[\sin(\alpha - \beta) + \sin(\alpha + \beta)] \qquad \text{EQ. 6.99}$$

a.)

$$t) \times \sin(\omega t)$$

$$s(\omega t) + \sin(\omega t)) \times \sin(\omega t)$$

$$(\omega t)\sin(\omega t) + \sin(\omega t)\sin(\omega t)$$

$$\left[in(\omega t - \omega t) + \frac{1}{2}\sin(\omega t + \omega t) \right] + \frac{1}{2}(1 - \cos(2\omega t))$$

$$\left[in(0) + \frac{1}{2}\sin(2\omega t) \right] + \frac{1}{2}(1 - \cos(2\omega t))$$

$$\cdot \sin(2\omega t) - \frac{1}{2}\cos(2\omega t)$$

b.)

$$t) \times \sin(\omega t + 90°)$$

$$s(\omega t) + \sin(\omega t)) \times \sin(\omega t + 90°)$$

$$(\omega t)\sin(\omega t + 90°) + \sin(\omega t)\sin(\omega t + 90°)$$

$$\left[in(\omega t + 90° - \omega t) + \frac{1}{2}\sin(\omega t + 90° + \omega t) \right] + \frac{1}{2}\left[\cos(\omega t - \omega t - 90°) - \cos(\omega t + \omega t + 90°) \right]$$

$$\left[in(90°) + \frac{1}{2}\sin(2\omega t + 90°) \right] + \frac{1}{2}\left[\cos(-90°) - \cos(2\omega t + 90°) \right]$$

$$\cdot \frac{1}{2}\sin(2\omega t + 90°) + \frac{1}{2}(0) - \frac{1}{2}\cos(2\omega t + 90°)$$

$$\frac{1}{2}\sin(2\omega t + 90°) - \frac{1}{2}\cos(2\omega t + 90°)$$

c.) After the LPFs

Therefore, the output of the I shaper is a logic 1 and the output of the Q shaper is a logic 0. So, I=1, Q=0.

PROBLEMS

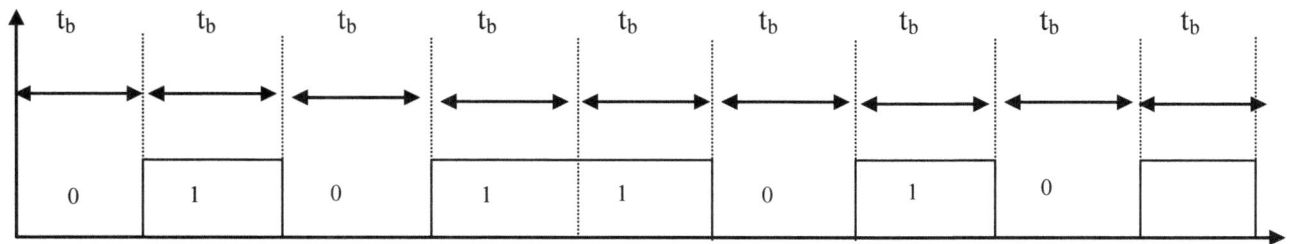

Figure 6.69. Serial data.

1. If t_b in Figure 6.69 is 1×10^{-3} seconds, what is the bit rate:
A. 1 bps
B. 10 bps
C. 100 bps
D. 1000 bps
E. 10000 bps

2. If t_b in Figure 6.69 is 1.0 μs, what is the bit rate:
A. 1000 bps
B. 10000 bps
C. 100000 bps
D. 1000000 bps
E. 100000000 bps

3. Given that t_b=100 μs and if every two bits in Figure 6.69 is encoded into a symbol and transmitted the corresponding symbol rate is:

A. 100 symbols/s
B. 1000 symbols/s
C. 5000 symbols/s
D. 10000 symbols/s

4. For an M-ary system, if 2 bits are encoded the number of possible symbols transmitted are:
A. 1
B. 2
C. 4
D. 8

5. For an M-ary system, if 3 bits are encoded the number of possible symbols transmitted are:

A. 3
B. 4
C. 6
D. 8
E. 16

6. For an M-ary system, if 4 bits are encoded the number of possible symbols transmitted are:
A. 5
B. 7
C. 9
D. 16
E. 32

7. For an M-ary system, a symbol that contains M=2 possible combinations consist of the following number of encoded bits:
A. 1
B. 2
C. 3
D. 4
E. 5

8. For an M-ary system, a symbol that contains M=4 possible combinations consist of the following number of encoded bits:
A. 2
B. 3
C. 4
D. 5
E. 8

9. For an M-ary system, a symbol that contains M=8 possible combinations consist of the following number of encoded bits:
A. 2
B. 3
C. 5
D. 8
E. 16

10 For an M-ary system, a symbol that contains M=32 possible combinations consist of the following number of encoded bits:
A. 2
B. 4
C. 5
D. 32
E. 64

11. For an M-ary system, a symbol that contains M=64 possible combinations consist of the following number of encoded bits:
A. 64
B. 32
C. 8
D. 7
E. 6

12. Which one of the following is a 2-ary system:
A. BPSK
B. FSK
C. ASK
D. All of the above
E. None of the above

13. Which one of the following is a 4-ary system:
A. BPSK
B. FSK
C. 8-PSK
D. QPSK
E. None of the above

14. Which one of the following is an 8-ary system:
A. 8-QAM
B. 8-PSK
C. 16-PSK
D. A and B above
E. None of the above

15. If the bit rate is 1000 Hz, the minimum bandwidth required for an M-ary (2 bits encoded) system is:
A.) 500 Hz   B.) 1000 Hz   C.) 5000 Hz   D.) 50 Hz

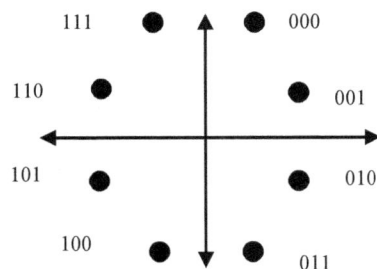

Figure 6.70.

16. The representation in Figure 6.70 is which one of the following:
A.) Truth Table   B.)Phasor diagram   C.) Constellation diagram   D.) None of the mentioned choices

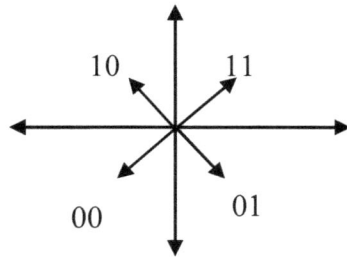

Figure 6.71.

17. The representation in Figure 6.71 is which one of the following:
A.) Truth Table   B.)Phasor diagram   C.) Constellation diagram   D.) None of the mentioned choices

18. Figure 6.70 represents which one of the following:
A.)8 QAM   B.)4 PSK   C.) FSK   D.) 8 PSK   E. None

19. Figure 6.71 represents which one of the following:
A.)4 QAM   B.)4 PSK   C.) 4 FSK   D.) ASK   E. None

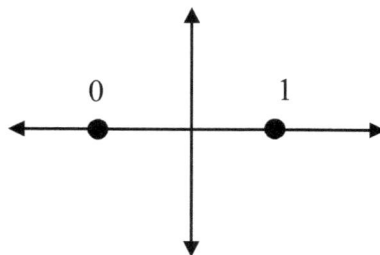

Figure 6.72.

20. Figure 6.72 represents which one of the following:
A.)4 QAM   B.)ASK   C.) 4FSK   D.) BPSK   E. None

21. The I bit in a QPSK system is out of phase with the reference carrier by 90 degrees.
A.) True   B.) False

22. The Q bit in a QPSK system is in phase with the reference carrier.
A.) True   B.) False

23. Time division multiplexing is when two or more channels of different carrier frequencies are combined but separately in the frequency domain.
A.) True   B.) False

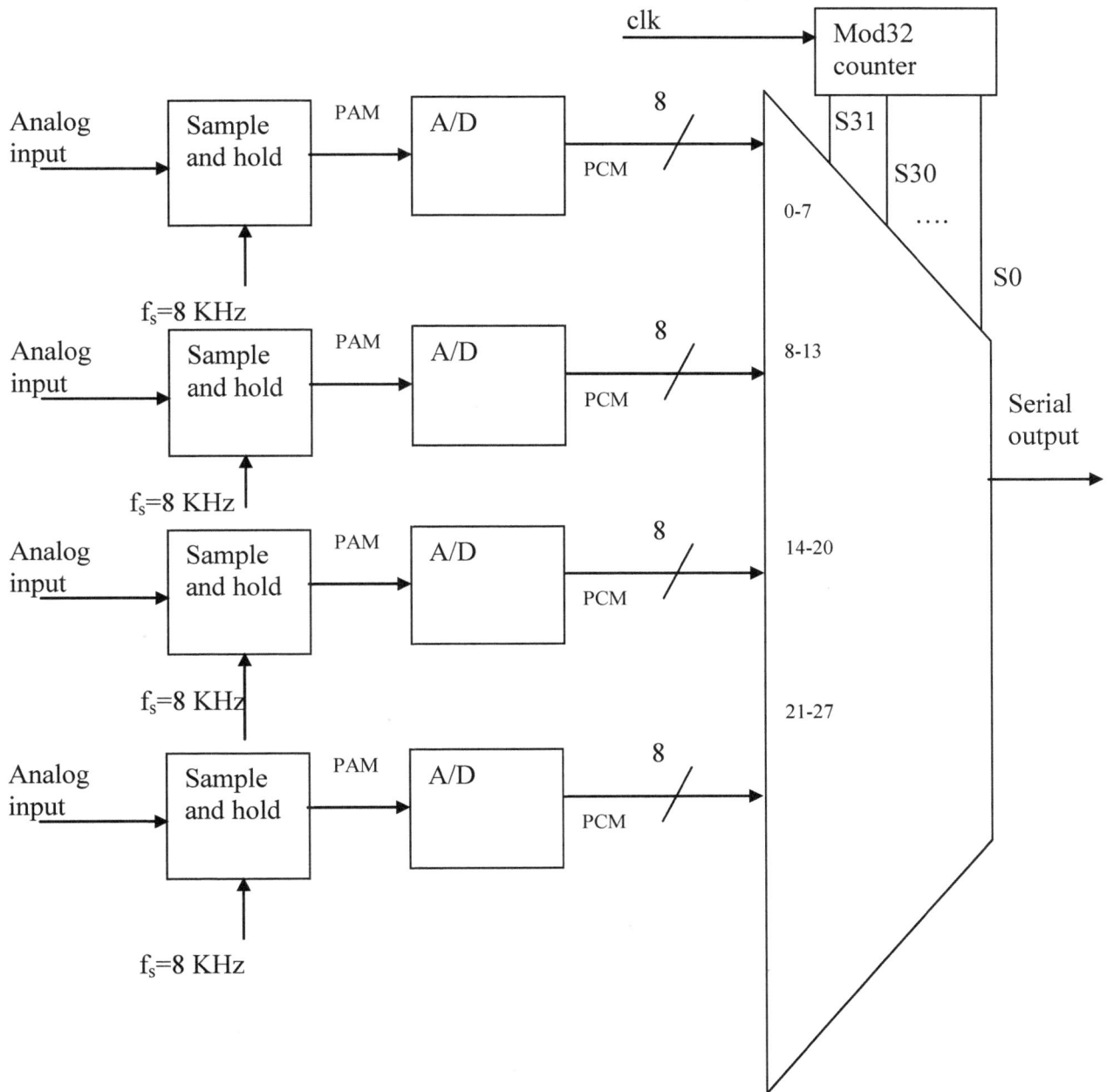

Figure 6.73. TDM system.

24. Four channels are multiplexed serially into 125μs frames as shown in Figure 6.73. The bit rate, f_b, at the output of the multiplexer is:

A.) 256Kbps   B.)64Kbps   C.)8Kbps   D.) 32Kbps

25. Problem 24 requires a multiplexer clock frequency (see clk in Figure 6.73.) of:

A.) 256KHz   B.)64KHz   C.)8KHz   D.) 32KHz

26. The bandwidth of an M-ary system is given by:

A. f_b/N

B. fmax-fmin

C. N/f_b

D. None

27. Sketch the frequency domain representation of $v(t) = 4\sin(2\pi1500t) + 3\sin(2\pi3000t) + 1.5\sin(2\pi500t)$ and label all axis and amplitudes appropriately.

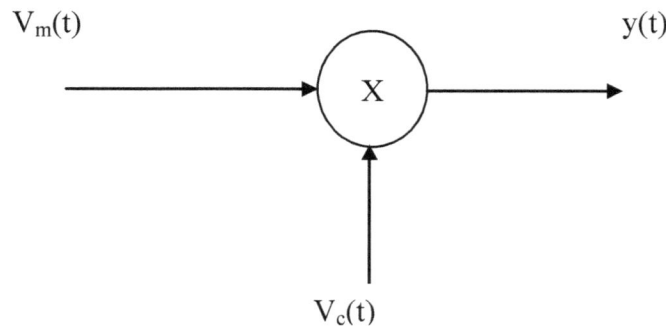

$V_m(t)$ $y(t)$

X

$V_c(t)$

Figure 6.74.

28. Using Figure 6.74, if $V_m(t)=10\sin(2\pi3000t)$ is the modulating signal and $V_c(t)=5\sin(2\pi12000t)$ is the carrier determine:

A. Time domain expression of y(t):

B. Frequency domain representation of y(t):

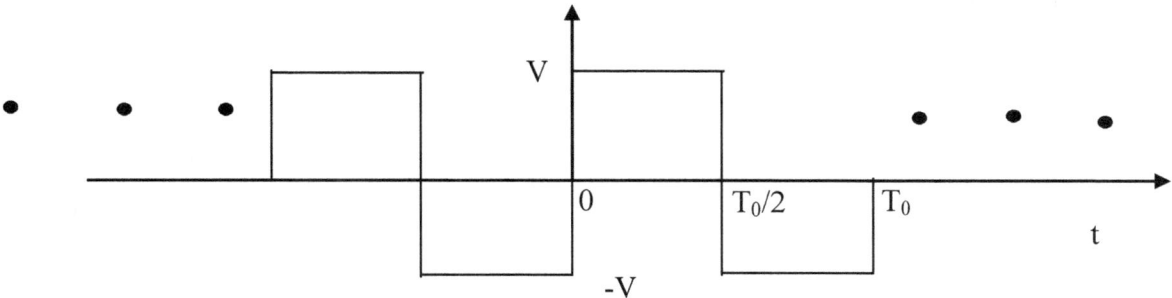

Figure 6.75.

29. The signal in Figure 6.74 has even, odd, or half-wave symmetry?

30. Using the signal in Figure 6.74 complete the table below for $T_0=1\times10^{-6}$ and $V=2$ volt.

Harmonic Number	Harmonic frequency	Harmonic Amplitude
0		
1		
2		
3		
4		
5		
6		
7		

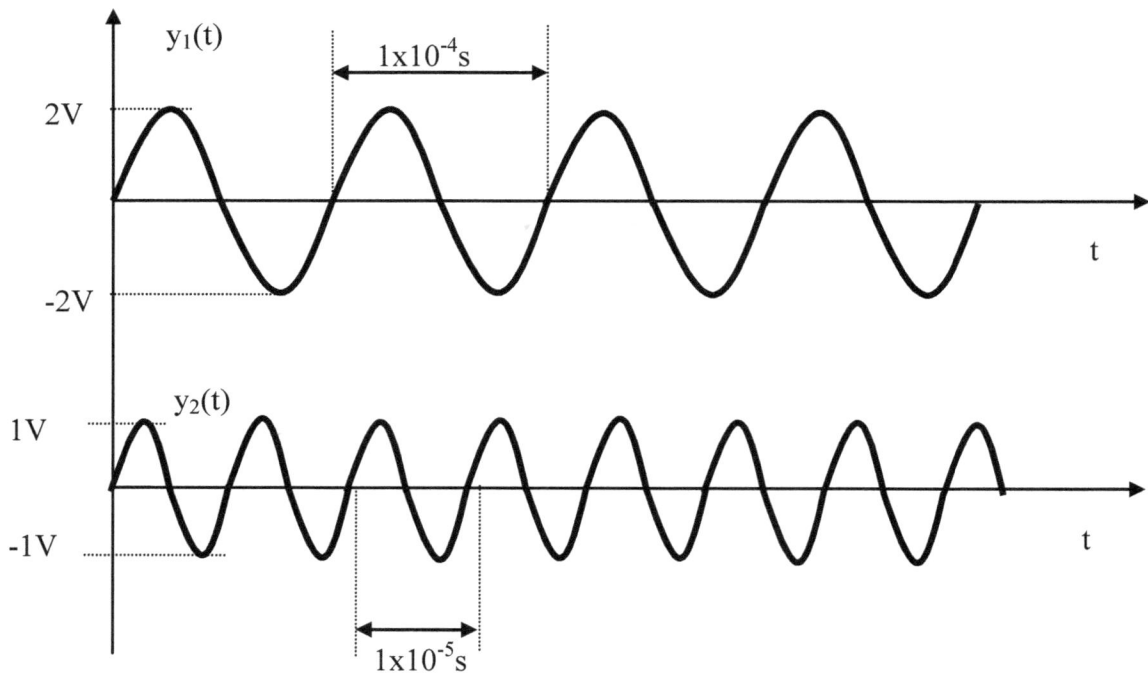

Figure 6.76. Periodic $y_1(t)$ and $y_2(t)$.

31. Using the waveforms in Figure 6.76, determine the reduced time domain equation and draw the corresponding frequency domain representation of each expression as indicated below:

	Reduced Time domain expression	Draw Frequency domain plot
A.	$y(t)=y_1(t) + y_2(t)$	
B.	$y(t)=10\times y_2(t)$	
C.	$y(t)=y_1(t)\times y_1(t)$	

$$L1=100\mu H,$$
$$C1=250 \text{ pF and}$$
$$C2=0.01 \ \mu F$$

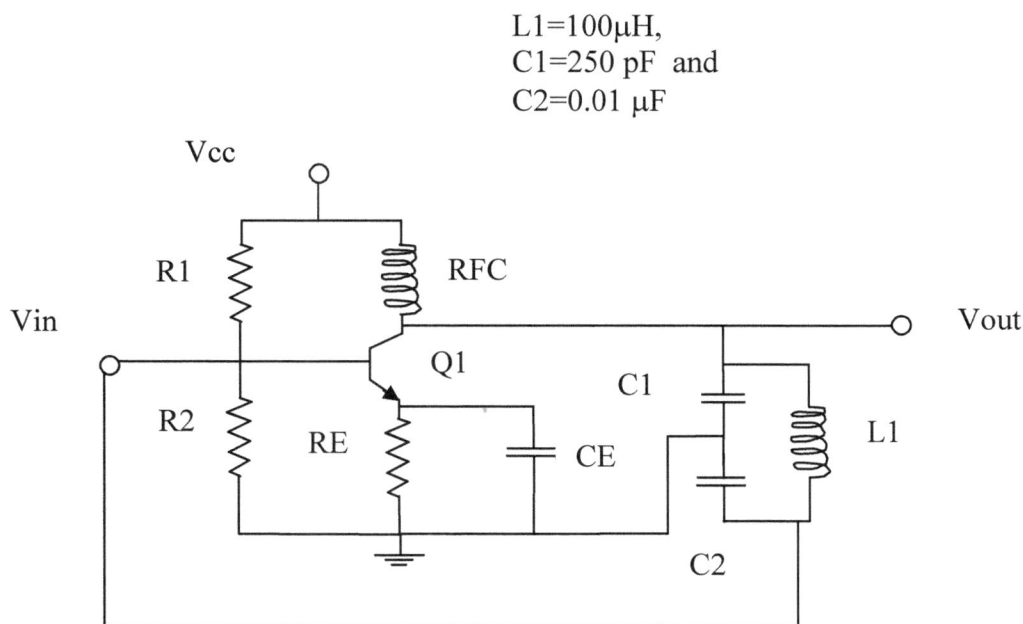

Figure 6.77.

Use Figure 6.77 to answer questions 31, 32, and 33 below:

32. Determine Ceq of the feedback circuit.

33. Determine the gain of the feedback circuit.

34. Determine the oscillating frequency of the oscillator circuit.

35. Write the sinusoidal equation for each of the vectors, A and B, in Figure 6.78 in terms of the positive sine function.

36. Write the sinusoidal equation for each of the vectors, A and B, in Figure 6.78 in terms of the negative sine function.

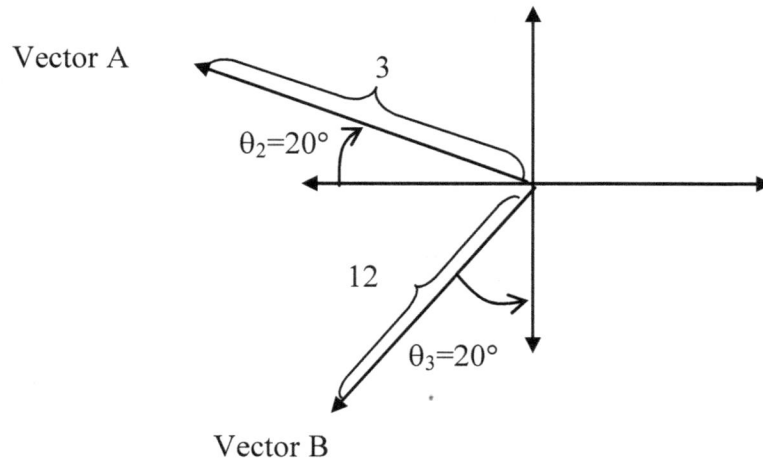

Figure 6.78.

37. The frequency of an electrical signal is 3000Hz. Determine its wavelength if it is radiated through free space (velocity factor k=1) from an antenna.

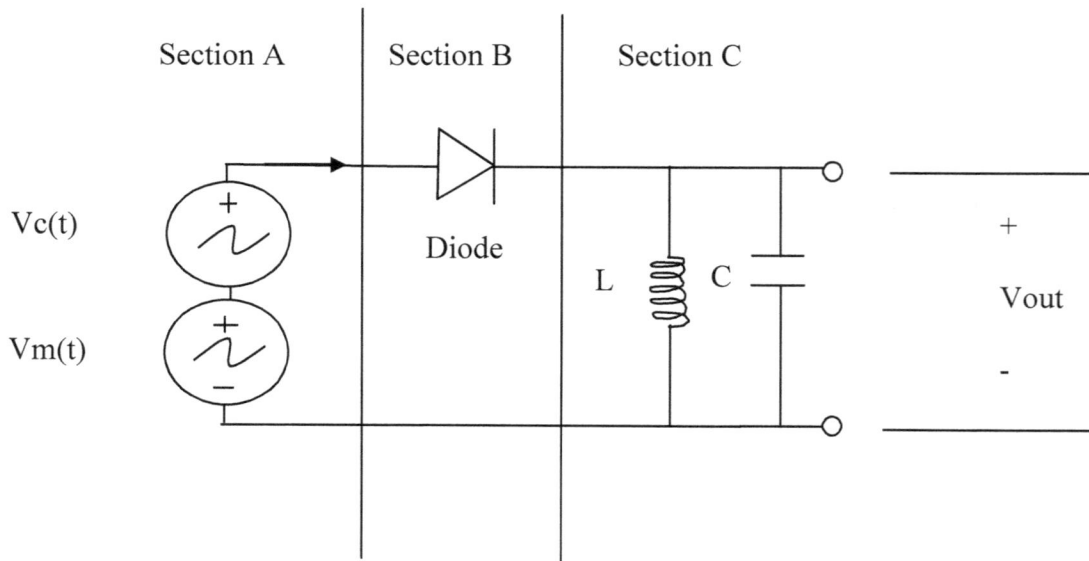

Figure 6.79. Vm(t) is the modulating signal and vc(t) is the carrier signal.

38. Completely describe the effect of each section, A, B, and C, of the circuit in Figure 6.79 on the input signal.

7. Basic VHDL

VHDL (VHSIC [Very High Speed Integrated Circuit] hardware description language) is a hardware description language used in electronic design automation to describe digital and mixed-signal systems. This Chapter is intended to serve as a basic course to get one started in VHDL for the robotic projects in this book. It is not intended for the advanced user. However, it can be used as a VHDL refresher for the advanced user. For more details on the VHDL language see Bhasker, 1995.

VHDL hardware systems include field-programmable gate arrays and integrated circuits, etc. VHDL was originally developed at the request of the U.S Department of Defense in order to document the behavior of the ASICs that supplier companies were including in equipment. During the original VHDL development, the Department of Defense required much of the syntax as possible to be based on Ada, to avoid re-inventing concepts that had already been thoroughly tested in the development of Ada. So, much of the VHDL language is similar to the Ada programming language in both concepts and syntax. The initial version of VHDL was based on the IEEE standard 1076-1987 from the United States Airforce, etc (IEEE, 1987). It included a wide range of data types, including numerical (integer and real), logical (bit and Boolean), character and time, plus arrays of bit called bit_vector and of character called string. Since 1987 it has undergone several revisions:

1. 1076-1993 had significant improvements due to several years of feedback. Thus, it has become the greatest vendor supported and most widely used version.
2. 1076-2000 introduces the use of protected types. This is a minor revision.
3. 1076-2002 introduces rules with regard to buffer ports that are relaxed. This is a minor revision of 1076-2000.
4. 1076-2008 consisted of a major revision released on January 26, 2009. Inclusive of other changes, this standard introduces the use of external signals.

VHDL, when used for system design, can be used to model (describe behavior) and simulate (verify) a given system before synthesis tools translate the design into real hardware (gates and wires). Another advantage of VHDL is that it allows a concurrent system to be modeled. Unlike other languages, VHDL is a dataflow language (a hardware decription language that can understood by the hardware which is concurrent), whereas computing languages such as BASIC, C, and assembly code, all run sequentially, one instruction at a time.

Several integrated design environment tools are available to perform VHDL designs. Examples for FPGA implementations are: Altera Quartus, Xilinx ISE, Synopsys Synplify or Mentor Graphics HDL Designer. These can be used to draw the RTL schematic of the desired circuit. Simulation software can then be used to verify the generated schematic which shows the waveforms of inputs and outputs of the hardware using the appropriate test bench.

The composite picture of VHDL is illustrated in Figure 7.1. A complete VHDL component description requires a VHDL entity and one or more VHDL architectures. The entity defines a component's interface, input and output signals, to its external environment such as other entities and the top level chip I/O. The architecture defines a component's functionality. An entity can only use one architecture of 1 to N at a time. That is, there can be many different architectures designated for an entity, however for each instantiation of the entity one must be selected. The architectures can have different VHDL components as shown. A configuration specifies the architecture that is to be used to simulate a design entity.

For Altera Quartus, a VHDL file must have the same name as the entity, if not the VHDL compiler will generate an error and thus not compile the code.

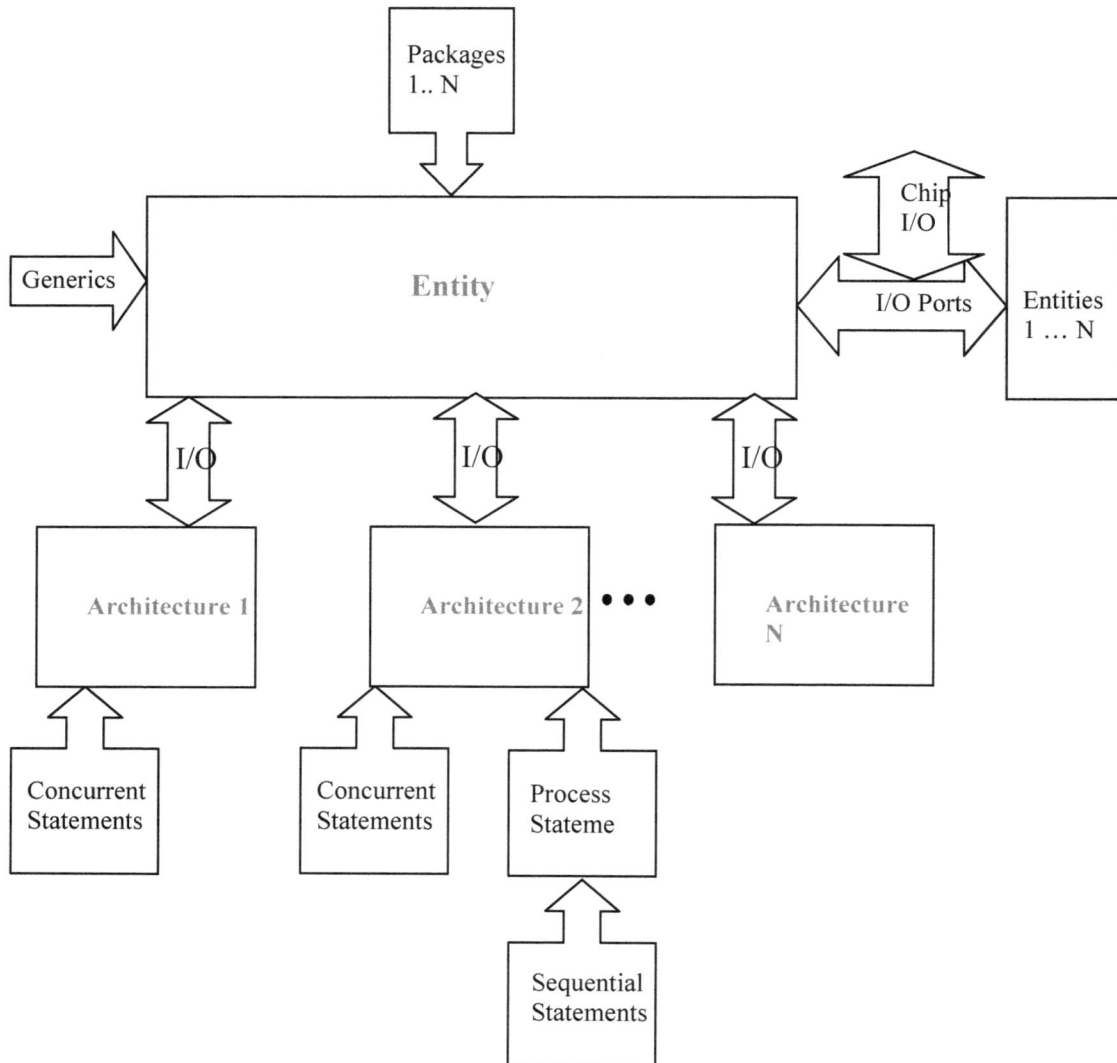

Figure 7.1. VHDL Composites.

Figure 7.2 shows an entity with its instantiated architecture. As indicated, process statements are concurrent.

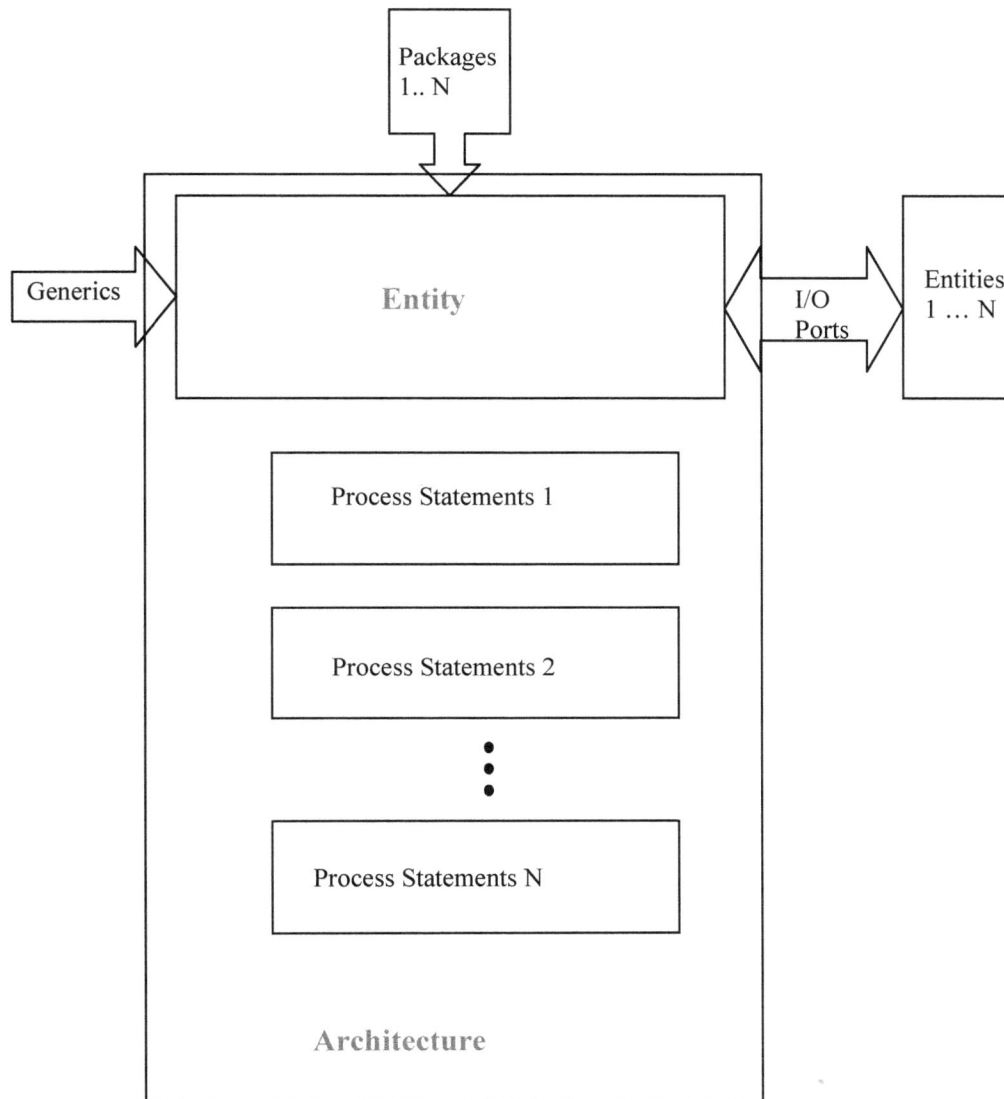

Figure 7.2. Relationship of VHDL entity, architecture, and processes.

Basic VHDL language elements consist of the following:
1.) Identifiers
2.) Delimiters, Literals, Operators and Precedence
3.) Types and Subtypes
4.) Scalar type - integer, enumerate (IEEE Std. 1164), physical, real
5.) Composite type - arrays, records
6.) Access type
7.) File
8.) Attributes

In summary, a VHDL description consists of the following main items:
1.) Libraries
2.) Packages
3.) Generics
4.) Interfaces
5.) Entities
6.) Architectures
7.) Configuration
8.) Processes
9.) Concurrent statements
10.)Sequential statements

This chapter will discuss the rules and syntax used for the various VHDL statements listed above using Backus–Naur Form (BNF). A BNF specification is a set of rules, using the following notation:

 <symbol> ::= __expression__

where <symbol> is a nonterminal, and the __expression__ consists of one or more sequences of symbols separated by the vertical bar, '|', indicating a choice. Symbols that never appear on a left side are terminals. Symbols that appear on a left side are non-terminals and are always enclosed between the pair <>. For example, the following represents an integer expressed relative to the Backus–Naur form:

<integer> ::= ['-'] <digit> {<digit>}
<digit> ::= '0' | '1' | '2' | '3' | '4' | '5' | '6' | '7' | '8' | '9'

In this example, the symbols (-,0,1,2,3,4,5,6,7,8,9) are terminal symbols and optionally signed <digit> and <integer> are nonterminal symbols.

7.1. Basic VHDL Language Elements
In this section the basic VHDL language elements shall be discussed. These are:
1.) Identifiers
2.) Delimiters, Literals, Operators and Precedence
3.) Types and Subtypes

4.) Scalar type - integer, enumerate (IEEE Std. 1164), physical, real
5.) Composite type - arrays, records
6.) Access type
7.) File
8.) Attributes

7.1.1. Identifiers

In VHDL there are two kinds of VHDL identifiers:
1.) Basic Identifier
2.) Extended Identifier

A basic identifier consists of a sequence of one or more characters. Legal characters are:
1.) A, B, C, …..X, Y, Z(upper case letters)
2.) a, b, c, ……x, y, z. (lower case letters)
3.) 0, 1, 2, ……8, 9 (numbers)
4.) _ (underscore)

Lower and upper case characters are considered to be identical, e.g., heaven, HEAVEN, Heaven, etc., are all equivalent. Two underscores cannot appear consecutively. VHDL keywords cannot be used as basic identifiers in VHDL. Examples of VHDL keywords are: process, architecture, entity, etc.

An extended identifier is a sequence of characters between two backslashes. Characters between back slashes are considered unique, so the same upper case and lower case, e.g., a, A, letters are not equivalent. Examples are: /earth/ and /Earth/ are unique.

7.1.2. Delimiter

A delimiter is a punctuation character or group of characters that indicates the separation of two names or two pieces of data. It also may be used to mark the beginning or end of a programming construct. Some delimiters are given in Table 7.1.

There are a few rules that apply to delimiters. No spacing is allowed before or after the following delimiters: 1) . 2) ' 3) #. Bit literals are bracketed with double quotes, e.g., "1001", with no spaces inside the quotes. Also, avoid using spaces before the following delimiters: 1) ; 2) (3)) 4) :.

Delimiter	Example(s)	Description
+	sum <= +10 + 2;	addition or positive
-	sum := -7 – 4;	subtraction or negative
/	div := 12/5; -- = 1	division
=	if sum = 12 then	equality
<	if count < max then	less than
>	if count > min then	greater than
&	y := '0' & "11"; -- = "011"	concatenator
|	when 'A' | 'a' =>	vertical bar
;	end if;	terminator
#	sumall := 16#3B7A#	enclosing based literals
(	type status is (ON, OFF);	left parenthesis
)	sum := 12 * (7 + y);	right parenthesis
:	signal tx_status: status;	separates data object from type
.	tx_status := state.idle;	dot notation
"	report "Hello There"	double quote
'	y := status'tx_status & ON;	single quote or tick
**	y := 3 ** 2; -- = 8	exponentiation
=>	when ON => sum := 7;	arrow, read as "then"
=>	eight_bit_register := (bit0 => 0, others => 1);	arrow, read as "gets"
:=	y:= sum + 12;	variable assignment
<=	count <= 3 after 12 ns;	less than or equal
<=	if count <= 9 then	signal assignment
>=	if count >= 12 then	greater than or equal to
/=	if count /= 15 then	inequality, not equal
<>	type mem_typ is array (integer range <>) of B32_typ;	box
--	-- this is a comment	comment

Table 7.1. Basic VHDL delimiters.

7.1.3. Literals

In a computer programming language, a literal is a value written exactly as it's meant to be understood. A name can be thought of as an identifier. Therefore, a variable is a name that can represent different values during the execution of the program. A constant is a name that represents the same value during the operation of a program. On the other hand, a literal is not a name, it is the value itself. Consider the VHDL expression y:= 12; y is the variable, and 12 is the literal. There are different VHDL literal types, these are:

1.) Literal numbers
2.) Literal characters
3.) Literal strings
4.) Literal bit strings

7.1.3.1.Literal Numbers

VHDL Literal numbers are represented using decimal (0,1,2,...9) or base (e.g., Octal use base 8, hexadecimal use base 16, binary use base 2) numbers. When the literal number includes a point it is a real number, otherwise it is an integer. In the foregoing paragraphs or the Sections, the notation ::= shall mean is defined by the given rules. Decimal literals follow the rules defined by:

$$\text{decimal\_literal} ::= \text{integer} \; [\; . \; \text{integer} \;] \; [\; \text{exponent} \;]$$

where:

$$\text{integer} ::= \text{digit} \; \{ \; [\; \text{underline} \;] \; \text{digit} \; \}$$

and where:

$$\text{exponent} ::= E \; [\; + \;] \; \text{integer} \; | \; E - \text{integer}$$

Some examples of number literals are given in Table 7.2.

Literal Type	Example(s)
integer	0
integer	1
integer	712E+2 or 712E2
integer	987_654_321
real	0.0
real	0.7
real	3.737E-3
real	7.5_103

Table 7.2. Examples of VHDL number literals.

Based literals follow the rules given by:

$$\text{based\_literal} ::= \text{base} \; \# \; \text{based\_integer} \; [\; . \; \text{based\_integer} \;] \; \# \; [\; \text{exponent} \;]$$

where:

$$\text{base} ::= \text{integer}$$

and where:

$$\text{based\_integer} ::= \text{extended\_digit} \; \{ \; [\; \text{underline} \;] \; \text{extended\_digit} \; \}$$

Some examples of base literals are given in Table 7.3. The base and the exponent are expressed in decimal.

Literal Type	Example(s)
base 2 (integer)	2#0011_1000#
base 2 (real)	2#1.1001_0001#E+2
base 8 (integer)	8#075#E1
base 8 (real)	8#3.50#E-3
base 16 (integer)	16#C3#
base 16 (real)	16#F.FA#E2
base 16	16#A1#E+2

Table 7.3. Examples of VHDL base literals.

7.1.3.2. Literal Characters

VHDL character literals are formed by enclosing ASCI characters in single quotes. Examples are given in Table 7.4.

Literal Type	Example(s)
character	'F'
character	'*'
character	'''
character (empty space)	' '
character	'C'
character	'+'
character	'0'

Table 7.4. Examples of VHDL character literals.

7.1.3.3. Literal Bit Strings

VHDL bit string literals are defined by the following rules:

$$bit\_string\_literal ::= base\_specifier\ "\ bit\_value\ "$$

where:

$$base\_specifier ::= B\ |\ O\ |\ X$$

and where:

$$bit\_value ::= extended\_digit\ \{\ [\ underline\ ]\ extended\_digit\ \}$$

For the base_specifier, the letter B stands for binary, O stands for Octal, and H stands for hexadecimal. Examples of bit strings are given in Table 7.5. To include a string within a string replace the inner string double quotes with a pair of double quotes.

Literal Type	Example(s)	Comments
bit string	B"1110010"	Length 7
bit string	O"3771"	Length 12, equivalent to B"_011_111_111_001"
bit string	X"1"	Length 4, B"0001"
bit string	X"37"	Length 8, B"0011_0111"

Table 7.5. Examples of VHDL bit string literals.

7.1.3.4. Literal Strings

VHDL literal strings are formed by enclosing a group of ASCI characters in double quotes. Examples of string literals are given in Table 7.6. To include a string within a string replace the inner string with a pair of double quotes.

Literal Type	Example(s)
string	"This is a string"
string	"This is a string "" in a string"".""
string	"Bit"
string (empty string)	""
string	"Status"
string	"Off"
string	"123"

Table 7.6. Examples of VHDL string literals.

7.1.4. VHDL Operators and Precedence

There are several predefined VHDL operators:

1. Logical operators
2. Relational operators
3. Shift operators
4. Adding operators
5. Multiplying operators
6. Exponentiation operator
7. Absolute value operator

These, along with examples, are given in Table 7.7. The VHDL operator precedence is given in Table 7.8. There are three basic forms of shift operators: Shift Logical Left (SLL), Shift Logical Right (SLR), and Shift Arithmetic Right (SAR). The logical and relational operators are similar to those in other languages. The concatenation operator &

joins two vectors together. Both vectors must be of the same type. For integer numbers and the exponentiation operator ** from the package STD, must use an exponent that is also an integer; no real exponents are allowed. Negative exponents are allowed only with real numbers.

Operator	Example(s)	Operator Description
and, or, xor, not, nand, nor, xnor	Y=A and B;	Logical
=, /=, <, >, <=, >=	Y<=A or B;	Relational
sll, srl, sla, sra, rol, ror	Y<="110010" sll 2;	Shift
+, -, &	Y <= A + B; Y <= '1' & '0'; -- "10"	Adding
*, /, mod, rem	Y <= 4 mod 3; -- 1	Multiplying
**	Y := 2.0 ** 3; -- 8	Exponentiation
abs	Y := abs A;	Absolute Value

Table 7.7. Basic VHDL operators.

Operator	Precedence
and, or, xor, nand, nor, xnor	Lowest
=, /=, <, >, <=, >=	
sll, srl, sla, sra, rol, ror	
+, -, &	
+, - (sign)	
*, /, mod, rem	
**, abs, not	Highest

Table 7.8. VHDL operator precedence.

7.1.5. Data Objects

VHDL data objects can be classified into one of the following categories:
1.) Constant
2.) Variable
3.) Signal
4.) File

Example declarations of the various data types are given in Table 7.9. These will be explained in more details in the following sections.

Type	Example(s)	Range
integer	signal count: integer:= 0;	$-(2^{31}-1)$ to $+(2^{31}-1)$
real	variable mult: real:= 1.0;	Implementation defined
Boolean	variable isdone: boolean:=TRUE;	TRUE, FALSE
bit	signal input_A: bit:= '0';	1. 1
bit_vector	variable register_A: bit_vector(15 downto 0)	Array with each element of type bit
character	variable display: character:= '@';	Exponentiation

Table 7.9. Basic VHDL data types.

More examples of the various data objects are given below:

```
constant FALL_TIME: TIME :=20 ns;
variable MULT: INTEGER range 0 to 50 := 12;
variable FLAG: BOOLEAN;
signal CLK: BIT;
signal ADD_BUS: BIT_VECTORS (0 to 15);

type BIT_FILE is file of BIT_VECTOR;
file VECTORS: BIT_FILE is
"/usr/home/troy/mult.vec";
```

7.1.5.1. Data Objects - Constants

A VHDL constant is given a value before the start of the simulation. It cannot be changed and maintains it value throughout the simulation. VHDL variables are similar to variables in other high level languages such as C, C++, Fortran, etc. VHDL signals are similar to wires in a circuit. A VHDL file contains a sequence of values and can be read from or written to.

A constant is an object which is initialized to a specified value when it is created. This value may not be modified during program execution. A constant declaration is defined by:

```
constant_declaration ::=
constant identifier_list : subtype_indication [ :=
expression ] ;
```

Some constants are predefined in the packages. The initial value must be given in the corresponding package body. Some examples are:

```
constant e : real := 2.71828;
constant delay : Time := 10 ns;
constant min_size : natural;
constant pi: real := 3.14159;
```

7.1.5.2. Data Objects - Variables
A VHDL variable is an object whose value may be changed after it is created. A variable is defined by the following rules:

```
variable_declaration ::=
variable identifier_list : subtype_indication [ :=
expression ];
```

If required, the initial value expression can be evaluated and assigned to the variable when it is created. If the initialization expression is missing, a default value is assigned when the variable is created. For scalar types, the default value is the left most value for the type, that is, the first in the list of an enumeration type, the lowest in an ascending range, or the highest in a descending range. If the variable is a composite type, the default value is the composition of the default values for each element, based on the element types. Some examples of variable declarations are:

```
variable count : integer := 0;
variable state : state_array;
```

7.1.5.3. Data Objects - Signals
Signals are analogous to wires in a circuit while variables are similar to variables in other programming languages such as C, C++, Fortran, etc. Signals are declared using the following rules:

```
signal_declaration ::=
signal signal_identifier : subtype_indication [ signal_kind
] [ := expression ] ;
signal_kind ::= register | bus
```

As indicated the signal_kind specification is optional. The absence of the signal_kind just results in a signal of the subtype_indication. In the declaration the expression is used to give the signal an initial value which is assigned at the start of the simulation. The absence of the expression results in a default initial assigned at the start of the simulation.

7.1.5.4.Data Objects - Files

Files provide a mechanism for VHDL to communicate with the host environment. Files are declared as follows:

```
type file_name is file of type_data;
```

The type_data is the type of data in the file. Examples are:

```
type vectors is file of BIT_VECTOR;
type input_names is file of STRING;
```

7.1.6. Data Types

VHDL data objects hold values that belong to a set of values. This is determined by using the type declaration. A type in VHDL has associated with it a set of values and operations. For example, BOOLEAN is a type that has the set of values TRUE or FALSE. Its operations are and, or, not, nor, nand, etc. Predefined types are given in the package STANDARD. All data types fall into one of the following categories:
1. Scalar Types
2. Composite types
3. Access types
4. Files types

7.1.7. Data Types – Scalar Types

The values of scalar types are ordered, for example, BIT is a scalar type with values '0' and '1', so the expression '1' > '0' is TRUE. Scalar types can be categorized into the following:
1. Enumeration
2. Integer
3. Physical
4. Floating point

7.1.7.1.Scalar Data Types – Enumeration Types

An enumeration type is said to be a discrete type since these types have discrete values. Enumeration types are user defined values of character literals and identifiers. Some examples are:

```
type OP is (DIV, MULT, ADD, SUB);
type MVL is ('U','0','1','Z');

subtype DSP_OP is OP range MULT to ADD;
```

MVL is represented by the set of ordered values U','0','1', and 'Z'. These are referred to as enumeration literals. DSP_OP is a subtype of the base type OP, it has a range constrained to include only MULT and ADD. The enumeration literals of OP are DIV, MULT, ADD, and SUB. The order of the enumeration literals determines their values. These values are used by relational operators to determine value of an expression. For example, DIV < MULT is FASLE while DIV > MULT is TRUE. The positional value of MULT is greater than the positional value of DIV. The value of the left most enumeration literal is 0. So, the positional value of DIV is 0, MULT is 1, ADD is 2, and SUB is 3.

There are predefined enumeration types. Some of these are defined as follows:

```
type character is (
NUL, SOH, STX, ETX, EOT, ENQ, ACK, BEL,
BS,  HT,  LF,  VT,  FF,  CR,  SO,  SI,
DLE, DC1, DC2, DC3, DC4, NAK, SYN, ETB,
CAN, EM,  SUB, ESC, FSP, GSP, RSP, USP,
' ', '!', '"', '#', '$', '%', '&', ''',
'(', ')', '*', '+', ',', '-', '.', '/',
'0', '1', '2', '3', '4', '5', '6', '7',
'8', '9', ':', ';', '<', '=', '>', '?',
'@','A', 'B', 'C', 'D', 'E','F', 'G',
'H', 'I', 'J', 'K', 'L', 'M', 'N', 'O',
'P', 'Q', 'R', 'S', 'T', 'U', 'V', 'W',
'X', 'Y', 'Z', '[', '\', ']', '^', '_',
'`', 'a', 'b','c', 'd', 'e', 'f', 'g',
'h', 'i', 'j', 'k', 'l', 'm', 'n', 'o',
'p', 'q', 'r', 's', 't', 'u', 'v', 'w',
'x', 'y', 'z', '{', '|', '}', '~', DEL);


type boolean is (false, true);
type bit is ('0', '1');
type severity_level is (note, warning, error, failure);
```

7.1.7.2. Scalar Data Types – Integer Types
A VHDL integer type is a type that falls within a specific integer range. Examples are:

```
--type declarations
type index is range 0 to 32;
type WORD_LENGTH is range 16 downto 0;
subtype MEMORY_WORD is WORD_LENGTH range 7
downto 0;

--object declarations
constant EEPROM_ADDRESS : INDEX :=3;
signal EEPROM_WORD: MEMORY_WORD;
```

Integer literals are values that belong to integer types. Examples are:
2E3   7   10_33_00

2E3 is the same as $2 * 10^3 = 2 * 1000 = 2000$. The 10_33_00 is the same as 103300. The underscore "_" in an integer value has no impact on the integer value.

7.1.7.3.Scalar Data Types – Floating Point Types

A VHDL floating point type is a type that falls within the real number range. Examples are:

```
--type declarations
type CIRCUIT1_VOLTAGE is range -10.5 to 3.5;
type MEASUREMENT is range 0.0 to 15.2;

--object declarations
variable CURRENT: MEASUREMENT is range 0.0 to 3.5;
variable VOLTAGE: CIRCUIT1_VOLTAGE is range 0.0 to 2.5;
```

Floating point literals are values that belong to floating point types. Examples are:

0.0   21.33   1_0.7_5   2.5E+3

Here, 0.0 is a floating point literal. 0 without the dot, ".", is integer literal. The dot, ".", distinguishes floating point literals from integer literals. 2.5E+3 is the same as $2.5 * 10^3$ $= 2.5 * 1000 = 2500.0$.

7.1.7.4.Scalar Data Types – Physical Types

An example of a VHDL physical type is a type that represents a measurement of a physical quantity such as voltage, current, time, length. These types are expressed as integer multiples of a base unit. Some examples are:

```
type length is range 0 to 1E9
units
um;    -- base unit
mm = 1000 um;
cm = 10 mm;
m = 1000 mm;
in = 25.4 mm;
ft = 12 in;
yd = 3 ft;
rod = 198 in;
chain = 22 yd;
furlong = 10 chain;
end units;
```

```
type resistance is range 0 to 1E8
units
ohms;   --base unit
kohms = 1000 ohms;
Mohms = 1E6 ohms;
```

The predefined physical type time is has the following declaration:

```
type time is range implementation_defined
units
fs;
ps = 1000 fs;
ns = 1000 ps;
us = 1000 ns;
ms = 1000 us;
sec = 1000 ms;
min = 60 sec;
hr = 60 min;
end units;
```

In the above examples, the base unit for length is um, the base unit for resistance is ohms, and the base unit for time is fs.

7.1.7.5.Scalar Data Types – Composite Types

VHDL composite types are represented by a collection of values such as array types and record types. Arrays have values of one type while records have values of multiple types.

Arrays can be one dimensional or two dimensional. Array types can be constrained or unconstrained. For constrained types the index of the array is defined when it is declared. For unconstrained types the index is defined later, after the array has been declared.

Examples of constrained array type declarations are:

```
type word is array (63 downto 0) of bit;
type ROM is array (0 to 31) of word;
type register_bank is array (byte range 0 to 132) of integer;
type decode_matrix is array (positive range 7 downto 0,
positive range 3 downto 0) of bit;
```

Examples of object declarations using these types are:

```
variable ROM_DATA: ROM;
signal ADDRESS_BUS: word;
variable DECODER_VALUE: decode_matrix;
```

An example of an unconstrained array type declaration is as follows:

```
type data_vector is array (integer range <>) of integer;
```

The VHDL symbol '<>', called a box, is referred to as a place-holder for the index range, which will be filled in later when the array type is used. For example, an object is declared to be a data_vector of 64 elements by using its type as follows:

data_vector(1 to 64)

Record types are composed of the same or different types. An example of a record type is as follows:

```
type WORD1 is range 31 downto 0;
type WORD2 is range 15 downto 0;
type MEMORY_DATA   is
record
size: integer range 0 to 1024;
memory1: WORD1;
memory2:WORD2;
prop_delay: TIME;
end record;
```

7.1.8. Access Types

A VHDL access type have values which are pointers to a dynamically allocated object of some type. Objects of access types can only belong to the variable class. These are similar to pointers in other languages such as C, etc. Examples of access types are:

```
type PTR is access MEMORY_DATA;

process
begin
......
variable MEMPTR1, MEMPTR2:PTR;   --default value is null
......
end process;
```

Here, PTR is an access type with values that point to objects of type MEMORY_DATA. When an access type has a **null**, it means that it does not point to an object. An access type object can be newly created and initial values assigned as follows:

```
MODPTR1:= new MEMORY_DATA(7, 3, 12, 12 ns);
```

7.1.9. File Types

Data can be saved to the host environment via file types. The syntax for the file type is as follows:

```
type file-type-name is file of type-name;
```

Some examples are as follows:

```
type MEM_DATA is file of WORD;
type LOG_DATA is file of string;
type ERROR_DATA is file of string;
```

Files can be opened, closed, read from, written to, and tested for an end of file condition. These statements can be included in procedures or functions (Bhasker, 1995).

7.1.10. Data Subtypes

AVHDL subtype is a subset of the type. It is a subset of type values. The type is called the base type of the subtype. Subtype declarations are used to declare subtypes. Examples of types and subtypes are:

```
type HEX is ('0','1','2','3','4','5','6','7','8','9','A','B',
'C','D','E','F');
subtype NUM is HEX range '0' to '9';
type TEEN is INTEGER range 13 to 19;
```

7.1.11. Attributes

In VHDL an attribute is associated with an entity name, architecture, signal, or label. A VHDL attribute is a value, function, type, range, signal, or constant. The attributes are categorized as follows:

1. User Defined
2. Predefined

User defined attributes are constants of any types. Attribute declarations have the following syntax:

attribute attribute-name : value-type;
attribute attribute-name **of** item-names: name-class **is** expression;

More details on attributes can be sought in Bhasker, 1995.

7.2. **VHDL Descriptions**

In summary, a VHDL description consists of the following main items:
1. Libraries
2. Packages
3. Configuration
4. Generics
5. Interfaces
6. Entities
7. Architectures
8. Processes
9. Concurrent statements
10. Sequential statements

These will be covered in the following Sections. For more details on any of the VHDL descriptions mentioned above please see Bhasker, 1995.

7.2.1. **Design Libraries**

A VHDL description is first written in a design file, then a compiler is invoked to analyze and insert them into a design library. A design library consists of the following:
1. Primary library units
2. Secondary library units

Primary library units are:
1. entity declarations,
2. package declarations and
3. configuration declarations.

The secondary library units are:
1. architecture bodies and
2. package bodies.

These library units depend on the specification of their interface in an associated primary library unit, therefore, the primary unit must be compiled before any corresponding secondary unit. Secondary library units must also reside in the same design library as primary units. For example, an entity declaration and all its architectural bodies must reside in the same library. Design units that reference other primary units can only be compiled after that primary unit has been compiled.

An architecture body, or architecture_name implicitly inherits all the declarations in the entity_name since it is called as follows:

architecture architecture_name **of** entity_name **is** ...

Similarly, VHDL package bodies also inherit all items declared in the package declaration as follows:

package body package_name **is** .....

VHDL design units can be explicitly called using the following two clauses:
1. **library** clause
2. **use** clause

The VHDL library clause has the following format:
library logical_name_list;

An example is:

library TTL, CMOS;

which makes the logical names TTL and CMOS visible in the design unit that follows. The following **use** clause in implicitly declared in every design file:

library STD, WORK;

After the **library** clause, the **use** clause must then be used to make the logical name present in the library visible to the following design unit. This is done using the **use** clause. The **use** clause has two forms:

1. **use** library_name.primary_unit_name;
2. **use** library_name.primary_unit_name.item;

The keyword **all** is used to specify all items in a design unit can be used. An example using the first use form is:

```
library TTL;
use TTL.AND3;
configuration ….. is
      …. use entity AND3(….);
end;
```

The second form of the **use** clause makes the item declared in the primary visible. This allows the item to be referenced from within the design unit as follows:

```
library ATTLIB;
use ATTLIB.SYNTH_PACK.MVL;
entity AND3 is
port (A, B, C: in MVL; Z: out MVL)
.
.
.
end;
```

To specify that all items in the design unit be used, the following is used:

```
library ATTLIB;
use ATTLIB.SYNTH_PACK.ALL;
entity AND3 is
port (A, B, C: in MVL; Z: out MVL)
.
.
.
end;
```

7.2.2. The Package

A VHDL package is the means to make available declarations across many design units. A package consists of the following

1. package declaration and
2. package body (optional)

The package declaration takes the following form:

```
package package_name is
--subprogram declarations
--type declarations
--subtype declarations
--constant declarations
--signal declarations
--variable declarations
--file declarations
--alias declarations
--component declarations
--attribute declarations
--disconnection specifications
--use clauses
end [package ] [package_name];
```

An example of a package declaration is given by:

```
package SYNTH_PACK is
    type MVL is ('U','0','1','Z');
    type MVL_VECTOR is array (NATURAL range <>) of MVL;
    constant HIGH2LOW: TIME := 10ns;
    type ALU_OP is (ADD, SUB, MUL, DIV);
    component AND3
        port(A,B,C:in MVL; Z: out MVL);
    end component;
end SYNTH_PACK;
```

A package body is not necessary if its package declaration does not have any subprogram or deferred constant declarations. However, a package body has the following syntax:

```
package body package_name is
--subprogram bodies
--complete constant declarations
--subprogram declarations
--type and subtype declarations
--file and alias declarations
--use clauses
end [package body ]
[package_name];
```

The package body name must have the same name as the package declaration. An example of a package declaration and its body is:

```
use WORK.SYNTH_PACK.all;
package ROBOT_TOP_PACK is
    constant PROP_DELAY: TIME;      -- deferred constant
    procedure LOAD (signal ARRAY_NAME: inout MVL_VECTOR;
        START_BIT,STOP_BIT, INT_VALUE: in INTEGER);
end package ROBOT_TOP_PACK;

package body ROBOT_TOP_PACK is
    use WORK.TABLES.all;
    constant PROP_TIME:TIME:=25ns;
    procedure LOAD(signal ARRAY_NAME: inout MVL_VECTOR;
        START_BIT,STOP_BIT, INT_VALUE: in INTEGER) is
        --local declarations
    begin
        -procedure behavior
    end LOAD;
end ROBOT_TOP_PACK;
```

7.2.3. The Configuration

VHDL configurations are used for several reasons:

1.   It may be necessary to define multiple views for a single entity. This can be done by specifying one architecture for each view. The configuration can be used to bind the architecture body. For example, there are three different architecture bodies: FA_BEH, FA_STR, FA_MIX, for entity FULL_ADDER. Any one of these can be selected by specifying a configuration.
2.   It may be necessary to define multiple entities for a single component.

So, in summary, a configuration is used to bind the following:

1.   An architecture body to an entity declaration.
2.   A component to an entity.

The configuration binding can be accomplished using two methods:

1.   Configuration specification.
2.   Configuration declaration.

7.2.3.1. Configuration Specification

VHDL configuration specifications are used to bind component instantiations to entities in a design library. These configuration specifications are included in the declaration section, along with the component declarations, of the architecture body. The configuration specification has the following syntax:

for list_of_component_labels: component_name binding_indication;
The binding_indication has several forms. One of the forms has the following syntax:

```
use entity entity_name [(architecture_name)]
[generic map (generic_association_list)]
[port map (port_association_list)]    -- form 1
```

An example of the specification binding is shown next using the structural VHDL description of the FULL_ADDER in Figure 7.3:

```vhdl
library HS_LIB, CMOS_LIB;
entity FULL_ADDER is
    port (A,B, CIN: in BIT; SUM, COUT: out BIT);
end;

architecture FA_STR of FULL_ADDER is
--Component declarations
component XOR2
    port (D1, D2: in BIT; DZ: out BIT);
end component;

component AND2
    port (Z: out BIT; A0, A1: in BIT);
end component;

component OR2
    port (N1, N2: in BIT; Z: out BIT);
end component;


--Configuration specifications
--Example of binding the entity with more than
--one instantiation of a component
for X1, X2: XOR2
    use entity WORK.XOR2(XOR2BEH)
--Example of binding the entity with a single
--instantiation of the component
for A3: AND2
    use entity HS_LIB.AND2HS(AND2STR)
    port map (HS_B=>A1, HS_Z=>Z, HS_A=>A0);
```

```
      --Example of binding the entity with all
      --instantiations of OR2 component
      for all: OR2
            use entity CMOS_LIB.OR2CMOS(OR2STR)
      --Example of binding the entity with all
      --unbound instantiations of AND2
      component
      for others: AND2
            use entity WORK.A_GATE(A_GATE_BODY)
            port map (A0, A1, Z);

      signal Y1, Y2, Y3, Y4, Y5: BIT;

      begin
            X1: XOR2 port map (A, B, Y1);
            X2: XOR2 port map (Y1, CIN, SUM);
            A1: AND2 port map (Y2, A, B);
            A2: AND2 port map (Y3, B, CIN);
            A3: AND2 port map (Y4, A, CIN);
            O1: OR2 port map (Y2, Y3, Y5);
            O2: OR2 port map (Y5, Y4, COUT);
      end FA_STR;
```

There are four configuration specifications shown in the architecture body of FA_STR. The first instantiation shows how an entity can be bound to more than one component. The instantiations X1 and X2 of component XOR2 are bound to the entity XOR2 (in library WORK) which is paired with architecture XOR2BEH. The second instantiation shows how an entity can be bound to one instantiation of the component. The instantiation A3 of component AND2 are bound to the entity AND2HS (in library HS_LIB) which is paired with architecture AND2STR. An association is used to map the port of entity AND2HS to the port of component AND2. For example, port HS_A of AND2HS is mapped to port A0 of the AND2 component, and so on. Please see Section 7.4.3 on page 302 for more on component instantiation and port associations. The third instantiation gives an example of binding the entity with all instantiations of OR2 component. The fourth instantiation gives an example of binding the entity with all unbound instantiations of the AND2 component.

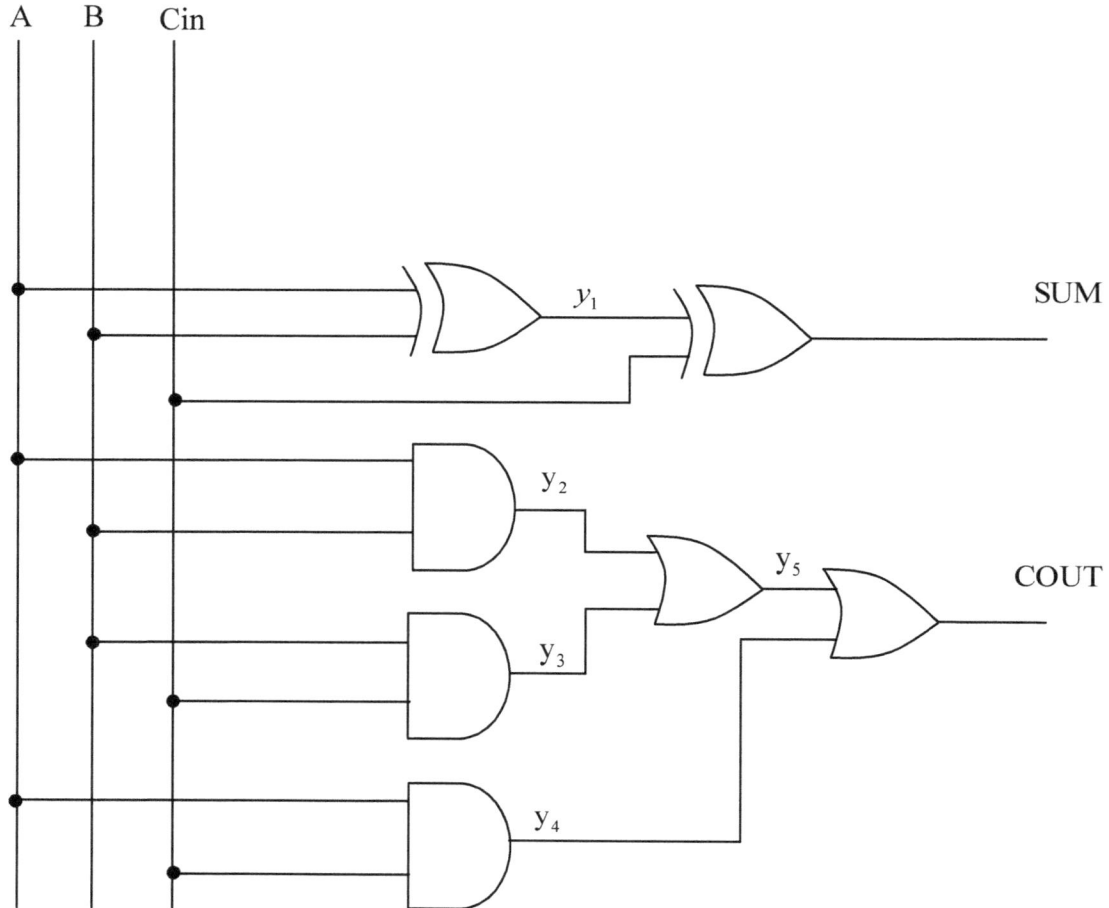

Figure 7.3. 1 bit full adder logic circuit.

7.2.3.2.Configuration Declaration

Sometimes it is not desired to change the architecture to implement a configuration specification and recompile it. An alternative is to use a VHDL configuration declaration. This is done using a separate design unit. Thus, configurations can be changed later without changing the architecture body. The configuration declaration can have the following syntax:

```
configuration configuration_name of entity_name is
    block_configuration
end [configuration][configuration_name];
```

More details on configuration declarations can be sought in Bhasker (1995) in the References on page 578.

7.2.4. Entities

The term entity refers to the VHDL construct in which a component's interface (which is visible to other components) is described. The first line in an entity declaration provides the name of the entity. The PORT statement indicates the actual interface of the entity: the signals in the component's interface, the direction of data flow for each signal listed, and type of each signal. Notice that if signals are of the same mode and type, they may be listed on the same line. No semicolon is required before the closing parenthesis in the PORT declaration (or GENERIC declaration, for that matter, which is not shown here). The VHDL entity declaration statement is closed with the END keyword, and the name of the entity is optionally repeated. The declaration of an entity is defined by the following rules:

```
entity_declaration ::=
entity identifier is
entity_header
entity_declarative_part
  [ begin
      entity_statement_part]
end entity [ entity_simple_name ] ;
```

where:

entity_header ::= [formal_generic_clause] [formal_port_clause]
generic_clause ::= generic (generic_list) ;
generic_list ::= generic_interface_list
port_clause ::= port (port_list) ;
port_list ::= port_interface_list
entity_declarative_part::= { entity_declarative_item }

entity_declarative_item ::=
subprogram_declaration
| subprogram_body
| type_declaration
| subtype_declaration
| contant_declaration
| signal_declaration
| shared_declaration
| file_declaration
| alias_declaration
| attribute_declaration
| attribute_specification
| use_clause

The port mode of the interface describes the direction in which data travels with respect to the component. The available port modes are:

1. In data comes in this port and can only be read
2. Out data travels out this port
3. Buffer data may travel in either direction, but only one signal driver may be on at any one time
4. Inout data may travel in either direction with any number of active drivers allowed; requires a Bus Resolution Function Linkage.

Some examples of entity declarations are as follows:

```
entity cpu is
      generic (max_clock_freq : frequency := 100 MHz);
      port (clock : in bit;
      address : out integer;
      data : inout word_16;
      control : out control_type;
      ready : in bit);
end cpu;

entity RAM is
      generic (width, depth : positive);
      port (enable : in bit;
      address : in bit_vector(depth - 1 downto 0);
      data : out bit_vector(width - 1 downto 0) );
end RAM;
```

7.2.5. Architectures

The architecture body contains the internal description of the entity. The descriptions may consist of one or a combination of the following:
1. Structural
2. Behavior
3. Dataflow

A structural description is when the body of the architecture describes composite subsystems. It consists of signal declarations, component instances, and port maps. A behavioral description is when the body of the architecture describes the algorithm performed. It consists of sequential statements in process statements. A Dataflow description is when the body of the architecture uses statements that define the actual flow of data. It consists of signal assignment statements. An architecture body is decfined by the following rules:

```
architecture_body ::=
architecture identifier of
entity_name is
      architecture_declarative_part
begin
      architecture_statement_part
end [ architecture_simple_name ] ;
```

where:

architecture_declarative_part ::= { block_declarative_item }
architecture_statement_part ::= { concurrent_statement }

block_declarative_item ::=
subprogram_declaration
| subprogram_body
| type_declaration
| subtype_declaration
| constant_declaration
| signal_declaration
| shared_variable_declaration
| file_ declaration
| alias_declaration
| component_declaration
| attribute_ declaration
| attribute_specification
| configuration_specification
| use_clause

where:

attribute_ declaration ::= **attribute** identifier : type_mark;
alias_declaration ::= **alias** alias_designator [:subtype_indication] **is** name [signature];
alias_designator ::= identifier | character_literal | operator_symbol

where:

concurrent_statement ::=
block_statement
| process_statement
| concurrent_procedure_call_statement
| concurrent_assertion_statement
| concurrent_signal_assignment_statement
| component_instantiation_statement

| generate_statement

where:

concurrent_signal_assignment_statement ::=
[label:][postponed] conditional_signal_assignment
| [label:][postponed] selected_signal_assignment

In summary, the architecture consists of two main parts:

1. declarations
2. statements.

For a structural architectural description the architecture declarations may consist of the following:

1. subprogram declaration
2. subprogram body
3. type declaration
4. subtype declaration
5. constant declaration
6. signal declaration
7. alias declaration
8. component declaration
9. configuration specification
10. use clause

The corresponding structural architecture statements may consist of one or more of the following concurrent statements:

1. block statement
2. component instantiation statement
3. process statement
4. concurrent procedure call statement
5. concurrent assertion statement
6. concurrent signal assignment statement
7. component instantiation statement
8. generate statement

The declaration rules of constants, variables, signals, and files have been discussed in 7.1.5.1(starting on page 270), 7.1.5.2, 7.1.5.3, and 7.1.5.4. The declaration rules of types and subtypes have been discussed starting with Section 7.1.6.

7.2.6. Processes

A VHDL process statement is used for all behavioral descriptions. Statements within a process are executed sequentially. Special attention may be required for signal assignment statements since they do not take effect immediately. All processes in a VHDL description are executed concurrently. Only variables can be declared in a process, and signals in and out of the process are used primarily for control. The rules for the process statement are defined by:

```
process_statement ::=
[ process_label : ]
process [ ( sensitivity_list ) ] [is]
      process_declarative_part
begin
      process_statement_part
end process [ process_label ] ;
```

process_declarative_items ::–
subprogram_declaration
| subprogram_body
| type_declaration
| subtype_declaration
| constant_declaration
| variable_declaration
| file_declaration
| alias_declaration
| attribute_declaration
| attribute_specification
| use_clause

process_statement_part ::= {sequential_statement}
sequential_statement ::=
wait_statement
| assertion_statement
| report_statement
| signal_assignment_statement
| variable_assignment_statement
| procedure_call_statement
| if_statement
| case_statement
| loop_statement
| next_statement
| exit_statement
| return_statement
| null_statement

The use of process_label at the beginning and end of a process is optional but recommended to enhance code readability. The sensitivity_list is optional in that a process may have either a sensitivity list, or it must include WAIT statements. However, a process cannot include both a sensitivity_list and WAIT statements. The process declaration_statement_part includes declarations for variables, constants, aliases, files, and a number of other VHDL constructs. The process_statement_part includes sequential statements such as report statements, variable assignment statements, signal assignment statements, procedure calls, wait statements, if statements, while loops, assertion statements, etc. The use of functions and procedures enables code compaction and enhances readability.

An example of a process statement is given here:

```
process (A, B, C, D)
variable V1, V2: BIT;
begin
      V1 := A and B;        --Statement 1
      V2 := C and D;        --Statement 2
      V1 := V1 or V2;       --Statement 3
      Z <= not V1;          --Statement 4
end process;
```

This process statement has four signals in its sensitivity list. Two variables are declared using the variable declaration statement. When an event occurs on any of the signals A, B, C, or D, the process is implemented. The statements are in the body of the process are then implemented in the order they occur, statement 1, statement 2, statement 3, and statement 4. After these are executed the process is then suspended until another event occurs on signals A, B, C, or D. However, the simulation continues to run.

7.3. Behavioral VHDL – Sequential Statements

In behavioral modeling the VHDL coding style uses sequentially executed statements. In other programming languages sequential statements are statements that are executed sequentially during the program run. In VHDL, the same concept holds, but with sequential statements that run inside processes. Some sequential statements are:
  1.) Variable statements
  2.) If statements
  3.) Case statements
  4.) Exit statements
  5.) Next statements
  6.) Loop statements
  7.) Null statement
  8.) Assertion statements
  9.) Wait statements

10.)   Sequential signal assignment statements
11.)   Report statements

These sequential statements are discussed more in the following sections.

7.3.1. The Variable Statement

Variable statements are statements that use variable assignment statements. A variable is given a new value using an assignment statement. A variable assignment statement is defined by the following rules:

variable_assignment_statement ::= variable_identifier := expression;

where:

variable_identifier ::= name | aggregate

The variable_identifier of the variable assignment is an object name. The value of the expression is given to the object name. The value and the object name must have the same base type.

Variables can be declared and used inside a process statement. The process statement shall be discussed in detail in Section 7.2.6 on page 290. An example of a process statement declaring a variable and executing a variable statement is as follows:

```
process (Y)
variable Y: integer := 0;
begin
     Y := Y+1;
end process;
```

The process is executed once at the start of the simulation. The variable Y is then initialized to 0 and incremented by 1; Afterwards, the process is implemented only when there's an event on Y at which time Y is again incremented by 1. So, the value of Y is retained until the next event on Y.

A variable can also be declared outside a process. These are called shared variables and will not be discussed here.

7.3.2. The If Statement

The if statement allows the selection of statements to execute depending on one or more conditions. The if statement uses the rules defined by:

```
if_statement ::=
if condition then
     sequential_statements
{ elsif condition then
     sequential_statements
 [ else
     sequential_statements
]
end if ;
```

The conditions represent Boolean expressions that result in Boolean values. The conditions are evaluated sequentially until one has a value of true. The sequential statements immediately following that condition are then executed. Otherwise, if the else clause is present, its sequence of statements list are executed.

7.3.3. The Case Statement
The case statement performs execution of the specified sequential statements when the value of the selection expression is true. The case statement follows the rules defined by:

```
case_statement ::=
 [case_label:]
case expression is

    when choices => sequence_of_statements

    when choices => sequence_of_statements
    .
    .
    .
    when choices => sequence_of_statements

end case [case_label];
```

where:

choices ::= choice { | choice }

and where:

choice ::=
choice_expression | discrete_range | element_simple_name | others

The expression must result in either a discrete type, or a one dimensional array of characters. If the expression results in an array then the choice_expression may result in a bit string or string. All the choices must be unique, that is, no values may be the same. Additionally, all values must be represented in the choice selection lists, or the special choice others must be included. Therefore, if no choice list includes the value of the expression, the others is selected. The sequence_of_statements are sequential statements which may include one of the following: wait statement, assertion statement, report statement, signal assignment statement, variable assignment statement, procedure call statement, if statement, case statement, loop statement, next statement, exit statement, return statement, or the null statement.

An example of a case statement is:

```
case cpu_opcode of
    when X"0000" => perform_add;
    when X"0001" => perform_subtract;
    when X"0010" => perform_sum;
    when X"0011" => perform_multiply;
    when others => cpu_illegal_opcode;
end case;
```

7.3.4. The Exit Statement

The exit statement can only be used inside a loop. It causes the execution to jump out of the current loop or jump to a specified label. It is defined by:

```
exit [loop_label] [when condition];
```

if no loop_label is specified, the current loop is exited. The optional when clause specifies the condition needed to exit the current loop, otherwise execution continues to the next statement. An example of the exit statement is presented in the following lines of VHDL code:

```
sum := 1; j:= 0;
SUM_LOOP: loop
        j:=j+12;
        sum := sum * 2;
exit when sum> 50;
end loop SUM_LOOP;
```

7.3.5. The Next Statement

The next statement can only be used inside a loop. It causes the execution to skip the remaining statements of the current loop or jump to a specified label. It is defined by:

```
next [loop_label] [when condition];
```

if no loop_label is specified, the current loop is exited. The optional **when** clause specifies the condition needed to exit the current loop, otherwise execution continues to the next statement. An example of the exit statement is presented in the following lines of VHDL code:

> sum := 1; j:= 0;
> SUM_LOOP: loop
> j:=j+12;
> sum := sum * 2;
> next when sum> 50;
>   sum := sum – 10;
> end loop SUM_LOOP;

When the next statement is executed, execution jumps to the end of the loop statement skipping the last statement "sum := sum -10;".

7.3.6. The Loop Statement

The loop statement is defined as follows:

```
loop_statement ::=
[ loop_label : ]

[ iteration_type ] loop
     Sequential_statements
end loop [ loop_label ] ;
```

where:

iteration_type ::= **while** condition | **for** *loop*_parameter_specification

where:

parameter_specification :: = identifier **in** discrete_range

Some examples are as follows:

```
loop
     statements;
end loop;
```

Here, the iteration type is omitted, so we get a loop which will repeat the enclosed statements indefinitely.

```
for index in 1 to N loop
    buffer(index) := 0;
end loop;
```

When the iteration type is given, it controls the number of iterations. In the above for loop, the loop_parameter_specification declares an object which takes on successive values from 1 to N for each iteration of the loop.

There are ways to control the iteration using statements inside the loop. These are:
1. Next statements
2. Exit statements

These are defined by:

next_statement ::= next [loop_label] [when condition] ;
exit_statement ::= exit [loop_label] [when condition] ;

Thus, modifying the previous VHDL loop statement example using the next statement gives the following:

```
for index in 1 to N loop
    buffer(index) := 0;
exit when buffer(index) = NUL;
end loop;
```

An example of the next statement is given in the following VHDL code:

```
outer_loop : loop
inner_loop : loop
i:=i+1;
temp:=buffer(i);
next outer_loop when
temp = 0;
.
end loop inner_loop;
end loop outer_loop;
```

7.3.7. The Null Statement

The null statement has no effect. Typically, it is used in case statements as shown in this example:

```
case robot_command is
when stop => motor_stop;
when forward => motor_forward;
when reverse => motor_reverse;
when left => motor_left;
when right => motor_right;
when idle => null;
end case;
```

All possible values of the case statement selection expressions must be listed in the "when choices". But for some choices no action is required which is where the NULL statement comes in handy.

7.3.8. The Assertion Statement

An assertion statement is used to identify a specific condition and to report if that condition has occurred. The assertion statement is defined by:

assertion_statement ::= assert condition [report expression][severity expression] ;

The report and severity and their expressions are optional otherwise their defaults are used. However, if a report is desired, the result of the expression must be a string. The string represents a message which will be reported if the assert condition is false. On the other hand, if the report is omitted, the default string message is "Assertion violation". If the severity and its expression are present the expression must be of the type severity_level. When the severity is omitted, the default is error. If an assertion violation occurs, a given simulator may terminate execution based on whether the severity value is greater than some implementation dependent threshold. Typically, the threshold is user controllable.

7.3.9. The Report Statement

A report statement is similar to an assertion statement, but without the assertion check. The report statement is simply used to display a message. The syntax of the report statement is:

report string_expression [severity expression];

The severity clause is optional. Its expression must be of the predefined type SEVERITY_LEVEL. If the severity clause is not used, a default level of NOTE is used. Some examples of the report statement are:

```
if  CLR  =  '0'  then
    report "Signal CLR has an logic LOW value.";
end if;


if  SYSTEM_RESET  =  '0'  then
    report "Signal SYSTEM_RESET has an logic LOW value!!!!!";
    severity ERROR;
end if;
```

7.3.10. The Wait Statement

The wait statement provides an alternative method for a process to suspend execution, otherwise, the sensitivity list is used to determine execution of the process. When a process has a sensitivity list, it always suspends after execution of the last sequential statement. There are three forms of the wait statement:

```
        wait on sensitivity_list;
        wait until boolean_expression;
        wait for time_expression;
```

Some examples are:

```
        wait on A, B, C, D;      --statement 1
        wait until A = D;    --statement 2
        wait for 12ns;       --statement 3
```

The execution of the wait statement in statement 1 results in a suspension of execution of the current process until an event on A, B, C, or D. The next event will cause the process to continue execution with the first statement after the wait statement. If the wait statement is at the end of the process, the execution will start at the first statement in the process.

The execution of the wait statement in statement 2 results in a suspension of execution of the current process until an event on A or D occurs. Then the condition A = D is checked. If the condition is not true, then process continues suspension, otherwise, the process continues execution at the next statement after the wait statement.

The execution of the wait statement in statement 3 results in a suspension of execution of the current process until after 10ns. If the current time is T, when the simulation time advances to time T + 10, the process continues execution at the next statement after the wait statement.

A process with a sensitivity list always suspends execution at the end of the process. It is continues execution at the first statement when there is an event on any signal in the sensitivity list. A process where the wait statement replaces the sensitivity list is given as follows:

```
process
      variable V1, V2: BIT;
begin
      V1 := A and B;          --Statement 1
      V2 := C and D;          --Statement 2
      V1 := V1 or V2;         --Statement 3
      Z <= not V1;            --Statement 4
wait on A, B, C, D;           --replaces
sensitivity list
end process;
```

This process is equivalent to one with a sensitivity list (A, B, C, D). The process statement is discussed in Section 7.2.6 on page 290.

7.3.11. The Sequential Signal Assignment Statement

There are two types of signal assignment statements:

1. Concurrent signal assignment statement
2. Sequential signal assignment statement

A signal assignment statement can occur within or outside a process. When it occurs within a process it is called a sequential signal assignment statement and is executed in sequentially as it appears with the other sequential statements. However, if the signal assignment appears outside the process it is called a concurrent signal assignment statement and is executed in concurrence with the other concurrent statements. A simple signal assignment statement can be represented by:

```
signal_name <= expression [ after delay_value];
```

The expression is executed when the signal assignment statement is executed (current simulation time). The computed value of the expression is assigned to the signal_name after the specified delay. If no after clause is specified the signal_name is updated after the default delta delay. Some examples of signal assignments are:

```
Y <= (A and B) or (C and D) after 7 ns;
Z <= A xor B after 12 ns;
address <= address + "10000";   -- address is assigned
                                -- after the default delta delay
```

There are differences between how variables and signals are treated in processes. This will be discussed next.

7.3.12. Delta Delays – Variables vs. Signals

A delta delay, Δ, is a very small delay. This delay models hardware where a minimum amount of time is required for a change to occur. Events always occur at integral multiples of the delta delay. For example, events occur at 20ns, 20ns + Δ, 20ns + 2Δ, etc. Consider the following signal declaration and process with a sequential signal assignment statement for signal Y:

```
signal A, Y: integer;
.

.

.
process (A)
      variable V1, V2: integer;
begin
      V1 := A + 1;           --Statement 1
      Y<=2* V1;              --Statement 2
      V2 := Y + V1;   --Statement 3
end process;
```

When an event occurs on signal A at simulation time T, statement 1 is executed, and V1 is assigned a value. Statement 2 is executed next. Y is scheduled to get updated with a value at time T + Δ. However, statement 3 is executed next, but V2 is using the old value of Y (value at time T), not the value computed by statement 2. This is so because the simulation time is still at T while the variable V2 is computed instantaneously as soon as statement 3 is executed, but statement 2 updates Y at T + Δ. In summary, variable assignments cause variables to get their values instantaneously while signal assignments get their values at a later time, a least a delta delay later.

7.4. Structural VHDL – Concurrent Statements

Structural VHDL consist mainly of concurrent statements which are executed at the same time. These will be explained next.

7.4.1. The Block Statement

Blocks are good for organizing sections of VHDL code into submodules. In particular, the submodules should represent a structural organization which may resemble the actual hardware. For example, a control unit and memory can be organized as separate blocks of code. Signals are the means to communicate between the blocks. Signals have been discussed in Section 7.1.5.3. The rules for defining a block are given by:

```
block_statement ::= block_label : block [ (
guard_expression ) ] [is]
[block_header]
[block_declarations]
begin
     block_statements
end block [ block_label ] ;
```

block_header ::= [generic_clause [generic_map_aspect ;]] [port_clause [
port_map_aspect ;]]
generic_map_aspect ::= generic map (generic_association_list)
port_map_aspect ::= port map (port_association_list)
block_declarations ::= { block_declarative_item}
block_statements ::= { concurrent_statement }

The block statement itself is concurrent. If present, the block_header describes the
interface of the block statement to its environment. Block declarations are known only to
statements within the body of the block, that is, between begin and end block. Any
number of concurrent statements can appear within body of the block. The block label
appearing at the beginning of the block statement is necessary. However, the block label
name appearing at the end is optional; if it is present, it must be the same block label
name used at the beginning. The guard_expression is used to disable signal drivers.

7.4.2. The Component Declaration

The component must first be declared before it can be used. The syntax of a component
declaration is defined by:

```
component_declaration ::=
component component_name [is]
     [ local_generic_clause ]
     [ port (list_of_interface_ports);]]
end component [component_name];
```

The component declaration declares the name and interface of the component. The
interface specifies the signals going in and out of the component, etc. The
component_name may refer to an existing entity name. The referred to entity can be
accessible by the component using the library clause. If the component_name does not
refer to an existing entity, it cannot be simulated. In this case, the configuration
statement is the means by which the binding can be done.

The names of the component port can be different than the name of the entity. These
then, need to be mapped using the configuration statement. Two examples of

components are given below. Assume that the component port names are the same as entity port names. The examples of component declarations are:

```
component XOR3
      generic (Tpd : Time := 1 ns);
      port (a, b, c : in BIT; y : out BIT);
end component;


component ROM
      generic (data_bits, addr_bits : positive);
      port (en : in bit;
      addr : in bit_vector(depth-1 downto 0);
      data : out bit_vector(width-1 downto 0) );
end component;
```

7.4.3. The Component Instantiation

A component instantiation defines a subcomponent of the entity in which it appears. The component associates the signals in an entity with the signals in the component port. The component must first be declared before it can be used. This is discussed in the previous section. The next step is to use the component by instantiating the component. The syntax of a component instantiation is defined by:

component_intance : component_name [port map (association_list)];

The component_instance represents any legal identifier and is the name of the component instance. The component_name is the name of the component declared using the declaration statement. The association list associates signals in the entity, referred to as actuals, with the ports of the component, referred as formals.

Two ways to map actual to formals are as follows:

1. Positional association
2. Named association

Positional notation is used in the following manner:

```
component component_name
    port (formal₁, formal₂, formal₃, ……., formalₙ);
end component;
component_instance: component_name port map (actual₁,
actual₂, actual₃, ……., actualₙ);
```

Here, the first actual$_1$ corresponds to the first formal$_1$, the second actual$_2$ corresponds to the second formal$_2$, and so on.

A named associated list is used as follows:

```
component component_name
    port (formal₁, formal₂, formal₃, ……., formalₙ);
end component;
component_instance: component_name port map
(formal₁=>actual₁, formal₂=>actual₂, formal₃=>actual₃,
……., formalₙ =>actualₙ);
```

Thus, in named association, the ordering of the associations is not important since the mapping is done explicitly.

It is also important to know that the names used in the associated name or positional notation must be of the same mode. That is, if the formal is a port of mode **in**, it must be associated with a formal of mode **in**, not **out** or **inout**. If the formal is a port of mode **out**, it must be associated with a formal of mode **out**, not **in** or **inout**. If the formal is a port of mode **inout**, it must be associated with a formal of mode **in**, **out**, or **inout**.

Let's look at two examples using the components declared in the previous Section 7.4.2. These are rewritten here for convenience:

```
component  XOR3
    generic (Tpd : Time := 1 ns);
    port (a, b, c : in BIT; y : out BIT);
end component;

XOR3: xor3_instance
port map (a => en1, b => en2, c => en3, y => error);

component ROM
    generic (data_bits, addr_bits : positive);
    port (en : in bit;
    addr : in bit_vector(depth-1 downto 0);
    data : out bit_vector(width-1 downto 0) );
end component;

ROM: rom0_instance
generic map (data_bits => 16, addr_bits => 8);
port map (en => rom_sel, data => sys_data_bus, addr =>
a(7 downto 0);
```

The first example shows an instance of component XOR3 which is xor3_instance. It uses a default value for the generic constant, Tpd, since no generic map specification is given. The next example shows an instance of component ROM, rom0_instance. Generic values are specified for the address and data port sizes. Here, the addr is represented by a slice of an array signal. Thus, a port does not have to use all of a signal.

The key word open may be used to signify that the port is not connected. For example,

```
XOR3: xor3_instance
    port map (en1, en2, open, error);
```

The 3rd input of the XOR3 component is not connected to any signal. The input port may be left open only if an initial value is specified in the declaration of the component. Thus, the component declaration for XOR3 must have an initial value expression as follows:

```
component  XOR3
    generic (Tpd : Time := 1 ns);
    port (a, b, c : in BIT:='0'; y : out BIT);
end component;
```

Here, inputs a, b, and c have been initialized to logic 0.

As another example, a structural representation of a four bit parity generator is illustrated in Figure 7.4. The structural VHDL code is given by:

```
enitity PARITY_4_BIT is
    port (D: in BIT_VECTOR(3 downto 0); EVEN: out BIT;
    ODD: buffer BIT);
end parity_4_BIT;

architecture PARITY_DATA of PARITY_4_BIT is
component  XOR2
    generic (Tpd : Time := 1 ns);
    port (a, b : in BIT:='0'; y : out BIT);
end component;

component  INV
    port (a : in BIT; y : out BIT);
end component;

signal Y1, Y2: BIT;
```

```
begin
    X0: XOR2 port map(D(0), D(1), Y1);
    X1: XOR2 port map(D(2), D(3), Y2);
    X2: XOR2 port map(Y1, Y2, ODD);
    IN0: INV port map(ODD, EVEN);
end PARITY_DATA;
```

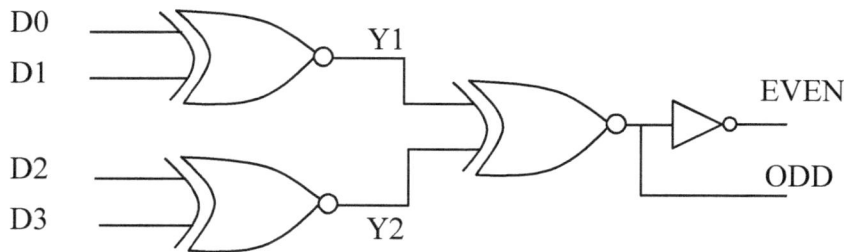

Figure 7.4. Four bit parity generator circuit.

Structural models such as the one given in Figure 7.4 can only be simulated after entities of the components have been added to a design library. The lowest level entities must be behavioral models. For example, the behavioral model for the entity of the component representing the XOR is:

```
enitity xor2 is
port (A, B: in BIT; Y: out BIT);
end xor2;
architecture xor_gate of xor2 is
begin

process
begin
    Y <= A xor B;
end process;

end xor_gate;
```

7.5. Dataflow VHDL

Concurrent statements occur at the same time. The concurrent signal assignment statement is the fundamental statement of dataflow modeling. Dataflow statements will be discussed next.

7.5.1. Concurrent Signal Assignment Statement

The concurrent signal assignment statement is a basic dataflow statement. Section 7.3.11, on page 299, discusses the two types of signal assignment statements – concurrent and sequential. Let's use the two input xor gate in Figure 7.5 to demonstrate dataflow VHDL code.

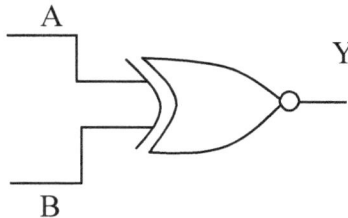

Figure 7.5. Xor circuit.

```
enitity xor2 is
port (A, B: in BIT; Y: out BIT);
end xor2;
architecture xor_gate of xor2 is
begin
     Y <= A xor B after 12ns;    --Dataflow Statement
end xor_gate;
```

The architecture body contains a single concurrent signal statement which represent the dataflow method as shown by the comment "--dataflow statement". The signals in the expression, A xor B, form the sensitivity list for the signal assignment statement. The concurrent signal assignment statement is implemented whenever there is an event on a signal, here, A and B. The expression on the right side is evaluated and its value is given to Y after a delay of 12ns.

There are other type of concurrent statements, these are:
1. Conditional signal assignment statement
2. Selected signal assignment statement

For each type of concurrent statement there is a corresponding process statement with the same meaning. The concurrent statements will be discussed in more details in the following Sections.

7.5.2. Conditional Signal Assignment Statement

The conditional signal assignment statement selects different values for the target based on multiple conditions. It is similar to an if statement. The conditional signal assignment statement can be thought of as the short version of a process containing signal assignments in an if statement. It has the following syntax:

conditional_signal_assignment ::= target <= options conditional_waveforms ;

options ::= [guarded] [transport]

conditional_waveforms ::=
{ waveform when condition else }waveform [when condition]

condition ::= Boolean expression

guard will not be discussed here. When an event occurs on any of the waveforms in the conditional assignment statement, the conditional assignment statement is executed by evaluating each condition one at a time. The first true condition found then results in an executed assignment. The following is an example:

```
Y <= A after 12 ns when S0='0' and S1='0' else
     B after 12 ns when S0='0' and S1='1' else
     C after 12 ns when S0='1' and S1='0' else
     D after 12 ns;
```

Here, the statement is executed when any event occurs on A, B, C, C, S0, or S1. The conditions are checked in order, first the condition " S0='0' and S1='0' ", if false then " S0='0' and S1='1' ", if false, then "S0='1' and S1='0' ", and so on. So, if "S0='1' and S1='0' " is true, the Y is assigned the value of C after 12 ns.

The equivalent process statement of the conditional assignment statement above is:

```
process
begin
    if S0='0' and S1='0' else
        Y <= A after 12 ns;
    elsif S0='0' and S1='1' else
        Y <= B after 12 ns;
    elsif S0='1' and S1='0' else
        Y <= C after 12 ns;
    else
        Y <= D after 12 ns;
    end if;
    wait on A, B, C, D, S0, S1;
end process;
```

7.5.3. Selected Signal Assignment Statement

The selected signal assignment statement is like a case statement. It selects different values for a target signal. The syntax is:

selected_signal_assignment ::=
with expression **select** target <= options selected_waveforms ;
selected_waveforms ::= { waveform **when** choices , }waveform **when** choices
choices ::= choice { | choice }

This syntax is somewhat similar to the conditional assignment statement just discussed. An example of the selected signal assignment statement is:

```
type OP is (MULT, DIV, SUB, ADD);
signal OP_CODE: OP;

with OP_CODE select
      Y <= A * B after MULT_PROP_DELAY when MULT,
           A / B after DIV_PROP_DELAY when DIV,
           A - B after SUB_PROP_DELAY when SUB,
           A + B after ADD_PROP_DELAY when ADD;
```

The above example results in the statement being executed whenever an event occurs on the signals A, B, or OP_CODE. For example, if ADD is true, then the expression "A + B" is executed and its value is assigned to Y after ADD_PROP_DELAY time. The equivalent process statement is:

```
        process
        begin
            case OP_CODE  is
                when MULT => Y <= A * B after
            MULT_PROP_DELAY;
                when DIV => Y <= A / B after
            DIV_PROP_DELAY;
                when SUB => Y <= A - B after
            SUB_PROP_DELAY;
                when ADD => Y <= A + B after
            ADD_PROP_DELAY;
            end case;
            wait on A, B, OP_CODE;
        end process;
```

7.6. **More VHDL Examples**

This section contains more VHDL examples for easy understanding of the concepts.

7.6.1. **The And Gate**

The basic building block of combinational logic is the and, or, and not gate. The VHDL for an AND Gate in Figure 7.6 is given in Listing 7.1.

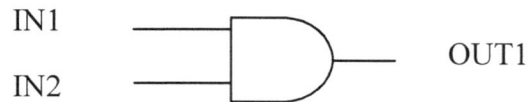

Figure 7.6. AND Gate.

```
-- AND Gate

-- import std_logic from the IEEE library
library IEEE;
use IEEE.std_logic_1164.all;

-- The entity
entity ANDGATE is
   port (
         IN1 : in std_logic;
         IN2 : in std_logic;
         OUT1: out std_logic);
end ANDGATE;

architecture RTL of ANDGATE is
begin

  OUT1 <= IN1 and IN2;

end RTL;
```

Listing 7.1. VHDL code for an AND gate.

7.6.2. **The D-Latch**

The D-Latch is shown in Figure 7.7. One way to implement the VHDL code for a D latch is given in Listing 7.2. Another way is shown in Listing 7.3.

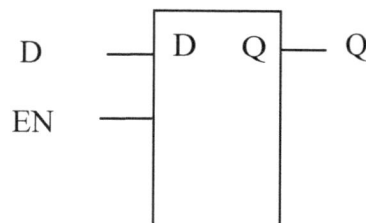

Figure 7.7. The D-latch.

```
-- D latch using when statement
Q <= D when EN = '1' else Q;
```

Listing 7.2. VHDL code for a transparent D-latch using when statement.

```
--D latch using process statement
process(D,EN)
begin
  if EN = '1' then
    Q <= D;
  end if;
end process;
```

Listing 7.3. VHDL code for a transparent D-latch using process statement.

In Listing 7.2 or Listing 7.3, the latch is basically one bit of memory which is updated when an active high enable (EN) signal is activated.

7.6.1. The D Flip-Flop

The D flip-flop is shown in Figure 7.8. One way to implement the VHDL code for a D flip-flop is given in Listing 7.4.

.

```
process(CLK, RESET)
begin
  if RESET = '1' then
    Q <= '0';
  elsif rising_edge(CLK)
then
    Q <= D;
  end if;
end process;
```

Listing 7.4. VHDL code for a D flip-flop.

In Listing 7.4, on the rising edge of the clock (CLK), the D flip-flop samples the incoming data. It has an active high asynchronous reset to set Q=0 when required.

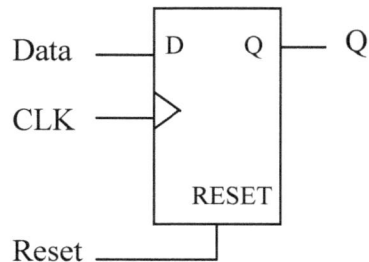

Figure 7.8. The D flip-flop.

7.6.2. The Configurable Width Counter

The configurable width counter is shown in Figure 7.9. The VHDL code for a configurable width counter is given in Listing 7.5.

```
library IEEE;
use IEEE.std_logic_1164.all;
use IEEE.numeric_std.all;    -- for the unsigned
type

entity counter is
generic ( WIDTH : integer := 32);
port (
CLK, RESET, LOAD : in std_logic;
DATA : in  unsigned(WIDTH-1 downto 0);
Q    : out unsigned(WIDTH-1 downto 0));
end entity counter;

architecture counter of counter is
signal cnt : unsigned(WIDTH-1 downto 0);
begin
process(RESET, CLK) is
begin
if RESET = '1' then
      cnt <= (others => '0');
elsif rising_edge(CLK) then
      if LOAD = '1' then
      cnt <= DATA;
      else
      cnt <= cnt + 1;
      end if;
end if;
end process;

Q <= cnt;

end architecture counter;
```

Listing 7.5. Counter VHDL Code.

The counter VHDL code in Listing 7.5 is an up-counter with asynchronous reset, parallel load and configurable width.

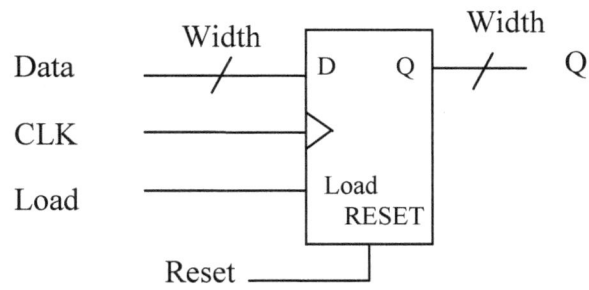

Figure 7.9. Configurable width counter.

8. Digital Design Development Kits

Programmable logic device (PLD) platforms of many types can be used for the code in this book. However, two PLD platforms considered for use in this book are: 1.) DeVry Electronic System on a Chip or eSOC (Waterman (2003), see References on page 578), 2.) and the Altera DE2 Board (see Appendix D). Note that PLD platforms with the supporting switches and LEDs are convenient to have for debugging purposes. The reference clock, on which most excersizes in this book is based on, is 4MHz. Other clock sources may be used as long as the same functionality is acheived.

8.1. ESOC Board

The VHDL code used in this book was programmed into an Altera programmable logic device (PLD, EPM7128SLC84) such as that shown in Figure 8.1. Many different robotic projects were successful.

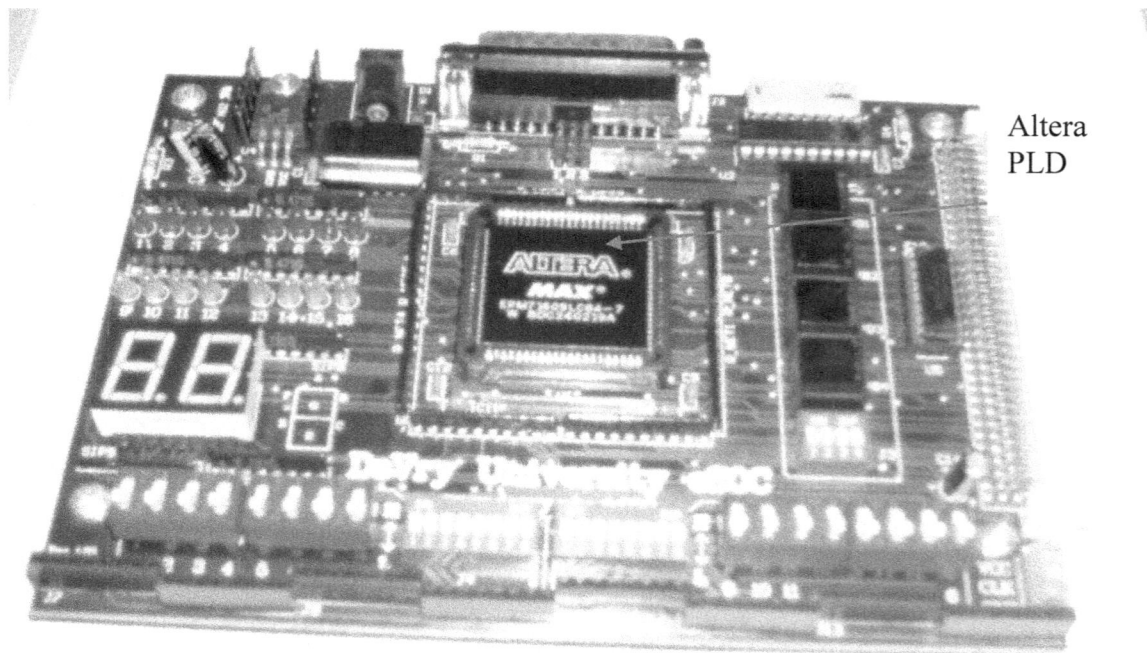

Altera
PLD

Figure 8.1. The DeVry Electronic System on a Chip (ESOC) platform with Altera Programmable logic Device.

8.2. **Altera DE2 Board**

Figure 8.2 shows the Altera DE2 board with Altera programmable logic device (PLD, EPM7128SLC84).

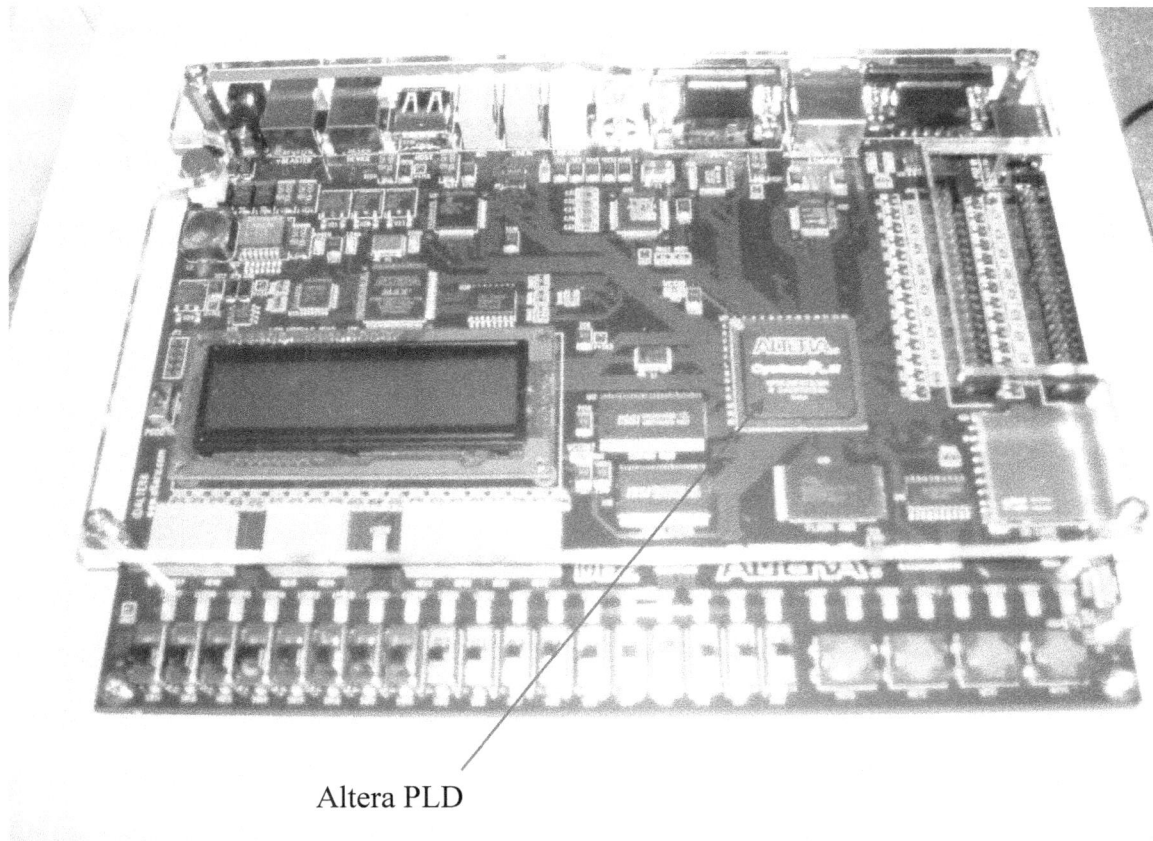

Altera PLD

Figure 8.2. Altera DE2 board with Altera Programmable logic Device.

9. Top Level and Low Level Robot Design

Robot top level codes are described in this chapter. The robot consists of two main portions: 1.) remote control transmitting PLD (or PIC platform, etc.) 2.) receiving PLD platform (or PIC platform, etc.) and supporting sensory circuits. Collectively, the robot is referred to as PLD-bot (or PIC-bot), which encompasses the transmitting PLD platform and receiving PLD platform.

9.1. Top Level Robot and Remote Control Design

The robot receiving PLD uses one top level graphic design file module called robot_top using the Altera Quartus II software. Currently, the Altera Quartus Web edition is free and can be downloadable from the following web site:

```
https://www.altera.com/download/software/quartus-ii-we
```

robot_top may contain the following components:

1. 74138 decoder
2. reg_ctl.vhd
3. pipo_ctl.vhd
4. a2d0804_ctl.vhd
5. d2a0808_ctl.vhd
6. compass_control.vhd
7. irmodule.vhd
8. servo_control.vhd
9. rxuart.bdf
10. txuart.bdf
11. count65536.vhd
12. sensorN_ct.vhd
13. loopback_ctl.vhd
14. tracktape.vhd
15. tracklight.vhd
16. speakerid_ctl.vhd

These are explained next and are shown in Figure 9.1. An example showing the typical top level wireless robot configuration for projects is shown in Figure 9.2. The robotic projects typically start testing in the manual mode first, thus eliminating all the bugs, the moving on to integrating the wireless mode and debugging. Figure 9.3 shows the block diagram of the remote control unit. It is used to send 8 bit commands to the robot base unit shown in Figure 9.2.

9.1.1. 74138 Decoder

The 74138 decoder module determines which VHDL module gets control of the servo_control module using the address sent to it from the address bus. It selects which module (tracklight_ctl, tracktape_ctl, irmodule, compass_control, etc.) controls the servo_control module.

robot_top.qfp

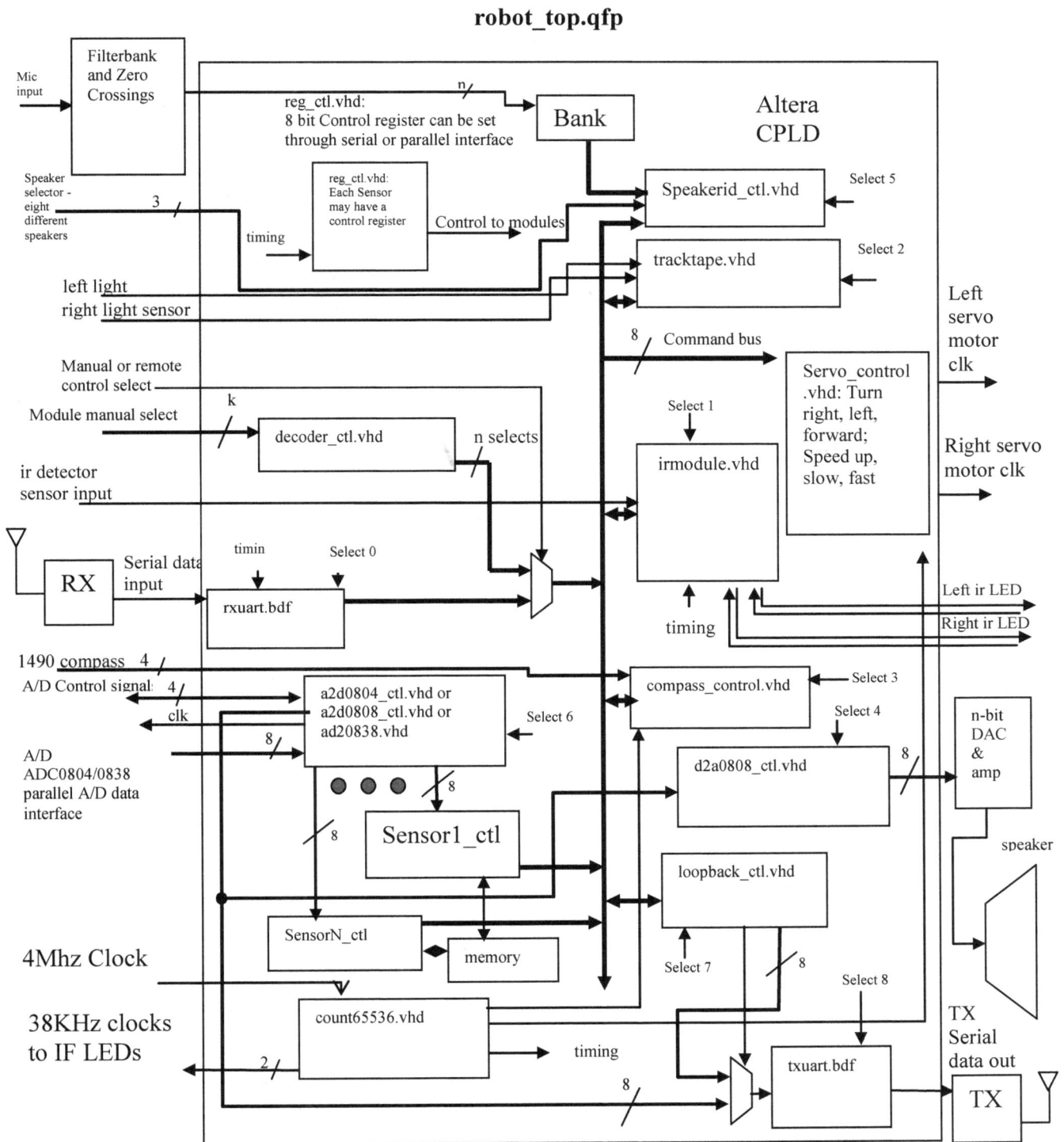

Figure 9.1. Top level view of robot module showing how VHDL modules are connected. Although, this block diagram shows all possible modules, projects typically consist of 2, 3, or even 4 sensor modules (e.g., ir_ctl.vhd, compass_ctl.vhd, tracktape.vhd, tracklight.vhd, etc.)

robot_top.qfp

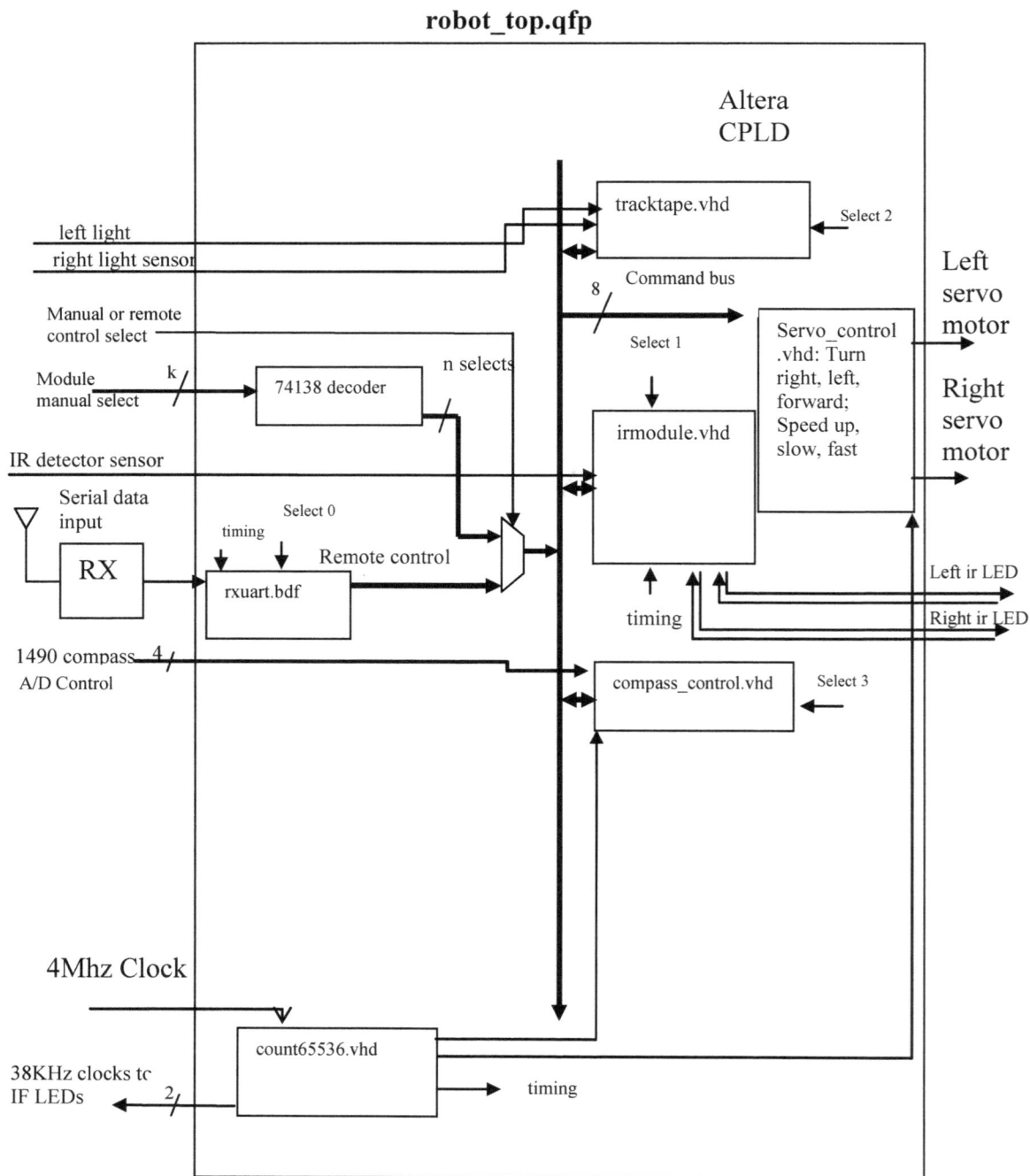

Figure 9.2. Remote controlled robot base block diagram using wireless receiver (RX). Typical 3 sensor based top level view of robot module showing how VHDL modules are connected.

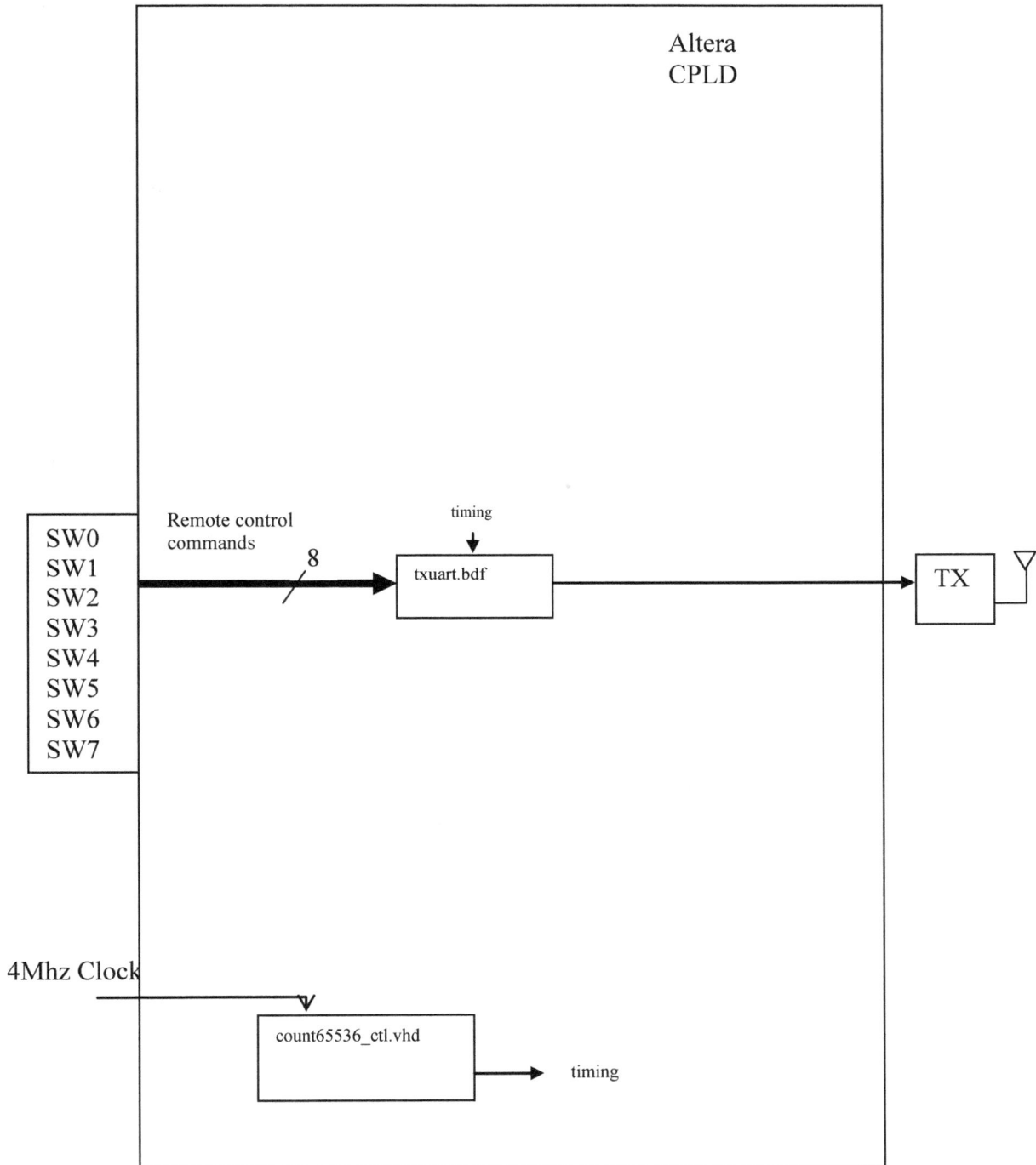

Figure 9.3. Block diagram of remote control transmitter unit. Commands are sent to the robot base unit in Figure 9.2. Commands are created manually using switches (SW0 – SW7) as shown. The 8 bit parallel commands are transmitted serially to the robot's base.

9.1.1. reg_ctl.vhd

The reg_ctl.vhd module holds the status and control information of each module. It is optional. If used, each module can then perform a task based on bits set in its register and can send status info to the register which can then be accessed by a microprocessor. This feature is not implemented here but can be done as an exercise by the student.

9.1.2. pipo_ctl.vhd

The pipo_ctl module determines whether a packet received from the serial data is an address packet or data packet. The easy method is split each received data bytes into 2 nibbles. The upper nibble of each serial byte sent is dedicated for the address bus and the lower nibble for the data bus. Then the upper nibble of the data bus and address bus is padded with zeros since each bus is 8 bits wide. Another method is to consider every 2 serial bytes as a packet. Let the most significant byte be the header byte and the least significant byte be the data byte. The received data byte is placed on the data bus depending on header info. . This feature is not implemented here but can be done as an exercise by the student.

9.1.3. a2d0804_ctl.vhd

The a2d0804_vhd module provides the control signals required for the ADC0804 A/D chips. The resulting 8 bit parallel data conversion is latched and placed on the data bus. This data can also be serialize by the txuart_ctl.vhd module and sent to a microprocessor. If the ADC0804 is used only one sensor module can be used unless used with an analog multiplexor. If the ADC0838 is used up to 8 sensors can be used. Details on how to construct the ADC0804 circuitry and test using the supporting VHDL code are given in Section 9.2 on page 321.

9.1.4. d2a0808_ctl.vhd

The digital to analog control module, d2a0808.vhd, send digital data to the 8-bit DAC0808. Digital data is converted to analog data using the correct timing. This data can be voice synthesis data, etc. Details on how to construct the ADC0804 circuitry and test using the supporting VHDL code are given in Section 9.3 on page 330. This feature is not implemented here but can be done as an exercise by the student.

9.1.5. compass_control.vhd

The compass_control.vhd goes in direction the robot is told to go in using a Dinsmore 1490 compass. See Section 9.10 on page 376 for more details.

9.1.6. sensorN_ctl.vhd

The sensorN_ctl.vhd modules represent multiple sensors, up to N sensors, connected to the a2d_ctl.vh module. Each sensor module determines the appropriate action for the servos using the A/D data from its respective analog sensor. For example, the sensor data can represent digitized temperature data, digitized voice data, digitized compass heading data, digitized infrared data, etc. Note that sensor1_ctl.vhd represents the VHDL code for sensor 1, sensor2_ctl.vhd represents the VHDL code for sensor 2, and so on, etc. Several modules have already been written for various sensor circuits as shown in Sections 9.10 through 9.14.

9.1.7. irmodule.vhd

The irmodule.vhd module receives data from the ir sensor and sends an appropriate command to the servo_control.vhd module. The ir_ctl module uses the received data to determine whether an object is the be avoided, if so, the appropriate command is sent to servo_control.vhd for servo action – turn left, turn right, reverse, forward, etc. The irmodule.vhd module is discussed in more details in Section 9.11 on page 379.

9.1.8.   servo_control.vhd

The servo_control module is used to control the servo motors. It receives an 8 bit control command from the various modules. The details of the servo_control module are given in Section 9.9 on page 370.

9.1.9.   rxuart.bdf

The rxuart.bdf module (Quartus schematic file: txuart.bdf) converts the received asynchronous serial data to parallel 8 bit data. The serial data contains one start bit, 8 data bits, and 2 stop bits. This data should be formatted according to the RS-232 UART protocol. The rxuart.bdf module is discussed in Section 9.7 on page 351.

9.1.10. txuart.bdf

The txuart.bdf module (Quartus schematic file: txuart.bdf) is used to convert parallel data to serial data using the RS-232 UART protocol, that is, one start bit, 8 data bits, and 2 stop bits. The choice of data is from the data bus or the a2d_ctl.vhd module. The txuart.bdf module is selected using the specified address via the decoder similar to the selection of other modules. Then the txuart.bdf module latches and decodes the data from the data bus and decides what to do. The txuart.bdf module is discussed in Section 9.6 on page 344.

9.1.11. count65536.vhd

The count65536.vhd module provides timing to all the modules and interfaces by dividing the 4 MHz clock. One example of a timing module is the count65536.vhd which divides the 4MHz clock into 16 different output frequencies which is used throughout the design for various clocking. The clocking module is discussed in Section 9.6.2 on page 349. Output clocks are provided in Table 9.4 and Table 9.5.

9.1.12. loopback_ctl.vhd

The loopback_ctl.vhd is used to put the robot_top.qpf in loopback mode. Data sent by a microprocessor, etc., can be returned for testing. This feature is not implemented here but can be done as an exercise by the student.

9.1.13. tracktape.vhd

The tracktape.vhd module maneuvers the robot to follow the black tape. When the robot's left sensor crosses the black tape it sends an active low signal to the tracktape.vhd control module as an indicator that darkness has been detected. The tracktape module responds by sending a left turn command to the servo control module. The servo control module (servo_control.vhd) then sends the correct frequencies to the left and right servo for a left turn. When the robot's right sensor detects darkness an active low is sent to the tracktape.vhd module. The module then sends a right turn command to the servo control module. The servo control module (servo_control.vhd) sends the appropriate frequencies to the servo motors causing the robot to turn right. The details of the tracktape module are discussed in Section 9.12 on page 384.

9.1.14. tracklight.vhd

The tracklight.vhd module maneuvers the robot to follow light. When the robot's left sensor senses light an active high signal is sent to the tracklight.vhd control module letting it know that light has been detected. The module responds by sending a left turn command to the servo control module. The servo control module (servo_control.vhd) then sends the correct frequencies to the left and right servo for a left turn. When the robot's right sensor detects light an active high is sent to the tracklight.vhd module. The module then sends a right turn command to the servo control module. The servo control module (servo_control.vhd) sends the appropriate frequencies to the servo motors causing the robot to turn right. The track tape and tracklight sensor circuits are exactly alike. The difference is in the VHDL code. For the track light module when a high is detected by the VHDL code for a sensor the robot performs a left or right turn movement. On the other hand, for the track tape module when a low is detected by the VHDL code for a sensor the robot performs a left or right turn movement. The tracklight module is discussed in Section 9.13 on page 387.

9.1.15. speakerid_ctl.vhd

Speakerid_ctl identifies a speaker before allowing the speaker to take control of the robot. The speaker ID module is given in Section 9.14 on page 390.

9.2. The Analog to Digital Converter Circuitry and VHDL Code for the ADC0804

This section gives the details of the analog to digital converter (ADC0804) circuitry along with the required testing to confirm the circuit functionality. Also given is the supporting VHDL code to test the circuitry. The ADC0804 chip will be explained next.

9.2.1. The ADC0804

The ADC0804 is an 8 bit successive approximation analog to digital converter. The pin configuration of the ADC0804 is shown in Figure 9.4. As shown the package type is the 20 pin dual in line (DIP) package. The analog input voltage range is 0 to VCC volts. The supply voltage input is represented by VCC. Its maximum is 6.5 volts. The ADC0804 has two grounds: analog ground (A GRD) and digital ground (D GRD). These two separate grounds insure that noise from analog circuits does not leak into the digital circuits within the chip. It consists of 4 digital control inputs: $\overline{CS}$, $\overline{WR}$, $\overline{INT}$, and $\overline{RD}$. The analog inputs, Vin+ and Vin, are differential. When the analog to digital conversion of the analog input is complete, the results can be read from the outputs, D0, D1, …D7.

9.2.2. Start Conversion

Conversion of an analog voltage on the differential inputs, Vin+ and Vin-, begin when the chip select, $\overline{CS}$, is low and the write, $\overline{WR}$, are both low. The chip select, $\overline{CS}$, must be low in order for the chip to be functional. When $\overline{WR}$ goes from being high to low, the SAR register is initialized and the interrupt, $\overline{INT}$, is set high. When the conversion is complete the interrupt goes active, that is, it goes from high to low indicating that the converted data is ready to be read from the outputs D0, D1, D2, D3, D4, D5, D6, and D7.

9.2.3. Read Converted Data

To read data when the conversion is complete the chip select must go low, followed by a high to low transition of the $\overline{RD}$ input. When the $\overline{RD}$ input goes low this resets the interrupt and causes the output data latches to be enabled so that the internal converted data appears on the data bus (D0, D1,…. D7).

The four digital control signals are summarized in the Table 9.1.

Digital control inputs	Active function
$\overline{CS}$/I	Chip select
$\overline{RD}$/I	Enable output
$\overline{WR}$/I	Start conversion
$\overline{INT}$/O	Data is ready

Table 9.1. The ADC0804 digital control inputs and their active function.

9.2.4. Continuous Conversion

For continuous conversion the chip select, $\overline{CS}$, and the read, $\overline{RD}$, are grounded, and the interrupt, $\overline{INT}$, and the write, $\overline{WR}$, are connected. In the continuous conversion configuration, the $\overline{WR}$ and $\overline{INT}$ pins must be brought low after power up to insure that the conversion process is started. See Figure 9.4 for the self-clocking configuration of the ADC0804.

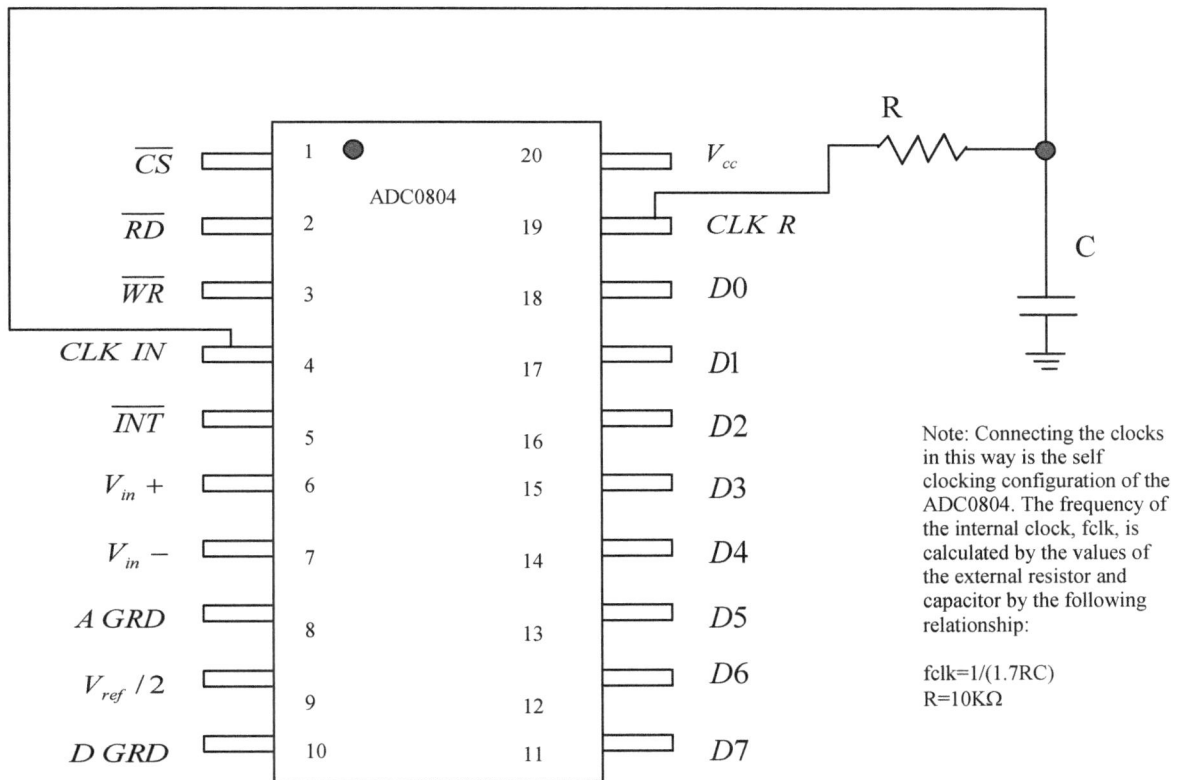

Note: Connecting the clocks in this way is the self clocking configuration of the ADC0804. The frequency of the internal clock, fclk, is calculated by the values of the external resistor and capacitor by the following relationship:

fclk=1/(1.7RC)
R=10KΩ

Figure 9.4. Pin configuration for the ADC0804 chip. Please see the ADC0804 datasheet for more information.

9.2.5.   Procedure for Testing and Controlling the ADC0804 using VHDL
The procedure for testing the ADC0804 is given below:

1. The ADC needs testing before interfacing it to the PLD. The test circuitry for the ADC0804 is shown in Figure 9.6. In this circuit the ADC0804 is configured in **free running** mode. This circuit consists of eight LEDs to verify the ADC0804's analog to digital conversion, a potentiometer to produce a test analog voltage for the ADC0804's analog input, a push button switch to start the analog to digital conversion process after power up, and the ADC0804 chip in the DIP package to do the analog to digital conversion. The LEDs are configured as active high. Therefore, a logic 1 on the digital output (pins 11, 12, 13, 14, 15, 16, 17, and 18) of an ADC0804 turns the corresponding LED on while a logic 0 turns the corresponding LED off. You may start by constructing the circuit in Figure 9.6 now.

2. Once the circuit is constructed power it up. Push the push button to start the analog to digital conversion process. Then turn the potentiometer to one extreme and observe the LEDs. Now turn the potentiometer to the other extreme and observe the LEDs. The LEDs should all turn off with one extreme and all turn on with the other extreme. If this is your observation you are ready to go to the next step if not you must troubleshoot the circuit until it is working. If all troubleshooting fails change the chip – you may have a bad chip.

3. The next step is to interface the working ADC0804 in the previous step to a programmable logic device (PLD). Figure 9.7 shows the schematic of the ADC0804 chip interfacing to a PLD, in this case, the Altera EMP7128SLC84. The ADC0804 can also be interfaced to a microprocessor, digital signal processor (DSP), or microcontroller, etc. Each serves the same purpose of processing the converted analog data digitally to perform a particular task, e.g., signal enhancements, noise removal, signal detection, digital transmission, etc. If you are interfacing to a PLD use Quartus to compile the VHDL code in Section 9.2.6 on page 329. Before compiling the code save it as a2d0804_ctl.vhd. This is the same name as the entity name. Most compilers such as the Altera VHDL compiler requires this to be done, else there will be an error in the compilation.

4. Create a Quartus project file called a2d0804_ctl.qpf. Insert the symbol a2d0804 into the graphic editor. The symbol is created when the VHDL file is compiled. Now complete the graphic design as shown in Figure 9.8. Once the graphic design file is finished compile and save it.

5. Now refer to Figure 9.5 for all pin assignments. You must now assign pins to the following outputs: cs, rd, wr, and write_led. Now assign pins to the following inputs: D0, D1, D2, D3, D4, D5, D6, D7, started, and int. The following outputs must be assigned to LEDs: Y0, Y1, Y2, Y3, Y4, Y5, Y6, and Y7. Compile the design after these signal names have been assigned. The net names are listed in Table 9.2 along with what pin they should be connected to on the ADC0804. The signal name startad is an input used to start the A/D conversion when configured in free running mode. This input should be connected to an active low push button. The input clock, clk, is not intended for the ADC0804 but is used for timing in the a2dcontrol4.vhd module. It was tested and works here for a 4MHz

clock. The input write_LED is used for testing purposes; it is connected to an active low LED. When an active write (either from a2d0804_ctl.vhd or startad) occurs the LED lights up. Note that the footprint for ADC0804 is given in to Figure 9.4. It shows all pins names and their numbers. Some of these pin names are used to control and accept data from the ADC0804. These pin names are shown in Figure 9.8 and has to be assigned pin numbers as shown in Figure 9.7. If you wish to use your own pin assignments please use Table 9.2. This Table is provided for you to assign and write your own pin assignments (numbers) in the column provided. Only the signals listed in the table need pin assignments, so ignore all other signals in Figure 9.8. Further information about the ADC0804's electrical characteristics, applications, and use, etc., can be sought in the ADC0804 datasheet.

ADC0804 pin name /chip input(I) or output(O)	ADC0804 pin number	Figure 9.8 signal name /PLD chip input(I) or output(O)	Your pin assignments
$\overline{CS}$ /I	1	cs/O	
$\overline{RD}$ /I	2	rd/O	
$\overline{WR}$ /I	3	wr/O	
$\overline{INT}$ /O	5	int/I	
D0/O	18	D0/I	
D1/O	17	D1/I	
D2/O	16	D2/I	
D3/O	15	D3/I	
D4/O	14	D4/I	
D5/O	13	D5/I	
D6/O	12	D6/I	
D7/O	11	D7/I	
N/A	N/A	write_LED/O	
N/A	N/A	startad/I	
N/A	N/A	clk/I	
N/A	N/A	Y0/O	
N/A	N/A	Y1/O	
N/A	N/A	Y2/O	
N/A	N/A	Y3/O	
N/A	N/A	Y4/O	
N/A	N/A	Y5/O	
N/A	N/A	Y6/O	
N/A	N/A	Y7/O	

Table 9.2. Assignment of names on Altera Quartus II graphic file to ADC0804 pin number. You must make sure that the indicated inputs and outputs in the Quartus II schematic are hardwired to the corresponding inputs and outputs on the ADC0804. Determine the pin assignments for each of the net names given in column 3 [signal name/PLD chip input (I) or output (O)] and write the assigned PLD pin number in the column provided (last column to right).

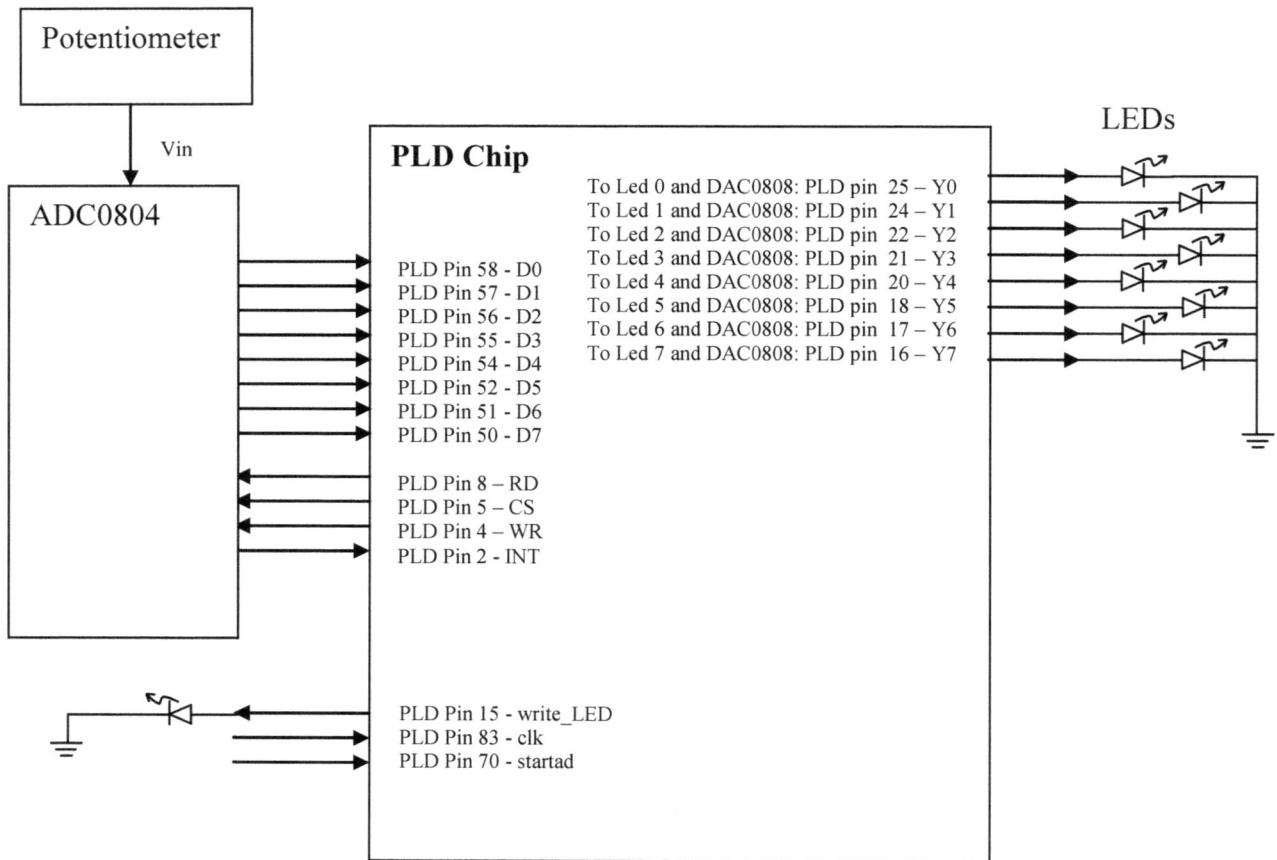

Figure 9.5. Pin assignments for the digital circuit in Figure 9.8. The PLDs meeting these pin assignments are: Altera EMP7128SLC84 or EMP7160SLC84. The ADC0804 and potentiometer circuits are external to the PLD platform while the LEDs, the 4MHz reference oscillator for generating the clk and pushbutton switch for started, which all reside on the PLD platform.

6. After the digital design has been compiled program it into the PLD.
7. Now power down the PLD board and connect the ADC0804 circuit to the PLD as shown in Figure 9.7. Disconnect all wires connected to the following pins on the ADC0804: 1, 2, 3, 5. This includes removing **R4** and the **push button** from the circuit. Let the remaining components remain as they are. Make sure the ADC0804 circuit shares the same ground as the PLD circuit. Also, if available connect the 5V source from the PLD board to the 5V inputs in the ADC0804 circuit.
8. Now power up the whole circuit. Push the push button on the PLD board to start the analog to digital conversion process. When this occurs the a2d0804_ctl module sends a signal to the ADC0804 chip via **int**. Now turn the potentiometer as you did in step 2 this time observing the LEDs on the PLD board. If the LEDs do not respond correctly troubleshoot the circuit.
9. When you are done have the instructor sign off on the working circuit.

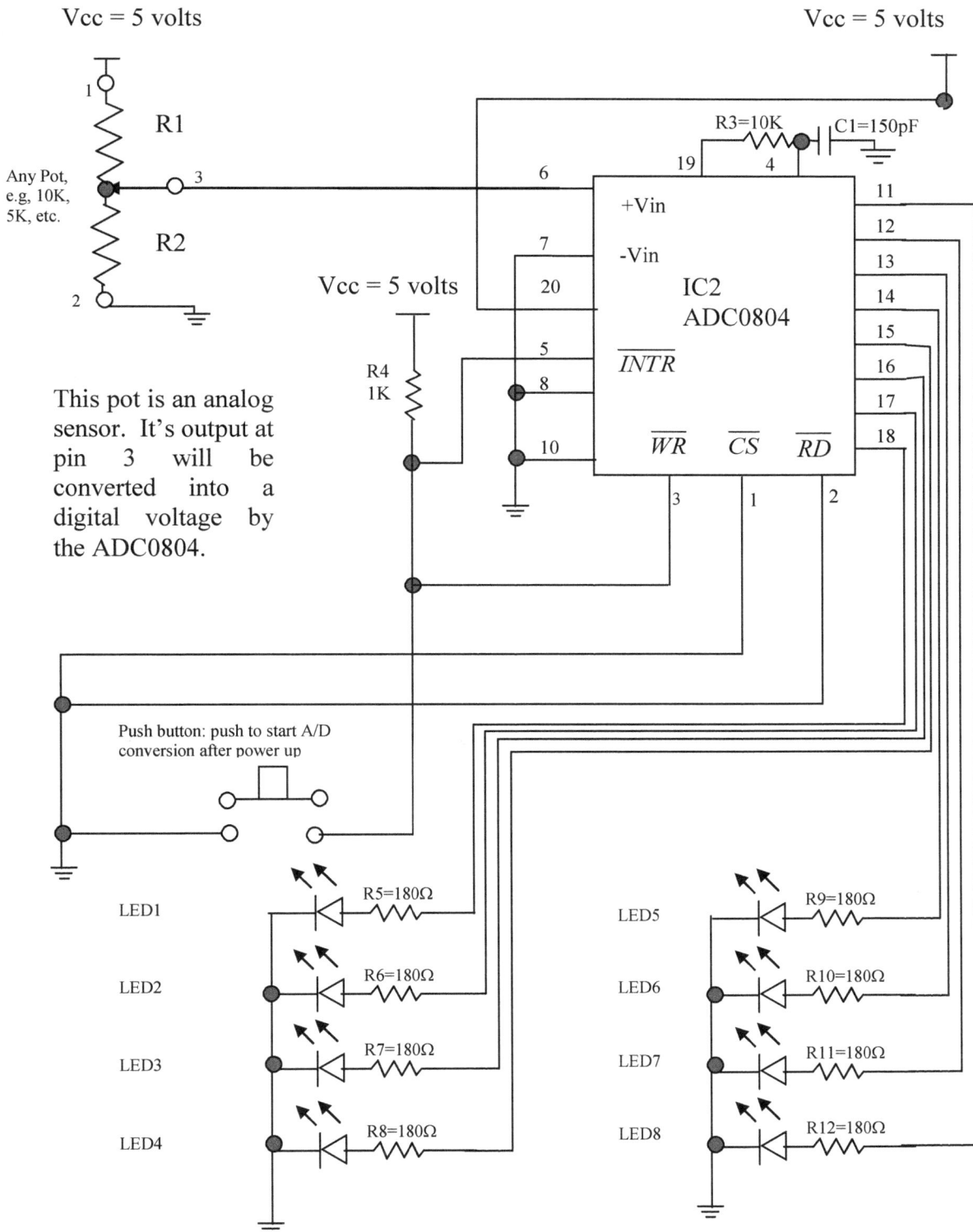

Figure 9.6. Schematic to test and verify the operation of the ADC0804 circuitry.

Figure 9.7 . Low level Schematic of ADC0804 and PLD.

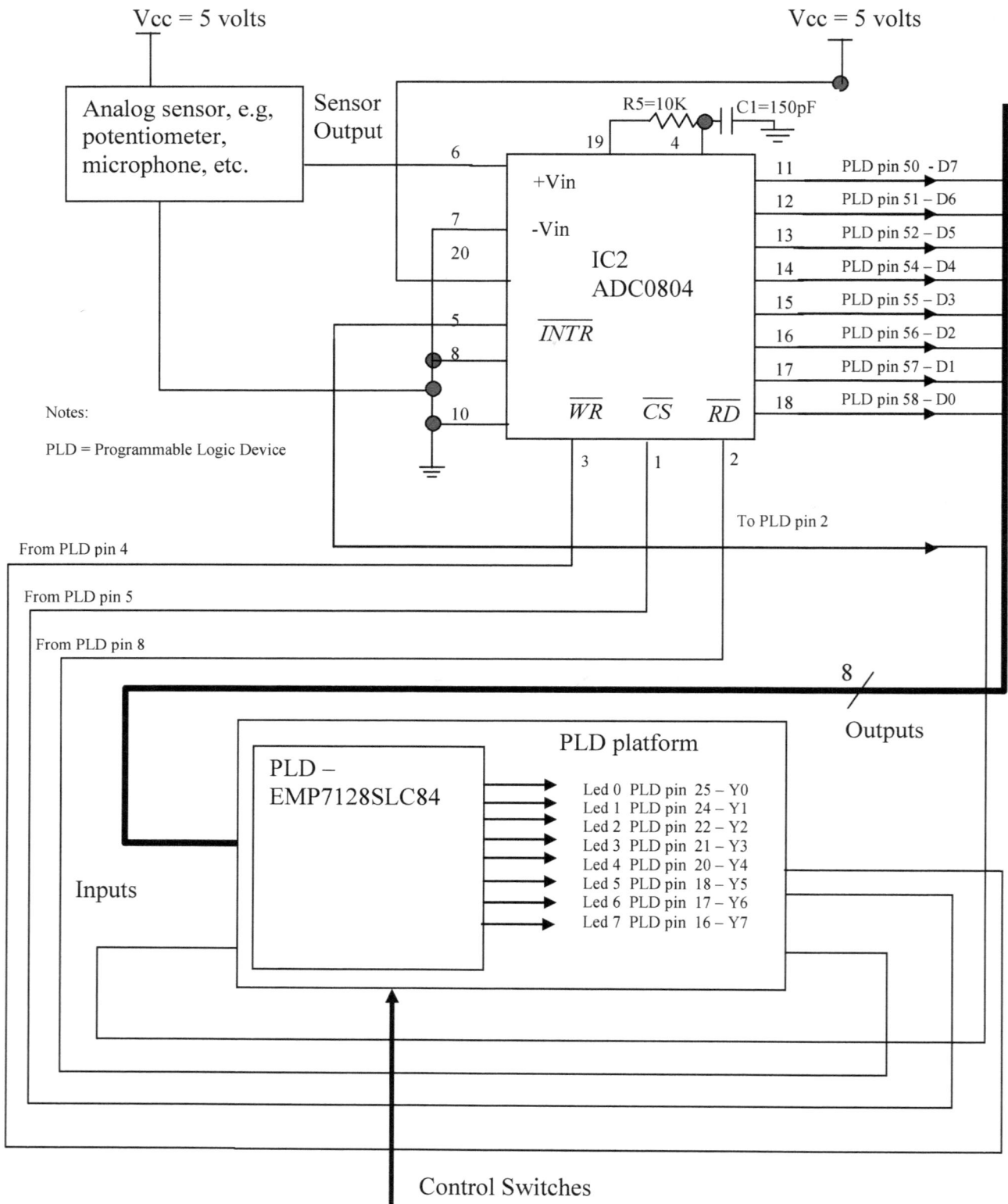

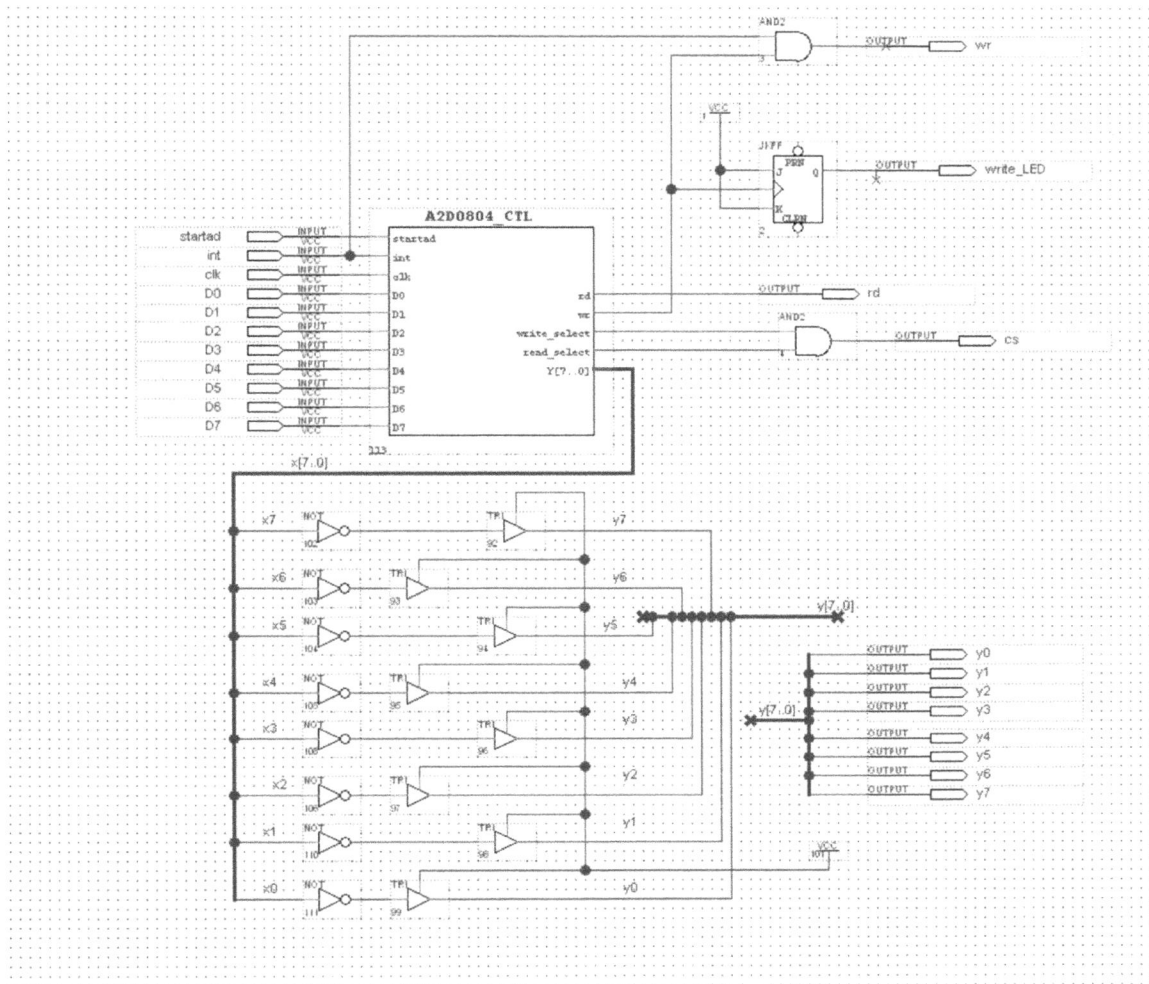

Figure 9.8. The analog to digital control module (Quartus file: a2d0804_ctl.qpf).

9.2.6.   VHDL code a2d0804_ctl.vhd for Controlling the ADC0804

The VHDL code for the A/D module (a2d0804_ctl.vhd) is given below.

```vhdl
library ieee;
use ieee.std_logic_1164.all;
-- A/D free running mode configuration, ADC0804
entity a2d0804_ctl is
port(
startad,int,clk: in std_logic;
rd,wr: out std_logic;
write_select,read_select: out std_logic;
D0, D1, D2, D3, D4, D5, D6, D7: in std_logic;
Y: out std_logic_vector (7 downto 0) );
end a2d0804_ctl;

architecture atod  of a2d0804_ctl is
begin

        get_data: process(int)
        begin
        if (int'event and int='1')then
              Y<=D7 & D6 & D5 & D4 & D3 & D2 & D1 & D0;
        end if;
        end process;

        assert_write: process(int,startad,clk)
        variable cnt: integer range 0 to 1023;
        begin
        --clk is assumed to be 4 MHz
            if (clk'event and clk='1')then
            if (startad = '1' or int='0')then
            cnt:=0;
            elsif (startad = '0' )then
            cnt:=cnt+1;
            end if;
            end if;

            -- startad is active high

            if (startad = '0' and cnt=1)then
            --initialize write signal to 1
            wr<='1';
            write_select<='1';
            elsif (cnt=2 and int='1') then
            write_select<='0';
            wr<='1';
```

```
                    elsif (cnt=3 and int='1') then
                wr<='0';
                write_select<='0';
                elsif (cnt=4 and int ='1') then
                wr<='1';
                write_select<='0';
                elsif (cnt=5 and int ='1') then
                write_select<='0';
                wr<='1';
                else
                wr<='1';
                write_select<='1';
                end if;
            end process;
-- Free running mode, grd rd and select
        read_select<='0';
        rd<='0';
end atod;
```

9.3. The Digital to Analog Converter Circuitry and VHDL Code for the DAC0808

The digital to analog converter circuitry is used for converting a digital signal to an analog signal. The dac0808_ctl.vhd module's main function is to capture the digital data and send it to the DAC at the right time. Let's start by building the DAC circuitry and testing it to insure that it is working properly before going on. This is explained in the next section.

9.3.1. Procedure for Building and Testing the DAC0808 Circuitry

The pin configuration of the DAC0808 is shown in Figure 9.9. This version of the DAC0808 is a 16 pin DIP. It is very important to know that pin 9 is the most significant bit and should be connected to the MSB of the 8 bit output data from the PLD. This will insure that the PLD output data is converted correctly. Pin 16 on the DAC0808 is the least significant bit (LSB)) and should also be connected to the LSB of the 8 bit output data from the PLD.

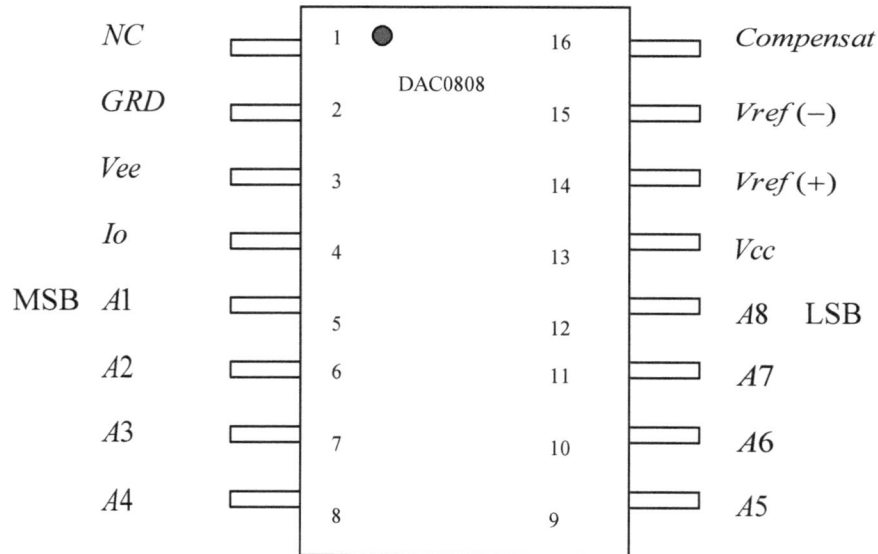

Figure 9.9. Pin configuration for the DAC0808 chip. Please see the DAC0804 datasheet for more information.

The procedure for constructing and testing the circuitry for a DAC0808 is given below:

1. Create the Quartus project dac0808_ctl.qpf . Compile the VHDL code for count256.vhd given in section 9.3.2 on page 335. Insert the symbol for count256 into a new Quartus design file, dac0808_ctl.qpf, and create the digital circuit as shown in Figure 9.10 below. Use the same pin assignments as given in Figure 9.11 on page 334. These pins are also shared by the LEDs to allow you to see any binary activity (if slow enough) of the counter outputs. Lastly, save and compile the graphic design file.

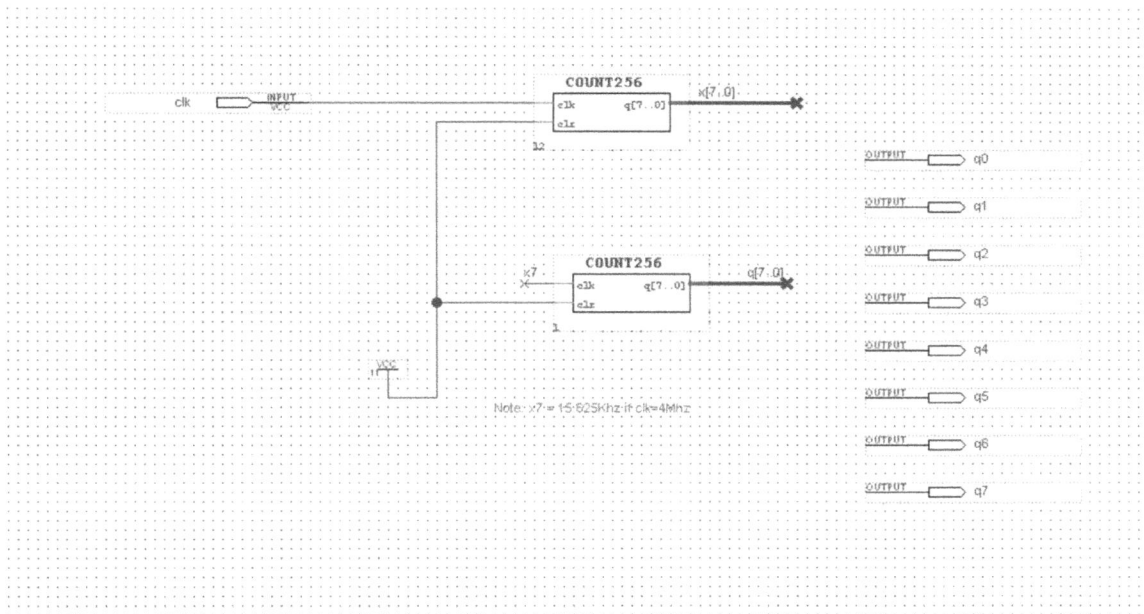

Figure 9.10. Illustration of two 8 bit counters used to drive the DAC0808 circuit (Quartus file: dac0808_ctl.qpf). The first 8 bit counter is used to derive the appropriate clock frequency for the second 8 bit counter while the second 8 bit counter is used to drive the DAC0808. The second 8 bit counter counts from 0 to 255 at a slower rate than the 1st. This creates a triangular waveform at the output of the DAC0808 circuitry.

2. Derive a 15.625khz clock (See Table 9.3below) from the 4Mhz clock signal by inserting a second count256.vhd in the current design. This should be in series with the previous count256. For more on cascading counters see Section 5.10 on page 137.

Mod256 Counter output	Frequency of divided clock, $4 \times 10^6/2^n$, n=1,2,3,…16; input counter clock frequency = 4Mhz
Q_0	$4\text{Mhz}/2^1 = 2\text{Mhz}$
Q_1	$4\text{Mhz}/2^2 = 1\text{Mhz}$
Q_2	$4\text{Mhz}/2^3 = 500\text{khz}$
Q_3	$4\text{Mhz}/2^4 = 250\text{khz}$
Q_4	$4\text{Mhz}/2^5 = 125\text{khz}$
Q_5	$4\text{Mhz}/2^6 = 62.5\text{khz}$
Q_6	$4\text{Mhz}/2^7 = 31.25\text{khz}$
Q_7	$4\text{Mhz}/2^8 = 15.625\text{khz}$

Table 9.3. Frequency division of the 4Mhz clock using the mod 256 counter (8 bit synchronous binary counter).

3. Program the PLD with the modified circuit created in step 1. Now, power down the PLD before going to the next step.

4. Build the DACC0808 circuitry as shown in Figure 9.11. Program the DC voltage of −12 V and 5V in the power supply and disable the outputs until needed. **Have your instructor sign off on the built circuit.** As can be seen in

5. Figure 9.11, the output of the DAC0808 is the current Io. Since the output of the DAC0808 is a current it can easily be converted into a voltage by using the 741 amplifier circuit as shown. The amplifier circuit is a typical inverting amplifier configuration. Therefore, the output voltage can be amplified by controlling the resistor R_f. The output voltage is computed by:

$$Vo = -Io \times R_f$$

where the current Io is calculated by:

$$Io = K\left(\frac{A_1}{2} + \frac{A_2}{4} + \frac{A_3}{8} + \frac{A_4}{16} + \frac{A_5}{32} + \frac{A_6}{64} + \frac{A_7}{128} + \frac{A_8}{256}\right)$$

and where K is

$$K \cong \frac{V_{ref}}{R_{ref}}$$

Once the DAC0808's surrounding circuit is built connect the digital inputs of the DAC0808 to the outputs of the 8 bit counter as shown.

6. Now, turn the power on and supply the −12V, +12V, and +5V to their respective inputs. Then power up the PLD circuit. Make sure the ground signal of the PLD circuit is tied to the ground signal of the DAC0808 circuit.

7. Use an oscilloscope and verify that the output waveform Vo is similar to the one shown in Figure 9.12. Print the waveform for your record. If you are not getting a similar waveform trouble shoot the circuit. **Have your instructor sign the waveform for approval.**

8. Repeat steps 2, 3, 5, and 6 for the following clock frequencies: 125KHz and 4MHz.

Note: Can use any PLD.

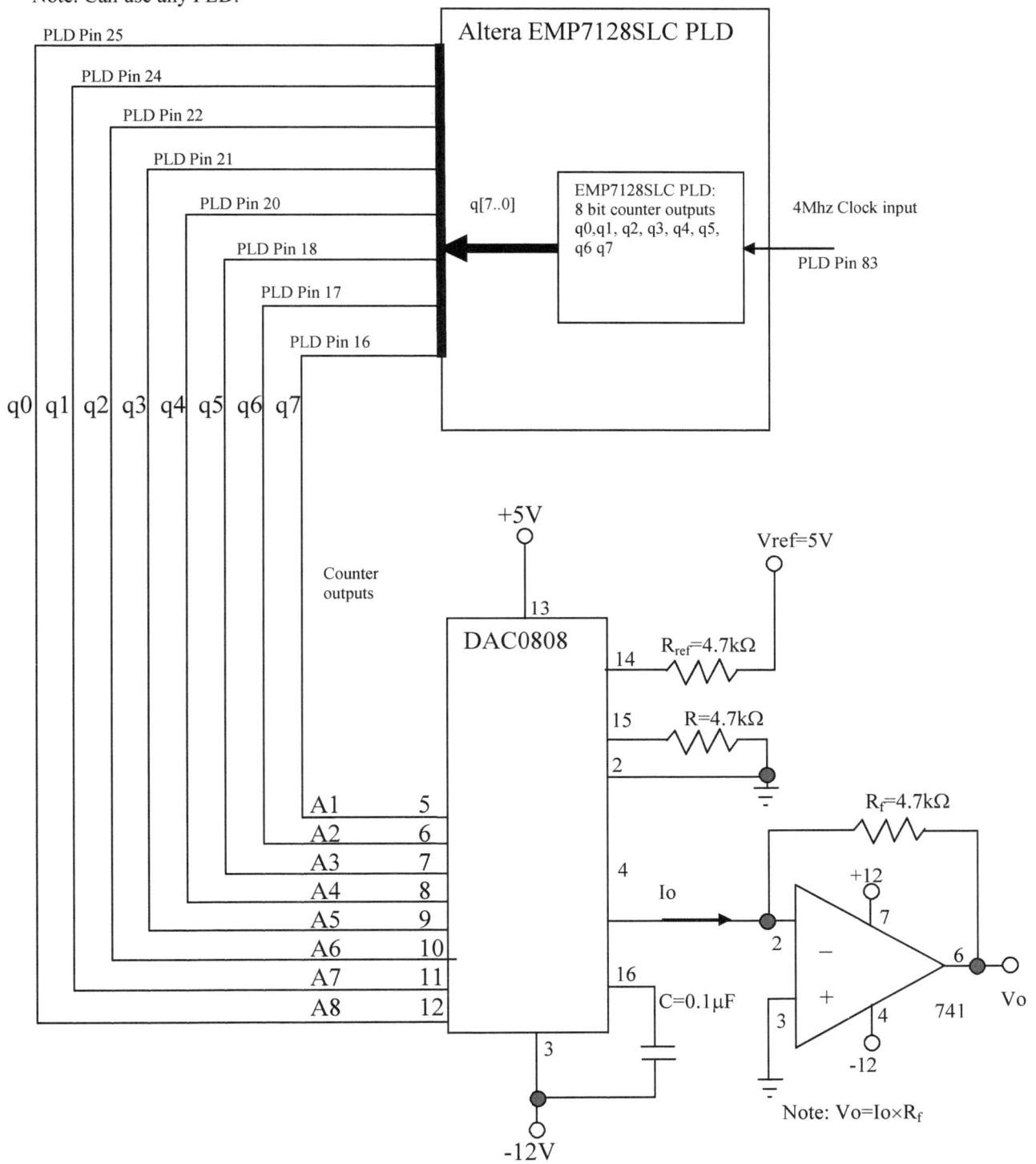

Figure 9.11. Circuitry for DAC0808 chip.

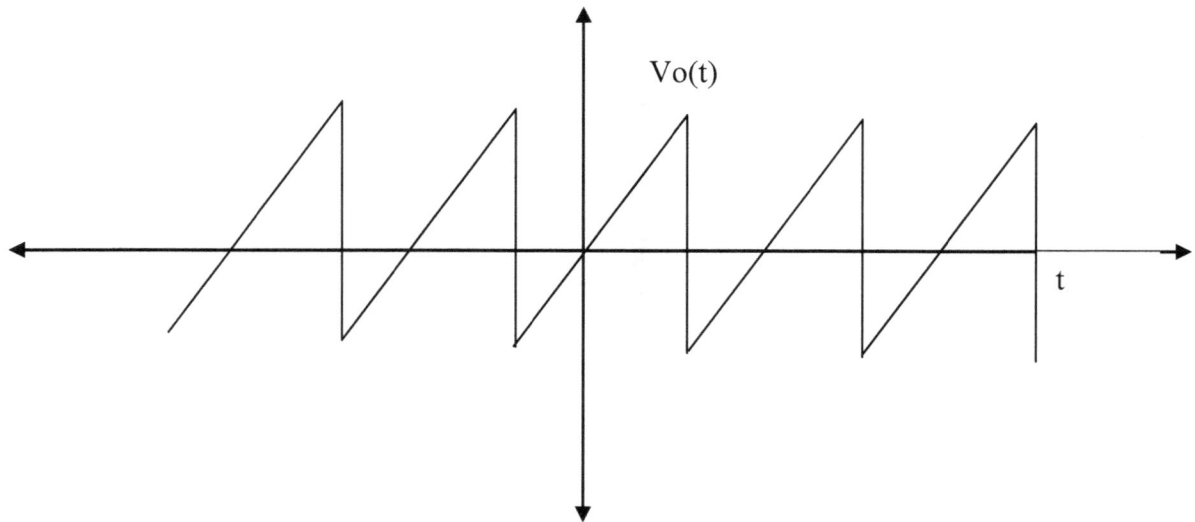

Figure 9.12. Expected output voltage waveform, Vo, for DAC0808 circuit shown in Figure 9.11.

9.3.2.   VHDL code for the 8 bit counter(count256.vhd)

The 2^8 provides the test data for the DAC0808. It counts from 0 to 255. The VHDL code for the 2^8 counter is listed as follows:

```
library ieee;
use ieee.std_logic_1164.all;

entity count256 is
port (

clk : in std_logic;
clr: in std_logic;
q: out integer range 0 to 255);

end count256;

architecture a of count256 is
begin

    process (clk)
    variable cnt: integer range 0 to 255;
    begin
        if (clr='0')then
            cnt:=0;
```

```
        else
            if(clk'event and clk='1')then
                cnt:=cnt+1;
            end if;
        end if;
        q<=cnt;

    end process;

end a;
```

9.4. Testing the ADC0804 and DAC0808 Together using a PLD and a Parallel to Parallel Interface

This section discusses the operation of the ADC0804 and DAC0808 as a complete A/D and D/A system. The block diagram of this system is shown in Figure 9.13. The function generator generates a sinusoidal signal in the amplitude range required by the ADC0804. This signal must be between 0-5 volts. This is half the full scale range of the ADC0804. Since a sinusoidal has both positive and negative values and the ADC0804 has a full scale range of 0-5V the sinusoidal should be offset by a DC voltage of 2.5V. This allows the sinusoidal to be converted without distortion or without losing the negative cycle in the A/D process. That is, the negative amplitudes will be converted to digital 0 if the DC offset is not applied. The procedure for performing this setup is given next.

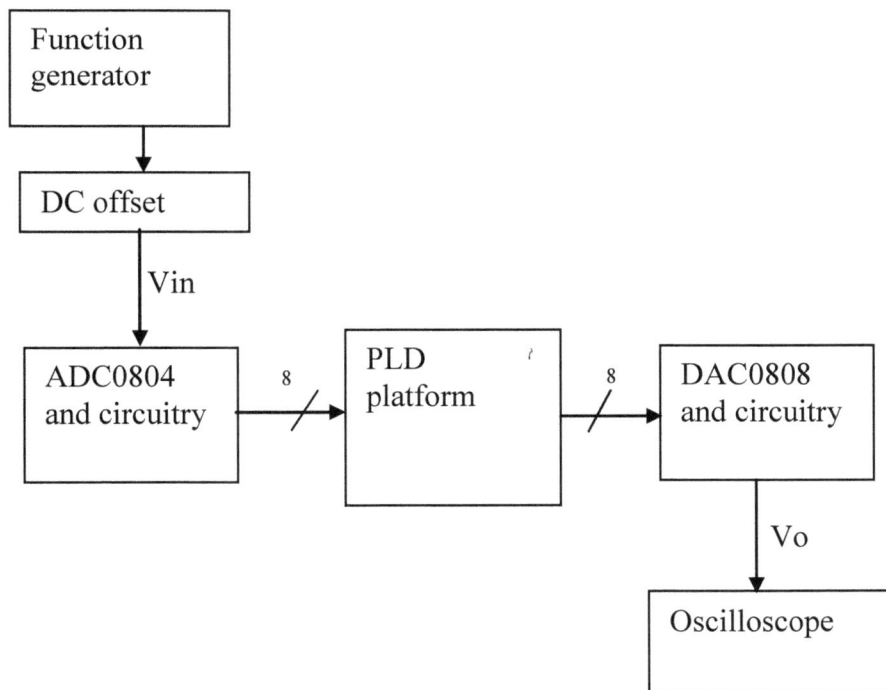

Figure 9.13. A complete A/D and D/A system using the ADC0804 and DAC0808.

9.4.1. Procedure for Building and Testing the ADC0804 and DAC0808 System

The procedure for building and testing the ADC0804 and DAC0808 system is given as follows:

1. Assuming that the ADC0804 and DAC0808 circuits are tested and working create a Quartus project file, a2d0804_2_d2a0808_ctl.qpf, and complete the digital circuit shown Figure 9.14 in the graphic editor. Compile and save the project.
2. Power down the ADC0804 and DAC0808 circuits. Program the PLD with the circuit in Figure 9.14.
3. Build the circuits in Figure 9.15 and Figure 9.16. Use a potentiometer to test the ADC0804 and DAC0808 together. You should see 0V or 5V on the DAC's output when the pot is turn to its extremes (see Section 2.11 on page 38 for OPAMPs)
4. Disconnect the potentiometer's output from the ADC0804's analog input. The output of the function generator should now be connected to the ADC0804's analog input. Test the DAC0808 using the following sinusoidal sources: 100Hz, 500Hz, 1000Hz, 5000Hz, 15000Hz, 50000Hz, 200KHz, and 600KHz. Make sure you apply the DC offset for all cases. Measure and print the waveforms showing the frequencies and peak-to-peak amplitudes of each waveform using the oscilloscope.
5. Have the instructor sign your work when done.

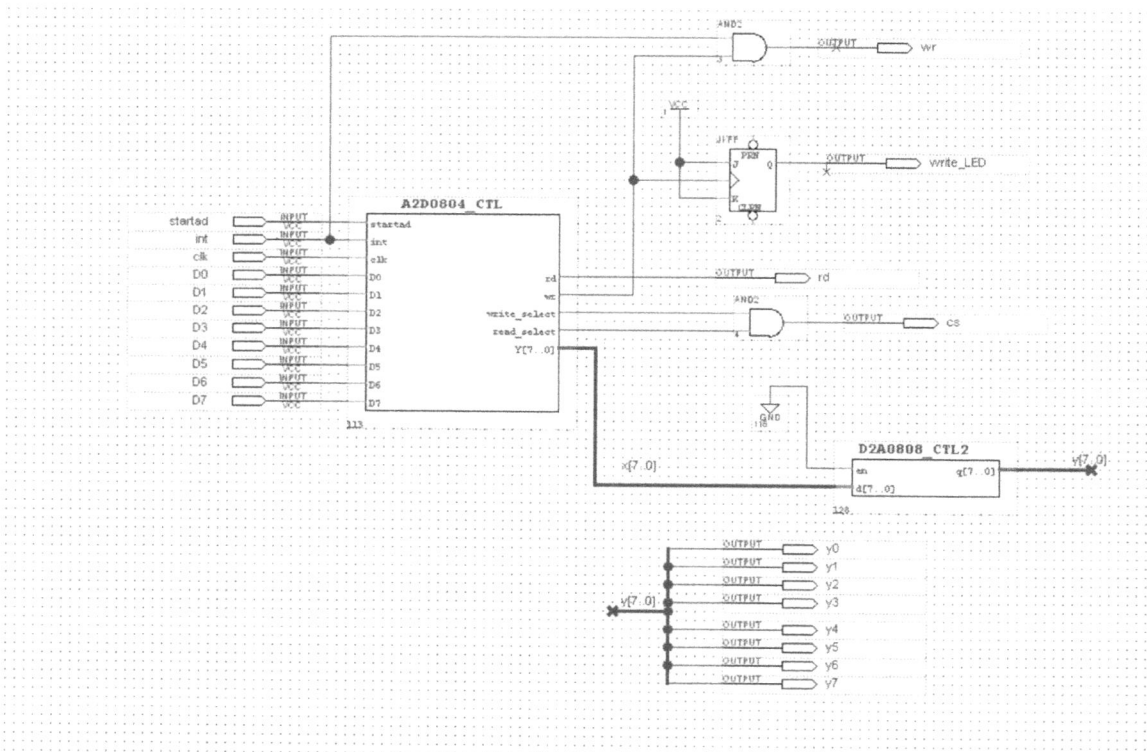

Figure 9.14. Digital circuit (Quartus file: a2d0804_2_d2a0808_ctl.qpf) for the ADC0804 and DAC0808 system.

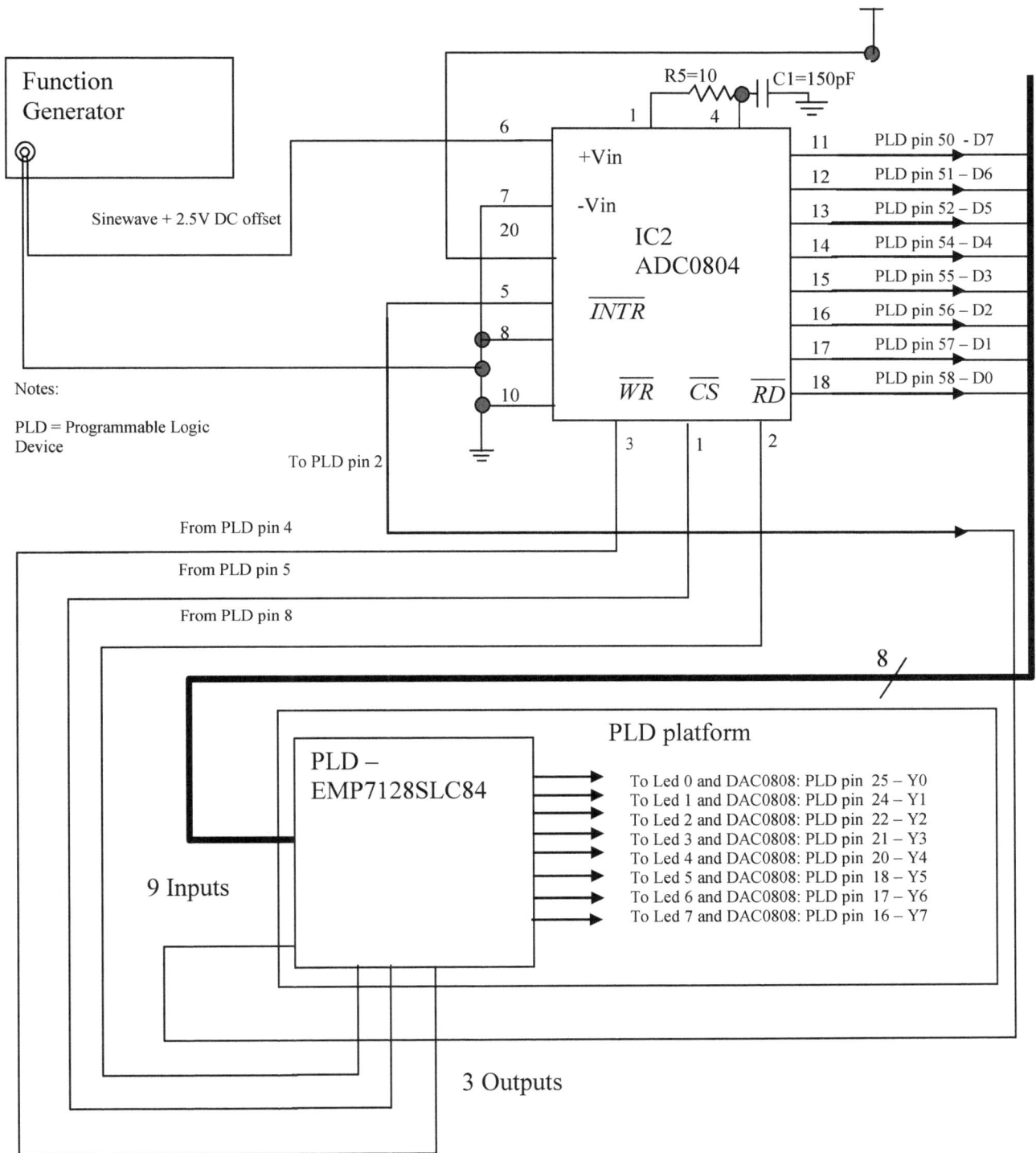

Figure 9.15. Analog transmit circuitry.

Note: Can use any PLD.

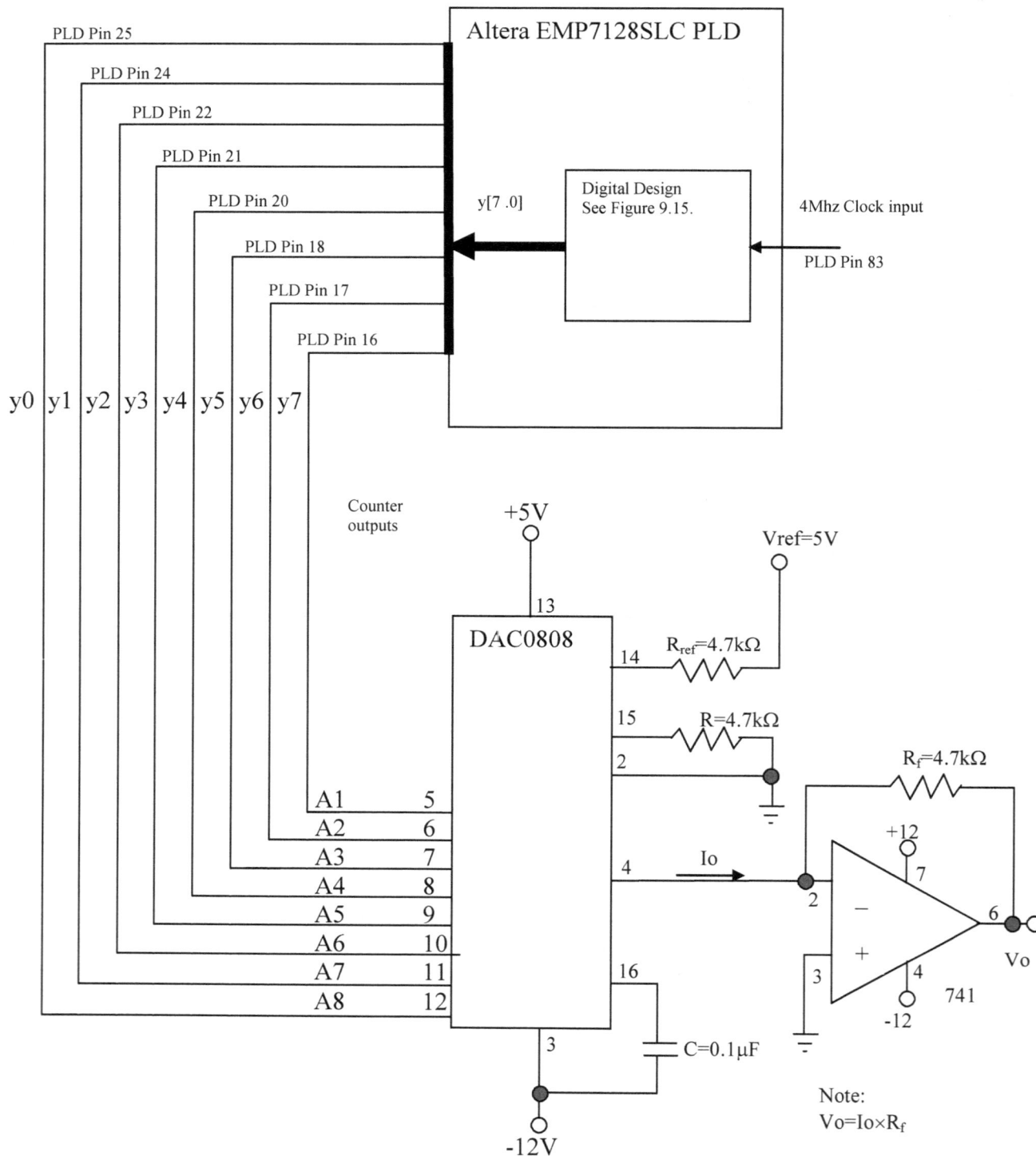

Figure 9.16. Analog receive circuitry for DAC0808 chip.

9.4.2. VHDL code for the d2a0808_ctl.vhd

The VHDL code for interfacing to the DAC0808 is listed below:

```
library ieee;
use ieee.std_logic_1164.all;

entity d2a0808_ctl2 is
    port(en:in std_logic;
         d: in std_logic_vector(7 downto 0);
       q: out std_logic_vector( 7 downto 0));
end d2a0808_ctl2;

architecture a of d2a0808_ctl2 is
begin
    process (en,d)
    begin
    if (en='0')then
         q<=d;
    end if;
    end process;
end a;
```

9.5. Testing the ADC0804 and DAC0808 Together Using a PLD and Using a Parallel to Serial Converter and Serial to Parallel Converter

This section is an extension of the previous section which dealt with an 8 bit parallel to parallel interface using an analog to digital converter's (ADC0804) parallel outputs that interfaced to the parallel inputs of a digital to analog converter (DAC0808). In this section the parallel output data of the ADC0804 is converted to serial data using a parallel in to serial output (PISO) converter. PISOs are discussed in Section 5.14.3 on page 162. This serial data is then converted back to parallel data for the DAC0808 parallel inputs using a SIPO. SIPOs are discussed in Section 5.14.2 on page 160. This is a pretest for a wireless voice data system that requires a serial bit stream for the transmitter's input data. Such systems are discussed in Chapter 17 on page 540. This serial data is then used to modulate the transmitter's carrier frequency using one of several modulation methods: ASK, FSK, PSK, QAM, etc. This modulated carrier is then sent to an antenna which allows the signal to be transmitted over a wireless channel to a receiver. The receiver demodulates received signal to recover the serial data. This serial data is then fed to a serial to parallel converter that converts the serial data back to parallel for the DAC0808's parallel input. The procedure for performing this experiment is given in the next section.

9.5.1. Procedure for Testing the ADC0804 and DAC0808 together using a PLD and using a Parallel to Serial Converter and Parallel to Serial Converter

The procedure is given as follows:

1. Create a Quartus project file, a2d0804_2ser_ser2_d2a0808_ctl.qpf. Add the digital circuits from the previous sections: txuart2.bdf, count65536.vhd, and rxuart.bdf. Begin by constructing the digital circuit for the txuart.bdf module is explained in Section 9.6, The txuart, on page 344. The txuart.bdf module converts parallel data to serial data using a 16 to 1 multiplexer. It consists of several VHDL files which are also given in Section 9.6. Finally the VHDL code for count65536.vhd is also given in Section 9.6.2, VHDL Code for the 2^{16} Counter (counter65536.vhd) on page 349. Finally, the rxuart.bdf module converts parallel data to serial data. It is explained in Section 9.7, The rxuart, on page 351.

2. The completed digital circuit should look like Figure 9.17. When the circuit is complete assign the correct PLD device and compile it. Then assign the input/output pins assignments as shown in Figure 9.18. Compile the circuit again and program it into the target PLD.

6. If not already done so, build the circuits in Figure 9.15 and Figure 9.16. Use a potentiometer to test the ADC0804 and DAC0808 together. You should see 0V or 5V on the DAC's output when the pot is turn to its extremes. OPAMPs are covered in Section 2.11 on page 38.

3. Disconnect the potentiometer's output from the ADC0804's analog input. The output of the function generator should now be connected to the ADC0804's analog input. Test the DAC0808 using the following sinusoidal sources: 100Hz, 500Hz, 1000Hz, 5000Hz, 15000Hz, 50000Hz, 200KHz, and 600KHz. Make sure you apply the DC offset for all cases. Measure and print the waveforms showing the frequencies and peak-to-peak amplitudes of each waveform using the oscilloscope.

4. Have the instructor sign your work when done.

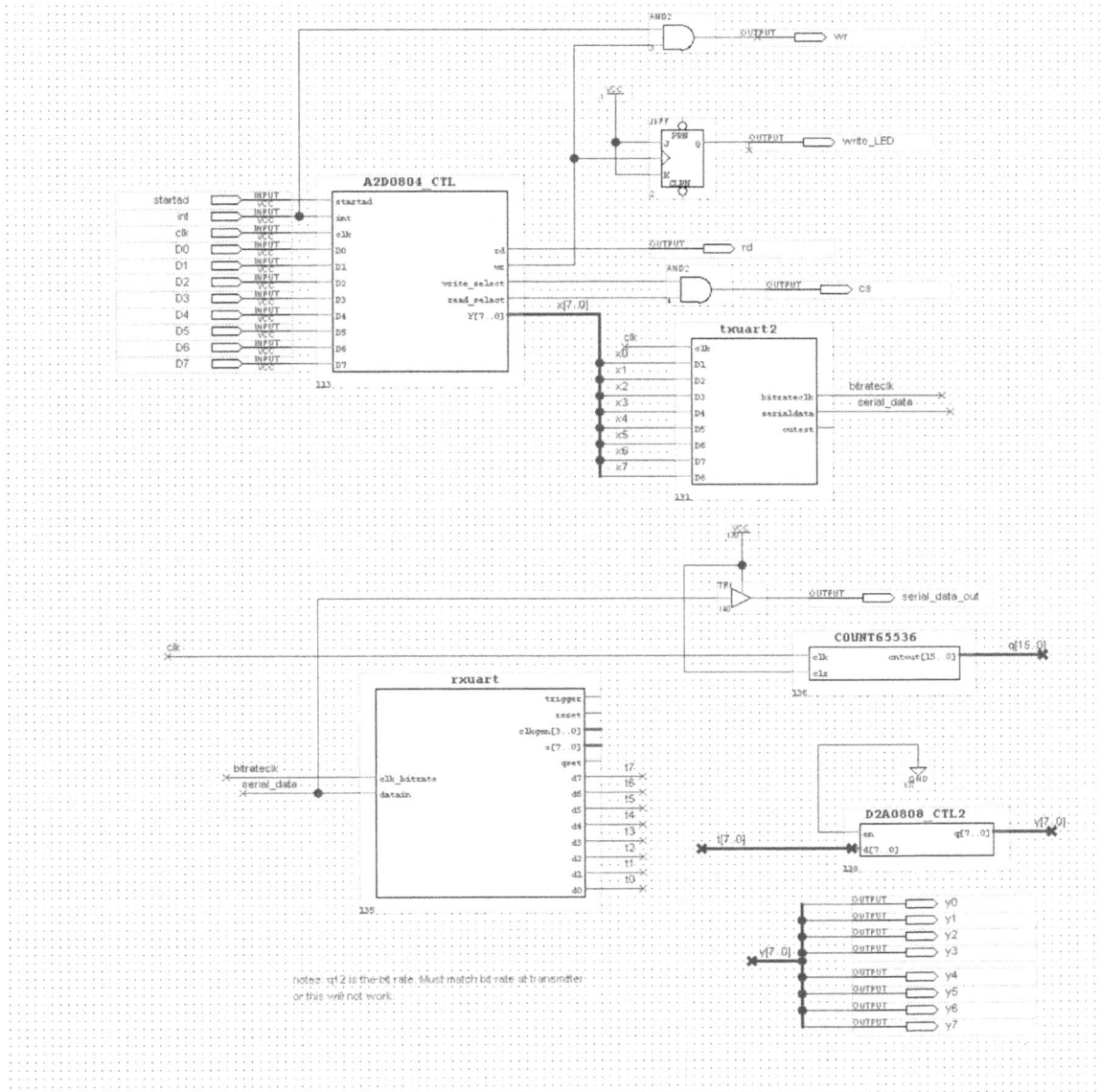

Figure 9.17. Parallel to serial and serial to parallel digital circuit for interfacing analog to digital and digital to analog conversion circuits (Quartus File: a2d0804_2ser_ser2_d2a0808_ctl.qpf). This shows how parallel digital data can be converted to serial data in preparation for modulation and transmission over a one link channel.

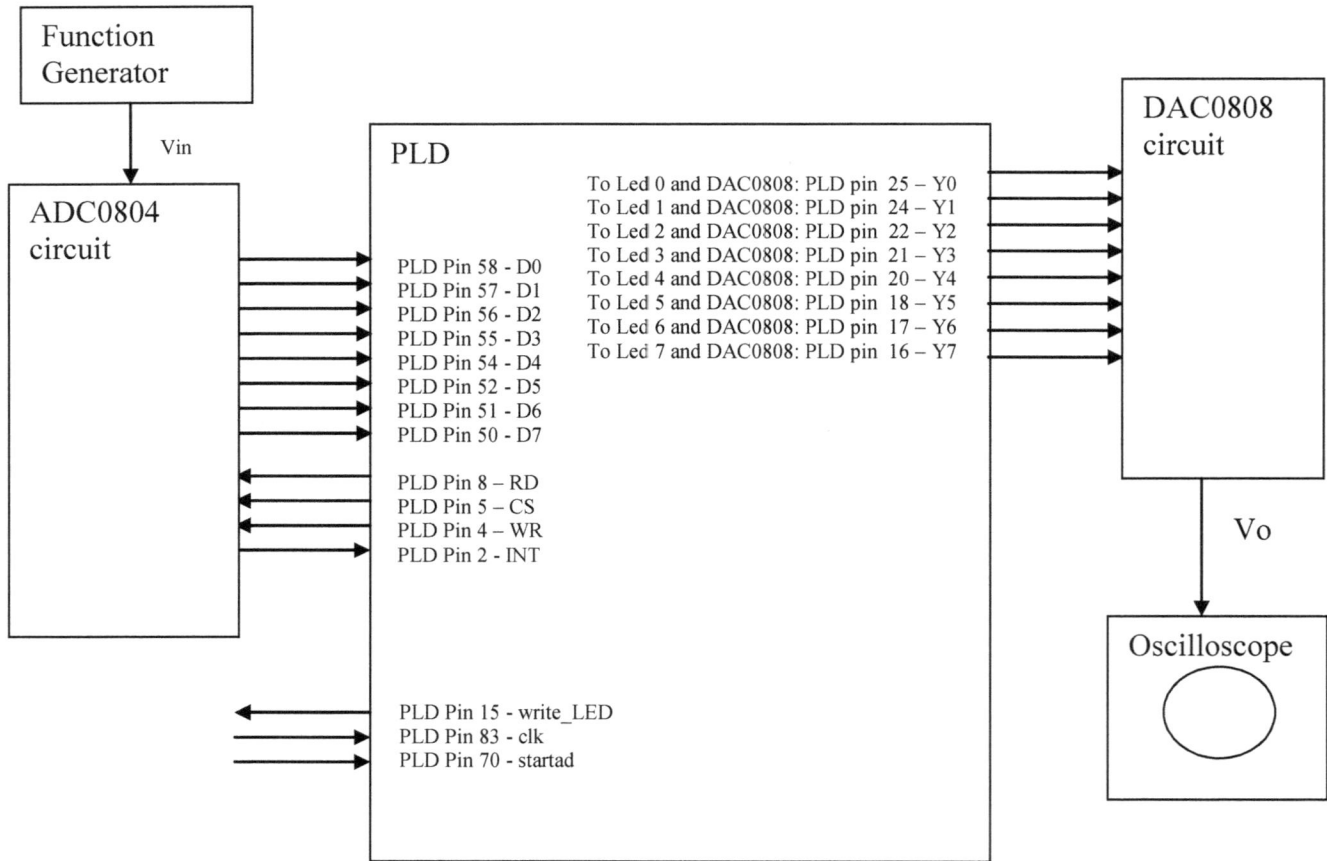

Figure 9.18. Pin assignments for the digital circuit in Figure 9.18. The PLDs with these pin assignments are: Altera EMP7128SLC84 or EMP7160SLC84.

9.6. **The txuart**

The low level description of the txuart module is shown in Figure 9.19.   The corresponding Quartus project file, txuart.qpf, of Figure 9.19 is shown in Figure 9.20. In Figure 9.19, **fb** is the bit rate of the txuart serial data. **fclk** is the 4MHz clock which is the frequency of the crystal oscillator which resides on the PLD platform (DeVry ESOC). **fb** represents the clock of the txuart output serial data which is derived by dividing down the 4Mhz clock. fb shown in Figure 9.19 is $4MHz/2^{12} = 976.56$ Hz. This is taken from Q_{11} of the divide by 2^{16} counter output (Q_{15}, Q_{14}, Q_{13}, Q_{12}, ..., Q_1, Q_0). See Table 9.4 for the actual frequencies of the divided down 4Mhz clock using a mod 65536 counter and Table 9.5 for further division of the 4Mhz clock using an additional mod 256 counter in cascade. For more on cascading counters see Section 5.10 on page 137.

The VHDL code for the mod 65536 counter is shown in Section 9.6.2 on page 349. The mod11 counter counts from 0 to 10 selecting only the first 11 inputs of the multiplexer and then recycling to 0 again. Note that only two stop bit is selected since the first 11 inputs of the 16 to 1 MUX are selected. The multiplexer in Figure 9.19 is a 16 to 1 multiplexer. Multiplexers are discussed in Section 4.11 on page 94. It is used to convert the incoming 8 bit parallel data into serial data.

Figure 9.19 yields serial data that follows the standard UART format. The details of the standard UART serial data format is shown below in Figure 9.21 with a total of 11 bits. Figure 9.21 shows the details of the standard 16 to 1 MUX serial output data format with a total of 11 bits. The start bit is transmitted first, followed by the least significant data bit, the remaining data bits, and lastly the two stop bits. The duration of one bit is called the bit period and is represented by **tb**. The inverse of the bit period is called the bit rate and is denoted by **fb**. It takes 11 bit periods to send all 11 bits. The VHDL code for the divide by 2^16 counter is shown in the next section. It is also called a mod 65,536 counter because it has 65,536 possible output states. It counts from 0 to 65,535. The mod11 counter has only 11 possible output states. It counts from 0 to 10. Additionally, a mod12 counter has only 12 possible output states. It counts from 0 to 11.

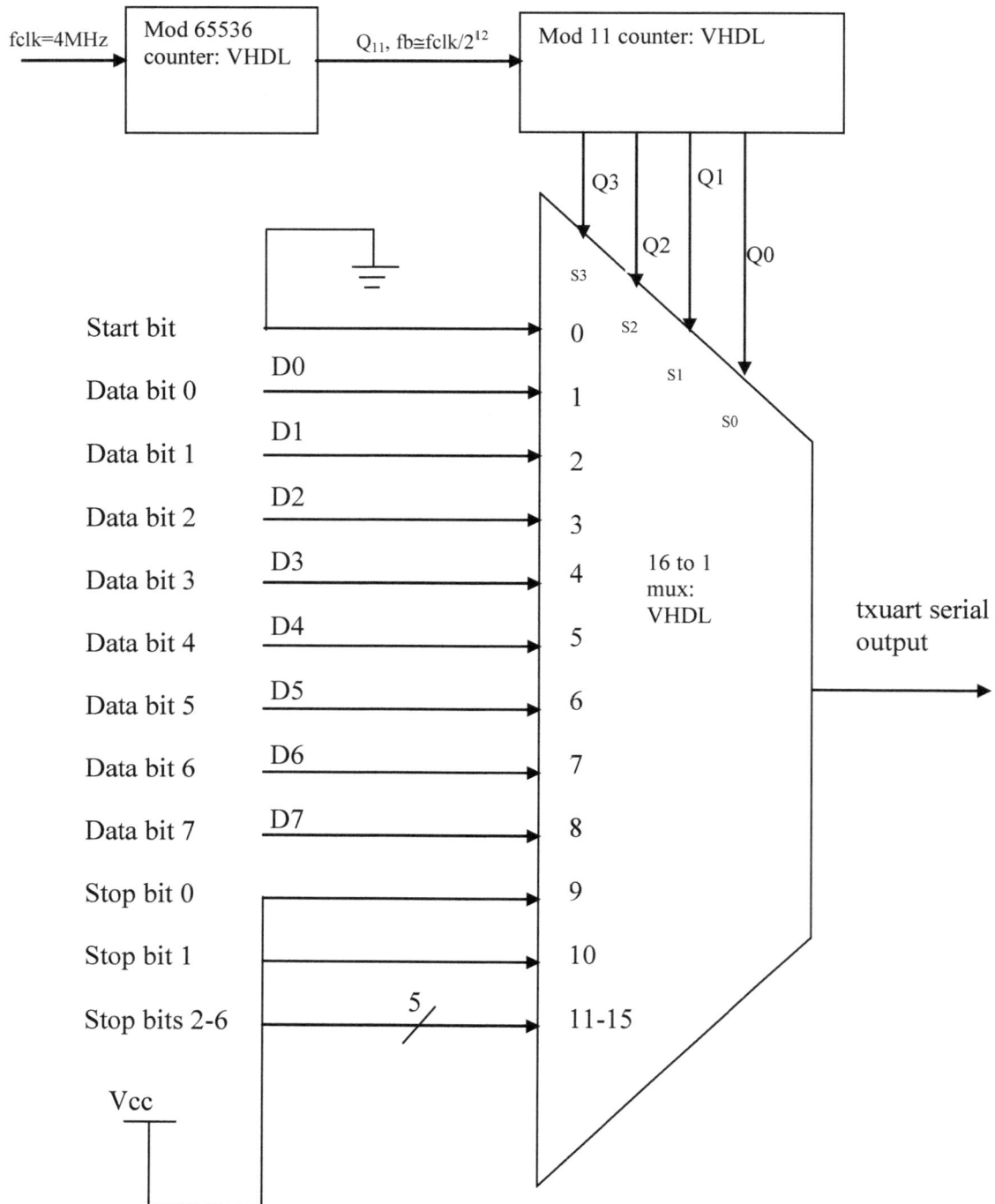

Figure 9.19. Universal serial asynchronous serial transmitter (UART) TX module low level representation.

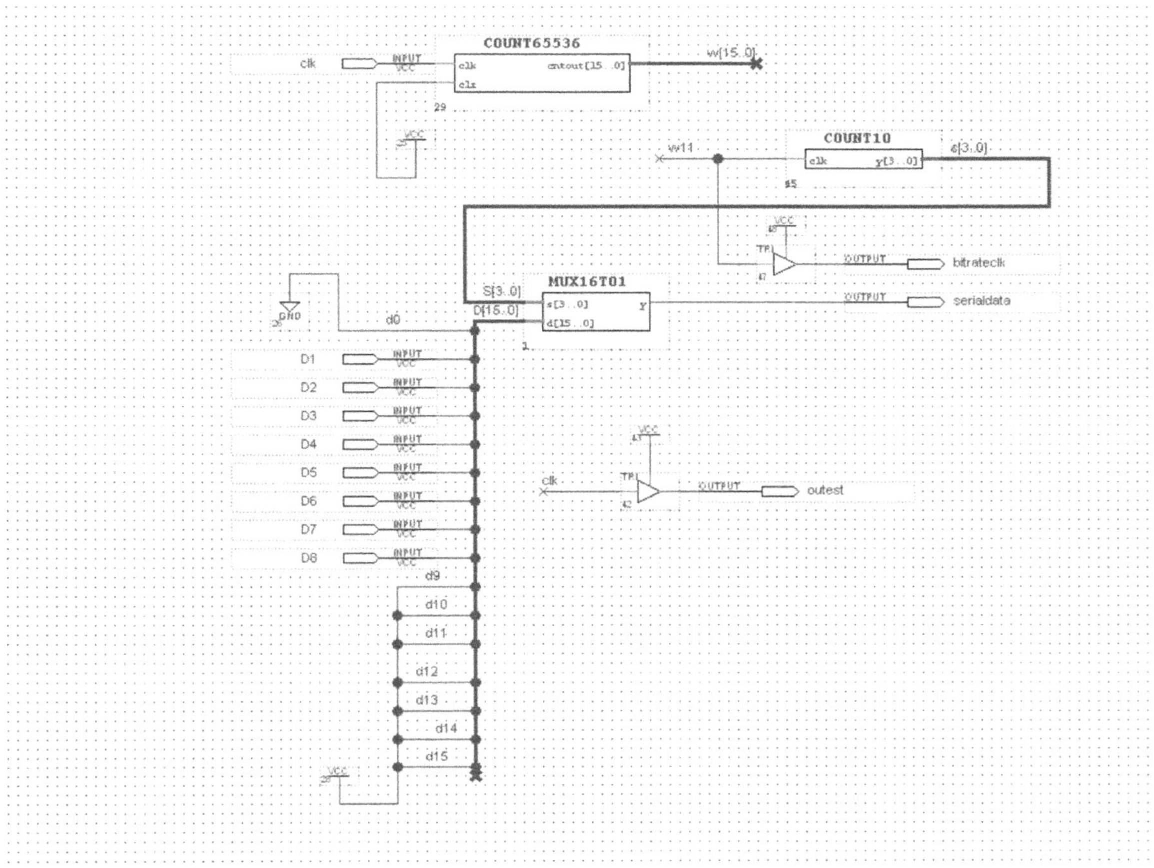

Figure 9.20. Quartus II version of txuart circuit (Quartus file: txuart.qpf) shown in Figure 9.19.

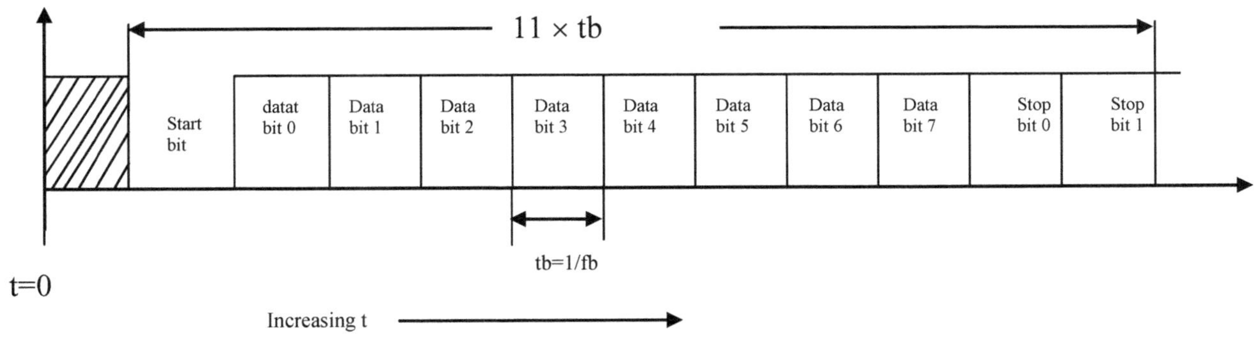

Notes: Start bit = logic 0; Data bit 0-7 = logic 0 or 1; Stop bit 0 = logic 1; Stop bit 1 = logic 1; tb = bit period (seconds); fb = bit rate = 1/tb (hertz).

Figure 9.21. Standard UART serial data format.

Mod65536 Counter output	Frequency of divided clock, $4 \times 10^6/2^n$, $n=1,2,3,\ldots 16$
Q_0	$4\text{Mhz}/2^1 = 2\text{Mhz}$
Q_1	$4\text{Mhz}/2^2 = 1\text{Mhz}$
Q_2	$4\text{Mhz}/2^3 = 500\text{khz}$
Q_3	$4\text{Mhz}/2^4 = 250\text{khz}$
Q_4	$4\text{Mhz}/2^5 = 125\text{khz}$
Q_5	$4\text{Mhz}/2^6 = 62.5\text{khz}$
Q_6	$4\text{Mhz}/2^7 = 31.25\text{khz}$
Q_7	$4\text{Mhz}/2^8 = 15.625\text{khz}$
Q_8	$4\text{Mhz}/2^9 = 7.8125\text{khz}$
Q_9	$4\text{Mhz}/2^{10} = 3.90625\text{khz}$
Q_{10}	$4\text{Mhz}/2^{11} = 1.953125\text{khz}$
Q_{11}	$4\text{Mhz}/2^{12} = 976.5625\text{hz}$
Q_{12}	$4\text{Mhz}/2^{13} = 488.28125\text{hz}$
Q_{13}	$4\text{Mhz}/2^{14} = 244.140625\text{hz}$
Q_{14}	$4\text{Mhz}/2^{15} = 122.0703125\text{hz}$
Q_{15}	$4\text{Mhz}/2^{16} = 61.03515625\text{hz}$

Table 9.4. Frequency division of the 4Mhz clock using the mod 65536 counter (16 bit synchronous binary counter).

Mod 256 Counter output (equivalent output due to cascading the 65536 with the mod 256 counter)	Frequency of divided clock, $61.03515625/2^n$, $n=1,2,3,\ldots 8$
Q_0 (Q_{16})	$4\text{Mhz}/2^{17} = 30.517578125\text{hz}$
Q_1 (Q_{17})	$4\text{Mhz}/2^{18} = 15.2587890625\text{hz}$
Q_2 (Q_{18})	$4\text{Mhz}/2^{19} = 7.62939453125\text{hz}$
Q_3 (Q_{19})	$4\text{Mhz}/2^{20} = 3.81469726563\text{hz}$
Q_4 (Q_{20})	$4\text{Mhz}/2^{21} = 1.90734863281\text{hz}$
Q_5 (Q_{21})	$4\text{Mhz}/2^{22} = 0.953674316406\text{hz}$
Q_6 (Q_{22})	$4\text{Mhz}/2^{23} = 0.476837158203\text{hz}$
Q_7 (Q_{23})	$4\text{Mhz}/2^{24} = 0.238418579102\text{hz}$

Table 9.5. Further division of the 4Mhz clock can be done by cascading the mod 65536 counter with a mod 256 counter. The mod 256 counter is an 8 bit synchronous counter. The Q15 output of the mod 65536 feeds the clock input of this mod 256 counter to give the frequencies indicated.

9.6.1. Procedure for the txuart

The procedure for constructing the txuart is listed as follows:

1. Create the project txuart.qpf and construct the circuit in Figure 9.20. You must first copy and compile the VHDL code for the following modules: count65536.vhd, mux16to1.vhd, and count10.vhd. See Table 9.6 below. Once the circuit is complete compile it. Notice that count11.vhd is not provided. You must modify a copy count65536.vhd to count from 0 to 10. That is, make a copy of count65536.vhd and rename it count11.vhd. Now edit the file, changing just three lines that will cause the counter to count from 0 to 10. The integer range should now go from **0 to 15** instead from **0 to 65535**. The if statement should be changed from **if (clr='0')then** to **if (clr='0' or cnt=11)then.**

VHDL File Name	Description
count65536.vhd	Counter that counts from 0 to 65535
mux16to1.vhd	16 to 1 multiplexer
count11.vhd (mod 11 counter)	Counter that counts from 0 to 10

Table 9.6. List of VHDL files needed for this procedure.

2. Now assign the correct device and corresponding pin assignments. If using the Altera EMP7128SLC84 or EMP7160SLC84 use the assign pins given in Figure 9.22 to interface the completed digital circuit to the outside world.
3. Compile the circuit. Now program the circuit into the PLD. Set the switches to the following: $D_8D_7D_6D_5D_4D_3D_2D_1=10101011$. Use an oscilloscope to verify that this data is being transmitted. The data transmitted follows the standard UART format shown in Figure 9.21. Also try using other values for $D_8D_7D_6D_5D_4D_3D_2D_1$.
4. Print the waveforms measured in step 3. Label the waveforms in detail indicating the start bit, eight data bits, and stop bits one by one. You must verify this to the instructor using the sign off below.

Instructor sign off_____ Date_____

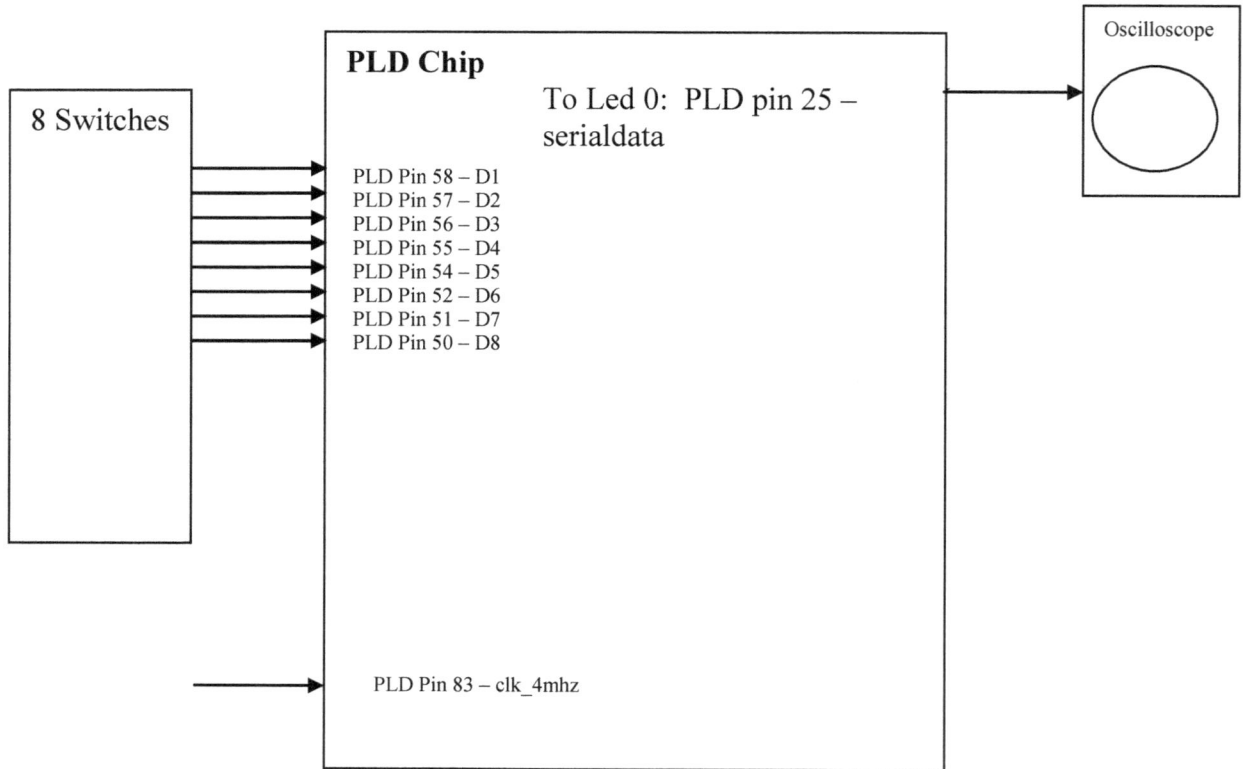

Figure 9.22. Pin assignments for the digital circuit Figure 9.20. The PLDs meeting these pin assignments are: Altera EMP7128SLC84 or EMP7160SLC84.

9.6.2. VHDL Code for the 2^{16} Counter (counter65536.vhd)

The 2^{16} counter provides the timing for all the VHDL modules. It counts from 0 to 65535. The VHDL code for the 2^{16} counter is listed as follows:

```
library ieee;
use ieee.std_logic_1164.all;

entity count65536 is
port (

clk : in std_logic;
clr: in std_logic;
cntout: out integer range 0 to 65535);

end count65536;

architecture a of count65536 is
begin
```

```
      process (clk)
      variable cnt: integer range 0 to 65535;
      begin
            if (clr='0')then
                  cnt:=0;
            else
                  if(clk'event and clk='1')then
                        cnt:=cnt+1;
                  end if;
            end if;
            cntout<=cnt;

      end process;

end a;
```

9.6.3.   VHDL Code for the 16 to 1 Multiplexor (mux16to1.vhd)
The VHDL code for the 16 to 1 multiplexer is given below:

```
library ieee;
use ieee.std_logic_1164.all;

entity mux16to1 is
port (
s: in integer range 0 to 15;
d: in std_logic_vector (15 downto 0);
y: out std_logic );
end mux16to1;

architecture a of mux16to1 is
begin

process (s)
begin

case s is
      when 0 =>
            y<=d(0);
      when 1 =>
            y<=d(1);
      when 2 =>
            y<=d(2);
      when 3 =>
            y<=d(3);
      when 4 =>
            y<=d(4);
```

```
          when 5 =>
                  y<=d(5);
          when 6 =>
                  y<=d(6);
          when 7 =>
                  y<=d(7);
          when 8 =>
                  y<=d(8);
          when 9 =>
                  y<=d(9);
          when 10 =>
                  y<=d(10);
          when 11 =>
                  y<=d(11);
          when 12 =>
                  y<=d(12);
          when 13 =>
                  y<=d(13);
          when 14 =>
                  y<=d(14);
          when 15 =>
                  y<=d(15);
end case;
end process;

end a;
```

9.7. The rxuart

The low level description of the rxuart module is shown in Figure 9.23 below. The corresponding Quartus project file, rxuart.qpf, is shown in Figure 9.24. The rxuart consists of five main components:

1. JK flip flop
2. clock generator
3. serial in parallel out (SIPO) register
4. parallel in parallel out (PIPO) register
5. one shot

The VHDL code for most of these components is shown in the following sections. The clock generator, clkgen.vhd, provides a four bit clock to the SIPO register. Shift registers are discussed in Section 5.14 on page 158. Specifically, the SIPO is discussed in Section 5.14.2 on page 160. The input clock of the clock generator must be the same as the serial data bit rate, fb. For example if the bit rate of the serial data is 973.5625 bps, then the input clk to the clock generator must be 973.5625 Hz. The clock generator is activated when it receives the set from the negative edge JK flip flop. The JK flip flop is in the set mode. For more on JK flip flops see Section 5.7 on page 126. When a high to low transition in the serial data occurs, which is the start bit equal to logic 0, the JK flip flop sends a set to the clk generator. This causes the clk generator to be enable. It then counts from 0 to 15. Counters are discussed in Section 5.9 on page 135. As long as the enable is high the clock generator counts 0 to 15 continuously. The SIPO register, ser2parsrg8_3.vhd, is used to convert the serial data into parallel data and is synchronized to the main clk and serial data using the four bit clock from the clock generator. When the conversion is complete an active low output, which is the trigger signal, is sent to the one-shot, oneshot_2.vhd and the PIPO register. PIPOs are discussed in Section 5.14.4 on page 162. The PIPO register loads the data from the SIPO register when the trigger becomes active low. When the one-shot receives the active low trigger from the SIPO register, a logic 0 is sent to the JK flip flop's asynchronous reset input causing it to reset to a logic 0. This cycle starts again upon detection of a high to low transition in the input serial data.

The simulation of the RX UART in Figure 9.24 is shown in Figure 9.25. The following test data bit groups are shifted in serially as shown in the table:

Data bit group	datain=0$D_1D_2D_3D_4D_5D_6D_7$11
1	01100111011
2	00111110111
3	00011000000

Table 9.7. The start bit (logic 0) is shifted in first followed by the data bits as follows: LSB=D0, then D1, D2, D3, ..., to the MSB=D7. Lastly, the two stop bits (logic 1) are shifted in.

Figure 9.25 shows that the data bit groups are shifted in serially as indicated in Table 9.7 and shifted out in parallel as can be seen by the output of the eight bit parallel in to parallel out (PIPO) register (par2parsrg8.vhd) for bits d0,d2,d3,d4,d5,d6,d7.

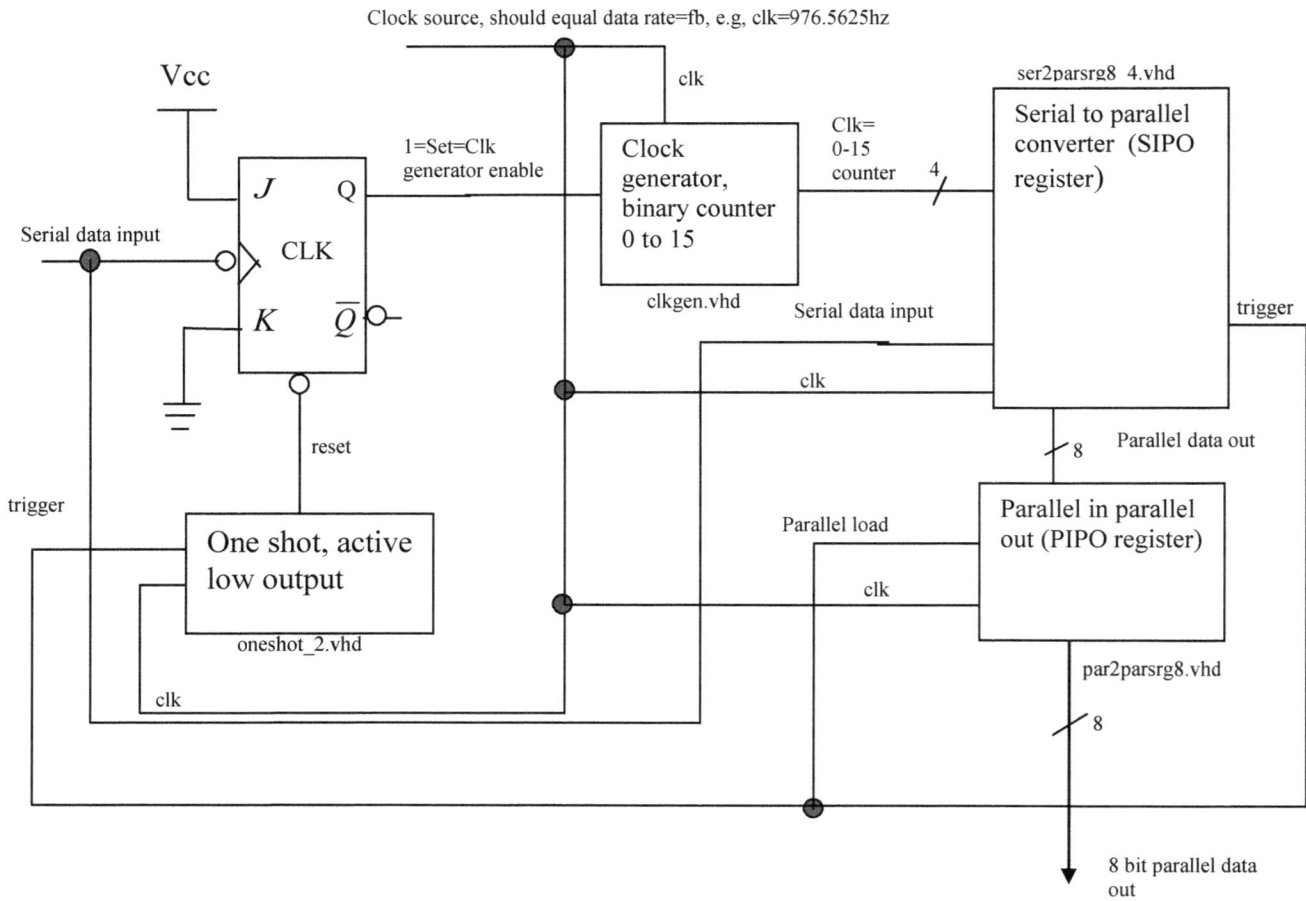

Figure 9.23. Receive UART architecture.

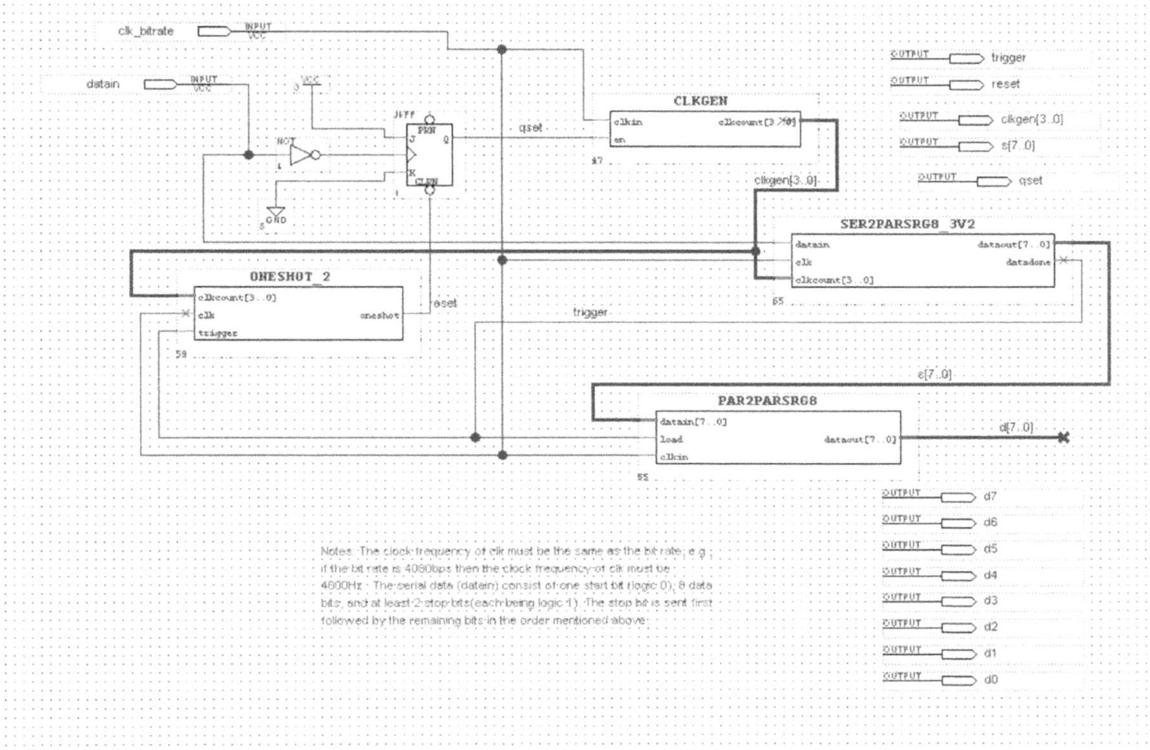

Figure 9.24. Receive UART architecture in Quartus II (Quartus file: rxuart.qpf).

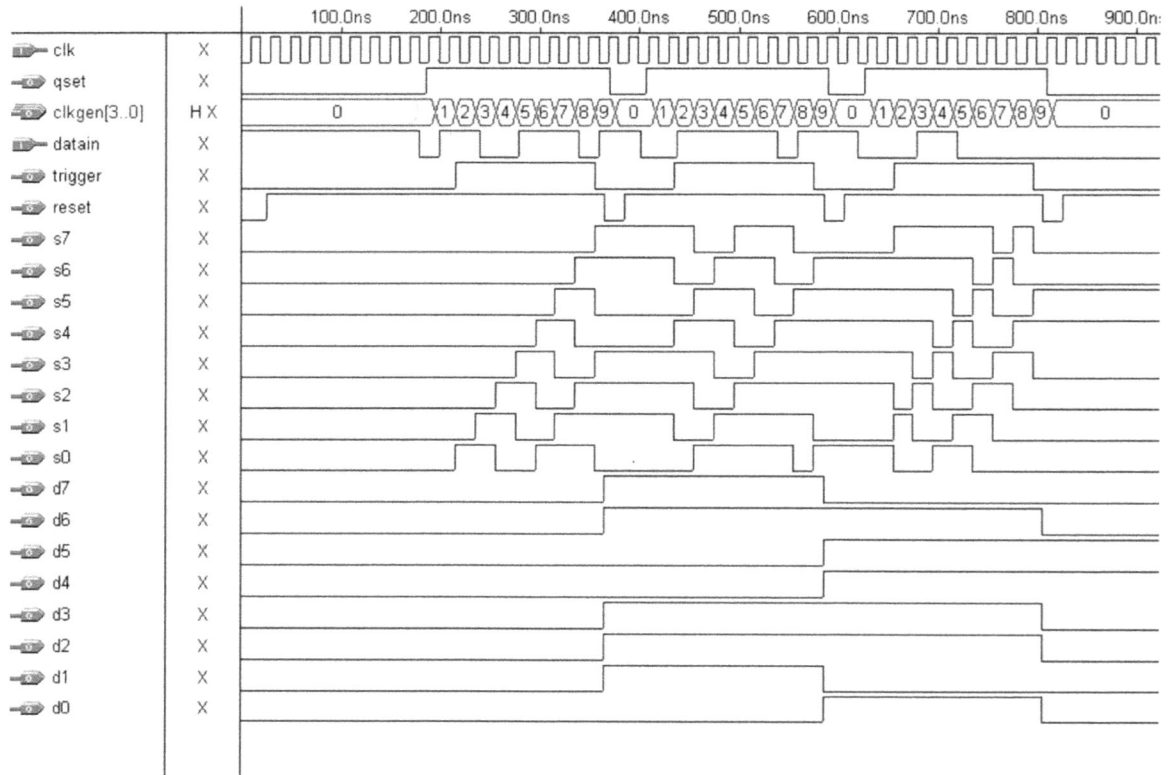

Figure 9.25. Simulation of Receive UART in Figure 9.24.

9.7.1. Procedure for rxuart

The procedure for the rxuart is as follows:

1. Copy the required files for building the rxuart digital circuit. These files are listed in Table 9.8 and with the corresponding VHDL codes given in the following sections.

VHDL File Name	Description
clkgen.vhd	Generates clock for serial to parallel register
ser2parsrg8_3.vhd	Serial to parallel register
par2parsrg8.vhd	Parallel to parallel register
oneshot_2.vhd	Generates reset signal for JK flip flop

Table 9.8. VHDL files needed for building the rxuart circuit.

Copy the codes to a VHDL text editor. Save each file using the given names. Compile each file. Open a project in your software (e.g., Quartus II) and save as **rxuart**. If Quartus is used, create a Quartus project file create rxuart.qpf. Now build the circuit shown Figure 9.24. Save, compile, and simulate the design when done.

2. When done show the instructor your design and that it compiles. The instructor sign off is below.

Instructor sign off_____ Date_____

9.7.2. VHDL Code for the Clock Generator (clkgen.vhd)

The VHDL code for the clock generator (clkgen.vhd) is given below. The graphic design file for the code is shown in Figure 9.26. The waveform simulation of clkgen.vhd is shown in Figure 9.27.

```
library ieee;
use ieee.std_logic_1164.all;

entity clkgen is
     port(
     clkin: in std_logic;
     en: in std_logic;
     clkcount: out integer range 0 to 15);
end clkgen;

architecture a of clkgen is
signal count16: integer range 0 to 15;
begin

     process(en,clkin)
     begin
     if (en='0')then
     count16<=0;
     elsif(clkin'event and clkin='1')then
     count16<=count16+1;
     end if;
clkcount<=count16;
     end process;
end a;
```

Figure 9.26. Graphic design file for clock generator VHDL code.

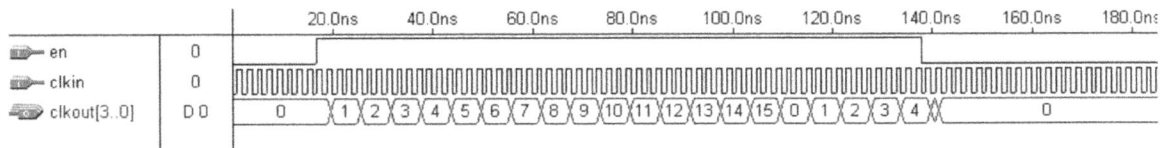

Figure 9.27. Waveform simulation of clock generator VHDL code.

9.7.3. VHDL Code for the SIPO Register(ser2parsrg8_3.vhd)

The VHDL code for the 8 bit SIPO register is shown below:

```vhdl
library ieee;
use ieee.std_logic_1164.all;

entity ser2parsrg8_3 is
port(

datain: in std_logic;
clk: in std_logic;
clkcount: in integer range 0 to 15;
dataout: buffer std_logic_vector(7 downto 0);
datadone:out std_logic);

end ser2parsrg8_3;


architecture shift_left of ser2parsrg8_3 is

begin

process(clkcount,clk)

begin

if (clk'event and clk='1')then
--shift the data left
case clkcount is
     when 1 =>
             datadone<='1';
             dataout(0)<=datain;
             dataout(1)<=dataout(0);
             dataout(2)<=dataout(1);
             dataout(3)<=dataout(2);
             dataout(4)<=dataout(3);
             dataout(5)<=dataout(4);
             dataout(6)<=dataout(5);
             dataout(7)<=dataout(6);
     when 2 =>
             dataout(0)<=datain;
             dataout(1)<=dataout(0);
             dataout(2)<=dataout(1);
             dataout(3)<=dataout(2);
             dataout(4)<=dataout(3);
             dataout(5)<=dataout(4);
             dataout(6)<=dataout(5);
```

```
          dataout(7)<=dataout(6);
when 3 =>
          dataout(0)<=datain;
          dataout(1)<=dataout(0);
          dataout(2)<=dataout(1);
          dataout(3)<=dataout(2);
          dataout(4)<=dataout(3);
          dataout(5)<=dataout(4);
          dataout(6)<=dataout(5);
          dataout(7)<=dataout(6);

when 4 =>
          dataout(0)<=datain;
          dataout(1)<=dataout(0);
          dataout(2)<=dataout(1);
          dataout(3)<=dataout(2);
          dataout(4)<=dataout(3);
          dataout(5)<=dataout(4);
          dataout(6)<=dataout(5);
          dataout(7)<=dataout(6);

when 5 =>
          dataout(0)<=datain;
          dataout(1)<=dataout(0);
          dataout(2)<=dataout(1);
          dataout(3)<=dataout(2);
          dataout(4)<=dataout(3);
          dataout(5)<=dataout(4);
          dataout(6)<=dataout(5);
          dataout(7)<=dataout(6);

when 6 =>
          dataout(0)<=datain;
          dataout(1)<=dataout(0);
          dataout(2)<=dataout(1);
          dataout(3)<=dataout(2);
          dataout(4)<=dataout(3);
          dataout(5)<=dataout(4);
          dataout(6)<=dataout(5);
          dataout(7)<=dataout(6);

when 7 =>
          dataout(0)<=datain;
          dataout(1)<=dataout(0);
          dataout(2)<=dataout(1);
          dataout(3)<=dataout(2);
          dataout(4)<=dataout(3);
```

```
        dataout(5)<=dataout(4);
        dataout(6)<=dataout(5);
        dataout(7)<=dataout(6);


    when 8 =>
        dataout(0)<=datain;
        dataout(1)<=dataout(0);
        dataout(2)<=dataout(1);
        dataout(3)<=dataout(2);
        dataout(4)<=dataout(3);
        dataout(5)<=dataout(4);
        dataout(6)<=dataout(5);
        dataout(7)<=dataout(6);
    -- finish reading 8 bits of data
        datadone<='0';
    when others =>

        dataout<=dataout;

    end case;
end if;

end process;

end shift_left;
```

9.7.4. VHDL Code for the PISO Register(par2parsrg8.vhd)

The VHDL code for the 8 bit PIPO register is given below. The graphic design file for the parallel to parallel register (par2parsrg8.vhd) is shown in Figure 9.28. The simulation for the para2parsrg.vhd is shown in Figure 9.29.

```
library ieee;
use ieee.std_logic_1164.all;

entity par2parsrg8 is
port( datain: in std_logic_vector(7 downto 0);
dataout: buffer std_logic_vector(7 downto 0);
load: in std_logic;
clkin: in std_logic);
end par2parsrg8;


architecture a of par2parsrg8 is
```

```
begin
process(clkin,load)
--data
begin
if (load='1') then
      dataout<=dataout;
elsif (clkin'event and clkin='0')then
      dataout<=datain;
end if;
end process;

end a;
```

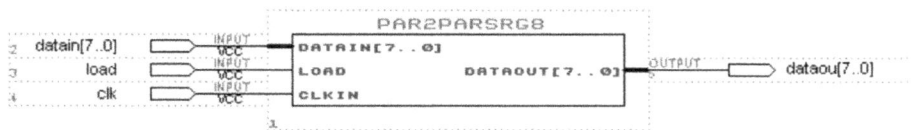

Figure 9.28. Graphic design editor of parallel to parallel register (**para2parsrg8.vhd**).

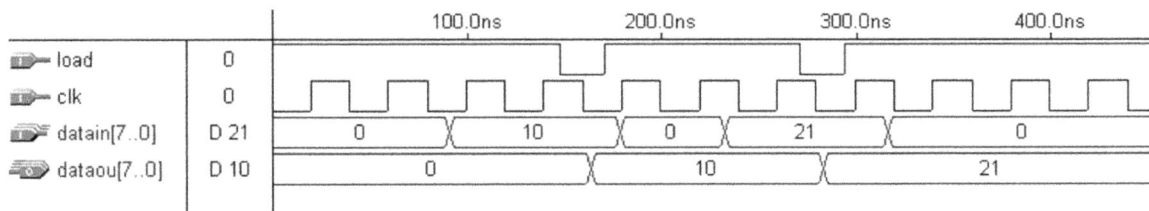

Figure 9.29. Simulation of parallel to parallel register (**para2parsrg8.vhd**).

9.7.5. VHDL Code for the Oneshot

The VHDL code (oneshot_2.vhd) for the active low trigger oneshot is given below:

```vhdl
library ieee;
use ieee.std_logic_1164.all;

entity oneshot_2 is
      port(
      clkcount: in integer range 0 to 15;
      clk:in std_logic;
      trigger: in std_logic;
      oneshot: out std_logic);
end oneshot_2;

architecture a of oneshot_2 is
begin

      process(clk)
      begin

      if (clk'event and clk='0')then

            if(trigger='0' and clkcount > 8)then
                  oneshot<='0';
            else
                  oneshot<='1';
            end if;

      end if;
      end process;

end a;
```

9.8. **VHDL Code for a More Robust rxuart to Reduce Bit Slippage**

The VHDL for a more robust rxuart is given in this section. The receive bit rate is selected to be twice as fast as the transmit bit rate. This allows the bits to latched more accurately which depends less on the startup clock conditions. Simulation shows that if the startup for the transmit bit rate clock is different, the bits will not be latched correctly. The design in this section reduces that error. The Quartus project file, ser2par_3.qpf, graphic representation is shown in Figure 9.30. Use the files in the following sections to build the digital circuit. Once complete compile and simulate.

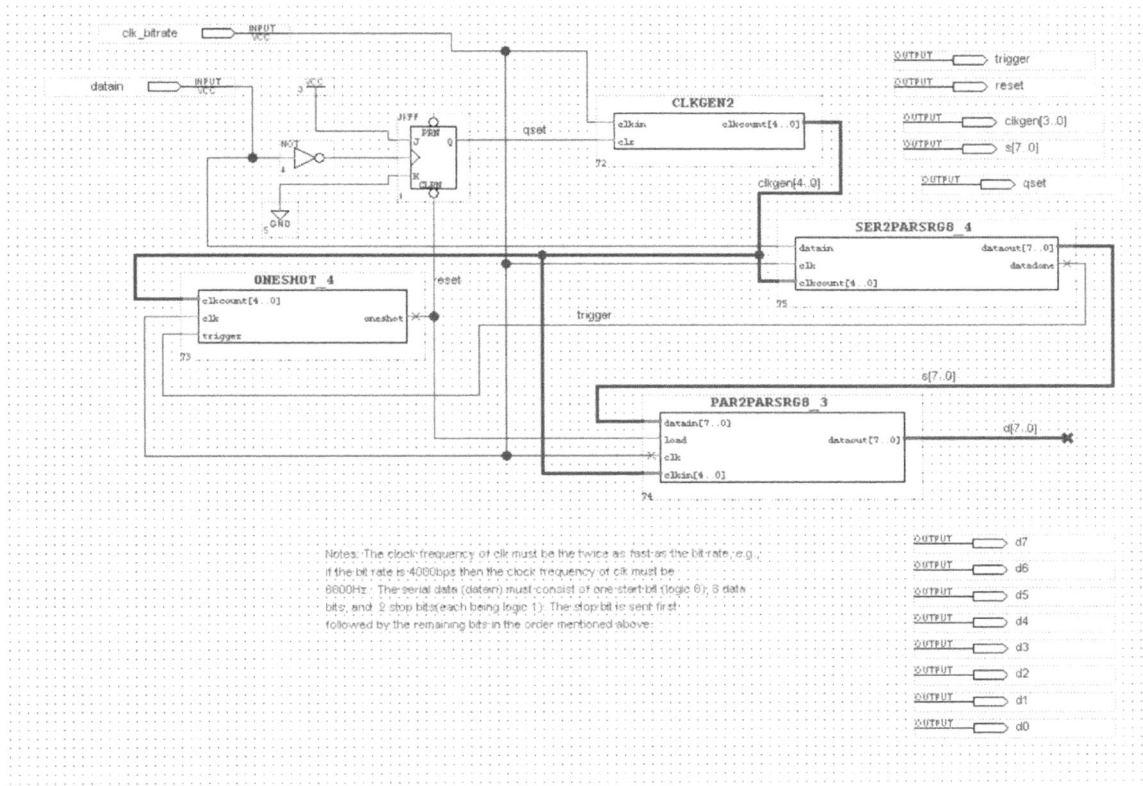

Figure 9.30. More robust rxuart using a receive clock that is twice the transmit bit rate clock (Quartus file: ser2par_3.qpf).

9.8.1. VHDL Code for the Clock Generator (clkgen.vhd)

The VHDL code for the clock generator (clkgen.vhd) is shown below. The graphic design file for the code is shown in Figure 9.26. The waveform simulation of clkgen.vhd is shown in Figure 9.27.

```
library ieee;
use ieee.std_logic_1164.all;

entity clkgen is
     port(
     clkin: in std_logic;
     en: in std_logic;
     clkcount: out integer range 0 to 15);
end clkgen;

architecture a of clkgen is
signal count16: integer range 0 to 15;
begin

     process(en,clkin)
     begin
     if (en='0')then
     count16<=0;
     elsif(clkin'event and clkin='1')then
     count16<=count16+1;
     end if;
clkcount<=count16;
     end process;
end a;
```

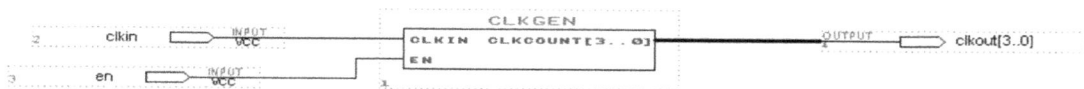

Figure 9.31. Graphic design file for clock generator VHDL code.

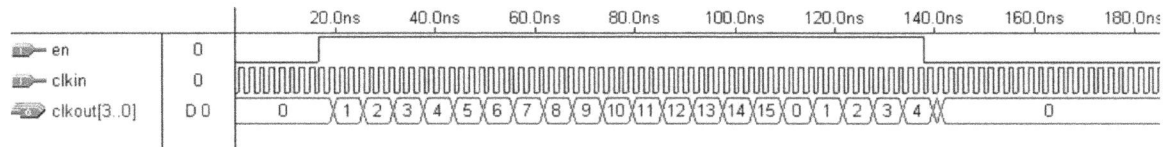

Figure 9.32. Waveform simulation of clock generator VHDL code.

9.8.2. VHDL Code for the SIPO Register

The VHDL code (**ser2parsrg8_4.vhd**) for the 8 bit serial input to parallel output (SIPO) register is shown below:

```
library ieee;
use ieee.std_logic_1164.all;

entity ser2parsrg8_4 is
port(

datain: in std_logic;
clk: in std_logic;
clkcount: in integer range 0 to 31;
dataout: buffer std_logic_vector(7 downto 0);
datadone:out std_logic);

end ser2parsrg8_4;


architecture shift_left of ser2parsrg8_4 is

begin

process(clkcount,clk)

begin
```

```
if (clk'event and clk='0')then
--shift the data right, assume LSB in first
case clkcount is
    when 1 =>
            datadone<='1';
            dataout(7)<=datain;
            dataout(6)<=dataout(7);
            dataout(5)<=dataout(6);
            dataout(4)<=dataout(5);
            dataout(3)<=dataout(4);
            dataout(2)<=dataout(3);
            dataout(1)<=dataout(2);
            dataout(0)<=dataout(1);
    when 3 =>
            dataout(7)<=datain;
            dataout(6)<=dataout(7);
            dataout(5)<=dataout(6);
            dataout(4)<=dataout(5);
            dataout(3)<=dataout(4);
            dataout(2)<=dataout(3);
            dataout(1)<=dataout(2);
            dataout(0)<=dataout(1);
    when 5 =>
            dataout(7)<=datain;
            dataout(6)<=dataout(7);
            dataout(5)<=dataout(6);
            dataout(4)<=dataout(5);
            dataout(3)<=dataout(4);
            dataout(2)<=dataout(3);
            dataout(1)<=dataout(2);
            dataout(0)<=dataout(1);
    when 7 =>
            dataout(7)<=datain;
            dataout(6)<=dataout(7);
            dataout(5)<=dataout(6);
            dataout(4)<=dataout(5);
            dataout(3)<=dataout(4);
            dataout(2)<=dataout(3);
            dataout(1)<=dataout(2);
            dataout(0)<=dataout(1);

    when 9 =>
            dataout(7)<=datain;
            dataout(6)<=dataout(7);
            dataout(5)<=dataout(6);
            dataout(4)<=dataout(5);
```

```
        dataout(3)<=dataout(4);
        dataout(2)<=dataout(3);
        dataout(1)<=dataout(2);
        dataout(0)<=dataout(1);

    when 11 =>
        dataout(7)<=datain;
        dataout(6)<=dataout(7);
        dataout(5)<=dataout(6);
        dataout(4)<=dataout(5);
        dataout(3)<=dataout(4);
        dataout(2)<=dataout(3);
        dataout(1)<=dataout(2);
        dataout(0)<=dataout(1);

    when 13 =>
        dataout(7)<=datain;
        dataout(6)<=dataout(7);
        dataout(5)<=dataout(6);
        dataout(4)<=dataout(5);
        dataout(3)<=dataout(4);
        dataout(2)<=dataout(3);
        dataout(1)<=dataout(2);
        dataout(0)<=dataout(1);


    when 15 =>
        dataout(7)<=datain;
        dataout(6)<=dataout(7);
        dataout(5)<=dataout(6);
        dataout(4)<=dataout(5);
        dataout(3)<=dataout(4);
        dataout(2)<=dataout(3);
        dataout(1)<=dataout(2);
        dataout(0)<=dataout(1);
-- finish reading 8 bits of data
        datadone<='0';
    when others =>
        NULL;
    end case;
end if;

end process;

end shift_left;
```

9.8.3. VHDL Code for the PISO Register

The VHDL code (**par2parsrg8_3.vhd**) for the 8 bit parallel input to parallel output (PIPO) register is shown below. The corresponding graphic file for the register is shown in Figure 9.28 while the simulation is shown in Figure 9.29.

```vhdl
library ieee;
use ieee.std_logic_1164.all;

entity par2parsrg8_3 is
port( datain: in std_logic_vector(7 downto 0);
dataout: buffer std_logic_vector(7 downto 0);
load: in std_logic;
clk: in std_logic;
clkin: in integer range 0 to 31);
end par2parsrg8_3;

architecture a of par2parsrg8_3 is

begin
process(load,clk)
--data
begin

if (clk'event and clk='1') then
     if (load = '0')then
     dataout<=datain;
--elsif (clk'event and clk='1')then
     --case clkin is
          --when 16 =>
          --dataout<=datain;
          --when others =>
          --dataout<=dataout;
     --end case;
end if;
end if;
end process;

end a;
```

9.8.4. VHDL Code for the Oneshot

The VHDL code (**oneshot_4.vhd**) for the active low trigger oneshot is shown below:

```vhdl
library ieee;
use ieee.std_logic_1164.all;
```

```
entity oneshot_4 is
    port(
    clkcount: in integer range 0 to 31;
    clk:in std_logic;
    trigger: in std_logic;
    oneshot: out std_logic);
end oneshot_4;

architecture a of oneshot_4 is
begin

    process(clk)
    begin

    if (clk'event and clk='1')then

        if(trigger='0' and clkcount >= 16 and clkcount <=
17)then
            oneshot<='0';
        else
            oneshot<='1';
        end if;

    end if;
    end process;

end a;
```

9.9. VHDL Code for Servo Control Module (servo_control.vhd)

The servo motors are driven by sending selected frequencies from the divide by 2^{16} counter. More information about the mod $(65536=2^{16})$ counter can be found in Section 9.6.2, VHDL Code for the 216 Counter (counter65536.vhd), on page 349. Each servo has its own set of frequencies from the mod 65536 counter that causes it to go forward, backward, reverse and stop depending on their configuration. If the servos are modified the same way the frequencies may be the same. These frequencies are found by trial and error using the last couple of outputs from the mod 65536 counter. The servo control module (servo_control.vhd) is a state machine (state machines are discussed in Section 5.15 on page 165) that responds with two output frequencies based on a command input. The correct set of frequencies for the left and right servo will cause the robot to turn right, turn left, go in reverse, or stop at a given command. The VHDL code for the servo module is given in the next section and its flowchart in the following section. Use the servo control VHDL code to create a project servo_ctl.qpf. Compile and simulate when done. The schematic representation of the servo_ctl module is shown in Figure 9.33 and its corresponding simulation in Figure 9.34. The servo control external circuit diagram is given in Section 9.9.3.

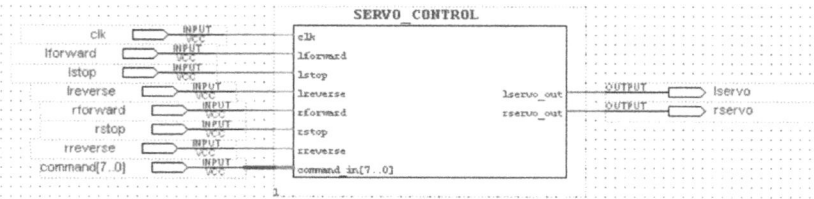

Figure 9.33. Quartus project file (servo_ctl.qpf) of servo control module (servo_control.vhd).

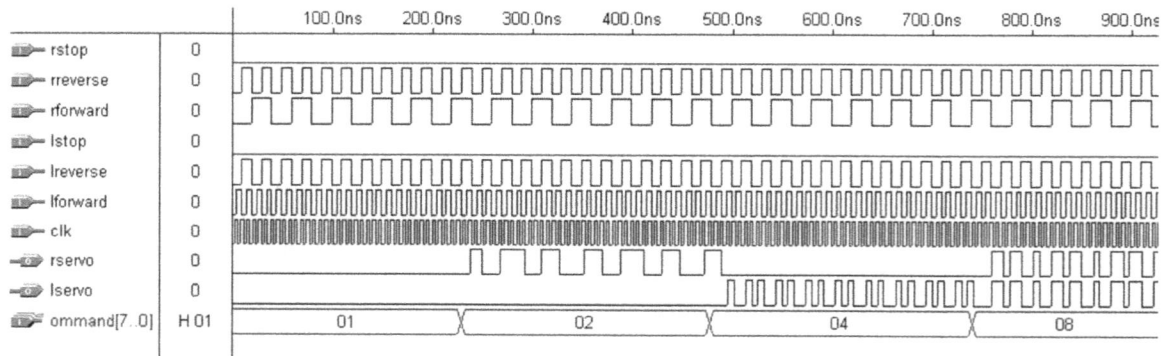

Figure 9.34. Simulation of servo control module (servo_control.vhd).

9.9.1. VHDL Code for Servo Control Module

The VHDL code for the servo control module (servo_control.vhd) is given below:

```
library ieee;
use ieee.std_logic_1164.all;

entity servo_control is
port (

clk: in bit;
lforward: in bit;
lstop: in bit;
```

```vhdl
lreverse: in bit;
rforward: in bit;
rstop: in bit;
rreverse: bit;
lservo_out: out bit;
rservo_out: out bit;
command_in: in bit_vector(7 downto 0));

end servo_control;

architecture a of servo_control is
begin


    process(clk)
    begin
        if (clk'event and clk='0')then
            case command_in is
                    --robot, stop command
                when "00000000" =>
                    lservo_out <= lstop;
                    rservo_out <= rstop;

                    -- robot, left turn command
                when "00000001" =>
                    lservo_out <= lreverse;
                    rservo_out <= rforward;

                    -- robot, right turn command
                when "00000010" =>
                    lservo_out <= lforward;
                    rservo_out <= rreverse;

                --robot, go forward command
                when "00000100" =>
                    lservo_out <= lforward;
                    rservo_out <= rforward;


                    -- robot, go reverse
                when "00001000" =>
                    lservo_out <= lreverse;
                    rservo_out <= rreverse;

                    --robot, stop!
                when others =>
                    lservo_out <= lstop;
```

```
                              rservo_out <= rstop;
                 end case;

        end if;

    end process;

end a;
```

9.9.2. Flowchart for the Servo Control Module

The flowchart of the servo control module is shown in Figure 9.35.

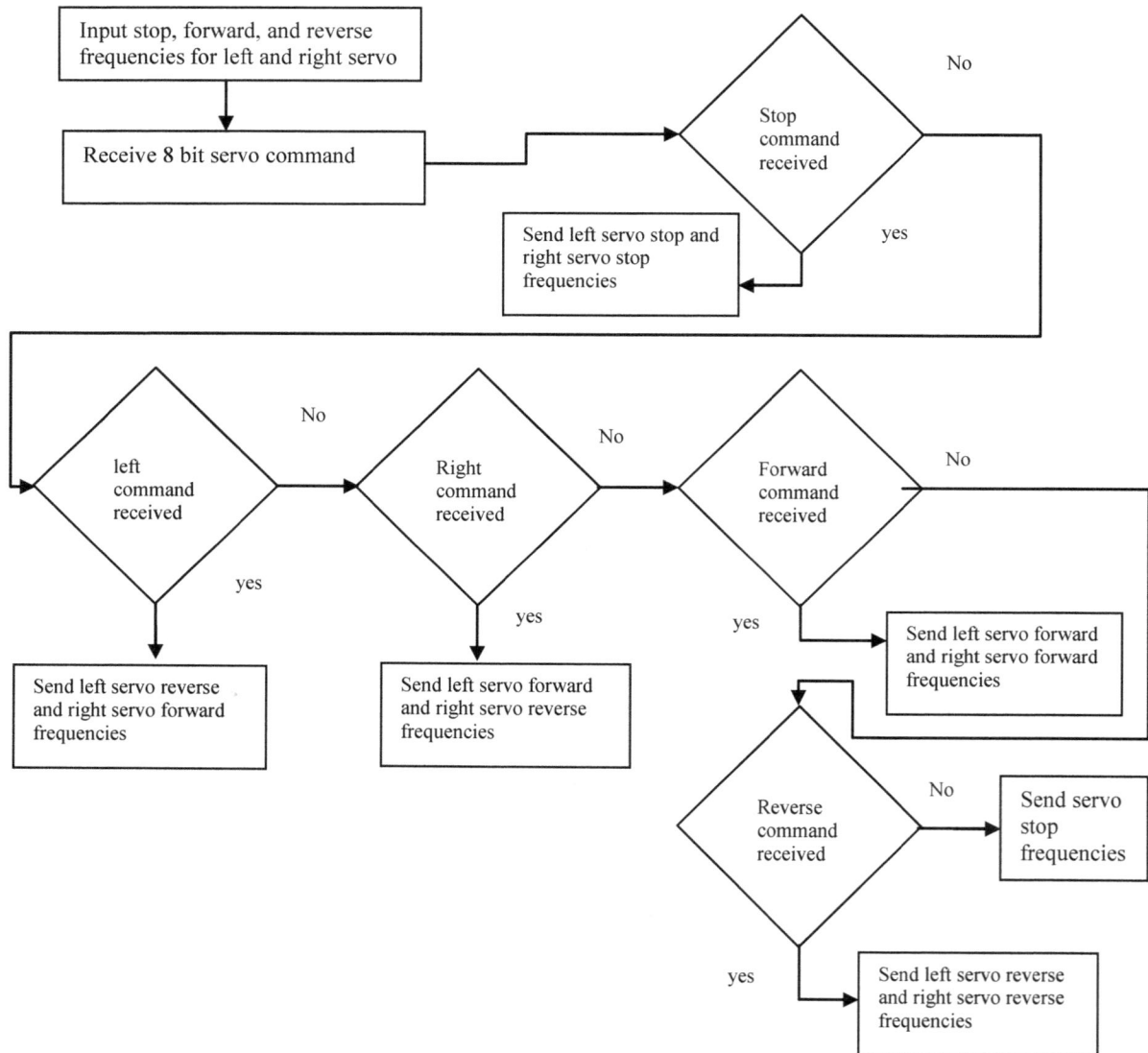

Figure 9.35. Flowchart of servo_control.vhd module.

9.9.3. External Circuitry for Servo Control Module

The external circuitry for the servo control module is given below in Figure 9.36. The servo control module (servo_control.vhd) receives an eight bit command from a sensor control module controlling the command bus. The following are sensor control modules: compass_control, irmodule, tracklight, tracktape, etc. The servo control module must receive six clock frequencies, three related to the left servo, and three related to the right servo. The three signals for each servo cause the servo to go forward, reverse, and stop. That is, the servo control module selects the right combination of clock frequencies that cause the robot to turn left, right, go forward, go reverse, or stop depending on the eight bit command it receives.

The sensor control module (compass_control, irmodule, tracklight, tracktape, etc.), based on sensory inputs, sends an eight bit command to the servo control module that tells the robot which direction to turn or go in. When the servo control module (servo_control.vhd) receives the eight bit command, a state machine immediately decodes it and determines the set of frequencies to send to the servos (left, right, forward, or reverse, stop). For example, when a sensor control module sends an eight bit command that is decoided as a left turn, a clock signal is sent to each servo. The clock signal sent to the left servo causes it to stop (or go reverse) and the clock signal to the right servo causes it to go forward. This dual action of the servos causes the robot platform to turn left.

As another example, suppose a sensor control module sends an eight bit command that is decoded by the servo control module's state machine as a right turn. Afterwards, a clock signal is sent to each servo. The clock signal sent to the left servo causes it to go forward and the clock signal to the right servo causes it to stop (or go forward). This dual action of the servos causes the robot platform to turn right.

Additionally, when an eight bit command is decoded as reverse, again, a clock signal is sent to each servo. The clock signal sent to the left servo causes it to go reverse and the clock signal to the right servo causes it to go reverse. This dual action of the servos causes the robot platform to go in reverse.

When the eight bit command is decoded by the servo control module's state machine as a forward command, a clock signal is sent to each servo. The clock signal sent to the left servo causes it to go forward and the clock signal to the right servo causes it to go forward. This dual action of the servos causes the robot platform to go forward.

Lastly, if any other eight bit command is sent to the servo control module a clock signal is sent to both servo causing them to stop. This dual action of the servos causes the robot platform to stand still. These are the suggested commands but can be redefined as needed per application.

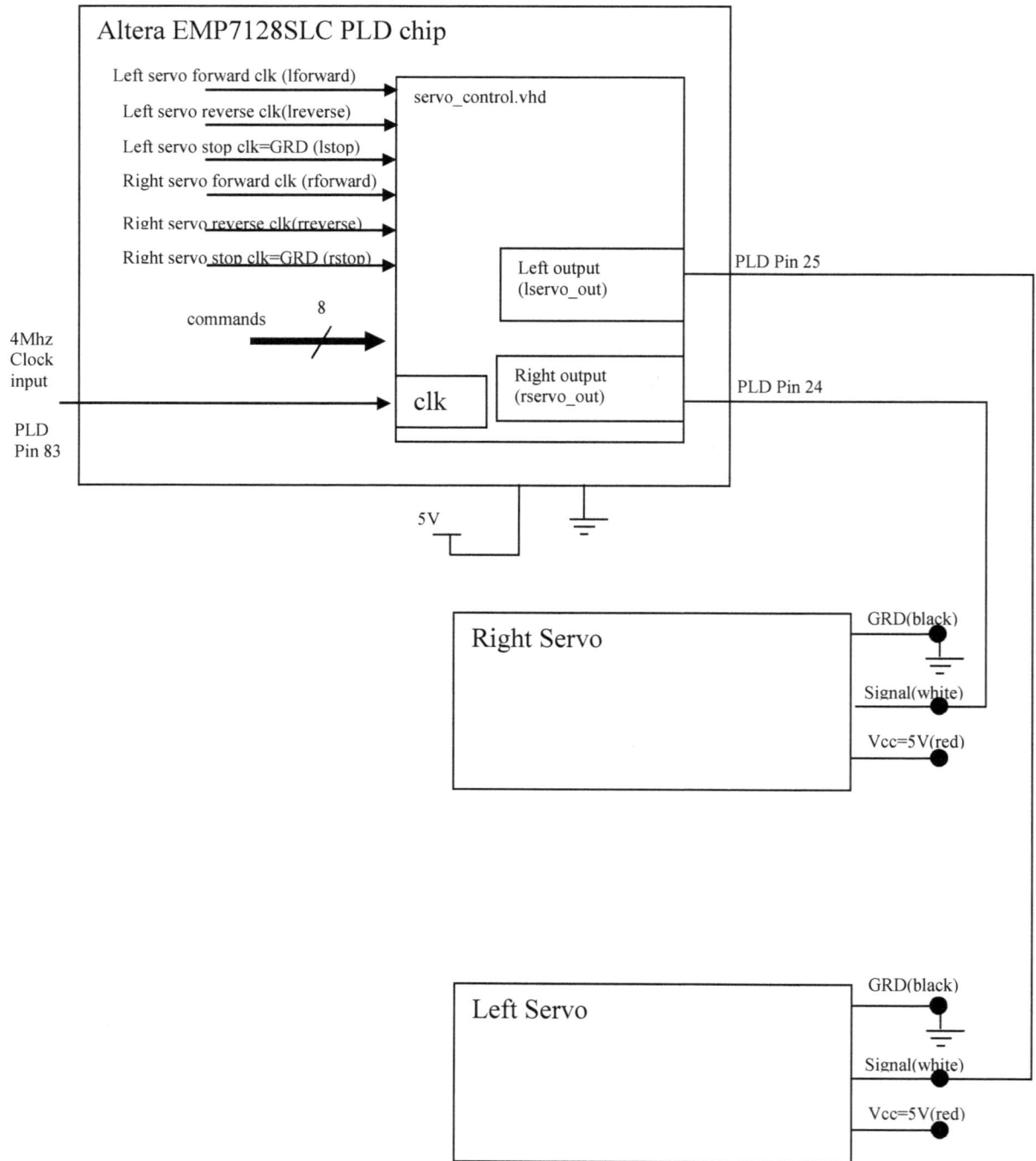

Figure 9.36. External circuitry for servo_control.vhd. See the servo datasheets for servo details and electrical characteristics. The color scheme for the servo wires given are from the Futaba S3003 servos.

9.10.   VHDL Code and Flowchart for the Compass Control Module

The compass control module determines the correct heading (north, south, east, west, etc.) for the robot and controls the robot to go in that direction. The VHDL code for this module is given in the next Section. Use the VHDL code to create the project compass_ctl.qpf . Compile and simulate when done.

9.10.1. VHDL code for the Compass Control Module

A bug has been left in compass_control2.vhd on purpose to allow the student to use his VHDL skills, to troubleshoot, find root cause, and provide a simple fix. The VHDL code for the compass control module is shown below.

```
library ieee;
use ieee.std_logic_1164.all;

entity compass_control2 is
port(
en: in bit;
clk: in bit;
north: in bit;
south: in bit;
east: in bit;
west: in bit;
direction: in bit_vector(3 downto 0);
command: buffer std_logic_vector( 7 downto 0));
end compass_control2;

architecture a of compass_control2 is
signal c:std_logic_vector(7 downto 0);
signal curdir: bit_vector(3 downto 0);

begin

process(en,c)
begin
--tri state command output when en is 1
    if(en='0')then
        command<=c;
    else
        command<="ZZZZZZZZ";
    end if;
end process;

-- direction code
--- 0001 north
--- 0010 east
--- 0100 west
```

```
--- 1000 south
--- 0011 north east
--- 0101 north west
--- 1100 south west
--- 1010 south east

curdir <= north & south & east & west;

process(direction,clk,north,south,east,west)
begin

     if(clk'event and clk='1')then

          case direction is
               when "0001" =>
                    -- going north
                    case curdir is
                    when "0111" =>
                         c <= "00000100";   --go forward
                    when others =>
                         c <= "00000010";   --turn right
                    end case;
               when "0010" =>
                    --going east
                    case curdir is
                    when "1101" =>
                         c <= "00000100";   --go forward
                    when others =>
                         c <= "00000010";   --turn right
                    end case;
               when "0100" =>
                    -- going west
                    case curdir is
                    when "1110" =>
                         c <= "00000100";   --go forward
                         when others =>
                         c <= "00000010";   --turn right
                    end case;

               when "1000" =>
                    -- going south
                    case curdir is
                    when "1011" =>
                         c <= "00000100";   --go forward
                    when others =>
                         c <= "00000010";   --turn right
                    end case;
```

```
                        when others =>
                            c <= "00000000";   --stop
                        end case;
            end if;
end process;
end a;
```

9.10.2. Magnetic Compass VHDL Flowchart

A simple flowchart for the magnetic compass module is shown in Figure 9.37.

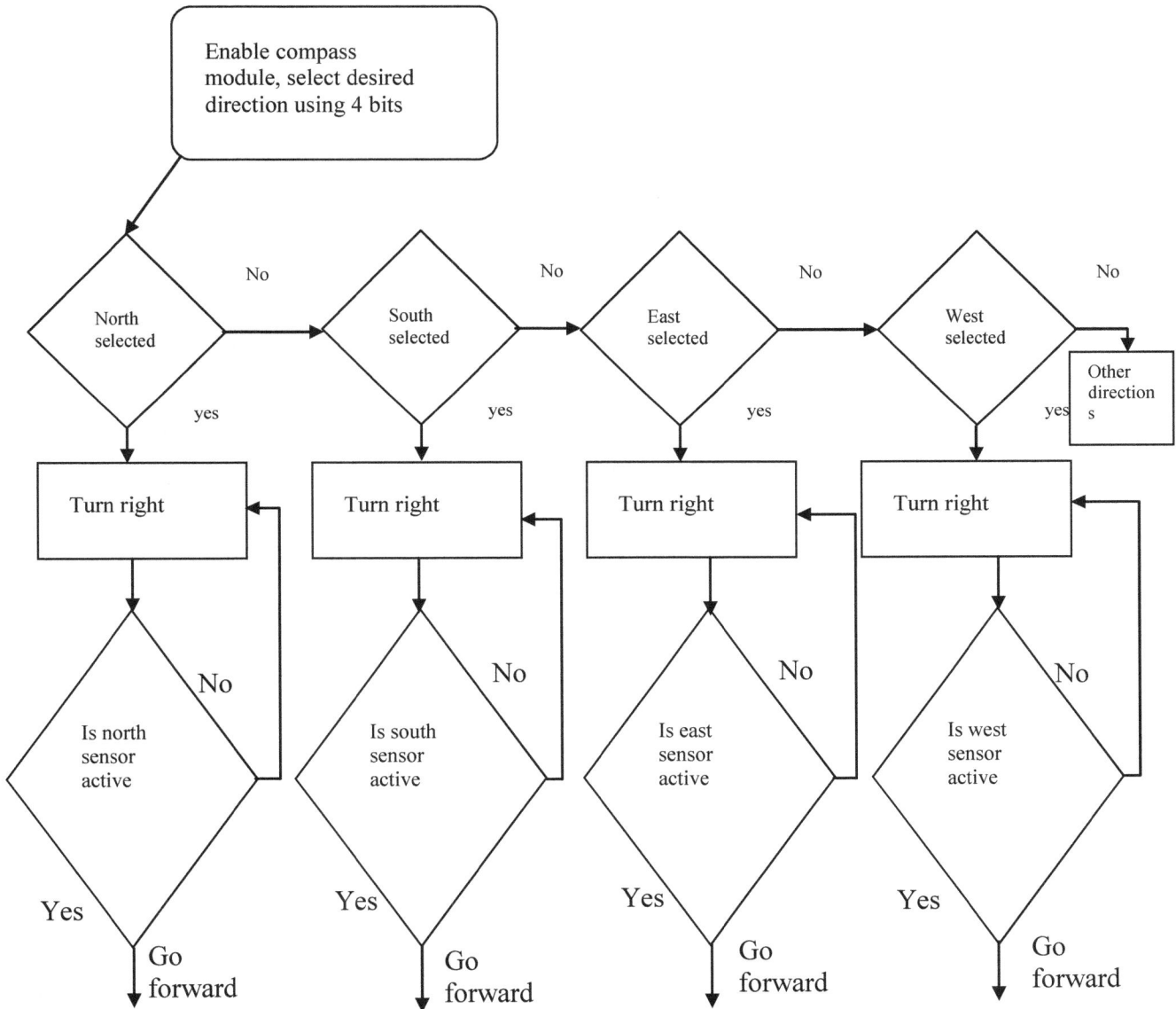

Figure 9.37. Flowchart of VHDL code for compass sensor.

9.11. VHDL Code and Flowchart for the Infrared Sensor Control Module

The infrared sensor module determines the appropriate action to take when an object is detected. The VHDL code and flowchart is given next. You will need to create the project irmodule_ctl.qpf for the ir module and VHDL code in the next Section. Compile and simulate when done.

9.11.1. VHDL Code for the Infrared Module

The VHDL code for the infrared module (`irmodule2.vhd`) is given below. The corresponding digital schematic and simulation are shown in Figure 9.38 and Figure 9.39, respectively.

```vhdl
library ieee;
use ieee.std_logic_1164.all;

entity irmodule is
port (en: in std_logic;
     irdet: in std_logic;
     clk: in std_logic;
     clk40khz:in std_logic;
     enleftir: buffer  std_logic;
     enrightir: buffer std_logic;
     leftirclk:out std_logic;
     rightirclk: out std_logic;
     command: out std_logic_vector(7 downto 0);
     cnt:buffer integer range 0 to 63);
end irmodule;

architecture a of irmodule is
signal d: std_logic_vector(2 downto 0);
signal x: std_logic_vector(1 downto 0);

begin
        d <= irdet & enleftir & enrightir;
        x <= enleftir & enrightir;


    process(en,clk40khz,d)
    begin
        if(en='1')then
            else
            if(clk40khz'event and clk40khz='1')then
            case d is
            when "001" =>
                -- object is on left
                -- move right
                command <= "00000010";  --turn right
```

```
            when "010" =>
                    -- object is on right
                    -- move left
                    command <= "00000001";   --turn left

            when "000" =>
                    -- object is in front
                    -- turn around
                    command <= "00000000";   --turn right

            when others =>
                    command <= "00000100";   --go forward

            end case;
            end if;
        end if;

end process;

process(en,x)
begin
        if(en='0')then
                case x is
                when "00" =>
                        --both ir leds are enabled
                        rightirclk<= clk40khz;
                        leftirclk<= clk40khz;
                when "10" =>
                        --right ir led is enabled
                        rightirclk<= clk40khz;
                        leftirclk<= '0';
                when "01" =>
                        --left ir led is enabled
                        rightirclk<= '0';
                        leftirclk<= clk40khz;
                when others =>
                        --neither left nor right ir led enabled
                        rightirclk<= '0';
                        leftirclk<= '0';
                end case;

        else

        leftirclk<='0';
        rightirclk<='0';

        end if;
```

```
        end process;


process(en,clk40khz,clk) is
begin
            if (clk40khz'event and clk40khz='1')then
            cnt <= cnt +1;
            end if;

if (en='0')then

        if(cnt >= 0 and cnt <= 21)then
                -- enable left ir led
            enleftir <= '0';
            enrightir <= '1';
        elsif (cnt>= 22 and cnt <= 42)then
                -- enable right ir led
            enleftir <= '1';
            enrightir <= '0';
        else
                -- enable both ir leds
            enleftir <= '0';
            enrightir <= '0';
        end if;
    else
            -- disable both ir leds
            enleftir <= '1';
            enrightir <= '1';
    end if;

end process;

end a;
```

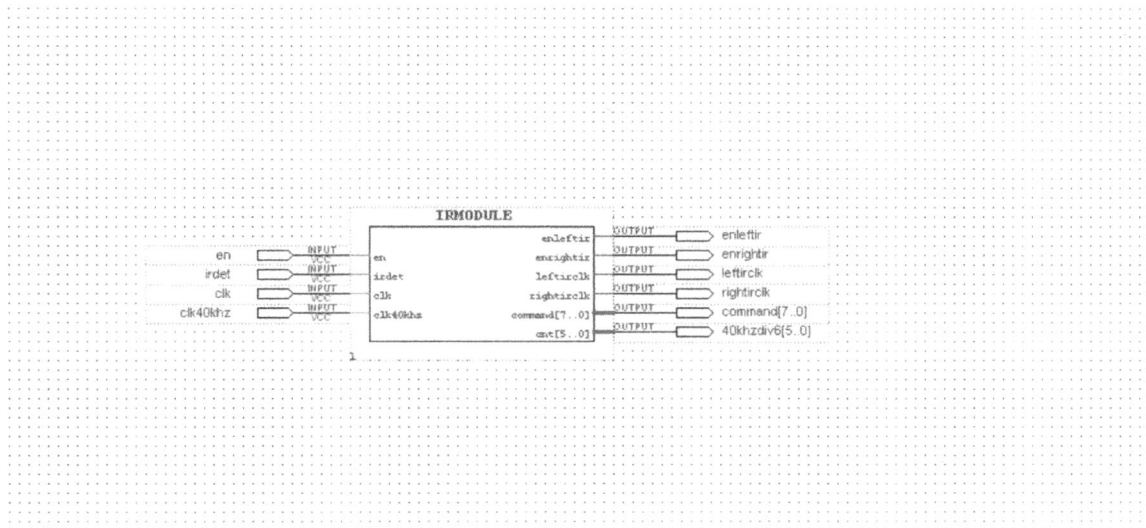

Figure 9.38. Quartus project file (irmodule_ctl.qpf) of infrared module (irmodule.vhd).

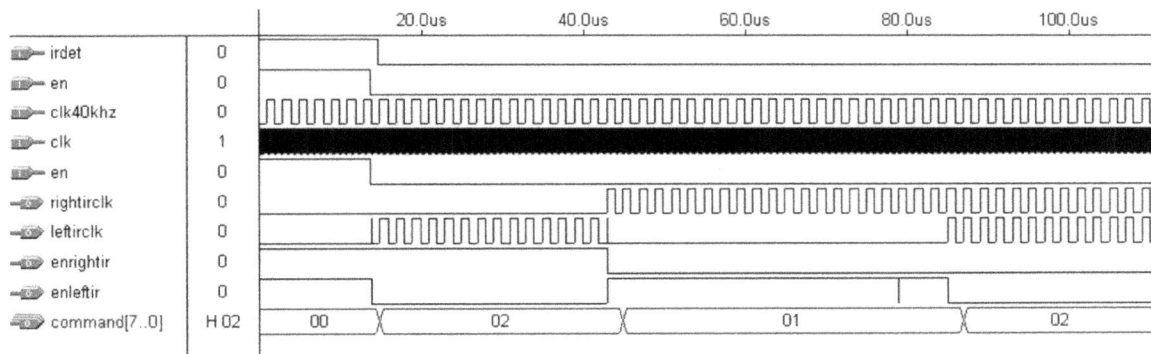

Figure 9.39. Simulation for infrared module (irmodule.vhd).

9.11.2. Infrared Sensor Module Flowchart

An example of the flowchart for the infrared sensor module and two infrared LEDs is shown in the Figure 9.40.

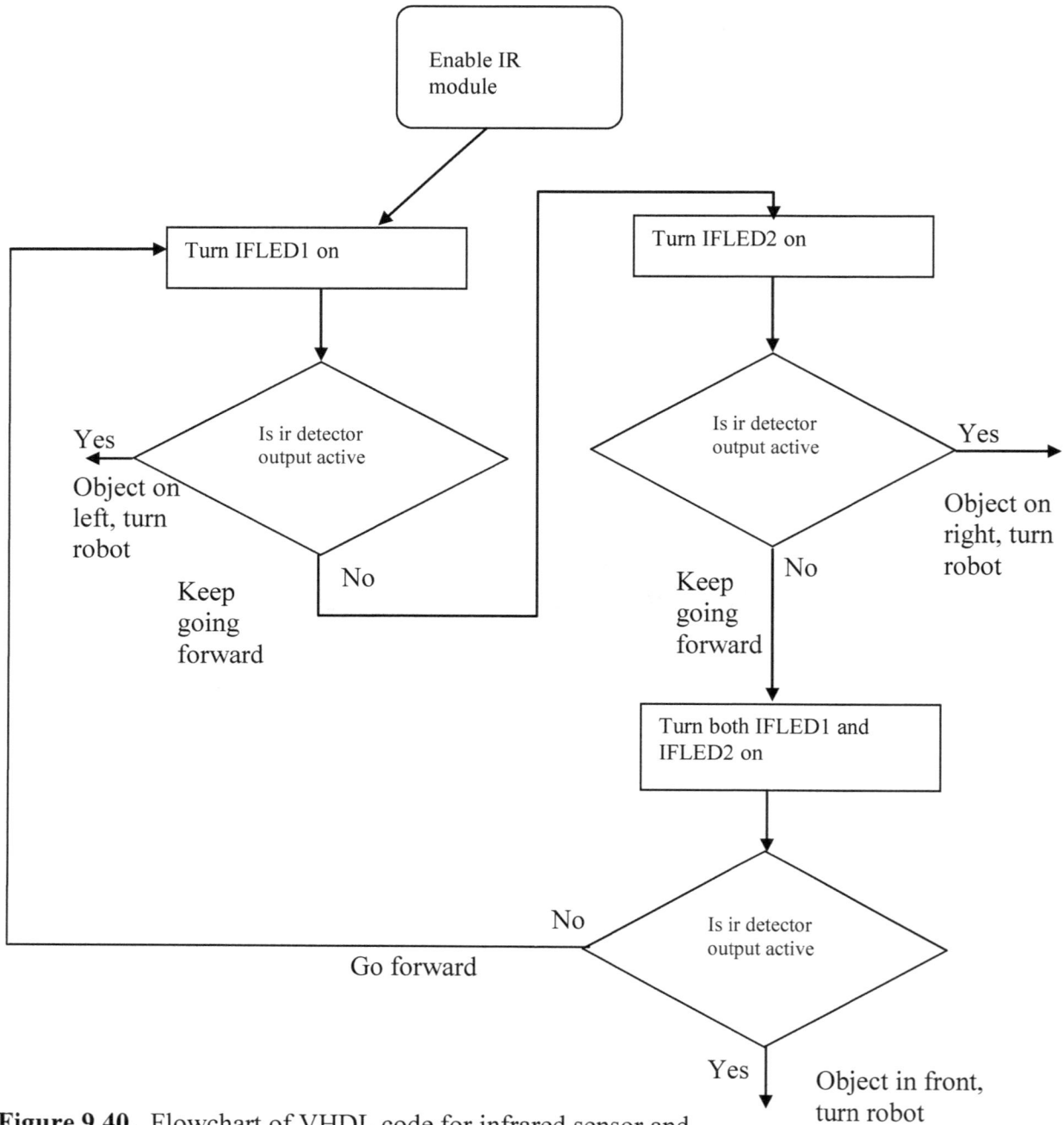

Figure 9.40. Flowchart of VHDL code for infrared sensor and IRLED circuitry.

9.12.   VHDL Code for the Tracktape Control Module

The VHDL code for the track tape module (tracktape.vhd) is given below. The Quartus design file for the tracking tape module is shown in Figure 9.41. You will need to create the project tracktape_ctl .qpf for the VHDL code below. Compile and simulate when done. The simulation is given in Figure 9.42.

```vhdl
library ieee;
use ieee.std_logic_1164.all;

entity tracktape is
port(
leftsen: in std_logic;
rightsen: in std_logic;
en: in std_logic;
command: out std_logic_vector(7 downto 0));
end tracktape;


architecture a of tracktape is
signal d:std_logic_vector(1 downto 0);
begin

--leftsen  => logic 0:senses darkness logic 1 :senses light
--rightsen => logic 0:senses darkness logic 1 :senses light

d<=leftsen & rightsen;

    process (en,d) is
    begin
        if(en = '1')then
            command<="ZZZZZZZZ";
        else
        case d is
            when "00" =>
                command <= "00000100";  --go forward
            when "10" =>
                command <= "00000001";  --turn right
            when "01" =>
                command <= "00000010";  --turn left
            when others =>
                command <= "00000100";  --go forward
            end case;

        end if;

    end process;
```

```
end a;
```

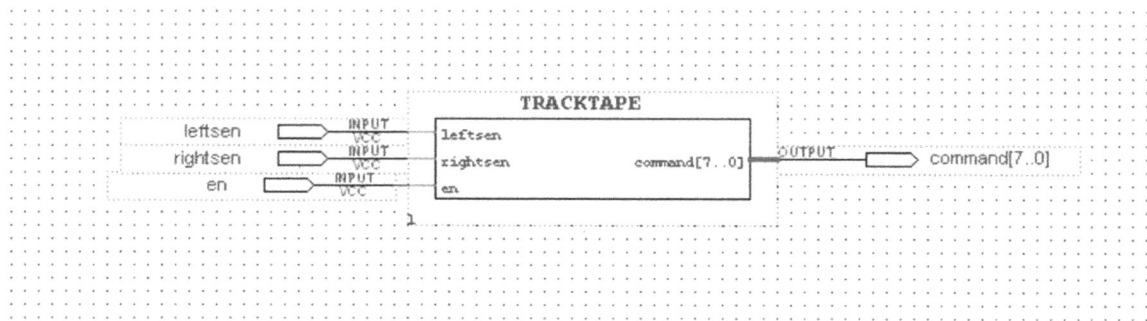

Figure 9.41. Quartus project file (tracktape_ctl.qpf) of track tape module (tracktape.vhd).

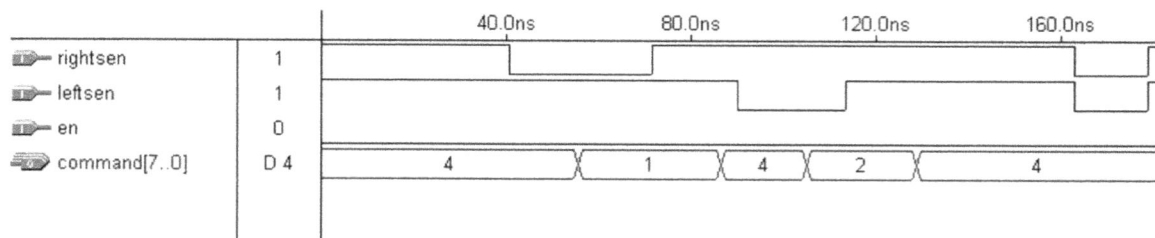

Figure 9.42. Simulation of track tape module (tracktape.vhd).

9.12.1. Tracktape Module External Circuitry

The Track tape external circuit is shown in Figure 9.49. It uses a comparator to sense a change in the input sensor resistance and corresponding voltage (Coughlin, 2001). The LM339 has four comparators. The pin configuration for the LM339 is shown in Figure 9.44. Adjust Pot1 and Pot2 such that Vo2 and Vo4 will go low when the sensor senses darkness. The sensor detects darkness using an active low output which is what the VHDL code will detect when the robot's sensor crosses the black tape.

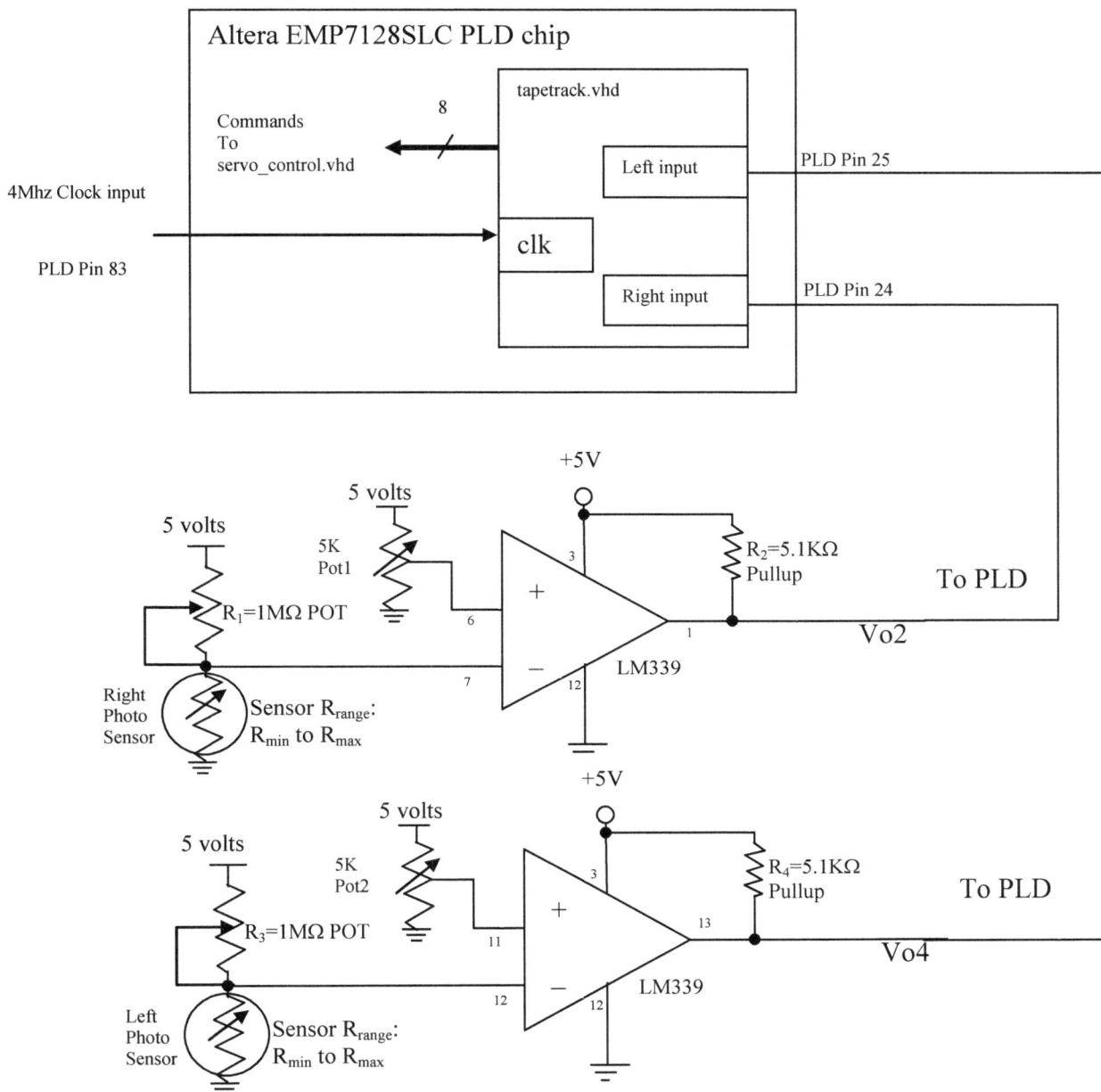

Figure 9.43. External circuitry for tracktape.vhd Photo sensors should be placed as close as possible to tape on ground. Assumed that when the sensor detects darkness the resistance will equal close to 1M. Also see photo sensor datasheet for R_{min} and R_{max}.

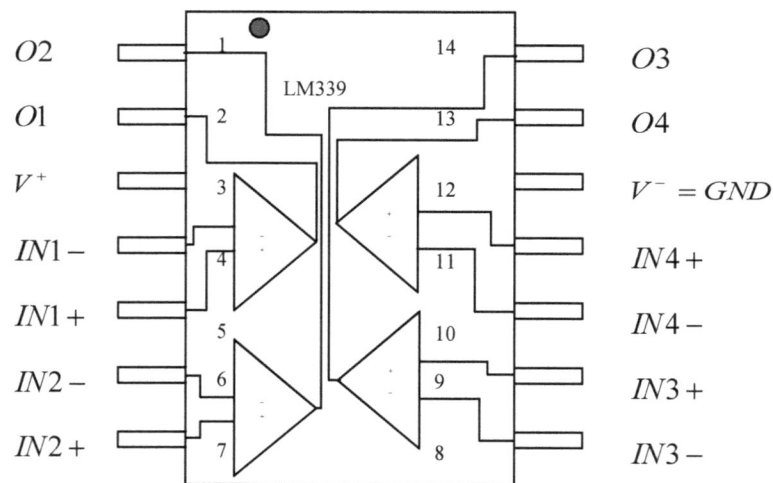

Figure 9.44. Pin configuration for the LM339 Quad comparator chip. Please see the LM339 datasheet for more information.

9.13. VHDL Code for the Tracklight Control Module

The VHDL code for the track light module (**tracklight.vhd**) is given below. The design file for the tracking light module is shown in Figure 9.49. Its simulation is given in Figure 9.46.

```
library ieee;
use ieee.std_logic_1164.all;

entity tracklight is
port(
leftsen: in std_logic;
rightsen: in std_logic;
en: in std_logic;
command: out std_logic_vector(7 downto 0));
end tracklight;


architecture a of tracklight is
signal d:std_logic_vector(1 downto 0);
begin

--leftsen  => logic 1: senses light          logic 0
:senses darkness
--rightsen  => logic 1: senses light          logic 0
:senses darkness
```

```
d<=leftsen & rightsen;

process (en,d) is
begin
    if(en = '1')then
        command<="ZZZZZZZZ";
    else
        case d is
        when "11" =>
            command <= "00000100";  --both sensors on,
                                    --  go forward
        when "10" =>
                command <= "00000001";--left sensor on,
                                    -- turn left
        when "01" =>
            command <= "00000010";  --right sensor on,
                                    --turn right
        when others =>
            command <= "00000000";  --stop
        end case;

    end if;

    end process;

end a;
```

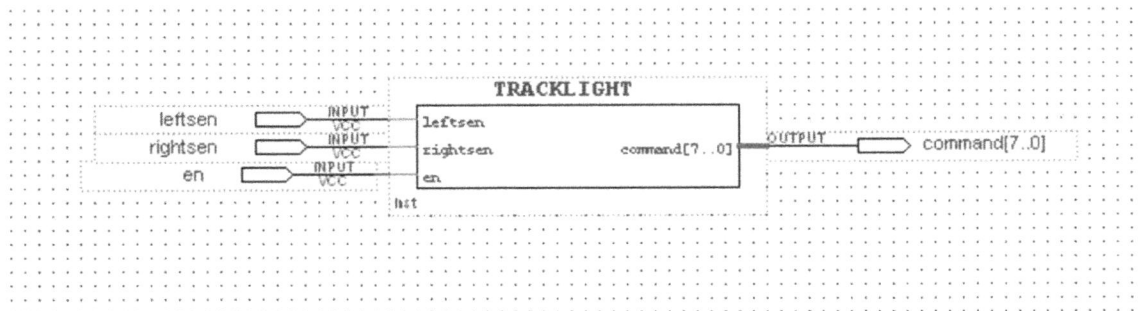

Figure 9.45. Design file for track light module (**tracklight.vhd**).

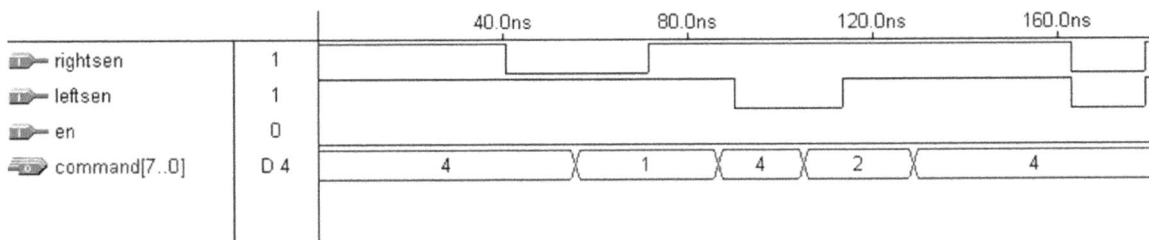

Figure 9.46. Simulation of track light module (**tracklight.vhd**).

9.13.1. Tracklight Module External Circuitry

The Track light external circuit is shown in Figure 9.47. It uses a comparator to sense a change in the input sensor resistance and corresponding voltage (Coughlin, 2001). The LM339 has four comparators. The pin configuration for the LM339 is shown in Figure 9.48. Adjust Pot1 and Pot2 such that Vo2 and Vo4 will go low when the sensor senses darkness. When the sensor detects light Vo2 and Vo4 should then go high, Thus, this is an active high input in the VHDL code indicating that the robot's sensor sees light.

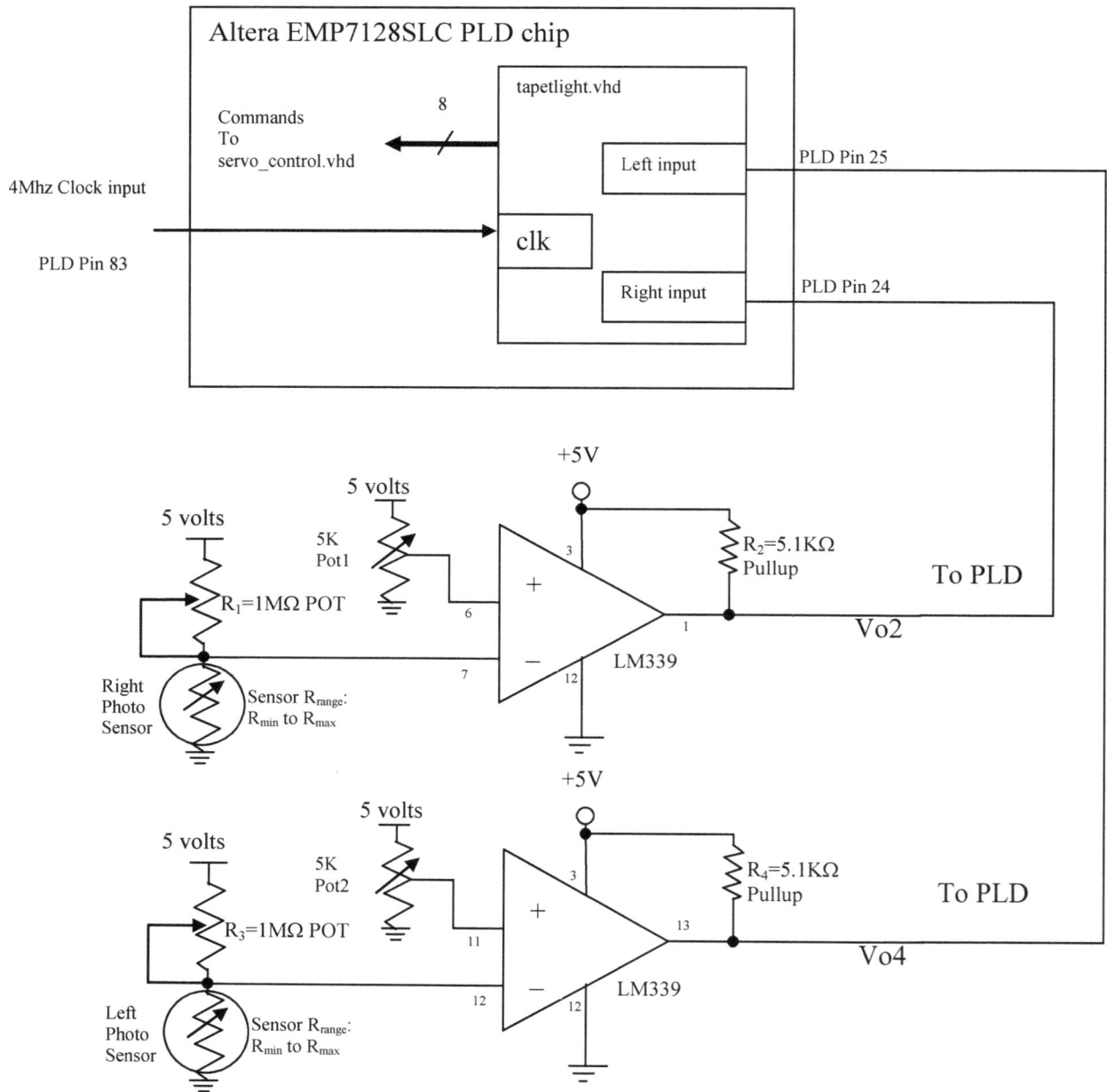

Figure 9.47. External circuitry for tracklight.vhd. It is assumed that when the sensor detects darkness the resistance will equal close to1M. Also see photo sensor datasheet for R_{min} and R_{max}.

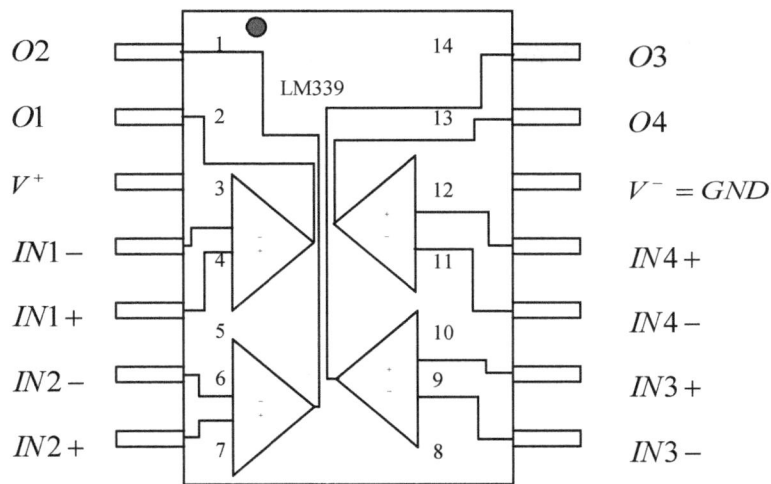

Figure 9.48. Pin configuration for the LM339 Quad comparator chip. Please see the LM339 datasheet for more information.

9.14. Speaker Identification Module

The design file for the speaker id module (speakerid_ctl.vhd) is shown in Figure 9.49. The simulation of the speaker id module is shown in Figure 9.50.

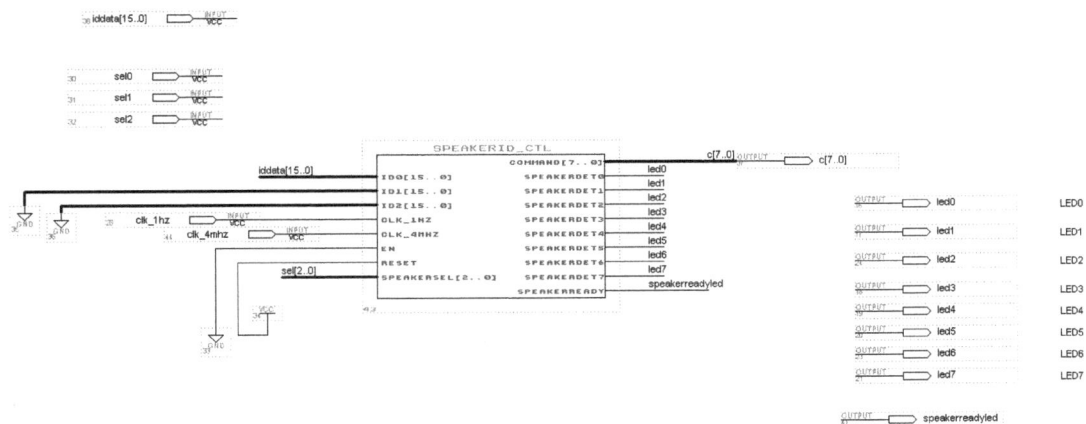

Figure 9.49. Quartus project file (speakerid_ctl.qpf) of speaker identification module (speakerid_ctl.vhd).

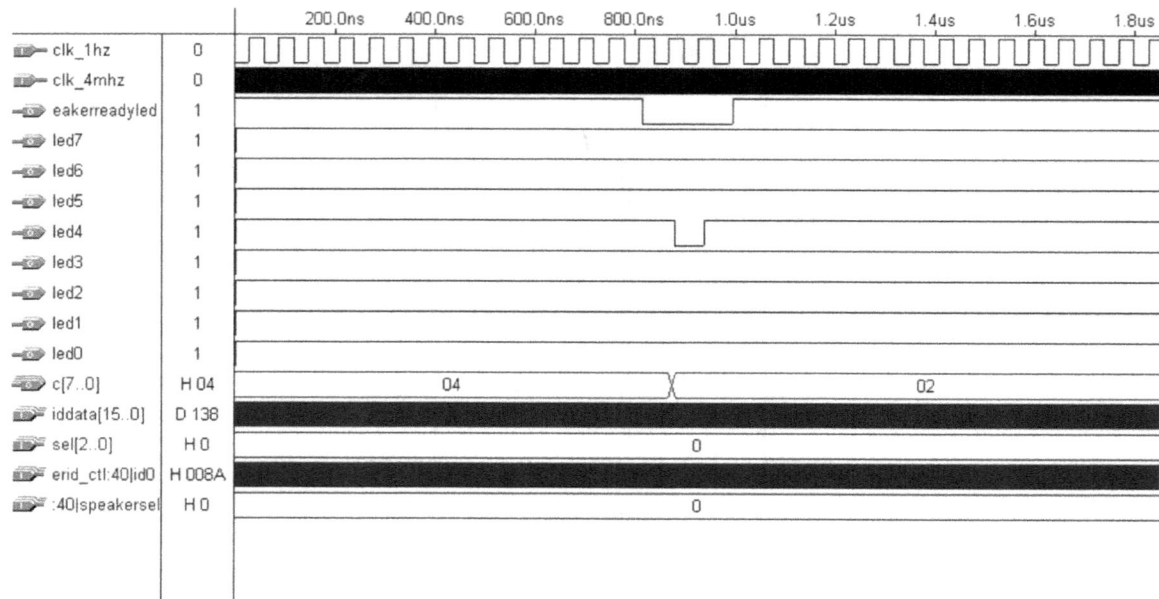

Figure 9.50. Simulation file of speakerid_ctl module.

9.14.1. VHDL Code for Speaker Identification Module (speakerid_ctl.vhd)

The VHDL for the speaker id (speakerid_ctl.vhd) control module is listed below:

```
library ieee;
use ieee.std_logic_1164.all;

entity speakerid_ctl2 is
port(id0: in integer range 0 to 65535;
id1:in integer range 0 to 65355;
id2:in integer range 0 to 65355;
clk_1hz: in std_logic;
clk_4Mhz:in std_logic;
en:in std_logic;
reset: in std_logic;
command: out std_logic_vector (7 downto 0);
speakersel: in integer range 0 to 7;
speakerdet0: buffer std_logic;
speakerdet1: buffer std_logic;
speakerdet2: buffer std_logic;
speakerdet3: buffer std_logic;
speakerdet4: buffer std_logic;
speakerdet5: buffer std_logic;
speakerdet6: buffer std_logic;
speakerdet7: buffer std_logic;
speakerready:buffer std_logic;
train_mode: in bit);
```

```
end speakerid_ctl2;


architecture a of speakerid_ctl2 is
signal thresholdmin0: integer range 0 to 65535;
signal thresholdmax0: integer range 0 to 65535;
signal thresholdmin1: integer range 0 to 65535;
signal thresholdmax1: integer range 0 to 65535;
signal thresholdmin2: integer range 0 to 65535;
signal thresholdmax2: integer range 0 to 65535;
signal thresholdmin3: integer range 0 to 65535;
signal thresholdmax3: integer range 0 to 65535;
signal thresholdmin4: integer range 0 to 65535;
signal thresholdmax4: integer range 0 to 65535;
signal thresholdmin5: integer range 0 to 65535;
signal thresholdmax5: integer range 0 to 65535;
signal thresholdmin6: integer range 0 to 65535;
signal thresholdmax6: integer range 0 to 65535;
signal thresholdmin7: integer range 0 to 65535;
signal thresholdmax7: integer range 0 to 65535;
signal speakerdet: integer range 0 to 15;
signal c: std_logic_vector(7 downto 0);
signal clk:std_logic;
signal cnt:integer range 0 to 31;
signal spd:std_logic_vector(7 downto 0);

begin


-- set speaker thresholds
                thresholdmin0 <= 0;        --speaker 0
                thresholdmax0 <= 100;

                thresholdmin1 <= 101;      --speaker 1
                thresholdmax1 <= 200;

                thresholdmin2 <= 201;      --speaker 2
                thresholdmax2 <= 400;

                thresholdmin3 <= 401;      --speaker 3
                thresholdmax3 <= 700;

                thresholdmin4 <= 701;      --speaker 4
                thresholdmax4 <= 1000;

                thresholdmin5 <= 1001;     --speaker 5
                thresholdmax5 <= 1100;
```

```
                thresholdmin6 <= 1101;    --speaker 6
                thresholdmax6 <= 1400;

                thresholdmin7 <= 1401;    --speaker 7
                thresholdmax7 <= 65535;

-- This process provides timing
    process (clk_1hz,reset)
    begin
        if (reset='0')then
            cnt<=0;
        elsif(clk_1hz'event and clk_1hz='1')then
            cnt<=cnt+1;
        end if;
    end process;

--This process provides the ready signal for speaker to
start speaking
    process (cnt, clk_1hz)
    begin

            if( cnt >13 and cnt <17)then
                speakerready<= '0';
            else
                speakerready<= '1';
            end if;

    end process;

--This process determines the speaker
    process(speakerready,reset)

    begin
        if (reset='0')then
            spd<="11111111";

    elsif(speakerready'event and speakerready='1')then

        if (id0 >= thresholdmin0 and id0 <=
thresholdmax0)then
            spd<="11111110";
            speakerdet<=0;

        elsif(id0 >= thresholdmin1 and id0 <=
thresholdmax1)then
            spd<="11111101";
```

```vhdl
                        speakerdet<=1;

            elsif(id0 >= thresholdmin2 and id0 <=
thresholdmax2)then
                    spd<="11111011";
                    speakerdet<=2;

            elsif(id0 >= thresholdmin3 and id0 <=
thresholdmax3)then
                    spd<="11110111";
                    speakerdet<=3;

            elsif(id0 >= thresholdmin4 and id0 <=
thresholdmax4)then
                    spd<="11101111";
                    speakerdet<=4;

            elsif(id0 >= thresholdmin5 and id0 <=
thresholdmax5)then
                    spd<="11011111";
                    speakerdet<=5;

            elsif(id0 >= thresholdmin6 and id0 <=
thresholdmax6)then
                    spd<="10111111";
                    speakerdet<=6;

            elsif(id0 >= thresholdmin7 and id0 <=
thresholdmax7)then
                    spd<="01111111";
                    speakerdet<=7;
            else
                    spd<="11111111";
            end if;
        end if;

speakerdet0<=spd(0);
speakerdet1<=spd(1);
speakerdet2<=spd(2);
speakerdet3<=spd(3);
speakerdet4<=spd(4);
speakerdet5<=spd(5);
speakerdet6<=spd(6);
speakerdet7<=spd(7);

    end process;
```

```
--set commamd if speaker is detected
     process(clk_4mhz,reset)
     begin
          if(clk_4mhz'event and clk_4mhz='1')then

          if (reset='0')  then
               c <= "00000010";  --turn right

          elsif ( (speakersel = speakerdet) )then
               c <= "00000100";  --go forward

          else
               c <= "00000010";  --turn right

          end if;
          end if;

     end process;

-- send command if not disabled
     process(en,clk_4mhz)
     begin
     if(clk_4mhz'event and clk_4mhz='1')then
          if (en='0')then
               command <= c;  --send command

          else
               command <= "ZZZZZZZZ";  --set to high Z
          end if;
     end if;

     end process;


end a;
```

10. WyJen Technologies Wireless Transmitters/Receivers

Wand yJen Technologies Incorporated designs and manufactures high quality, high performance, wireless devices for wireless applications. The WyJen digital wireless modules are listed in Table 10.1 and used in the robotic experiments in this book. Other similar wireless modules may be used. Basic digital communication concepts such as FSK, ASK, and PSK are discussed in Section 6.7 on page 232.

Device Part Number	Description
GTX315M-3V-A00S06A-M27A	315MHz ASK transmitter module, 3V
GTX433M-3V-A00S06A-M27A	433MHz ASK transmitter module, 3V
GRX315M-3V-A00S12B-M07A	315MHz ASK receiver module, 3V
GRX433M-3V-A00S12B-M07A	433MHz ASK receive module, 3V
GTR315M-M13	315MHz FSK transceiver module, 3 or 5V

Table 10.1. Digital WyJen Wireless Devices.

10.1. WyJen TransmitterGTX315M-3V-A00S06A-1 Description

The GTX315M-3V-A00S06A-1 transmitter module operates at bit rates up to 100Kbps. This module implements amplitude shift keying. Amplitude shift keying is discussed in Section 6.7.2 on page 237. It is ideal for battery applications because of its low power usage. The GTX315M-3V-A00S06A-1 transmitter module input/output pin information is shown in Figure 10.1 and their corresponding descriptions are given in Table 10.3. As shown in Figure 10.1, the GTX315M-3V-A00S06A-M27A is a 6 pin SIP package.

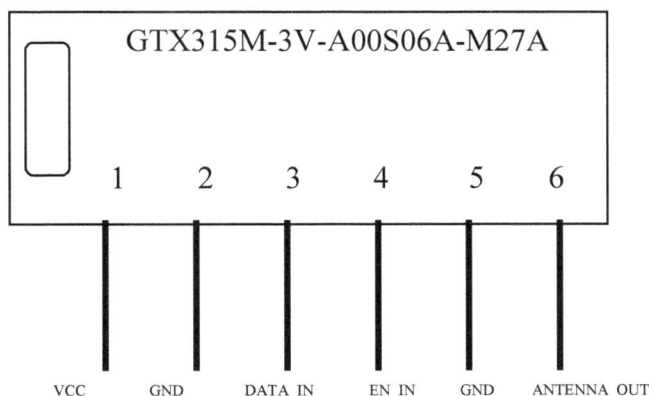

GTX315M-3V-A00S06A-M27A

| 1 | 2 | 3 | 4 | 5 | 6 |

VCC   GND   DATA_IN   EN_IN   GND   ANTENNA_OUT

Figure 10.1. Illustration of **GTX315M-3V-A00S06A-M27A** transmitter module showing inputs and outputs.

10.1.1. Key Features

The key features of the transmitter module are:

- Carrier frequency of 315 Mega Hertz
- Supports Amplitude Shift Keying (ASK) modulation with 90dB modulation depth
- Supports 3.3V input digital data (0 –3.3V)
- Supports two modes of operation: power down and power up
- Transmit bit rate up to 100Kbps
- Supply voltage range of 2.1 to 3.6 for 3.3V technology
- Supports a low 5.3mA supply current
- Operating temperature range of -40 to 125 degrees
- Antenna matching input of 50 ohms
- Low power consumption
- SIP package of 6 pins with 100 mil spacing between pins
- Module dimensions are 1.125 inches wide by 0.75 inches height

10.1.2. Targeted Applications

Some of the targeted applications of the GTX315M-3V-A00S06A-M27A transmitter are:

- Remote controlled garage door openers
- Remote controlled robotics
- Remote keyless entry systems
- Remote controlled toys
- Wireless sensors
- Wireless computer peripherals
- Wireless security systems
- Wireless game consoles
- Wireless pressure monitoring systems

10.1.3. Ordering Info

The transmitter order information is provided in Table 10.2. A similar transmitter may also be used.

Manufacturer/Distributor	Part Number	Online Store	Sales
WyJen Technologies Incorporated	GTX315M-3V-A00S12B-M27A	www.wyjen.com	1-800- 490-4165

Table 10.2. Transmitter ordering information.

10.1.4. Input/Output Pin Description
The function of each pin is described in Table 10.5.

Name	Pin Number	Description
VCC	1	3.3 V power supply.
GND	2	GROUND.
DATA_IN	3	Digital Data input. Power amplifier is on when data input is high (3.3 volts) and off when data input is low (0 volts). Should be 3.3 volt peak to peak..
EN_IN	4	Enable input. Module is in standby mode when this pin is low (0 volts), otherwise a high (3.3 volts) on this input puts the module is in power up mode.
GND	5	GROUND.
ANTENNA_OUT	6	50 ohm antenna output.

Table 10.3. Pin description table for the GTX315M-3V-A00S06A-M27A transmitter module.

10.2.   WyJen Receiver GRX315M-3V-A00S12B-M07A Description
The GRX315M-3V-A00S12B-M07A receiver is shown in Figure 10.2.   The major components of the module are a low noise amplifier (LNA), a Superherodyne receiver, peak detector, and power up/down circuitry.   The GRX315M-3V-A00S12B-M07A receiver is a radio frequency module that operates with a 315MHz carrier frequency using amplitude shift keying (ASK) modulation.   Amplitude shift keying is discussed in Section 6.7.2 on page 237.   The module is based on 3.3V CMOS technology that permits it to operate at low power usage.   Furthermore, the module uses power up/down circuitry which allows more power savings.   It is small and can easily be incorporated into existing system that transmits digital data over a physical cable or wire up to 10,000 bits per second.   This is done by replacing the wire link with a wireless link using the GTX315M-3V-A00S12A-M27A transmitter and GRX315M-3V-A00S12B-M07A receiver.   The module uses a SIP style connector that has strength and durability over other RF modules which allows it to withstand the ruggedness of the surrounding environment more.   The output of the module is 3.3V raw digital data or non return – to –zero (NRZ) serial data when the input transmitter data is also NRZ.   An appropriate length antenna can be used to optimize the maximum transmitting distance.   The module is easy to use and does not require RF knowledge to install.   The GRX315M-3V-A00S12B-M27A transmitter module is ideal for various educational as well as industrial projects where one wishes to transmit digital data over a wireless transmission link.   These projects often involve wireless sensors systems, remote control robotics, wireless garage door openers, wireless automotive systems, wireless microprocessor systems, wireless digital signal processing systems, wireless microcontroller systems, wireless security systems, wireless lighting

systems, wireless safety systems etc. The data transmitted in these systems should be limited to 10Kbps for maximum performance.

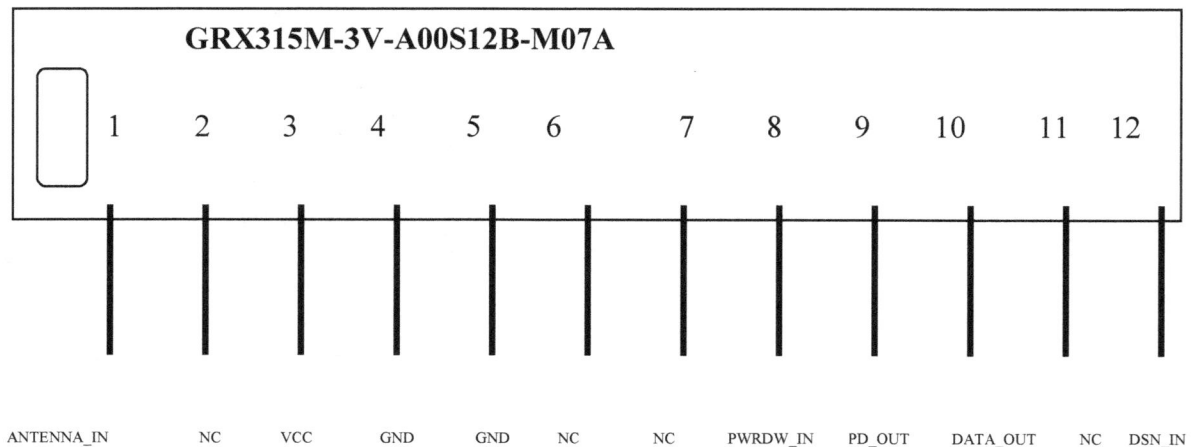

GRX315M-3V-A00S12B-M07A

| 1 | 2 | 3 | 4 | 5 | 6 | 7 | 8 | 9 | 10 | 11 | 12 |

ANTENNA_IN   NC   VCC   GND   GND   NC   NC   PWRDW_IN   PD_OUT   DATA_OUT   NC   DSN_IN

Figure 10.2. Illustration of GRX315M-3V-A00S12B-M07A receiver showing inputs and outputs.

10.2.1. Key Features
The key features of the GRX315M-3V-A00S12B-M07A receiver module are:

- Carrier frequency of 315 Mega Hertz
- Supports Amplitude Shift Keying (ASK) modulation
- Supports -115dB receive sensitivity
- Uses a superheterodyne receiver with a Low Noise Amplifier
- Supports 3.3V output data (0 –3.3V)
- Supports a low 5.5mA supply current
- Supports two modes of operation: power down and power up
- Supports a supply voltage range of 3.0V to 3.6V for 3.3V technology
- Supports operating temperature range of -40 to 85 degrees Celsius
- Antenna matching input of 50 ohms
- Low power consumption
- SIP package of 12 pins with 100 mil spacing between pins
- Module dimensions are 2 inches wide by 0.75 inches height

10.2.2. Targeted Applications
Some of the targeted applications of the GRX315M-3V-A00S12B-M07A receiver module are:

- Remote controlled garage door openers

- Remote controlled robotics
- Remote keyless entry systems
- Remote controlled toys
- Wireless sensors
- Wireless computer peripherals
- Wireless security systems
- Wireless game consoles
- Wireless pressure monitoring systems

10.2.3. Ordering Info

The receiver order information is given in Table 10.4. A similar receiver may also be used.

Manufacturer/Distributor	Part Number	Online Store	Sales
WyJen Technologies Incorporated	GRX315M-3V-A00S12B-M07A	www.wyjen.com	1-800- 490-4165

Table 10.4. Receiver ordering information.

10.2.4. Functional Block Diagram

A functional block diagram of the GRX315M-3V-A00S12B-M07A receiver is shown in Figure 10.3.

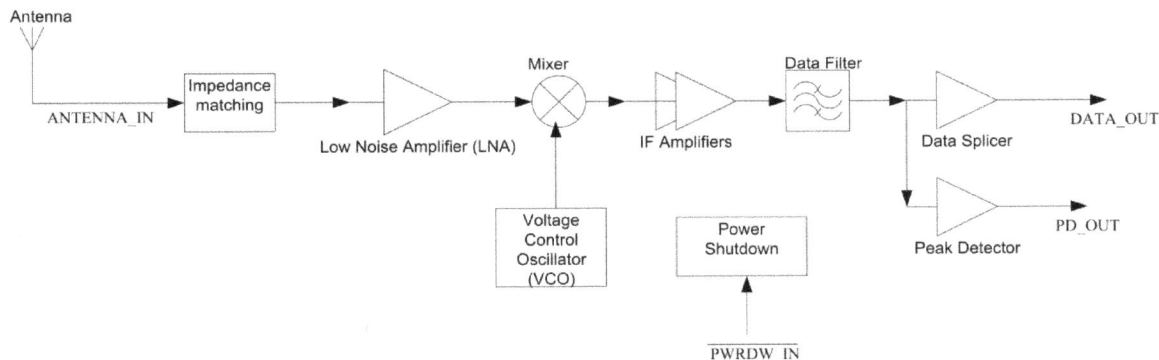

Figure 10.3. Illustration of GRX315M-3V-A00S12B-M07A receiver functional block diagram.

10.2.5. Input/Output Pin Description

A diagram of the GRX315M-3V-A00S12B-M07A receiver module with the input/output information is shown in Figure 10.2. The function of each pin is described in the Table 10.5.

Name	Pin Number	Description
ANTENNA_IN	1	50 ohm antenna input
NC	2	No connect
VCC	3	3.3 V power supply
GND	4	GROUND
GND	5	GROUND
NC	6	No connect
NC	7	No connect
PWRDW_IN	8	Power down input. Module is in power down mode when this pin is low, otherwise when this pin is high the RX module is in power up mode.
PD_OUT	9	Peak detector output
DATA_OUT	10	Digital data output
NC	11	No connect
DSN_IN	12	Data slice negative input, leave unconnected if no input

Table 10.5. Pin description table for the GRX315M-3V-A00S12B-M07A receiver.

10.3. **WyJen Transceiver GTR315M-M13 Description**

A top view of the GTR315M-M13 transceiver module is shown in Figure 10.4. The GTR315M-M13 transceiver module uses frequency shift keying (see Section 6.7.3 on page 237) with data rates up to 33000bps using Manchester encoded data or 66000bps using non return to zero encoded data. The module operates at two popular voltage source levels: 3V or 5V. In particular this module easily targets battery wireless applications using 3.3V. It also operates at the 5 volt supply which is typical for university laboratories. This module can also be used for non-battery applications with low noise sources.

The module integrates a transmitter and receiver section that uses a phase locked loop, low noise amplifier, and power amplifier. The receive section of the GTR315M-M13 module consists of a low noise amplifier with good sensitivity as low as -110dB. The transmit section of the GTR315M-M13 module consists of a power amplifier with a typical output power of 10dBm. Both the transmit and receive sections use a phase locked loop to modulate and demodulate the digital data, respectively. The module also provides a shutdown feature for low power consumption with a typical current of $0.8\mu A$.

The GTR315M-M13 module provides a transmit/receive input and bidirectional data which can be easily interfaced to and controlled by a baseband processor or programmable logic device. Other features of the module is that it is robust and rugged and can fit in small areas.

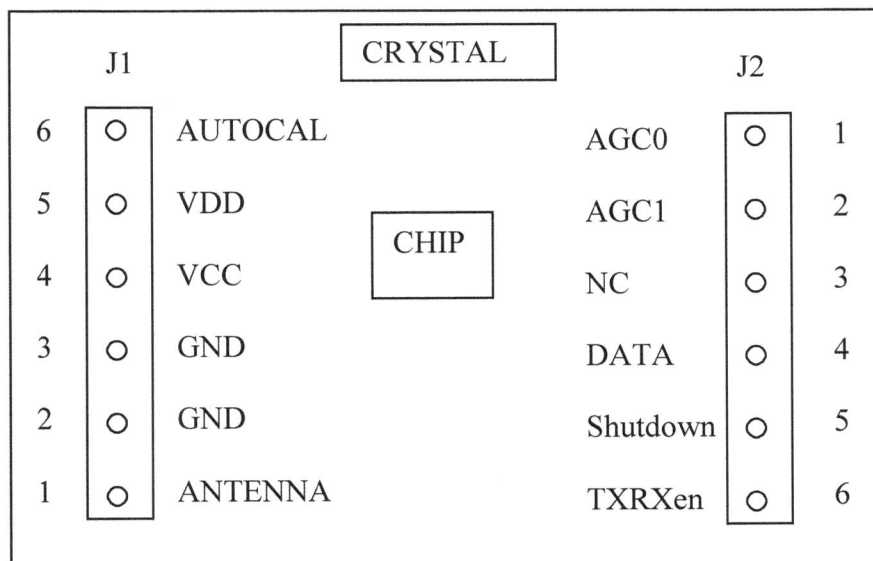

J1		CRYSTAL		J2	
6	○ AUTOCAL		AGC0	○	1
5	○ VDD		AGC1	○	2
4	○ VCC	CHIP	NC	○	3
3	○ GND		DATA	○	4
2	○ GND		Shutdown	○	5
1	○ ANTENNA		TXRXen	○	6

Figure 10.4. Illustration of GTR315M-M13 transceiver module showing pins. Note: Top view is from component side.

10.3.1. Key Features

The key features of the transceiver module are:

- Carrier frequency of 315 Mega Hertz
- Supports Frequency Shift Keying (FSK)
- Supports 3.3V or 5V input digital data
- Supports two modes of operation: power-down and power-up
- Transmit bit rate up to 66Kbps using NRZ line encoding or 33Kbps using Manchester line encoding
- Selectable Supply voltage range of 3.3V or 5V
- Selectable transmit/receive mode
- Operating temperature range of -40 to 125 degrees
- Antenna matching input of 50 ohms for TX or RX mode
- Low power consumption
- Provides TX/RX data via a 6 pin header with 100 mil spacing between pins and power and ground on separate 6 pin header with same spacing
- Module dimensions are 1.6 inches wide by 1 inch height
-

10.3.2. Targeted Applications

Some of the targeted applications of the GTR315M_M13 transceiver are:

- Remote controlled garage door openers
- Remote controlled robotics
- Remote keyless entry systems
- Remote controlled toys
- Wireless sensors
- Wireless computer peripherals
- Wireless security systems
- Wireless game consoles
- Wireless pressure monitoring systems

10.3.3. Ordering Info

Table 10.6 provides the transceiver ordering information. A similar transceiver may also be used.

Manufacturer/Distributor	Part Number	Online Store	Sales
WyJen Technologies Incorporated	GTR315M-M13A	www.wyjen.com	1-800- 490-4165

Table 10.6. Transceiver ordering information.

10.3.4. Input/Output Pin Description

The function of each pin for the GTR315M_M13 transceiver module is described in Table 10.7.

Name	J1 Pin Number	Input/Output	Description
ANTENNA	1	Bidirectional	50 ohm antenna input/output. When TX/RXen is HIGH this pin acts as a 315MHz RF output and when TX/RXen is HIGH this pin acts as an input to an incoming 315MHz RF signal.
GND	2	Input	GROUND.
GND	3	Input	GROUND.
VCC	4	Input	5V power supply.
VDD	5	Input	3.3V power supply configuration.
AUTOCAL	6	Input	Set to HIGH to auto-calibrate the internal FSK demodulator Phase Locked Loop to adjust to variations in temperature and voltage every minute, otherwise set to zero to calibrate only on power up.

Name	J2 Pin Number	Input/Output	Description
AGC0	1	Input	Automatic gain control 0
AGC1	2	Input	Automatic Gain control 1
NC	3	NC	No Connection
DATA	4	Bidirectional	When in receive mode DATA is an output (receive data), when in transmit mode DATA is an input (transmit data). See pin 6 of J2.
SHUTDOWN	5	Input	Shutdown input. When HIGH the module is in normal operation, when LOW the module is in shutdown mode.
TX/RXen	6	Input	Transmit/receive enable. To put the module in transmit mode set TX/RXen HIGH, to put it in receive mode set TXRXen LOW.

Table 10.7. Pin description table for the GTR315M-M13 transceiver module.

10.3.5. GTR315M_M13 5V or 3.3V Power Supply Configuration Table

Table 10.8 shows how to put the module in 5V or 3.3 power supply operation.

VCC Voltage	VDD Voltage	Power Supply Operation
NC	5V	5V operation
3.3V	NC	3.3V operation

Table 10.8. GTR315M_M13 power supply configuration.

10.3.6. GTR315M_M13 Auto-Calibration of Automatic Gain Control Table

Table 10.9 shows how to set the automatic gain control operation of the GTR315M_M13 module. The AGC is disabled when AGC0=0 and AGC1=0. The dwell time is used after the AGC has been enabled. The dwell time is the time that that the module waits before adjusting the gain. When the AGC is enabled using any of the three on states in Table 10.9 the strength of the incoming RF signal is monitored. When the strength is – 55dB (strong RF signal) the module waits for a certain amount of time (dwell time) before reducing the gain of low noise amplifier (LNA). When the RF signal strength is - 59dB the module waits a certain amount of time (dwell time) before increasing the LNA gain. If after the dwell time the RF signal changes back to the original RF level the LNA gain will remain unchanged.

AGC1	AGC0	Automatic Gain Control Operation
0	0	AGC disabled, High Gain
0	1	Dwell time is 162us
1	0	Dwell time is 1.3ms
1	1	Dwell time is 83ms

Table 10.9. GTR315M_M13 AGC configuration.

10.4. WyJen Transmitter/Receiver Part Codes

WyJen wireless transmitters and receivers follow the naming convention described in this section. The part code convention for the WyJen transmitter and receiver is given next.

GAABBBC-XX-YYYZWWT-VVVU

G	Genesis Engineering Design Group
AA	TX stands for Transmitter
	RX stands for Receiver
	TR stands for Transmitter and Receiver
BBB	3 digital number representing carrier frequency
	2 digit number representing carrier frequency with decimal point represented by D
C	M stands for mega Hertz
	G stands for Giga Hertz
XX	5V stands for 5 volt technology
	3V stands for 3.3V technology
YYY	A00 stands for ASK modulation
	F00 stands for FSK modulation
	P00 stands for PSK modulation
	Q00 stands for QAM modulation
	AF0 stands for ASK and FSK modulation
	AFP stands for ASK, FSK, and PSK modulation
	AM0 stands for amplitude modulation
	AFM stands for ASK and FM modulation
	FFM stands for FSK and FM modulation
Z	S stands for SIP package
	D stands for DIP package

	M stands for surface mount package
T	A stands for 1.125 inches width by 0.75 inches height
	B stands for 2 inches width by 0.75 inches height
WW	2 digit number representing number of input and output pins in the package
VVV	3 character series code
U	1 character feature code

Notes:

1.   T Field: For SIP package width is side with connector
2.   BBB and C Fields are used together: Some example values are

315M
433M
868M
915M
2D5G
5D8G

Examples

The part number GTX315M-3V-A00S06A-M27A corresponds to a WyJen Technologies transmitter, 315 mega Hertz operating frequency, that requires a 3.3 volt power supply, uses amplitude shift keying modulation with a 6 pin SIP package, series M27 transmitter, with version A features. The dimensions of the package are 1.125 width by 0.75 height.

The part number GTX2D5G-3V-F00S12B-M08B corresponds to a WyJen Technologies transmitter, 2.5 Giga Hertz operating frequency, that requires a 3.3 volt power supply, uses frequency shift keying modulation with a 12 pin SIP package, series M08 receiver, with version B features. The dimensions of the package are 2 width by 0.75 height.

11. Wyjen Transmitter, Receiver, and Transceiver Experiments

Wireless devices are fun to work with. The goal of this Chapter is to help you become more familiar with a simple wireless communication system as discussed in Section 6.3 on page 198. First, you will learn how to test a wireless transmitter/receiver module and a transceiver module followed by several experiments.

11.1. Testing the GTX315M-3V-A00S06A-M27A Transmitter and GRX315M-3V-A00S12B-M07A Receiver or GTX433M-3V-A00S06A-M27A Transmitter and GRX433M-3V-A00S12B-M07A Receiver modules

The GTX315M-3V-A00S06A-M27A is an ASK transmitter with a 315MHz carrier frequency while the GRX315M-3V-A00S12B-M07A is an ASK receiver. The two devices can operate together to make a complete simplex wireless ASK system. The GTX433M-3V-A00S06A-M27A transmitter and GRX433M-3V-A00S12B-M07A receiver are the same type of pair but operate at the 433MHz carrier frequency. These devices also require a 3.3V supply voltage and 3.3V interfacing signals. The procedure for testing these devices is given next.

11.1.1. Procedure for Testing the 315/433MHz Transmitter and Receiver Pair

The procedure for testing the GTX315M-3V-A00S06A-M27A transmitter and GRX315M-3V-A00S12B-M07A receiver pair is the same for testing the GTX433M-3V-A00S06A-M27A transmitter and GRX433M-3V-A00S12B-M07A receiver pair. The procedure to follow is given below:

1. Acquire the necessary test equipment listed in Table 11.1.

Equipment/Component	Quantity	Function	Description
Transmitter: GTX315M-3V-A00S06A-M27A or GTX433M-3V-A00S06A-M27A	1	TX	315MHz Transmitter, 3.3V, www.wyjen.com
Receiver: GRX315M-3V-A00S12B-M07A or GRX433M-3V-A00S12B-M07A	1	RX	315MHz Receiver, 3.3V, www.wyjen.com
Function Generator	1	Signal Generation	Function generator for 3.3V square wave with 1.65V offset
Oscilloscope	1	Waveform measurement	Oscilloscope to measure transmitted and received data signal
Power Supply	1	3.3V Supply	3.3V source for transmitter and receiver test circuit

Table 11.1. Equipment and Electronic part list for the WyJen transmitter and receiver testing.

2. Build the transmitter/receiver test circuit shown in Figure 11.1. It is recommended that two different circuits boards be used, one for the transmitter circuit and one for the receiver circuit. This will allow the transmitter and receiver to be separated various distances apart if needed.

3. Now setup the function generator to output a test waveform which is a square wave with 3.3V peak-to-peak and with a 1.65V offset. This mimics 3.3V digital data (with amplitudes between 0V to 3.3V, also called unipolar Non Zero Return or UNRZ) which is what the WyJen transmitter wants to see. Set the frequency to 500 Hz. The square waveform now represents a serial bit stream with a 1000bps data rate ($f_b=2f$, where f_b is the data rate and f is the frequency of the function generator, i.e., $f_b =2*500=1000bps$). Separate the transmitter and receiver circuits by a distance of about 1 foot. The WyJen transmitter module, GTX315M-3V-A00S06A-M27A or GTX433M-3V-A00S06A-M27A, can transmit digital data at rates up to 100,000 bps. The WyJen receiver module GRX315M-3V-A00S12B-M07A or GRX433M-3V-A00S12B-M07A can receive at rates up to 10,000bps. Thus, the bit rate of this wireless communication system is limited by the receiver at 10,000bps.

4. Now, using the oscilloscope, compare the transmitted waveform (function generator output) to the received waveform (receiver data output). Compare their amplitudes and frequencies. What similarities or differences do you see. Do you see any noise, etc. Demonstrate your results to the instructor. Make sure that the oscilloscope displays the frequency and amplitude measurements. Save and print the waveform for your records and your report.

5. Now repeat steps 3 to 4 using the following bit rates: 1000bps, 2000bps, 3000bps, 7000bps, 10000bps, 12000bps, 14000bps, and 20000bps.

6. If a spectrum analyzer is available measure the output of the transmitter module using one of the bit rates above. Determine the amplitudes and frequencies of the highs and valleys in the spectrum. Save and print your results.

7. Have your instructor sign off your work:

Instructor_____Date_____

Note: Function Generator Output - 3.3V peak-to-peak square waveform with 1.65V offset.

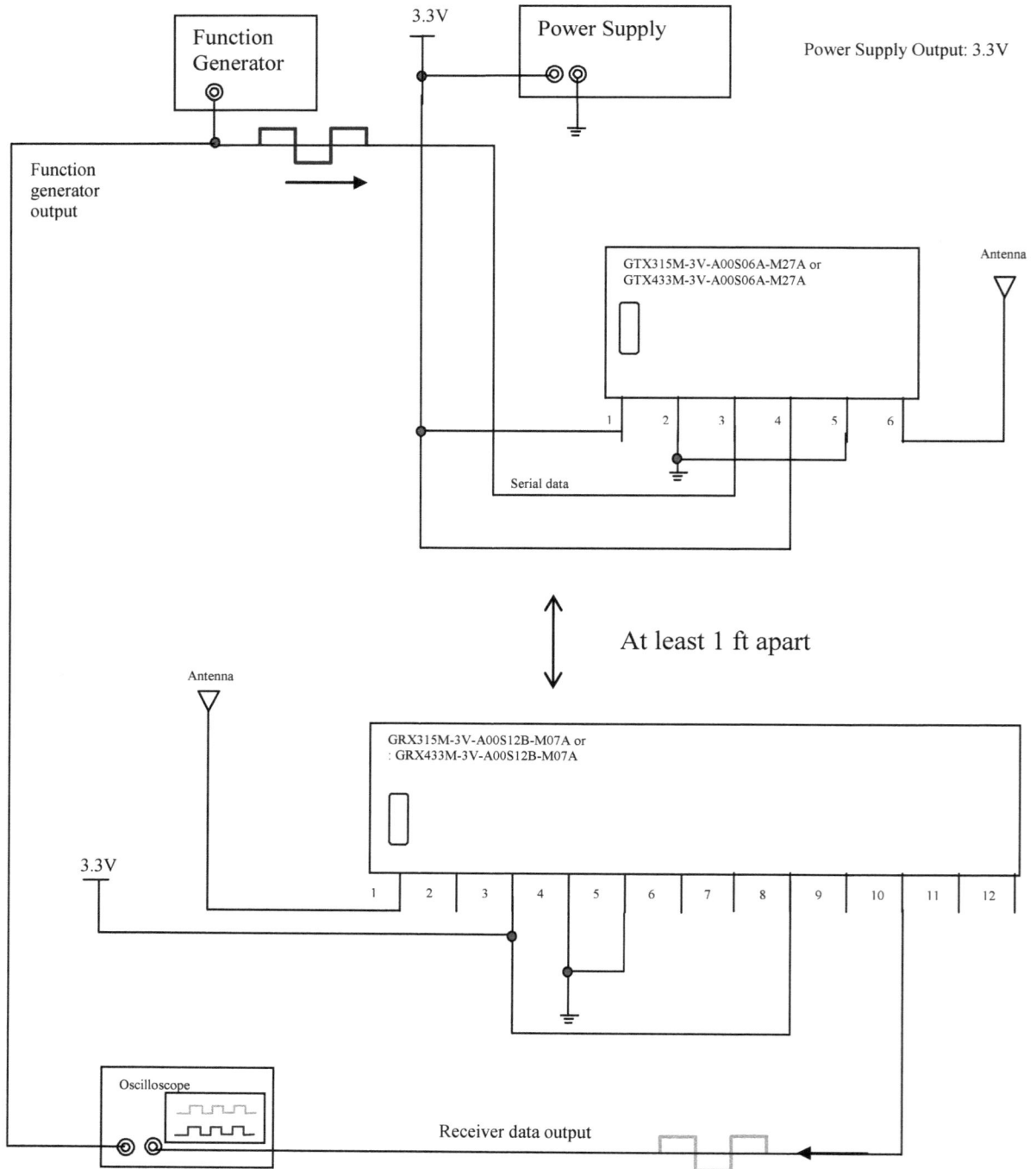

Figure 11.1. WyJen 3.3V 315/433 MHz transmitter and receiver pair test circuit.

11.2. Procedure for Testing a 315MHz Transceiver Pair

The procedure for testing the GTR315M_M13 is given as follows:

1. Acquire the necessary test equipment listed in Table 11.1.

Part	Quantity	Ref designator	Description
Transceiver: GTR315M_M13	2	TXRX	315MHz transmitter/receiver, 5V, half duplex transmission, www.wyjen.com
Capacitor	6		10pF capacitor
Antenna	2	wires	Antenna wire, strip wire of insulation for self made antenna.
Bread boards	2		Small bread boards
Oscilloscope	1	Oscilloscope	Oscilloscope to measure transmitted and received data signal
Power Supply	1	Power	5V source for transmitter and receiver test circuit
Function Generator	1	Signal Generation	Function generator for 5V square wave with 2.5V offset

Table 11.2. Equipment and Electronic part list for the WyJen GTR315M_M13 transceiver testing.

Build the transceiver test circuit shown in Figure 11.2. It is recommended that two different circuits boards be used, one for each transceiver circuit. This will allow each transceiver to be separated various distances apart if needed.

2. Now setup the function generator to output a test waveform which is a square wave with 5V peak-to-peak and with a 2.5V offset. This mimics 5V digital data (with amplitudes between 0V to 5V, also called unipolar Non Zero Return or UNRZ) which is what the WyJen transceiver wants to see. Set the frequency to 500 Hz. The square waveform now represents a serial bit stream with a 1000bps data rate ($f_b=2f$, where f_b is the data rate and f is the frequency of the function generator, i.e., $f_b =2*500=1000$bps). Separate the transceiver circuits by a distance of about 1 foot. The WyJen transceiver module, GTR315M_M13, can transmit and receive digital data at rates up to 66,000 bps.

3. Now, using the oscilloscope, compare the transmitted waveform (function generator output) to the received waveform (receiver data output). Compare their amplitudes and frequencies. What similarities or differences do you see. Do you see any noise, etc. Demonstrate your results to the instructor. Make sure that the oscilloscope displays the frequency and amplitude measurements. Save and print the waveform for your records and your report.

4. Now repeat steps 3 to 4 using the following bit rates: 1000bps, 2000bps, 3000bps, 7000bps, 10000bps, 12000bps, 14000bps, and 20000bps.

Note: Function Generator Output - 5V peak-to-peak square waveform with 2.5V offset.

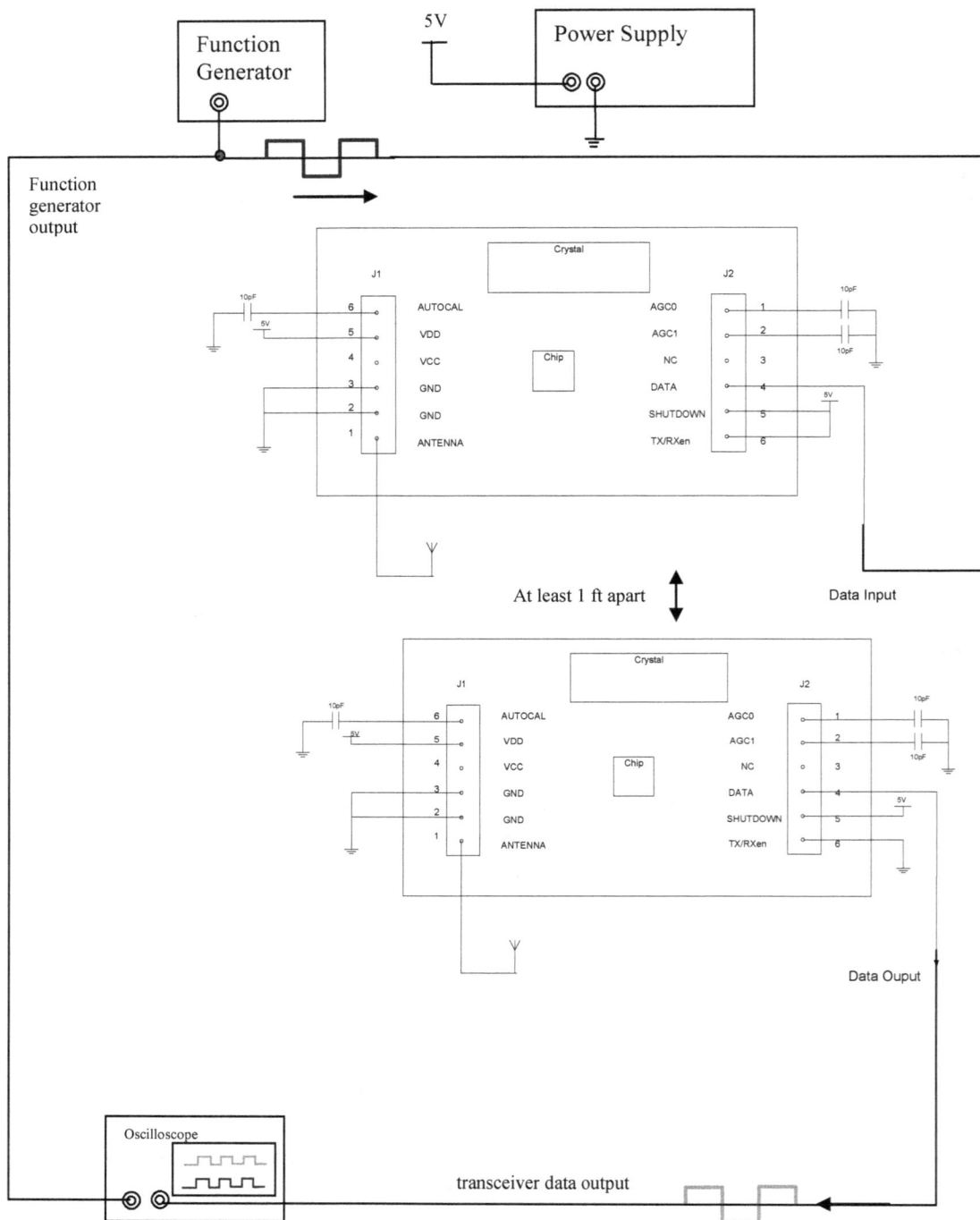

Figure 11.2. WyJen GTR315M_M13 Transceiver 5V test circuit.

5. If a spectrum analyzer is available measure the output of the transmitter module using one of the bit rates above. Determine the amplitudes and frequencies of the highs and valleys in the spectrum. Save and print your results.

6. Have your instructor sign off your work:

Instructor_____Date_____

11.3. A Simple Simplex Communication System using a 315/433MHz Transmitter and Receiver Module Pair

The procedure for constructing a simple communication system is given next.

11.3.1. Part A - Preparing Digital Data for Wireless Transmission - from txuart to rxuart

1. Acquire the necessary equipment and components for the experiment which are listed in Table 11.1. During this exercise you will create a Quartus project file to build a digital circuit - the txuart circuit. This will be compiled and programmed into the PLD. A few tests will be performed. Afterwards, the rxuart circuit will be added next and then programmed into the PLD. To perform these tasks see the following steps.

Part	Quantity	Ref designator	Description
Transmitter: GTX315M-3V-A00S06A-M27A or GTX433M-3V-A00S06A-M27A	1	TX	315MHz Transmitter, 3.3V, www.wyjen.com
Receiver: GRX315M-3V-A00S12B-M07A or GRX433M-3V-A00S12B-M07A	1	RX	315MHz Receiver, 3.3V, www.wyjen.com
MM74HC244 or SN74LVC244AN	2	IC1	5V to 3.3V converter/buffer, www.digikey.com or Devry Lab
AA Batteries	4		Any Store, Very clean voltage source for TX and RX. Put 2 AA batteries in series for 3.0 V sources. One pair for the TX and one pair for the RX
TPS7133QP, Digikey (optional)	3	IC2	3.3V Voltage Regulator
PLD Platform	2	PLD	Serial data generation, DeVry ESOC or Altera
Oscilloscope	1	Oscilloscope	Oscilloscope to measure transmitted and received data signal
Power Supply	1	Power	3.3V source for transmitter and receiver test circuit

Table 11.3. Equipment and Electronic part list for the WyJen transmitter and receiver testing.

2. Open Quartus II. Create the project file **txuartrxuart_ctl.qpf**. Insert the symbol for the txuart file. Complete the txuart digital circuit portion shown in Figure 11.3. Verify that the device assigned to the design is the EMP7128SLC or an equivalent pin compatible PLD device Assign pins to the txuart main inputs and outputs. Assign switches to the txuart **D inputs**. These switches will be used to set the bits in the serial data. Also, assign a pin to the **serial data output** and the **main clock input**. The other inputs and outputs are not important at this time and do not require a pin assignment. Now compile the program.

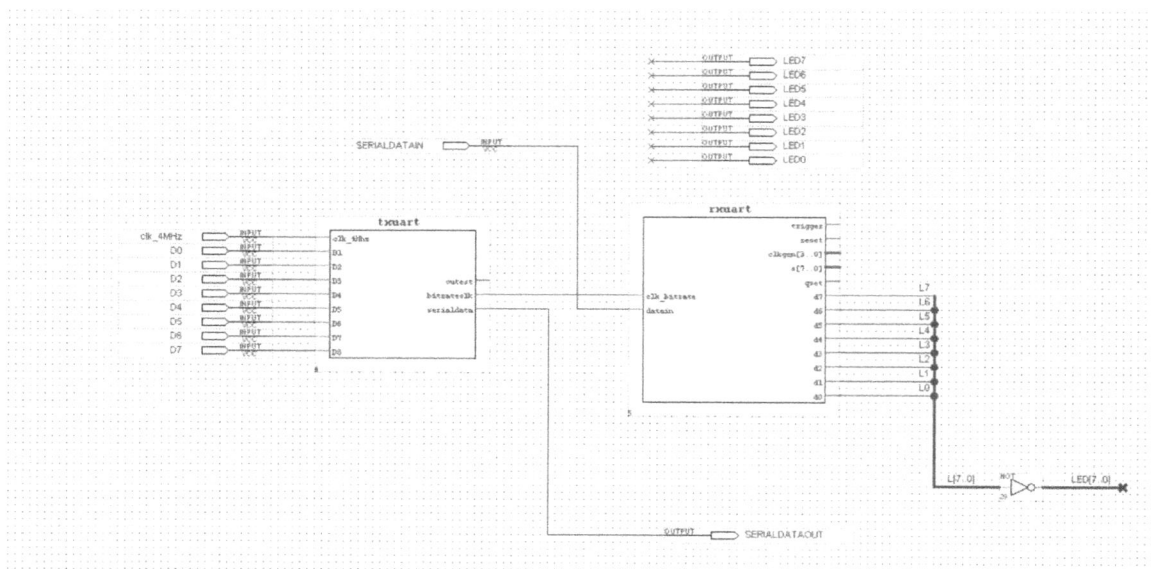

Figure 11.3. Quartus project file (**txuartrxuart_ctl.qpf**) of txuart and rxuart digital circuit.

3. Now, program the PLD with the txuart circuit of **txuartrxuart_ctl.qpf** of the project file.

4. Now use an oscilloscope to measure the serial data output. Set the serial data to alternating zeros and ones. Measure the output using 3 different bit rates by selecting the appropriate counter output clock. Observe the stop and start bits. How many **stop** bits are displayed? _____ How many **start** bits are displayed? _____ How can you change the number of stop bits? Please explain:

_____
_____
_____
_____

Save and print your results. Indicate the start, data, and stop bits using a pencil. Indicate on the print out what each bit is: Start, D7, D6, D5, D4, D3, D2, D1, D0, Stop1, Stop1, etc. Do this for one period of the data on the print out.

5. Now select two desire serial data bit patterns using the switches and measure, save, and print each showing the stop and start bits using a pencil. Show which data bits are the MSB (D7) and LSB(D0).

6. Open Max Plus II or Quartus. Open the rxuart file. Make a symbol for the rxuart circuit. Now, insert the rxuart symbol into the design file txuartrxuart_ctl.qpf. Assign pins to the rxuart's main inputs and outputs. Assign the bit rate clock used for the txuart to the rxuart's **bit rate input.** Assign the 8 data outputs of the rxuart to 8 LEDs on the PLD platform. The other inputs and outputs are not important at this time and do not require a pin assignment. Now compile the design and program it into the PLD again with the updated design file **txuartrxuart_ctl.qpf.** The digital circuit should now be complete and is shown in Figure 11.3.

7. Now physically connect a wire from the **serialdataout** output to the **serialdatain** input. Use the user-guide to determine where the PLD pins are on the board. Select a serial data pattern to transmit using the switches. Observe the LEDs. What do you see? _____

_____

8. Now increase the number of stop bits from your current number (1) to a number between 2 and 6. The number of stop bits may be changed by implementing the changes to count10.vhd as indicated in Table x below. Keep increasing the number of stop bits until the LEDs match the switches. How many stop bits did you use to get the match?_____

Number of Stop bits	Change the count10.vhd VHDL statement: if (clr='0' or cnt=10)then. to:	Count sequence
1	if (clr='0' or cnt=10)then	0,1,2,3,4,5,6,7,8,9
2	if (clr='0' or cnt=11)then	0,1,2,3,4,5,6,7,8,9,10
3	if (clr='0' or cnt=12)then	0,1,2,3,4,5,6,7,8,9,10, 11
4	if (clr='0' or cnt=13)then	0,1,2,3,4,5,6,7,8,9,10, 11,12
5	if (clr='0' or cnt=14)then	0,1,2,3,4,5,6,7,8,9,10, 11,12,13
6	if (clr='0' or cnt=15)then	0,1,2,3,4,5,6,7,8,9,10, 11,12,13,14
7	if (clr='0')then	0,1,2,3,4,5,6,7,8,9,10, 11,12,13,14,15

Table 11.4. Number of stops bits verses VHDL statement changed to. The count10.vhd if statement: if (clr='0' or cnt=10)then, changes the number of stop bits selected by making count10 a truncated counter. When the number of stop bits is 7 the counter is no longer truncated.

9. Have your instructor sign this part off:

Instructor_____ Date _____

11.3.2. Part B - Sending Data from the txuart to a Transmitter and over a Wireless Link to a Receiver and the rxuart

This section continues from the previous section:

10. Construct the analog/digital circuits shown in Figure 11.1 to interface the 3.3V transmitter and 3.3V receiver to the 5V PLD. When a 5V chip drives or sinks a 3.3 chip a translator chip is commonly used to match the voltage and current levels between the two different circuits. In this case, a 5V to 3.3V buffer translator chip. SN74LVC244AN or MM74HC244 is required between the transmitter serial data input and the PLD and between the PLD and the receiver serial data output.

11. Use the switches to send the following bit patterns: A, B, C, D, E, and F, in Table 11.6 from the transmitter to the receiver (over a wireless link) using the indicated bit rate. In the txuart the bit rate clk is the input clock to coun10. The bit rate can be changed by selecting clk input to count10 in txuart: a net name of W11 gives a bit rate of 976.5625bps , W12 gives a bit rate of 488.28125bps, W13 gives a bit rate of $4Mhz/2^{14}=244.140625bps$, and so on as shown in Table 11.5 for various counter outputs (count65536). Note that the frequency of the counter output is equal to the bit rate because a bit is placed on the serialdataout on every positive edge of the bit rate clock. Observe the LEDs and write down the LEDs that are off/on. Are the LEDs consistently on or off or blinking on and off.

count65536 Counter output	Frequency of divided clock, $4 \times 10^6/2^n$, n=1,2,3,...16
W_0	$4Mhz/2^1=2Mhz$
W_1	$4Mhz/2^2=1Mhz$
W_2	$4Mhz/2^3=500khz$
W_3	$4Mhz/2^4=250khz$
W_4	$4Mhz/2^5=125khz$
W_5	$4Mhz/2^6=62.5khz$
W_6	$4Mhz/2^7=31.25khz$
W_7	$4Mhz/2^8=15.625khz$
W_8	$4Mhz/2^9=7.8125khz$
W_9	$4Mhz/2^{10}=3.90625khz$
W_{10}	$4Mhz/2^{11}=1.953125khz$
W_{11}	$4Mhz/2^{12}=976.5625hz$
W_{12}	$4Mhz/2^{13}=488.28125hz$
W_{13}	$4Mhz/2^{14}=244.140625hz$
W_{14}	$4Mhz/2^{15}=122.0703125hz$
W_{15}	$4Mhz/2^{16}=61.03515625hz$

Table 11.5. Frequency division of the 4Mhz clock using the mod 65536 counter (16 bit synchronous binary counter).

Bit Pattern Number	Bit Rate	Bit Pattern The MSB is to the far left	Observed LED pattern, put MSB to far left
A	61.035bps	10101010	
B	488.28kbps	11001100	
C	976.56kbps	00110011	
D	7.8125kbps	10010010	
E	15.625kbps	00010000	
F	15.625kbps	11111111	

Table 11.6. Bit patterns to send over wireless link.

12. What is the WyJen transmitter Baud rate for each bit pattern given in Table 11.6. Record these in the Table below:

Bit Pattern from Table 11.6	Bit Rate	Baud Rate
A	61.035bps	
B	488.28kbps	
C	976.56kbps	
D	7.8125kbps	
E	15.625kbps	
F	15.625kbps	

Table 11.7. Baud Rate vs. bit rate of the WyJen ASK transmitter.

13. Demonstrate and have your instructor sign off on your results:

Instructor_____ Date _____

A. MM74HC244 or SN74LVC244AN pin configuration. The pin configuration of the MM74HC244 is compatible to that of the SN74LVC244AN. Shown below is the PDIP package, or N package, for the SN74LVC244A. The complete part number is SN74LVC244AN.

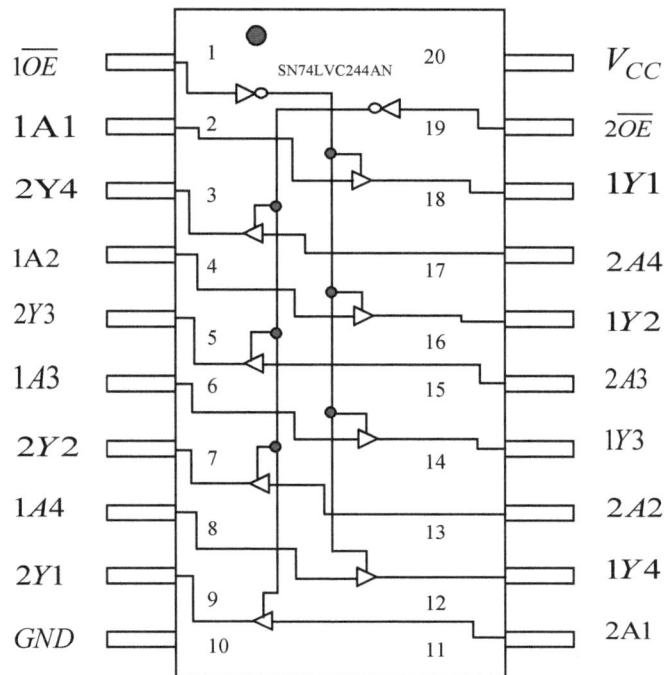

Pin	Signal		Pin	Signal
1	$1\overline{OE}$		20	V_{CC}
2	$1A1$		19	$2\overline{OE}$
3	$2Y4$		18	$1Y1$
4	$1A2$		17	$2A4$
5	$2Y3$		16	$1Y2$
6	$1A3$		15	$2A3$
7	$2Y2$		14	$1Y3$
8	$1A4$		13	$2A2$
9	$2Y1$		12	$1Y4$
10	GND		11	$2A1$

SN74LVC244AN

B. SN74LVC244A truth table.

Inputs		Output
$\overline{OE}$	**A**	**Y**
L	H	H
L	L	L
H	X	Z

Figure 11.4. A.) Pin configuration and B.) truth table, for the SN74LVC244AN or MM74HC244 DIP package.

MM74HC244/SN74LVC244AN

3.3V

Power Supply or battery supply

Power Supply Output: 3.3V or two AA batteries in series at 3.0V

Antenna

GTX315M-3V-A00S06A-M27A or
GTX433M-3V-A00S06A-M27A

Output: serialdata to TX

PLD Platform

8 LEDs

Serial data

1 2 3 4 5 6

At least 1ft apart

Input: serialdata from RX

Antenna

GRX315M-3V-A00S12B-M07A or
: GRX433M-3V-A00S12B-M07A

3.3V

1 2 3 4 5 6 7 8 9 10 11 12

MM74HC244/SN74LVC244AN

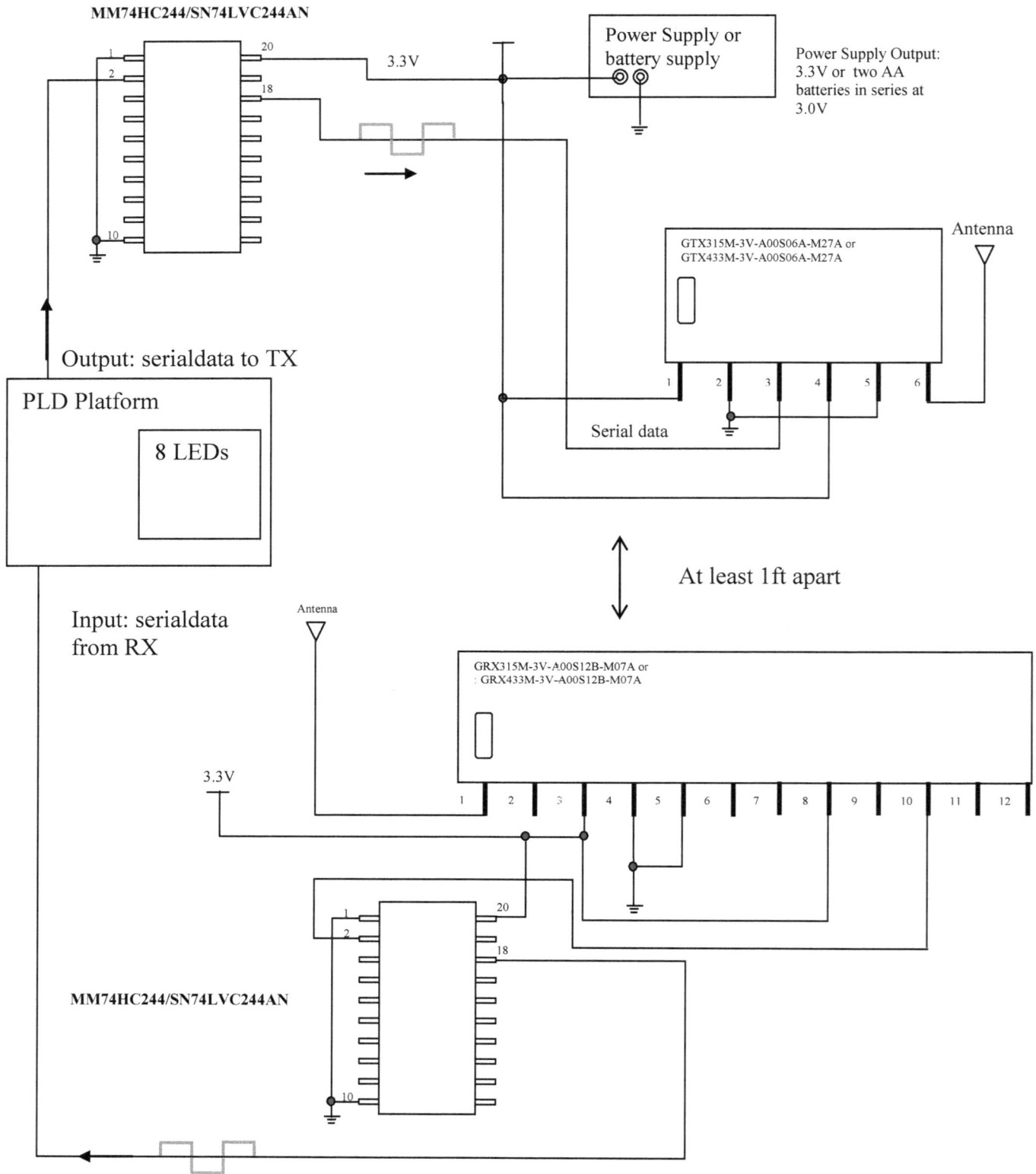

Figure 11.5. WyJen 3.3V 315/433 MHz transmitter and receiver pair test circuit.

11.4. A Simple Simplex Communication System using a 315 MHz Transceiver Module Pair

The procedure for constructing a simple communication system is the same as the transmitter/receiver experiment in Section 11.3. A brief procedure is given here.

1. Acquire the necessary test equipment listed in Table 11.1.

Part	Quantity	Ref designator	Description
Transceiver: GTR315M_M13	2	TXRX	315MHz transmitter/receiver, 5V, half duplex transmission, www.wyjen.com
Capacitor	6		10pF capacitor
Antenna	2	wires	Antenna wire, strip wire of insulation for self made antenna.
Bread boards	2		Small bread boards
Oscilloscope	1	Oscilloscope	Oscilloscope to measure transmitted and received data signal
Power Supply	1	Power	5V source for transmitter and receiver test circuit
Function Generator	1	Signal Generation	Function generator for 5V square wave with 2.5V offset
PLD platform	1	PLD	Programmable logic device

Table 11.8. Equipment and Electronic part list for the WyJen GTR315M_M13 transceiver communication system.

2. Repeat the experiment in Section 11.3 using the circuit in Figure 11.6.

3. Lastly, have the instructor check the results.

Function Generator Output: 5V peak-to-peak square waveform
with 2.5V offset

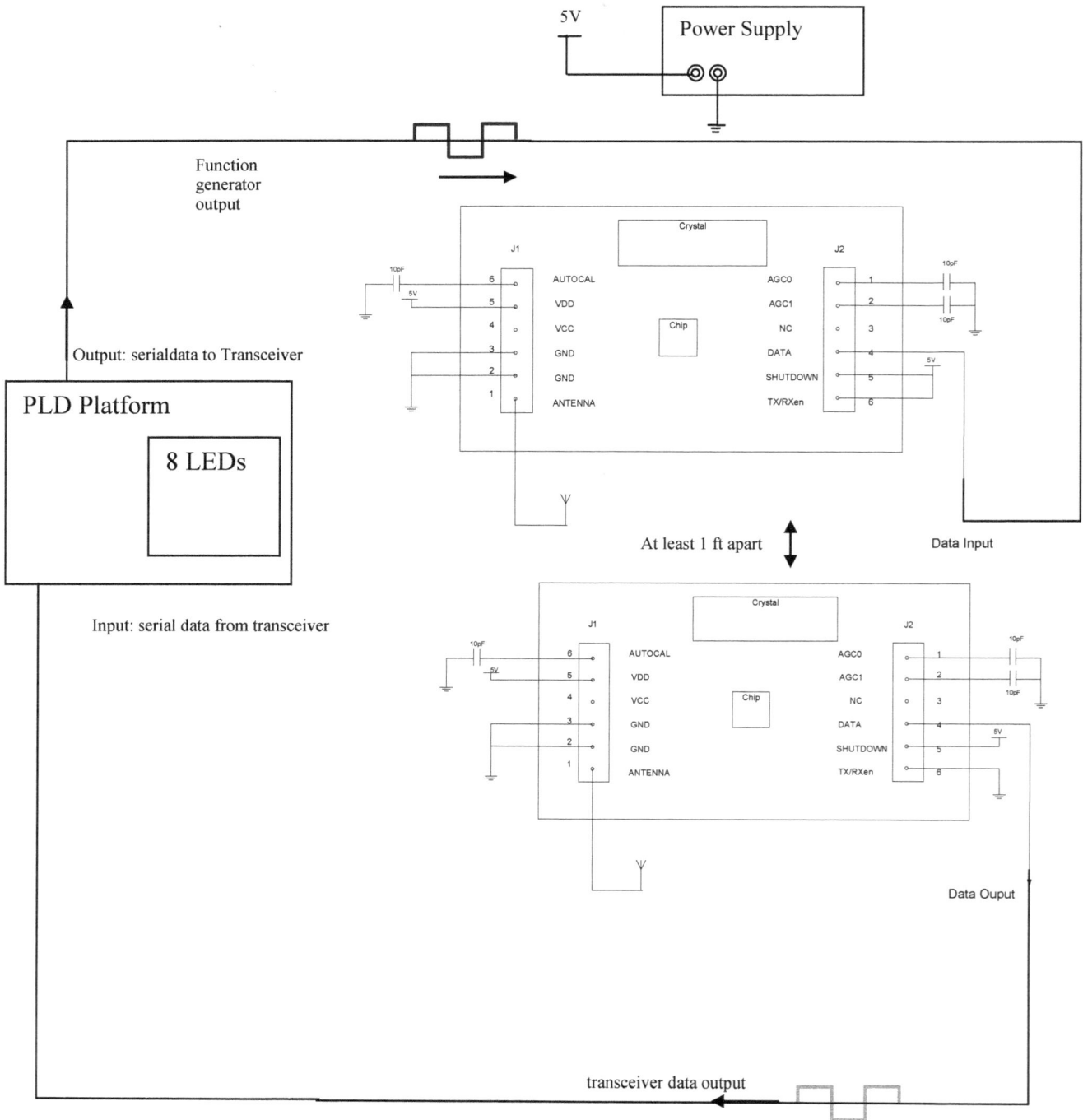

Figure 11.6. Simple communication system using WyJen Transceiver.

11.5. Implementing a Noise Improvement Method for Wireless Transmitter Receiver Systems

There are digital data encoding techniques called line codes that can use to improve the transmission quality of digital signals in digital transmission over a cable medium (wireline) and digital wireless systems. Some line codes can also be used for ease of clock recovery. Common line codes are:

1. Unipolar return to zero (URZ)
2. Bipolar return to zero (BRZ)
3. Unipolar nonreturn to zero (UNRZ)
4. Bipolar nonreturn to zero BNRZ)
5. Alternate Mark Inversion (AMI)
6. Manchester

An example of each line code is shown in Figure 11.7 below (Tomasi, 2004). One in particular we have chosen here is the Manchester line code. This is discussed next.

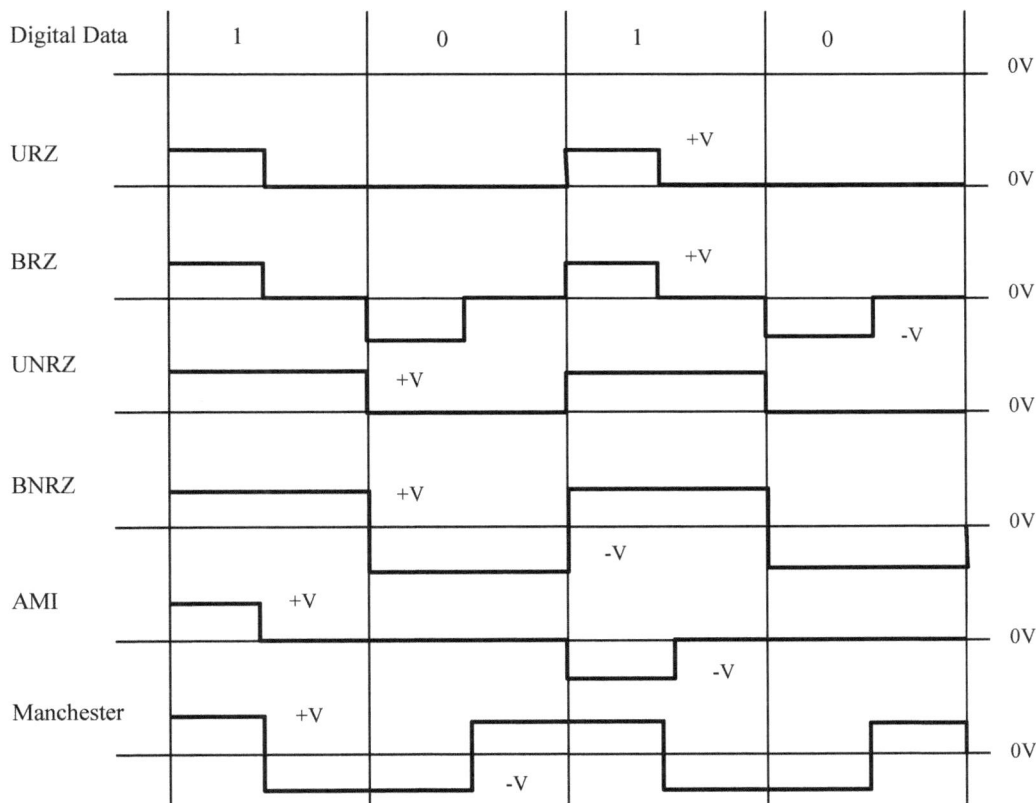

Figure 11.7. Some common lines codes used for encoding digital data.

11.5.1. Manchester Encoding Implementation

Manchester offers ease of clock recovery and noise reduction. It also offers a simple and convenient way to encode the transmitted data using an XOR gate as shown in Figure 11.8. The corresponding input and output waveforms of the digital circuit shown in Figure 11.8 are shown in Figure 11.9. This can easily be done in a programmable logic device (PLD) or microcontroller (μC). The block diagram for Manchester encoding in a radio frequency (rf) system is shown in Figure 11.10.

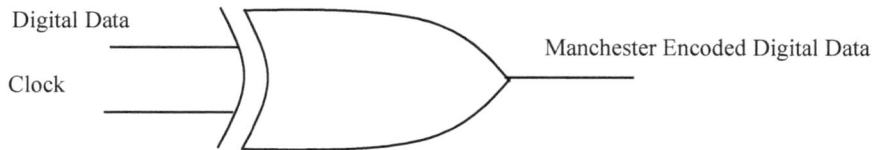

Figure 11.8. Method to perform Manchester encoding on digital data.

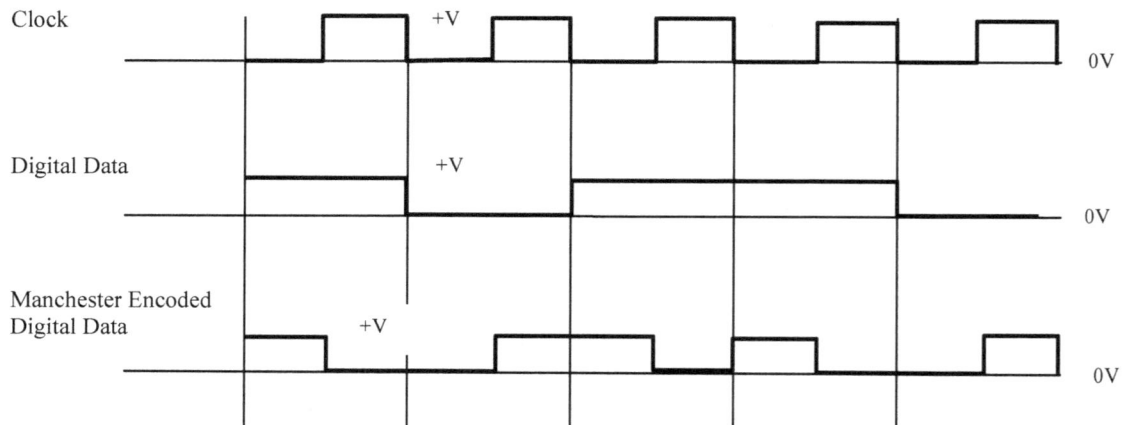

Figure 11.9. Input and output waveforms for Figure 11.8.

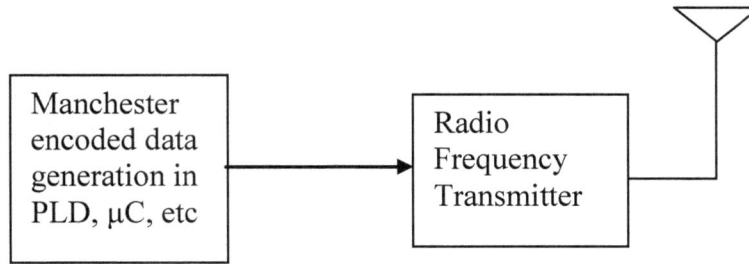

Figure 11.10. Manchester encoding in a radio frequency system.

A circuit for converting the received Manchester encoded data back to digital logic levels prior to being processed by a Manchester decoder is shown in Figure 11.11. The output of this circuit is then fed to a PLD or μC for decoding the Manchester data back into digital data as shown in Figure 11.12. The Manchester decoder logic determines when a positive to zero transition occurs and outputs a logic one. Similarly the Manchester decoder logic outputs a logic zero when a zero to positive transition occurs.

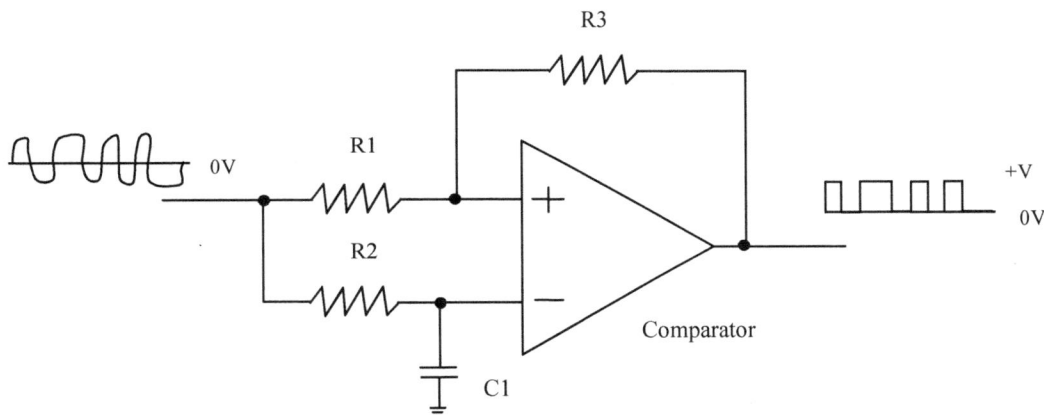

Figure 11.11. Data slicer circuit to convert incoming Manchester encoded data into digital logic levels.

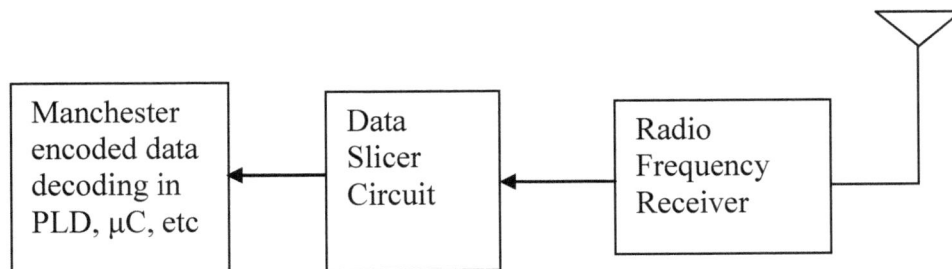

Figure 11.12. Manchester decoding in a radio frequency system.

11.6.   A Simple Simplex Communication System using a 315/433MHz Transmitter, Encoder, Decoder, and Receiver Module Pair

This section helps one to become more familiar with and how to test the Holtek HT12E encoder and Holtek HT12D decoder. The next step then is to include the encoder and decoder in a wireless simplex system using a WyJen transmitter/receiver.

11.6.1. Holtek HT12E Encoder and Holtek HT12D Decoder Test Procedure

1. Acquire the necessary equipment and components for the experiment which are listed in Table 11.9.

Part	Quantity	Ref designator	Description
HT12E	1	IC1	Holtek encoder, www.holtek.com can order from www.holmate.com
HT12D	1	IC2	Holtek decoder, www.holtek.com can order from www.holmate.com
Resistor	2		For transmitter and receiver oscillator circuits
Bread boards	2		Small bread boards
SPST or DPST switches	8		On/off switches for inputs
LEDs	4		LEDs for output
Oscilloscope	1	Oscilloscope	Oscilloscope to measure transmitted and received data signal
Power Supply	1	Power	3.3V source for transmitter and receiver test circuit

Table 11.9. Equipment and Electronic part list for the simplex wireline communication system.

2. The pin configuration for the Holtek HT12E encoder and HT12D decoder is shown in Figure 11.13 and Figure 11.14, respectively. Download a copy of the datasheet for details and your reference. Study the pinout, whether it's an input or output, etc., and understand what is required by each pin. You are now ready to construct the circuit.

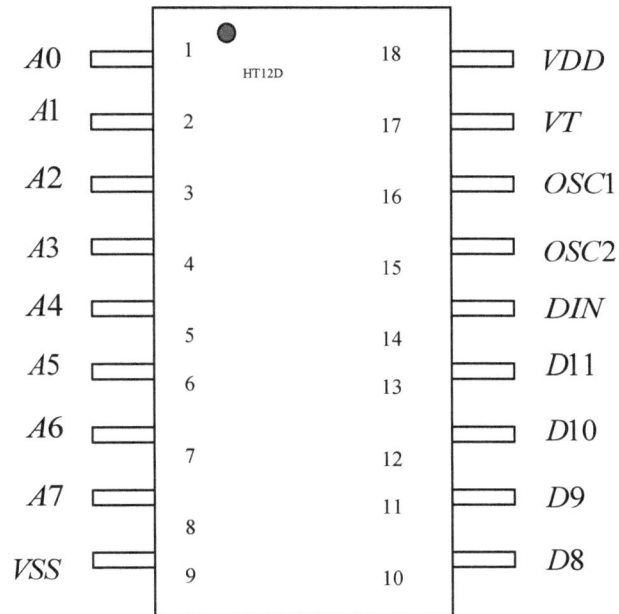

Figure 11.13. Pin configuration for the Holtek HT12D decoder 18 pin DIP package. See datasheet for details.

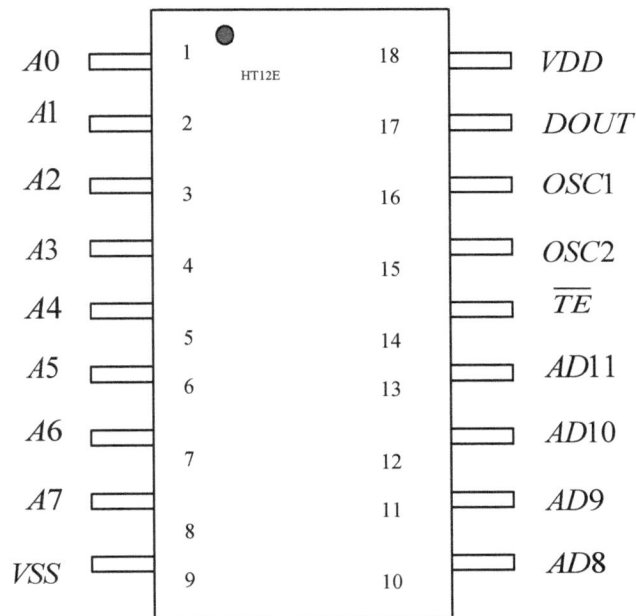

Figure 11.14. Pin configuration for the Holtek HT12E encoder 18 pin DIP package. See datasheet for details.

3. Connect the encoder and decoder circuit together as shown in Figure 11.15.

4. Select a resistor for the encoder and the decoder circuit. The decoder oscillator frequency is required to be 50 times the oscillator frequency of the encoder. What values did you chose. What is the bit rate, fb, of the encoder and decoder as a result of the resistor values (see datasheet for Rosce and Roscd selection). Use an oscilloscope to measure the bit period, Tb, of the encoder output, DOUT. Note that fb=1/Tb. Select a bit rate that is within the required bit rate of the transmitter and receiver. It may be helpful to use a potentiometer to select the resistance of the encoder. Turn the potentiometer until the desired bit rate is achieved. Print the output waveform, DOUT, for two different bit rates.

5. Connect the address bits of the encoder and decoder to ground. Push the buttons to see if the corresponding LEDs come on. Record your results.

6. Connect the address bit 7 on the decoder to GND (Ground). Push the buttons to see if an LED comes on. Record your results.

7. Connect the Address bit of the decoder back to ground. Push the buttons. Does the LEDs come on. Record your results. Explain why the LEDs come on relative to A0-A7.

Instructor sign off _____ Date:_____

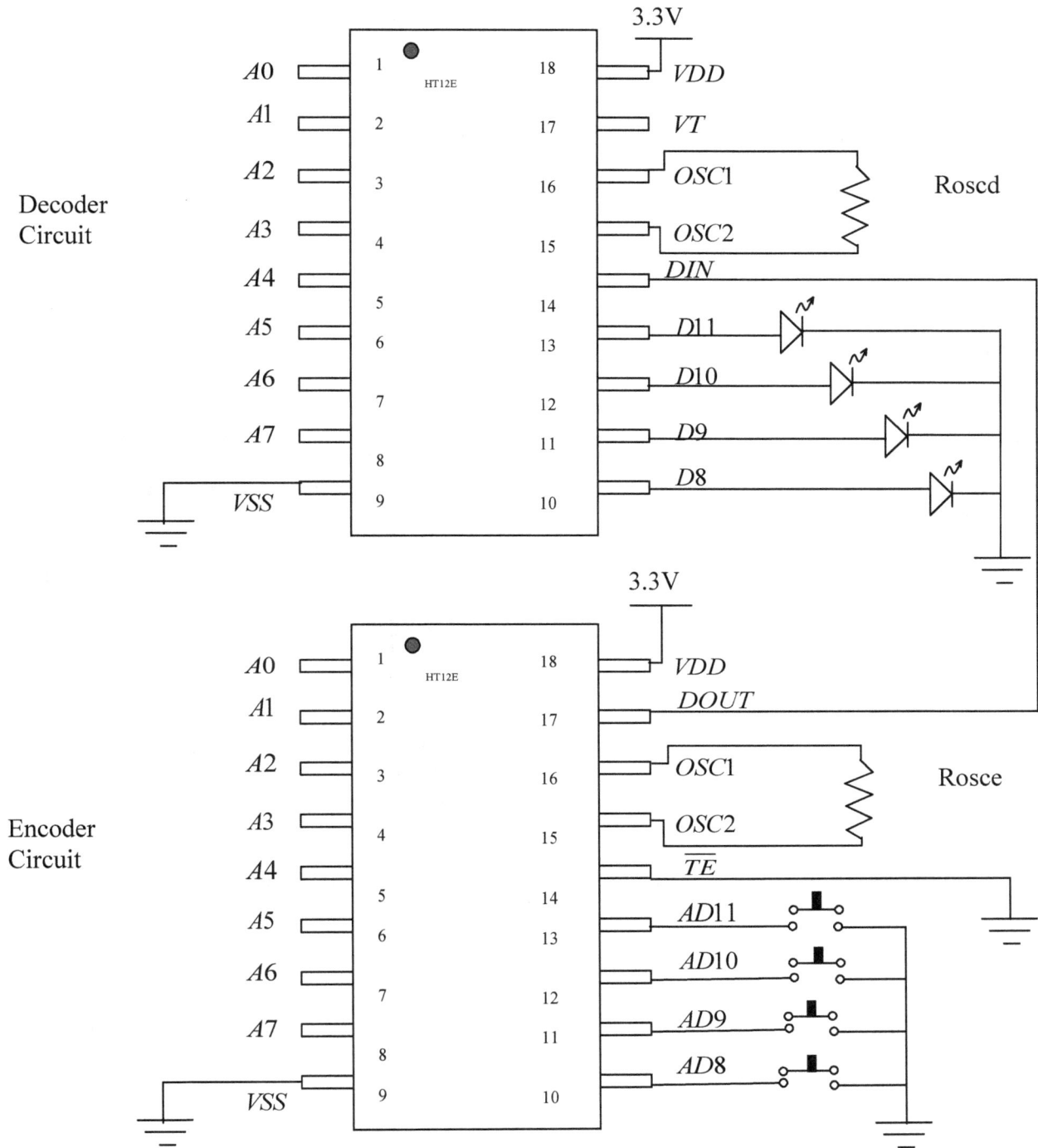

Figure 11.15. Schematic for the Holtek HT12E and Encoder and HT12D circuit.

11.6.2. A Simplex Wireless Communication System using the Holtek HT12E Encoder and HT12E Decoder

The previous Section discussed the Holtek encoder and decoder and a simple experiment. The experiment in this section uses the same encoder and decoder combination in a simplex wireless communication system. The goal of this experiment is become more familiar with encoders and decoders in wireless communications. To implement this experiment perform the following steps:

1. Acquire the parts listed in Table 11.10.

Part	Quantity	Ref designator	Description
Transceiver: GTR315M_M13	2	TXRX	315MHz transmitter/receiver, 5V, half duplex transmission, www.wyjen.com
HT12E	1	IC1	Holtek encoder, www.holtek.com can order from www.holmate.com
HT12D	1	IC2	Holtek decoder, www.holtek.com can order from www.holmate.com
Resistor	2	Rosce, Roscd	For encoder and decoder oscillator circuits
Capacitor	6		10pF capacitor
Antenna	2	wires	Antenna wire, strip wire of insulation for self made antenna.
Bread boards	2		Small bread boards
SPST or DPST switches	8		On/off switches for inputs
LEDs	4		LEDs for observing outputs
Oscilloscope	1	Oscilloscope	Oscilloscope to measure transmitted and received data signal
Power Supply	1	Power	5V source for transmitter and receiver test circuit

Table 11.10. Equipment and Electronic parts list for the simplex wireless communication system.

2. Construct the circuit shown in Figure 11.16 and Figure 11.17.
3. Push the push buttons on the encoder and verify the corresponding LEDs on the decoder light up.

Instructor sign off_____ Date_____

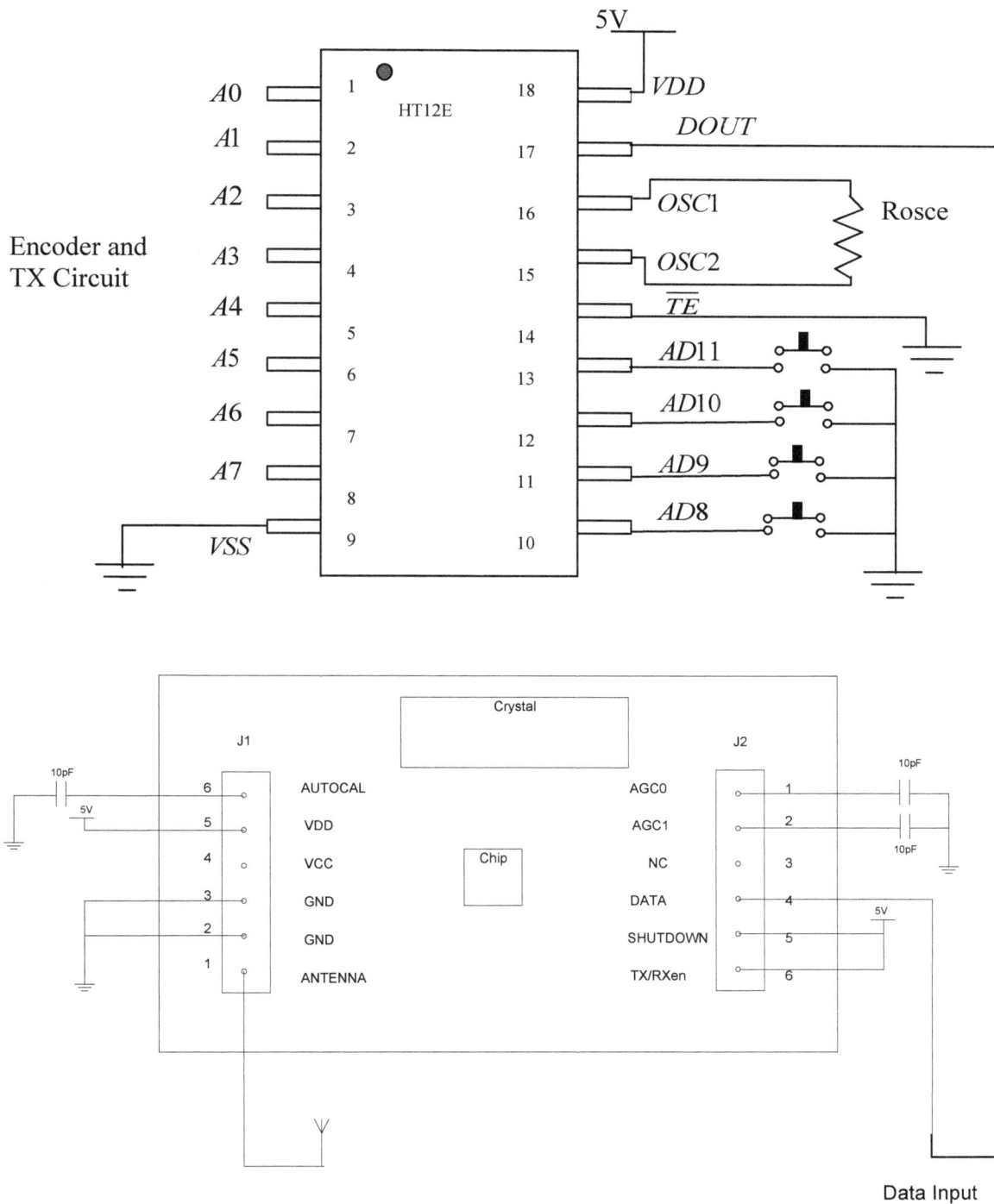

Figure 11.16. Schematic for the WyJen **GTR315M_M13** transceiver and Holtek HT12E encoder circuit.

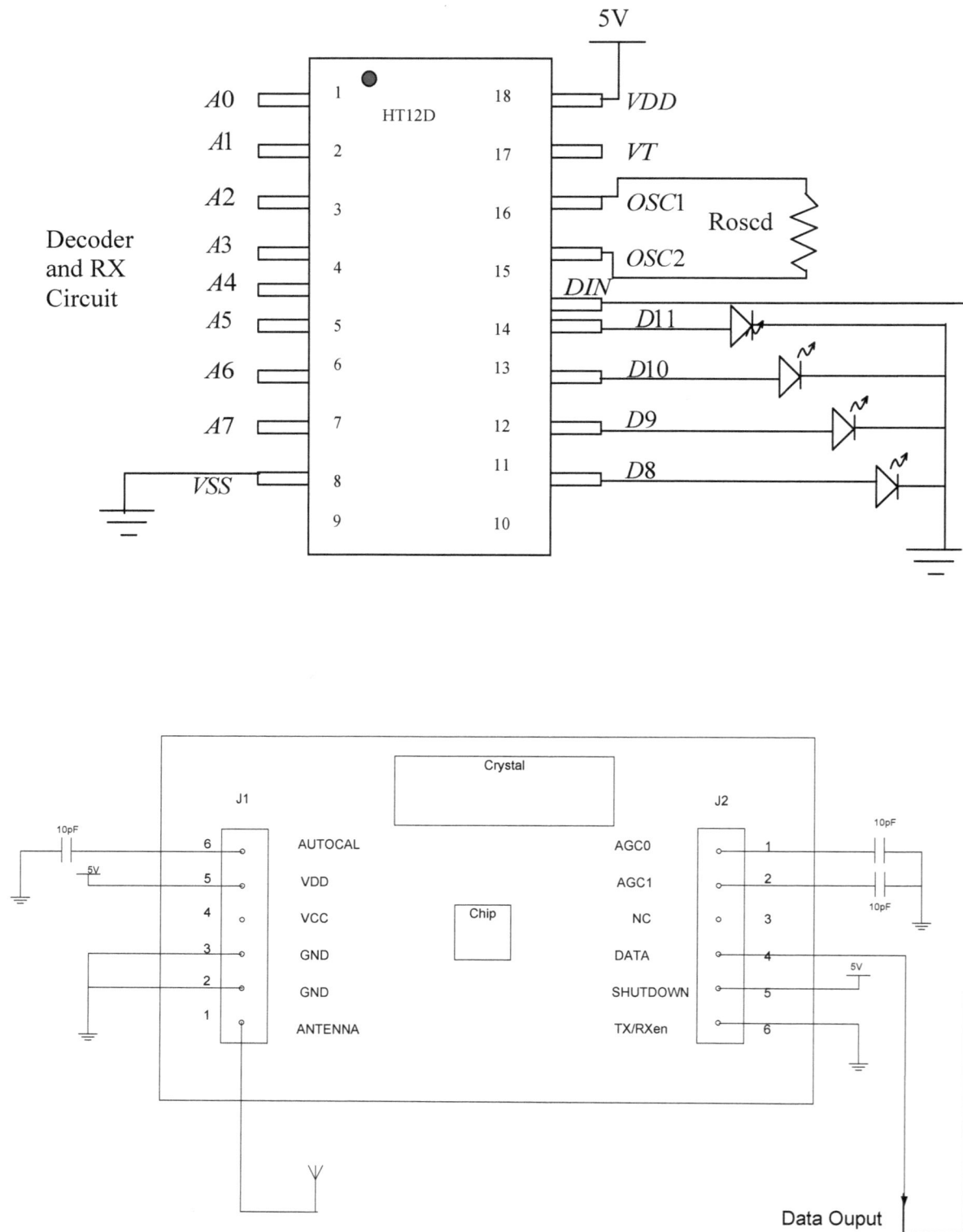

Figure 11.17. Schematic for the WyJen GTR315M_M13 transceiver and Holtek HT12D decoder circuit.

12. Transmission Lines and Antenna Design

Antennas are simply electrical conductors that radiate electrical energy into space. The simplest antenna is a piece of wire. Two common antennas are 1.) vertical antenna in Figure 12.1, and 2.) horizontal antennas in Figure 12.2. A vertical antenna is simply a wire that is vertical. It is a quarter wave dipole antenna and is sometimes called a Marconi antenna. A horizontal antenna is simply a wire that is horizontal. The horizontal antenna is referred to as a half wave dipole antenna. Antennas are connected to the radio frequency (RF) output of a transmitter and the radio frequency input of a receiver. The output impedance of the transmitter or the input impedance of the receiver should match the impedance of the antenna for maximum transmission and maximum reception, respectively. More about the different antennas can be referenced in Tomasi (2004) on page 578.

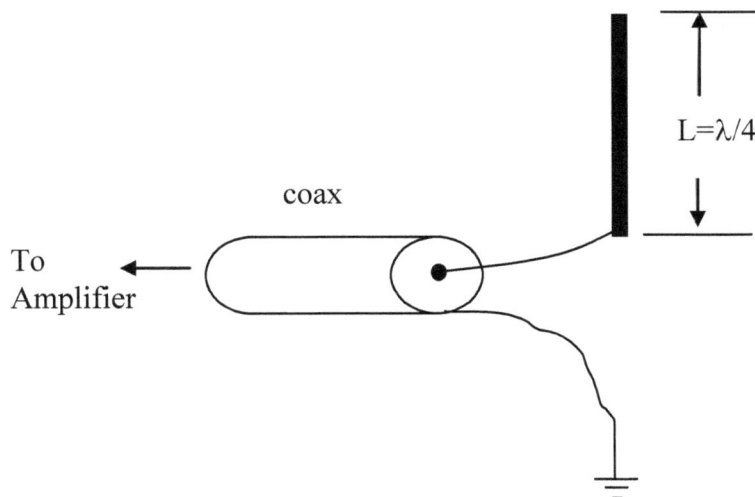

Figure 12.1. Vertical antenna.

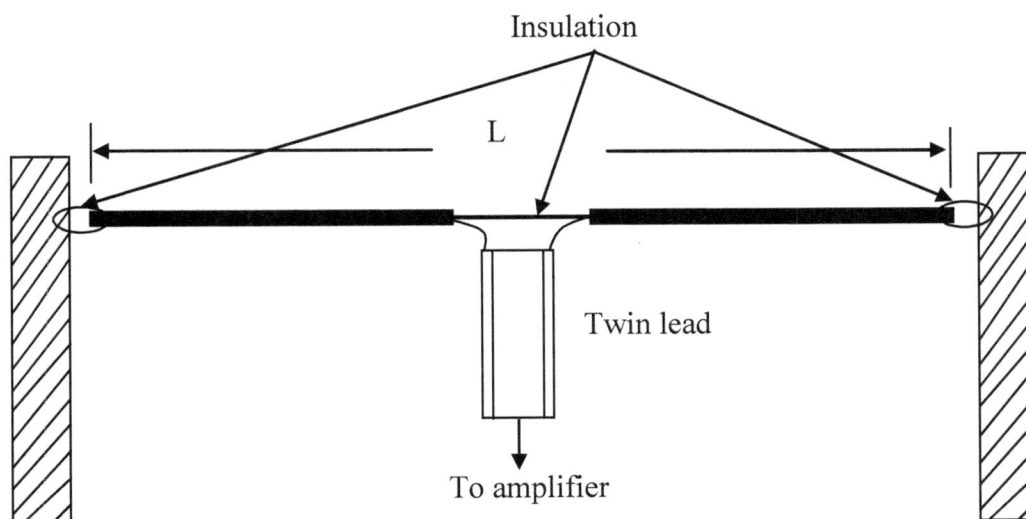

Figure 12.2. Horizontal antenna.

An antenna is used for radiating waves into space or receiving them from space. When dealing with radio waves, it is common to use units of **wavelengths** rather than frequency. **Wavelength** is the distance that one cycle of an electromagnetic wave occupies in a medium, e.g., free space, transmission line, etc. An antenna is an open ended transmission line as shown in Figure 12.3. The transmission line is $(1/4)\lambda$ or a quarter wavelength. This insures that the voltage is maximum at the open end of the line and the current is minimum.

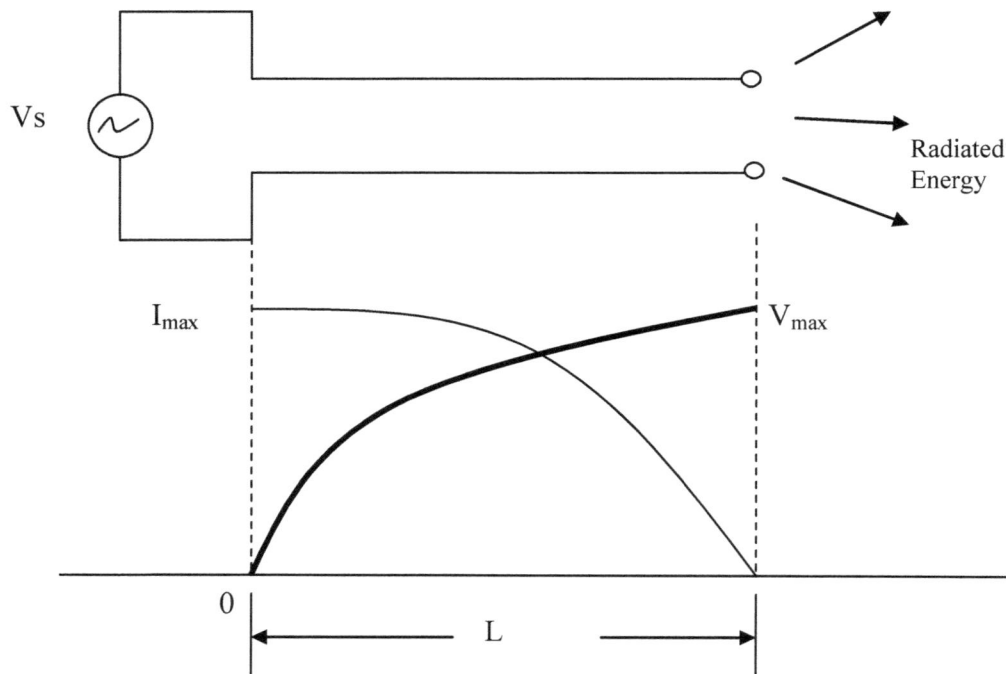

Figure 12.3. Modeling the vertical antenna as an open ended transmission line.

For a vertical antenna, the length of the wire should be about ¼ the carrier frequency wavelength. First the wavelength is calculated by

$$\lambda = \frac{(k \cdot c)}{f} = \frac{c}{f}$$

EQ. 12.1

EQ. 12.1 assumes the medium is free space where the velocity factor k=1. Therefore, the length of the antenna, L, should be

$$L = \left(\frac{1}{4}\right)\lambda$$

EQ. 12.2

The radiation pattern for a vertical antenna is shown in Figure 12.4. A top and side view are illustrated.

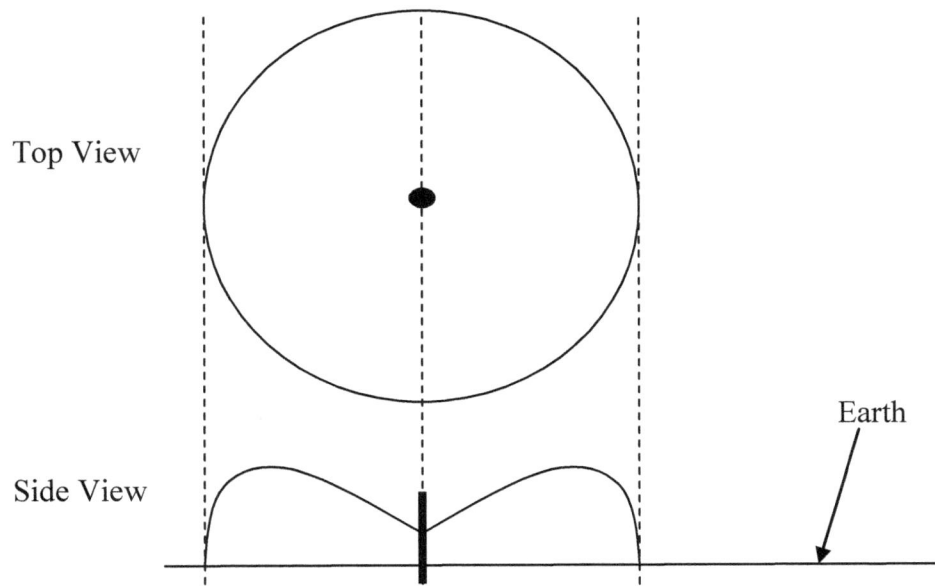

Figure 12.4. Radiation pattern of vertical antenna.

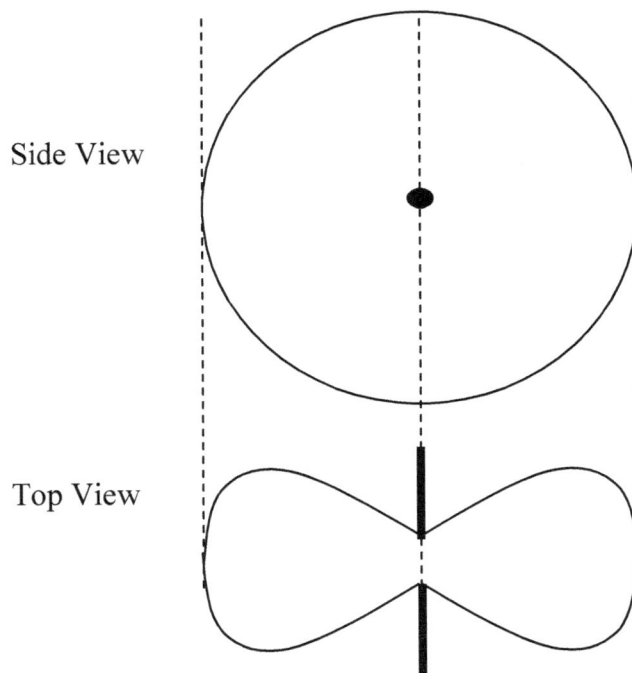

Figure 12.5. Radiation pattern of the horizontal antenna.

If a horizontal antenna is used the wire should have a length of about ½ the carrier frequency wavelength. So the wavelength is calculated using EQ. 12.1 and the antenna length, L, as indicated in Figure 12.6, is determined by

$$L = \left(\frac{1}{2}\right)\lambda$$

EQ. 12.3

because $1/4\lambda + 1/4\lambda$ is $1/2\lambda$ for maximum voltage at the ends of the open ended transmission line.

The radiation pattern for a horizontal antenna is shown in Figure 12.5. A top and side view are illustrated.

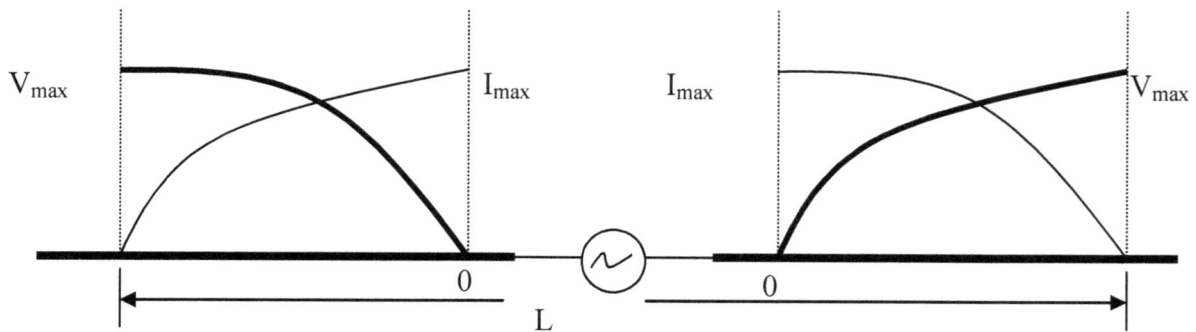

Figure 12.6. Modeling the horizontal antenna as an open ended transmission line.

12.1. **Antenna Radiation Resistance**

The radiation resistance is the fictitious resistance of the antenna or open ended transmission line. It is defined as

$$R_{rad} = \frac{P}{I^2}$$

EQ. 12.4

where R_{rad} is the radiation resistance of the antenna, P is the total power converted into radiated energy, and I is the antenna current in RMS.

The radiation resistance of a vertical antenna is 36.5Ω. The radiation resistance of a horizontal antenna is 73Ω. An antenna can be modeled by inductive, capacitive, and resistive (radiation resistance) components as shown in Figure 12.7. Thus, an antenna can mathematically be represented by:

$$Z_T = j\omega L + \frac{1}{j\omega C} + R_{rad} \qquad \qquad \textbf{EQ. 12.5}$$

When the antenna is at resonance, with radiation resistance, R_{rad}, then the antenna is correctly matched to the transmitter. If the frequency of operation does not match the antenna length, the antenna circuit will not resonate and may be more capacitive and inductive. In such cases an inductive coil or capacitance is needed to give the correct matching. This is referred to as **antenna tuning**.

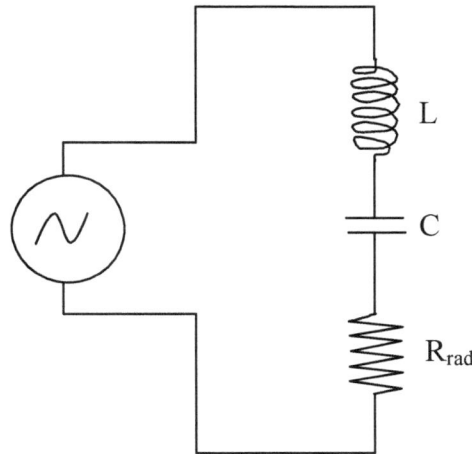

Figure 12.7. Antenna Model.

For optimal antenna transmission or maximum power transfer the antenna must meet the following conditions:

    A. Antenna resonate

    B. Transmitter match antenna

A series RLC circuit, as shown in Figure 12.7, is at resonance when

$$f_r = \frac{1}{2\pi\sqrt{LC}} \qquad \qquad \textbf{EQ. 12.6}$$

For an antenna that is shorter than the required length, an inductance is needed to compensate for capacitance. This is called loading the antenna. The inductance is calculated by:

$$f_r = \frac{1}{2\pi\sqrt{LC}} \quad \Rightarrow \quad L = \frac{1}{(2\pi f)^2 C} \qquad \qquad \textbf{EQ. 12.7}$$

where C is the antenna capacitance.

Once the inductance is determined the coil parameters of a single layer coil can be determined from

$$L = \frac{n^2 d^2}{459d + 1009\ell}$$ **EQ. 12.8**

where n is the number of turns, d is the coil diameter in mm, ℓ is the coil length in mm. Equation EQ. 12.8 is valid if the wire spacing is equal to the wire diameter.

The capacitance of a vertical wire is given by

$$C = \frac{24h}{\log\left(\dfrac{1.15h}{d}\right)}$$ **EQ. 12.9**

where h is the height of the antenna in meters, d is the diameter of the wire, and C is the capacitance of the vertical antenna. In most cases C = 6pF/m is accurate enough. The radiation resistance of vertical antenna is calculated by

$$R = \frac{40\pi^2 h^2}{\lambda^2}$$ **EQ. 12.10**

where h is the height of the vertical antenna and λ is the wavelength of the transmitting signal in meters.

EXAMPLE 12.1

A vertical antenna with a height of 10m and wire diameter of 3mm is used to transmit a 136kHz signal. Determine
   A. the radiation resistance
   B. capacitance
   C. the inductance of the loading coil.

Solution:
A.

$$\lambda = \frac{c}{f} = \frac{3\times10^8}{136\times10^3} = 2205.88m$$

$$R = \frac{40\pi^2 h^2}{\lambda^2} = \frac{40\pi^2 10^2}{2205.8^2} = 8.1m\Omega$$

B.

$$C = \frac{24h}{\log\left(\frac{1.15h}{d}\right)} = \frac{24(10)}{\log\left(\frac{1.15(10)}{3\times10^{-3}}\right)} = 66.97 \text{pF}$$

C.

$$L = \frac{1}{(2\pi f)^2 C} = \frac{1}{\left(2\pi136\times10^3\right)^2 66.97\times10^{-12}} = 20.45 \text{mH}$$

12.2. Antenna Gain

Antenna gain refers to improved power in a specific direction. That is, it does not refer to an increase in power. The antenna gain is given by:

$$A_{dB} = 10\log\left(\frac{P_t}{P_{in}}\right)$$
EQ. 12.11

where A_{dB} is the antenna gain in decibels, P_{in} is the actual input power of the antenna in Watts (W), and P_t is the total transmitted power in W. Power gains of 10 or more can be easily accomplished, especially at higher RF. Thus, a 100W transmitter can be made to perform like a 1000W transmitter with the right antenna gain.

12.3. Transmitter and Power Density

The power density at a given point r distance away from an isotropic radiator is given by:

$$P_d = \frac{P_t}{4\pi r^2}$$
EQ. 12.12

where P_d is the power density of the signal in Watts per square meters (W/m^2), P_t is the total transmitted power in Watts (W), r is the radius distance away from the point source in meters (m). Isotropic refers to an imaginary sphere. The distance, r, is the radius of the imaginary sphere, enclosing the point source where $4\pi r^2$ is the area of a slice of the sphere. For practical purposes, antennas are not truly isotropic, they are directional in terms of signal strength. So EQ. 12.12 must be modified slightly to:

$$P_d = \frac{A_t P_t}{4\pi r^2}$$
EQ. 12.13

where A_t is the antenna gain. A dipole has an antenna gain of 1.64 or 2.15dB.

EXAMPLE 12.2

If the transmitting antenna is a dipole where the total transmitter power is 30 Watts, determine the power density at a distance of 30 miles (48,300m) from the antenna.

Solution:
The power density is calculated by:

$$P_d = \frac{A_t P_t}{4\pi r^2} = \frac{1.64(30)}{4\pi(48300)^2} = 1.68 \times 10^{-9} \, W/m^2 = 1.68 nW/m^2$$

The next section talks about how the received power can be calculated which is very useful to know at the receiver.

EXAMPLE 12.3

A transmitting antenna has a power gain of 10 and an input power of 100W. Determine the power density at a distance of 10km from the antenna.

Solution:
The power density is calculated by:

$$P_d = \frac{A_t P_t}{4\pi r^2} = \frac{10(100)}{4\pi(10000)^2} = 0.796 \, \mu W/m^2$$

The next section talks about how the received power can be calculated which is very useful to know at the receiver.

12.4.   Receiver Power

The receiver power density at a given point r distance away from an isotropic radiator is given by:

$$P_r = \frac{P_t A_t A_r \lambda^2}{4\pi r^2}$$ **EQ. 12.14**

where P_r is the received power in W, P_t is the total transmitted power in W, A_t is the gain of the transmitting antenna expressed as a power ratio, A_r is the gain of the receiving antenna expressed as a power ratio, r is the radius or distance from the transmitter in meters (m), and λ is the signal wavelength in meters (m).

EXAMPLE 12.4

A transmitter is operating at 118MHz with a power of 3W into a quarter wavelength vertical antenna. The receiver is 30 miles (48,300m) away and has an antenna with a gain of 8dB. What is the received power?

Solution:

Using EQ. 12.1, the wavelength at 118MHz is $\lambda = c/f = 300/118 = 2.54m$. The gain of the quarter wave antenna is the same as the dipole. With a gain of 1, we must multiply this by 1.64 to get the gain of the isotropic source.

The gain of the receiving antenna is 8dB. Since the gain is typically expressed as the gain over a dipole, we need to convert to a gain relative to an isotropic source by adding 2.15dB. This is the same as multiplying the power ratio by the gain, 1.64, in decibels. Thus, the gain is 8 + 2.15=10.15dB. Now, converting to an actual power ratio using:

$$\frac{P_{out}}{P_{in}} = \log^{-1}\left(\frac{dB}{10}\right) = \log^{-1}\left(\frac{10.15}{10}\right) = \log^{-1}(1.015) = 10.35$$

where P_{in} is the input power of the antenna and P_{out} is the output power of the antenna. Thus, the received power can be calculated by:

$$P_r = \frac{P_t A_t A_r \lambda^2}{4\pi r^2} = \frac{3(1.64)(10.35)(2.54)^2}{16(9.87)(48300)^2} = \frac{328.52}{36.84 \times 10^{10}} = 8.92 \times 10^{-10}\,W = 89.2nW$$

In EXAMPLE 12.4, we can further calculate the input voltage at the receiver by

$$P = \frac{V^2}{R} \Rightarrow V = \sqrt{PR}$$ **EQ. 12.15**

EXAMPLE 12.5

Calculate the voltage at the receiver using the result in EXAMPLE 12.4.

Solution:

If the antenna, transmission line, and front end input impedance are 50Ω then the receiver input voltage is

$$V = \sqrt{PR} = \sqrt{(89.2 \times 10^{-9})(50)} = 2.11mV$$

This is a strong voltage, most good narrowband receivers can detect receive signals down to 1μV or less.

12.5. Impedance Matching

Impedance matching can be done using LC networks, transformers, balums, Q sections, antenna tuners, etc. The basic concept of impedance matching networks are to provide maximum power transfer from one circuit to another. The basic impedance matching problem is illustrated in Figure 12.8. There are three basic types of antenna matching networks(see Frenzel (1998) in Reference Section on page 578):

A. L network
B. T network
C. π network

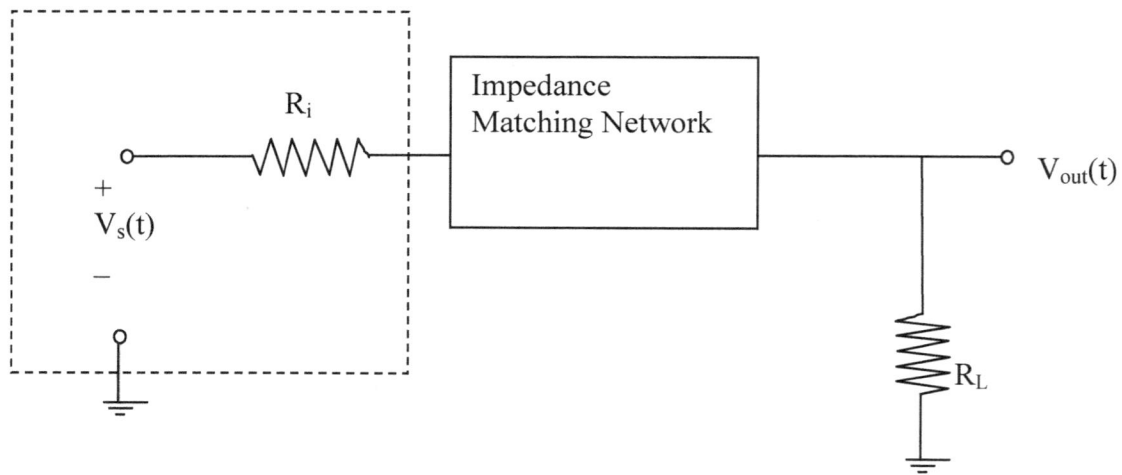

Figure 12.8. Impedance matching block diagram.

The L- network (impedance matching network in Figure 12.8) can be one of the circuits in Figure 12.9. Figure 12.9A and Figure 12.9B are lowpass filters while Figure 12.9C and Figure 12.9D are highpass filters. Thus, the L network consists of an inductor and capacitor connected in several L like arrangements as shown Figure 12.9.

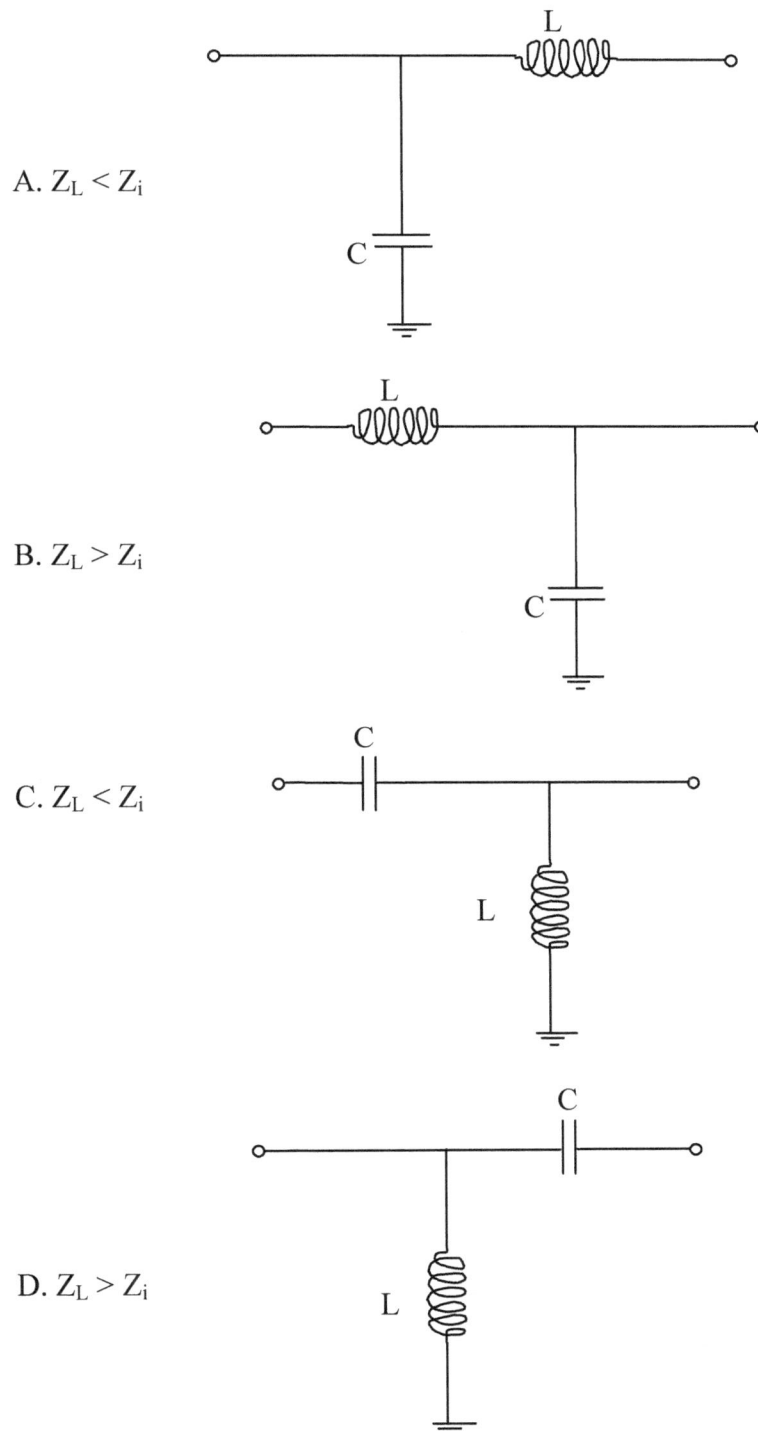

Figure 12.9. L network configurations.

When the source impedance and load impedance are resistive in Figure 12.9 we can apply the following equations for the case $Z_L < Z_i = R_L < R_i$:

$$X_L = \sqrt{R_i R_L - R_L^2}$$ EQ. 12.19

$$Q = \sqrt{\frac{R_i}{R_L} - 1}$$

$$X_C = \frac{R_i R_L}{X_L}$$

These apply to Figure 12.9A and Figure 12.9C. Additionally, for resistive internal source and load impedances for the case when $Z_L > Z_i = R_L > R_i$, use the following formulas:

$$X_L = \sqrt{R_i R_L - R_i^2}$$ EQ. 12.20

$$Q = \sqrt{\frac{R_L}{R_i} - 1}$$

$$X_C = \frac{R_i R_L}{X_L}$$

These apply to Figure 12.9B and Figure 12.9D.

EXAMPLE 12.6

Suppose we wish to match a 6Ω transistor amplifier to a 50Ω antenna load at 155MHz. Determine the L network.

Solution:

$Z_L > Z_i$ or 50Ω > 6Ω, so use Figure 12.9B formulas and circuit as follows:

$$X_L = \sqrt{R_i R_L - R_i^2} = \sqrt{6(50) - 6^2} = 16.25\Omega$$

$$Q = \sqrt{\frac{R_L}{R_i} - 1} = \sqrt{\frac{50}{6} - 1} = 2.7$$

$$X_C = \frac{R_i R_L}{X_L} = \frac{50(6)}{16.25} = 18.46\Omega$$

Now find L and C using the basic reactance formulas:

$$X_L = 2\pi f L$$

$$\Rightarrow L = \frac{X_L}{2\pi f} = \frac{16.25}{6.28(155 \times 10^6)} = 16.7\text{nH}$$

$$X_C = \frac{1}{2\pi fC}$$

$$\Rightarrow C = \frac{1}{2\pi fX_C} = \frac{1}{6.28\left(155\times 10^6\right)18.46} = 55.65pF$$

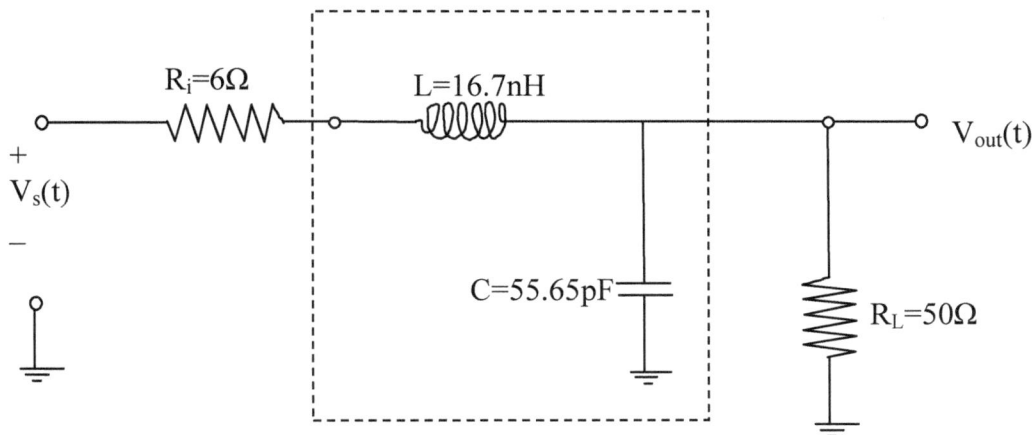

Figure 12.10. Impedance matching block diagram. Beware of stray inductance in series with L and stray capacitance in parallel with C.

12.6. Transmission Lines

A transmission line is a pair of wires or electrical conductors that transport signals form one point to another point. The transmission line connects the source, V_s, to the load impedance, Z_L, as shown in Figure 12.11. The source impedance, Z_i, is not shown but is assumed to match the transmission line impedance, Z_o. In this section , we will focus mainly on the transmission line impedance matching the load impedance. There are two types of transmission lines:

   A. Balanced
   B. Unbalanced

In a balanced transmission line current flows in both wires in the wire pair. However, the current is 180 degrees out of phase. In an unbalanced transmission line the current flows in one wire while the other is grounded.

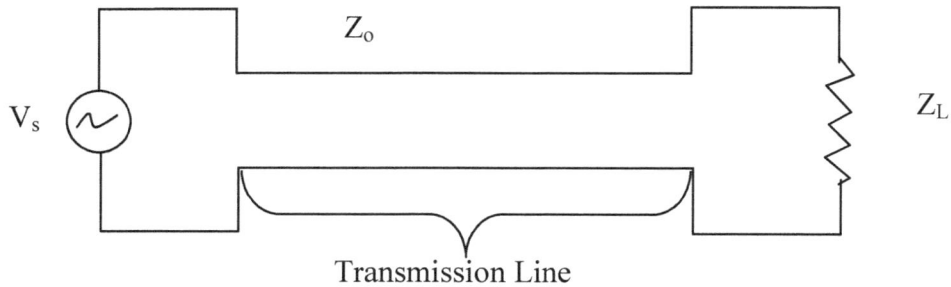

Figure 12.11. Transmission line.

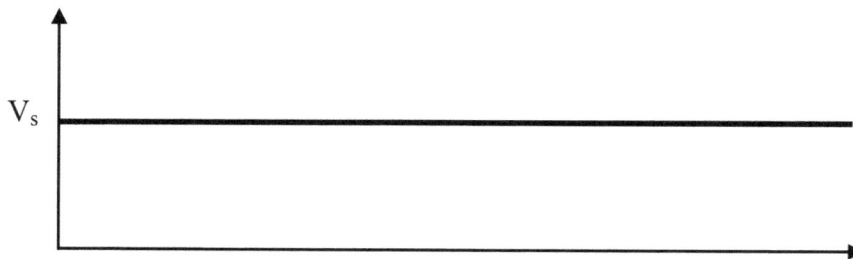

Figure 12.12. Ideal transmission line voltage.

For an ideal transmission line which has zero resistance, the voltage or current will be the same everywhere. In Figure 12.12 the x axis represents the distance while the y axis represents the voltage or current with distance.

The amount of time it takes a signal to travel a transmission line is given by

$$t = \frac{d}{v}$$ **EQ. 12.22**

where t is the time, d is the distance of the transmission line, and v is the velocity of the electromagnetic wave in a medium. When the medium is free space $v = 3 \times 10^8 \text{m/s}$.

The transmission line impedance, more commonly called the **characteristic impedance,** Z_0, of the transmission line is the impedance seen looking into the line from the input side using an infinite length of the transmission line. When the load matches the characteristic impedance of the transmission line maximum power is transferred.

12.6.1. Standing Waves

When the transmission line is not terminated with a load that matches the characteristic impedance reflections occur. That is, the original signal is reflected back to the source. The reflected wave is called the **reflected wave.** The original wave is called an **incident wave.** The original and reflected wave are 180 degrees out of phase. The phase relationship of the waves are called **standing waves.**

The **standing wave ratio** (SWR) is a ratio of the characteristic impedance, Z_0, to the load impedance Z_L. The ratio is determined by putting the value of the largest impedance, Z_0 or Z_L, in the numerator, and the smallest value, Z_0 or Z_L, in the denominator. This is expressed as

$$SWR = \frac{Z_L}{Z_0} \Rightarrow Z_L > Z_0 \qquad \text{EQ. 12.23}$$

or

$$SWR = \frac{Z_0}{Z_L} \Rightarrow Z_0 > Z_L \qquad \text{EQ. 12.24}$$

The standing wave ratio for the voltage is denoted by **VSWR** while the standing wave ratio for the current is denoted by **ISWR**. The various standing wave ratios are related by

$$SWR = VSWR = ISWR \qquad \text{EQ. 12.25}$$

12.6.2. The Reflection Coefficient

The reflection coefficient, Γ, is another quantity for determining the characteristic of the matched or unmatched transmission lines. It is given by

$$\Gamma = \frac{V_{ref}}{V_{inc}} = \frac{I_{ref}}{I_{inc}} \qquad \text{EQ. 12.26}$$

The SWR can be expressed in terms of Γ as follows

$$SWR = \frac{\Gamma + 1}{1 - \Gamma} \qquad \text{EQ. 12.27}$$

The reflection coefficient can be expressed in terms of Z_0 or Z_L as follows:

$$\Gamma = \left| \frac{Z_L - Z_o}{Z_L + Z_o} \right|$$ **EQ. 12.28**

EXAMPLE 12.7

A 72 Ω transmission line is connected to a 50Ω load. Determine:
  A. SWR
  B. Γ
  C. V_{ref} if V_s=5V

Solution:
A.

$$SWR = \frac{Z_o}{Z_L} = \frac{72}{50} = 1.44$$

B.

$$\Gamma = \left| \frac{Z_L - Z_o}{Z_L + Z_o} \right| = \left| \frac{50 - 72}{50 + 72} \right| = 0.180$$

C.

$$\Gamma = \frac{V_{ref}}{V_{inc}} \quad \Rightarrow \quad V_{ref} = \Gamma V_{inc} = .180(5) = 0.9V$$

Digital Mobile Robotic Projects

13. Overview of Digital Robotic Projects

PLD-bot is a small moving robot controlled by a remote controlled PLD platform. It can be programmed by a PLD that resides on the PLD platform . The PLD is programmed to control two servomotors, using pulse width modulated waveforms, to move in a specified direction in response to sensory input. A wide variety of sensors can be attached to the PLD - bot. Several project variations of the PLD-bot based on the electronic discipline, e.g., electronic controls, electronic communication, signal processing, etc., of the students are given next.

13.1. Introduction of Scenarios

This section discusses several potential robotic projects based on three technological areas:

4. Control electronics using sensors
5. Communication electronics using wireless technologies
6. Signal processing electronics and data acquisition using sensors and signal processing

The control electronics based scenarios focuses on how a mobile robot can be controlled using various sensors, programmable devices, servo motors, etc.

The communication electronics scenarios focuses on how a mobile robot can be controlled using remote control electronics such as Amplitude modulation (AM), frequency modulation (FM), amplitude shift keying (ASK), frequency shift keying (FSK), phase shift keying (PSK), etc. A transmitter, and received is used. The receiver is mounted on the robot while a transmitter is hand held.

The signal processing and data acquisition scenario focuses on signal processing software algorithms that respond to sensory input data. This data is sampled and converted to digital data for spectral analysis. The spectral analysis may be done in real time using a digital signal processor (DSP), programmable logic device (PLD), etc. It can also be done in real time by sending data from the PLD-bot to the personal computer via the parallel port.

The next few chapters will discussed these projects in detail. Earlier chapters introduced VHDL and the various VHDL modules used to drive these projects.

13.2. Project Expectations

This section discusses several projects, but before considering a project is very important that you are organized and have duties and expected accomplishment dates set. So the next few sections will discuss what is expected from you during and at the completion of your project. In general you will have **project deliverables** and a **project schedule**.

13.3. **Project Deliverables**

The deliverables are due at the end of the term for a total of 100 points. These are listed below as follows:

1. Oral Presentation (50 points): TBA. You must be present to receive credit for the project.

2. Project Trouble Shooting Manual
    a. Cover page – Title Trouble shooting manual. Must be 1 page.
    b. Number and list possible problems and corresponding solution to each problem using the following format:
        Problem
        Solution
        Must be at least 1 page.

3. Project Written Report (50 points - includes trouble shooting manual)
    b. Cover page – shows Title (Industrial Controls Robotic Project: PLD-bot), date, professor, list individual team members showing duties performed. Must be 1 page.
    c. Table of Contents. Must be a least 1 page.
    d. Introduction – Must give brief overview of project. Show main diagrams. Must be 1 –2 pages.
    e. Body of Report. Must discuss each section of the robot (1. Each sensor used, 2. Platform used, 3. Servomotor/motors used, 4. the remote control, 5. VHDL code (for PLD-bot) or C code (for PIC-bot or DSP-bot) and applicable simulation results, etc.) using a heading for each section. Must be at least 7 pages.
    f. Conclusion. Must summarize main points. Must be a maximum of 1 page.
    g. Appendix –Contains all Data Sheets, VHDL code, C code, Schematics, etc.

Notes:
1. All pages must be numbered - except cover page
2. All Figures must be clear (text readable, graphics must look professional) and numbered, e.g., Figure 1.
3. Use size 12 or 10 for all fonts including headings.
4. Each section of the report must have a separate heading that stands off from the rest of the text following it.

13.4. **Project Schedules**

You will be expected to meet certain major deadlines during the duration, development, and completion of your project. When these deadlines are mapped to dates it is called a project development schedule. At the end of the project you will be given a time slot to discuss your project accomplishments.

13.4.1. Tentative Development Schedule

The basic 15 week schedule for implementing the robotic projects is shown in Table 13.1. To be successful, you must try to follow it.

Week	Things to do
Week 1,2,3	Get ideas and teams together, assignment of a leader
Week 4,5	High-level layout of robot complete. Devices selected (e.g., sensors, communication, etc.), start getting parts, platform selected, assignment of job duties, analog designer, digital designer, programmer, project report, power point presentation, weekly status reports due(sign off required)
Week 6,7	Parts Due, platform complete, batteries acquired (or substitute power source), weekly status reports due (sign off required)
Week 7-12	Test all circuits, sensors, etc. weekly status reports due (sign off required)
Week 7-10	VHDL code simulation, weekly status reports due (sign off required)
Week 9-13	Assemble robot and integrate all circuits, sensors, etc., working with VHDL code, weekly status reports due (sign off required)
Week 14	Power point presentations and written reports due

Table 13.1. Tentative 15 week schedule

13.5. Oral Presentation Schedule

The class projects are split into groups. These groups and their corresponding oral presentation time slots will be assigned as listed in **Table 13.2**.

Time Slot Number and Time of Oral Presentations	Group Number
Time Slot 1	Group 1
Time Slot 2	Group 2
Time Slot 3	Group 3
Time Slot 4	Group 4
Time Slot 5	Group 5
.	.
.	.

Table 13.2 Oral presentation schedule.

14. Control Electronics and Robotics

Control electronics is the focus of this Chapter. The following sections present design variations on how to build a mobile robot using control electronics. Control Electronics is rich in sensor and feedback circuits (see Bateson (2002) in Reference Section on page 578). The control robot consists mainly of the PLD platform with supporting sensory circuits.

14.1. Project 1: Robotic Vision and Direction Determination using Photoelectric and Hall Effect Sensors

Project 1 consists of at least two light sensors/light emitters(LEDs) for the robot's vision and a digital magnetic compass for the robot's orientation (see Figure 1). The light sensors are photoresistors while the light emitters are LEDs. The light sensors operate as a switch when the correct voltage is applied. When the LED is on, the photoresistor (switch) turns on, otherwise, the photoresistor is off. Black tape can be used to test the light sensor. It does not reflect the LED light, therefore, no infrared light is returned to the phototransistor. The control PLD-bot has the following requirements:

Requirement 1: The PLD-bot must go north, south, east, or west at the command of the controller using the magnetic compass sensor.

Requirement 2: The PLD-bot must follow or track black tape on the floor using photoresistors.

Requirement 3: The PLD-bot must follow/track a light source using photoresistors.

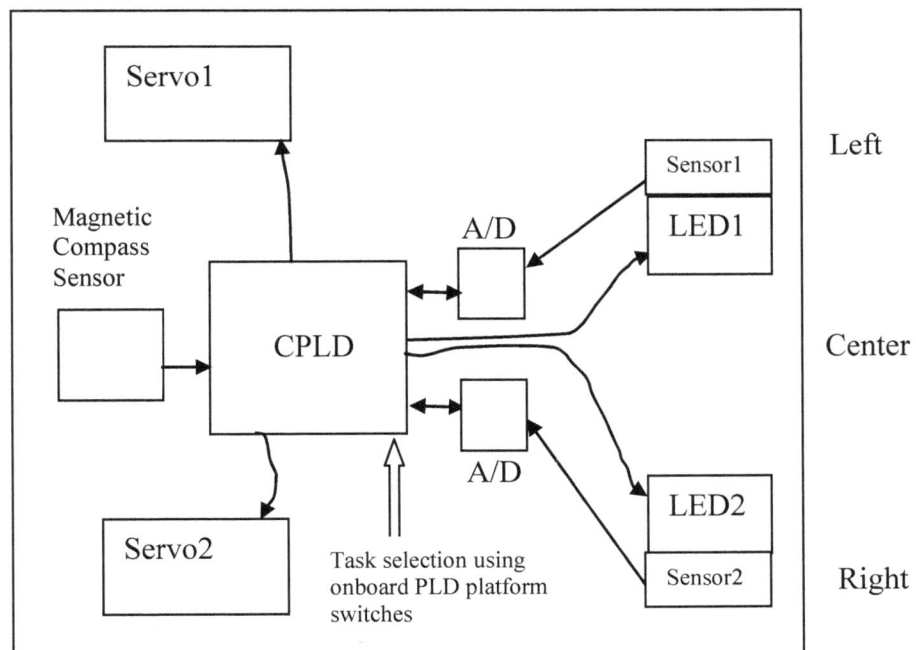

Figure 14.1. Project 1 of the PLD-bot. LED1 is light sensor 1 (on the left), and LED2 is light sensor 2 (on the right).

14.1.1. Project 1.0 Electronic Parts List

The starter electronic part list for Project 1.0 is given in Table 14.1.

Item	Quantity	Ref designator	Description
Altera Software	1		Software used to program the EPM7128SLC84. See website: http://www.altera.com/products/software/quartus-ii-switching/max-plus-ii/max-plus-ii-users.html
A/D converter, ADC0838	1	IC2	8 channel input, analog to digital converter, serial output
Resistor, 100K ohms photo	2	R1, R3	Input: Light Output: Resistance In the dark the output resistance is typically from 100K to 1M, this becomes very low when the photoresistor is exposed to light.
Resistor, 1K ohms	2	R2, R4	Resistor
Resistor, 10K ohms	6	R5, R6, and R7, R8, R9, R10	Resistor
Capacitor, 470PF	2	C1, C2	Capacitor
Capacitor, 33uF	2	C3, C4	Capacitor
LED, 276-143C, Radio Shack	2	IFLED1, IFLED2	Infrared light emitting diode
Altera CPLD, EPM7128SLC84-15	2	CPLD, IC1	Devry ESOC board, Controller. The pin connections are shown for this device. See http://www.elexp.com/tst_pld2.htm
Altera CPLD, FLEX10K20 or FLEX10K70 (optional)	2	CPLD, IC1	Altera UP1 or UPX board. The pin connections are not shown for this device
Magnetic Compass, Dinsmore model 1490	1	IC3	Magnetic compass, 4 bit, active low digital outputs
Futaba servos S3003 (Tower TS53J), or High Tech servos HS300	2	servo1, servo2	Servo motors, www.towerhobbies.com
Platform I	1	Platform I	Platform 1 uses two servo motors, a caster wheel, etc. See Figure 15.16 on page 485
Circuit board	3	CB1, CB2, and CB3	Circuit board for TX, RX, and compass circuit.
20 gauge solid wire, Radio Shack part 278-1222			Black, red, green. Wire for connecting various electronic parts and components
20 gauge solid wire			Need other colors such as white, blue, yellow, etc. Wire for connecting various electronic parts and components

Table 14.1. Electronic starter parts list for Project 1.0. For more details on where to order parts see Appendix D on page 577.

14.1.2. Project 1.0 Procedure

For the steps below save all Altera and VHDL files in the same directory. Preferably a directory called **work**. Follow the steps below to complete this project. Acquire all the necessary parts for constructing the electronic portion of project 1. Most of these parts are listed in Table 14.1.

1. Also see Section Appendix D on page 577 for more information on purchasing the various parts.
2. Modify the servo motors for continuous rotation by a.) clipping or removing the potentiometer, b.) clipping away the stopper.
3. It is very important that you record your results using Table 15.3 and Table 15.4 in this section for later referral! Test the servo motors using the PLD platform's 4Mhz on board clock as follows. This clock frequency must be divided down below 1000 Hz for proper operation of the servos. A divide by 2^{16} counter can achieve this. The VHDL code for a divide by 2^{16} (mod65536) counter is given in Section 9.6.2 on page 349. Send the last six outputs of the counter to the output ports as shown in Figure 14.2. Connect each of these outputs to the signal input of the servo motor. The ground wire of the servo must be grounded to the PLD Platform ground. Similarly, the power supply wire of the servo must be connected to the PLD platform board's VCC. An example of the external test circuit is shown in Figure 14.2. Record the outputs and their respective frequencies that cause the servo to turn clockwise (forward), counter clockwise (reverse), and stop. This information may be unique for each servo and must be recorded separately for each servo motor. Use Table 14.2and Table 14.3 below to record the effect of each output from the counter on the servos. Put a check mark in the column labeled slow speed, medium speed, and fast speed. Slow speed is when the servo turns very slow. Fast speed is the highest speed the servo will go. Medium speed is somewhere in between. You have to use your best judgment on what is a slow, medium, or fast speed.

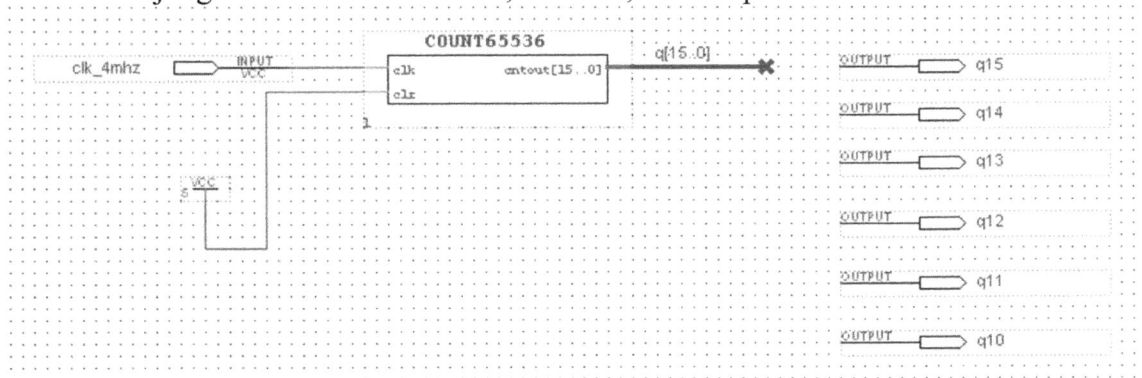

Figure 14.2. Digital test circuit for a servo motor using a mod 65536 counter to divide down a 4MHz clock signal. The output ports for q10, q11, q12, q13, q14, and q15 are individually connected to the signal input of the servo motor.

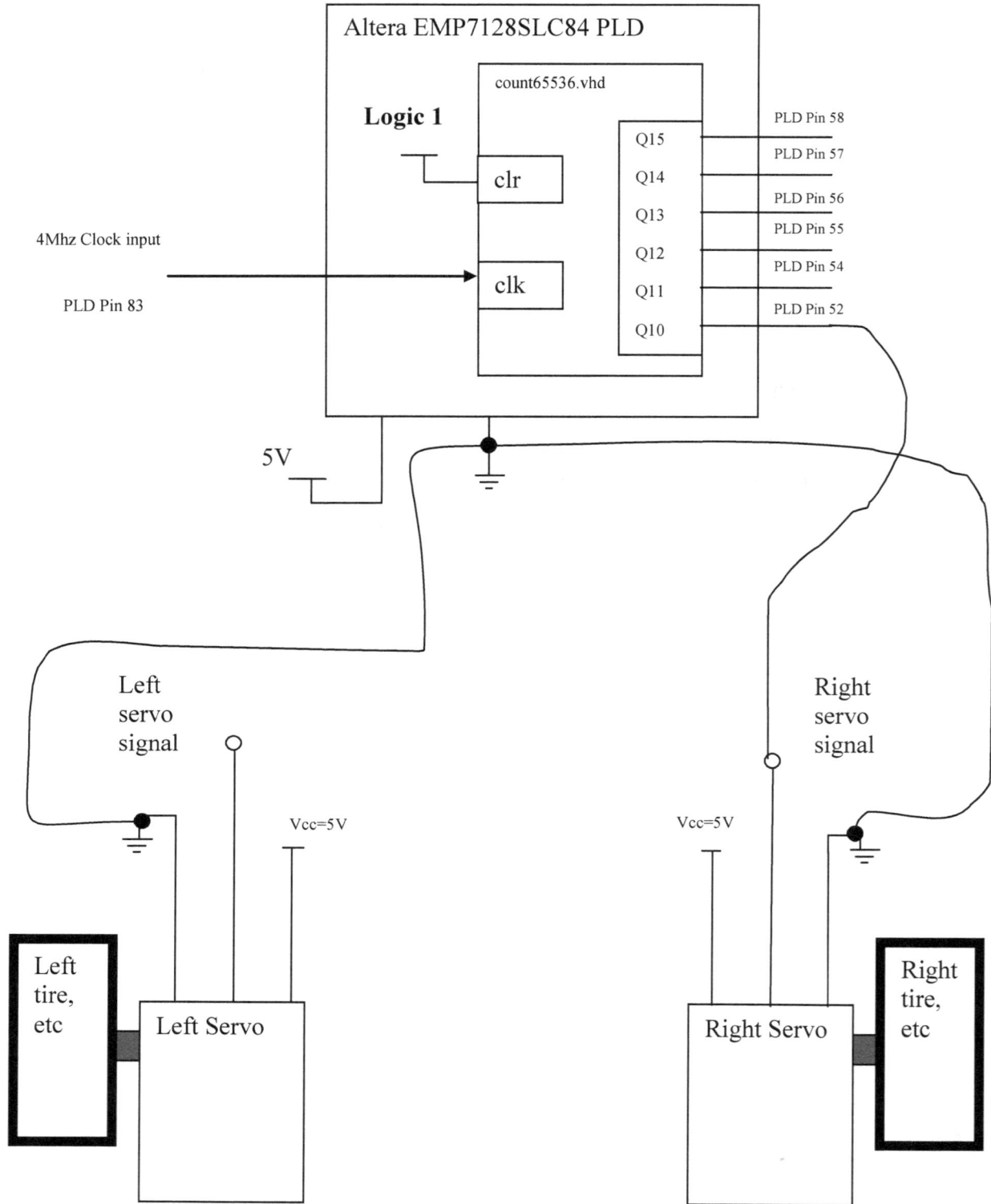

Figure 14.3. Testing of left servo and to right servo using the mod 65536 counter.

Left Servo Clock Frequency (Speed Characteristics and rotation direction)						
Counter Output	Frequency	Slow Speed	Medium Speed	Fast Speed	Forward	Reverse
Q10						
Q11						
Q12						
Q13						
Q14						
Q15						

Table 14.2. Left Servo clock frequency verses speed and rotation direction.

Right Servo Clock Frequency vs. (Speed Characteristics and rotation direction)						
Counter Output	Frequency	Slow Speed	Medium Speed	Fast Speed	Forward	Reverse
Q10						
Q11						
Q12						
Q13						
Q14						
Q15						

Table 14.3. Right Servo clock frequency verses speed and rotation direction.

Before you begin recording the servo clock frequency verse speed and rotation direction in Table 16.4 and Table 16.5 the servos should be identified by a label as the right servo and the left servo. The servos should be positioned as shown in Figure 14.4. The shaft and tire of the servos should be pointed in opposite directions as shown.

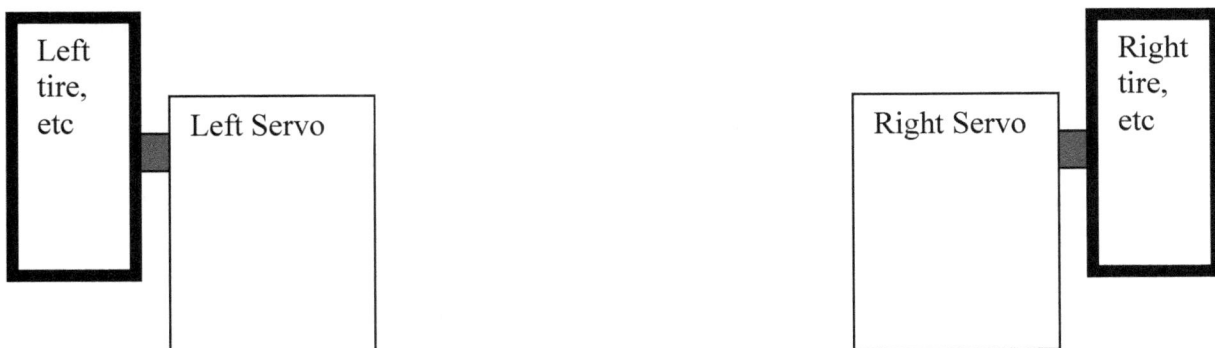

Figure 14.4. Position of left servo relative to right servo.

It should be noted that clockwise is forward and counter clockwise is reverse. While each servo is positioned differently, the robot will go forward when both servos are going forward. The robot will go backward or reverse when both servos are in reverse. The robot will turn left when the left servo is stopped and the right servo is in forward and the robot will turn right when the right servo is stopped and the left servo is forward. For the robot to turn left in place the left servo needs to be in reverse and the right servo need to be in forward and for the robot to turn right in place the left servo needs to be in forward while the right servo needs to be in reverse. Keep in mind that when the robot is moving forward the left servos and the right servo are actually going in opposite directions. This is so because they are positioned on opposite sides of the robot as well as faced in opposite directions.

4. Open a new Quartus project file. Name this project file robottop.qpf.
5. Add the following symbols: servo_control, count65536. You must assign the frequencies determined in step 3 to the servo control module. For this particular design, w14 causes the left servo to turn clockwise, w13 causes the left servo to turn counter clockwise, w13 causes the right servo to turn clockwise, and w15 causes the right servo to turn counter clockwise. This will be different for each servo which may depend on the servo modification and the potentiometer position, etc. Assign pins to the left (lservo), and right servo output (rservo) signals. Compile and program the PLD on the PLD platform with this design.
5. Program the PLD platform with the digital circuitry that controls the SERVOCONTROL. You will need to write the VHDL code to interface the SERVOCONTROL with each paired sensors (left and right) at a time. The robot should go right if the left sensor is active and left if the right sensor is active.
6. Now, completely build the robot's digital circuitry for the compass sensor and track tape sensor. You can change this design according to your design requirements. The VHDL code for the compass sensor can be obtained from Section 9.10 on page 376. Compile the design.
7. Construct the hardware platform for the robot as shown in Section 15.1.3 on page 484. The RX PLD platform needs to stand above the plexiglass platform. This is done here with 1 and ½ inch standoffs. The platform is about 10 inches in diameter. The weight of the ESCO board along with the other components needs to be evenly distributed to prevent the robot from easily turning over.
8. Complete the project as shown in Figure 14.5, Figure 14.6 or Figure 14.7. Lastly, program the design into the PLD or FPGA. Troubleshoot and test the robot.
9. Incorporate the battery supply into the project. The receive robot may now feed off the battery supply. The battery will be drained in a few minutes demonstration time. Make sure you have a spare battery pack or a fast charger and a power outlet nearby for uninterrupted demonstrations.

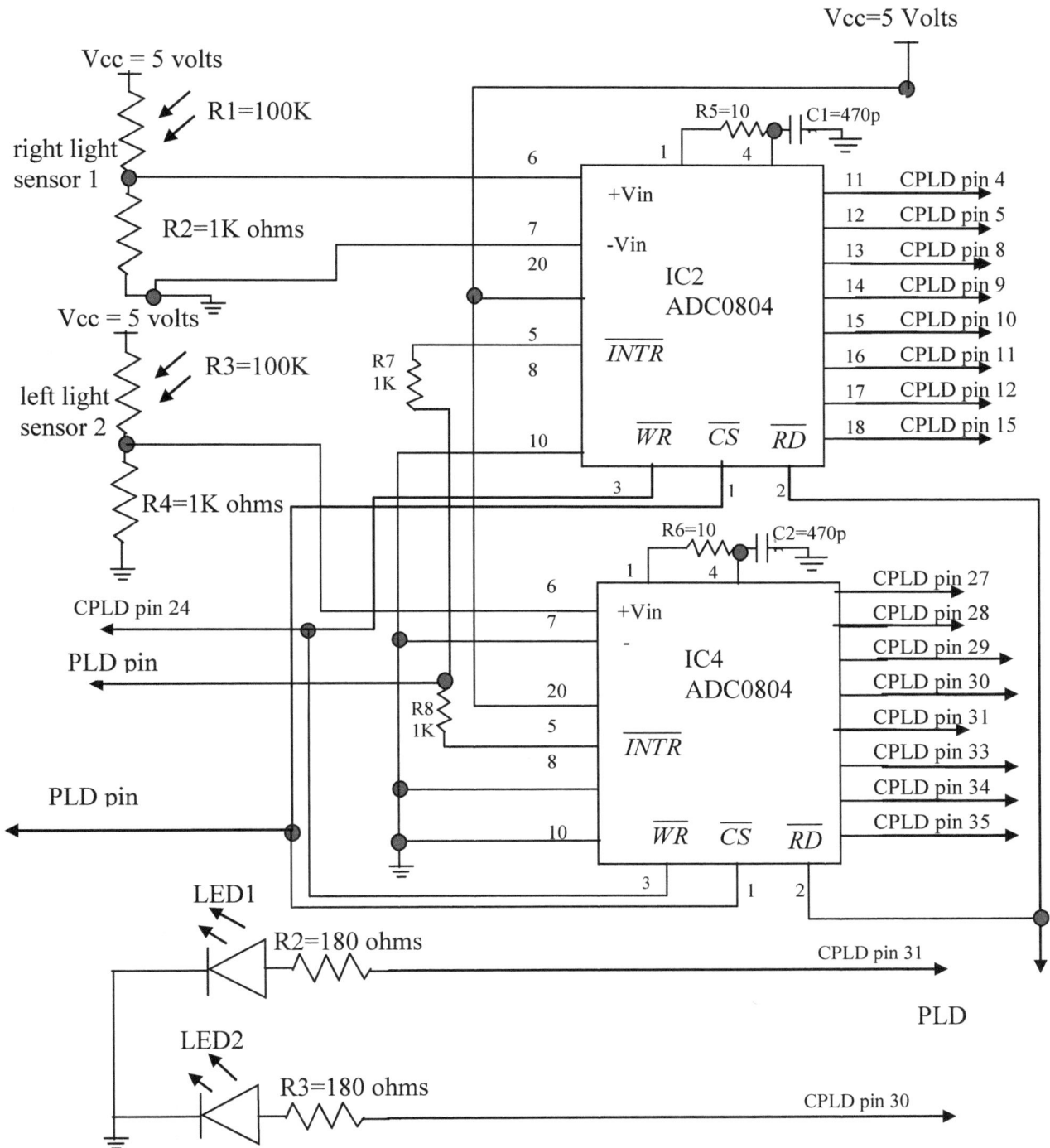

Figure 14.5. Schematic page I using 2 or more A/D converters: Schematic of Project 1, photo sensor input circuitry. Two 1 channel analog to digital converters are used here. The ADC0803 has 1 analog input channel and an 8 bit parallel output bit stream. Therefore, one analog sensor can be used with each A/D. The digital data sent to the CPLD over 8 wires. This is not very convenient when CPLD pins are scarce. However, it is easier to implement relative to VHDL coding in the CPLD.

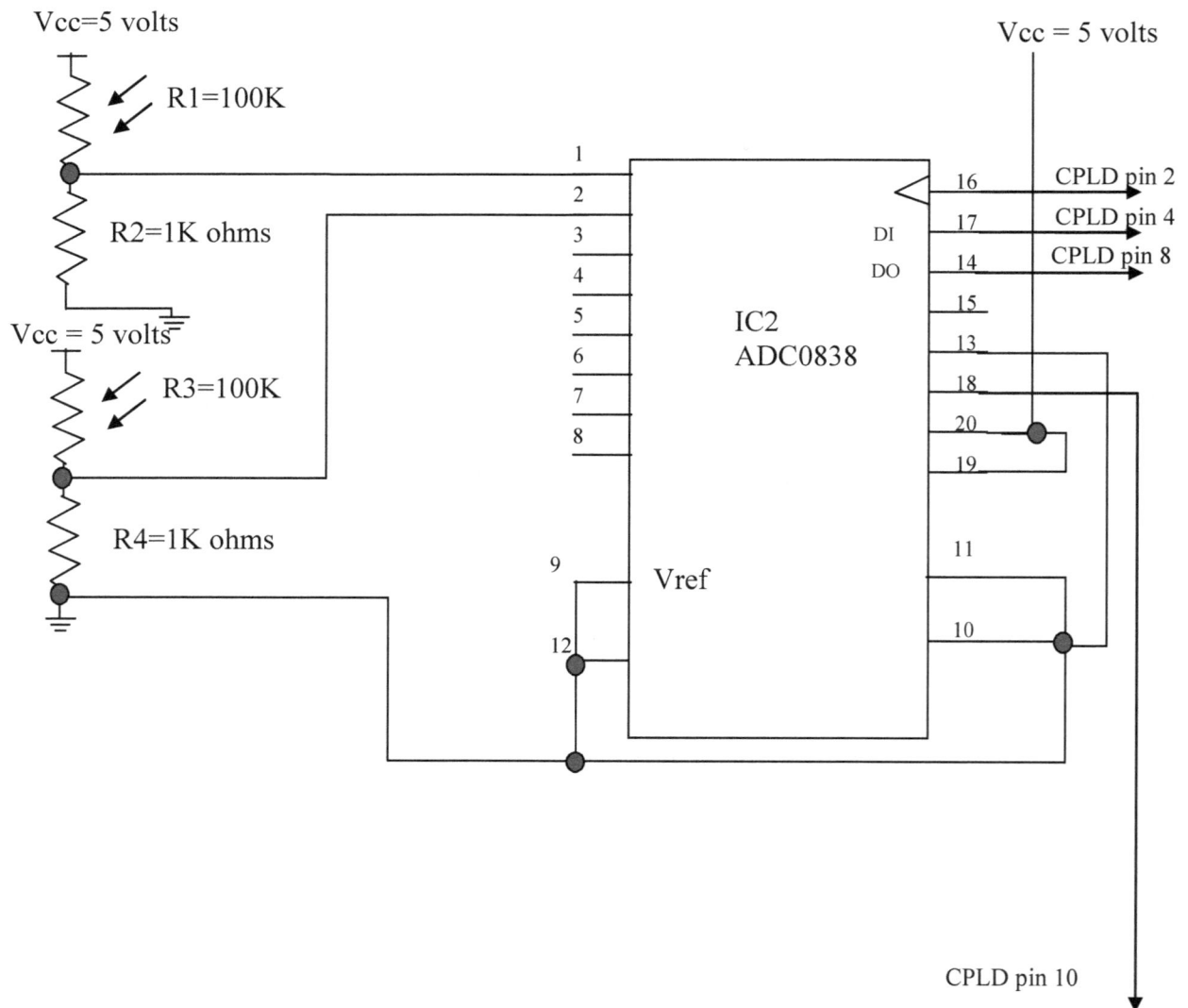

Figure 14.6. Schematic page I using one A/D converter: Schematic of Project 1, photo sensor input circuitry. An 8 channel analog to digital converter is used here. The ADC0838 has 8 analog input channels and a serial output bit stream. Therefore, up to 8 analog sensors can be used with the digital data sent to the CPLD over one wire. This is convenient when CPLD pins are scarce.

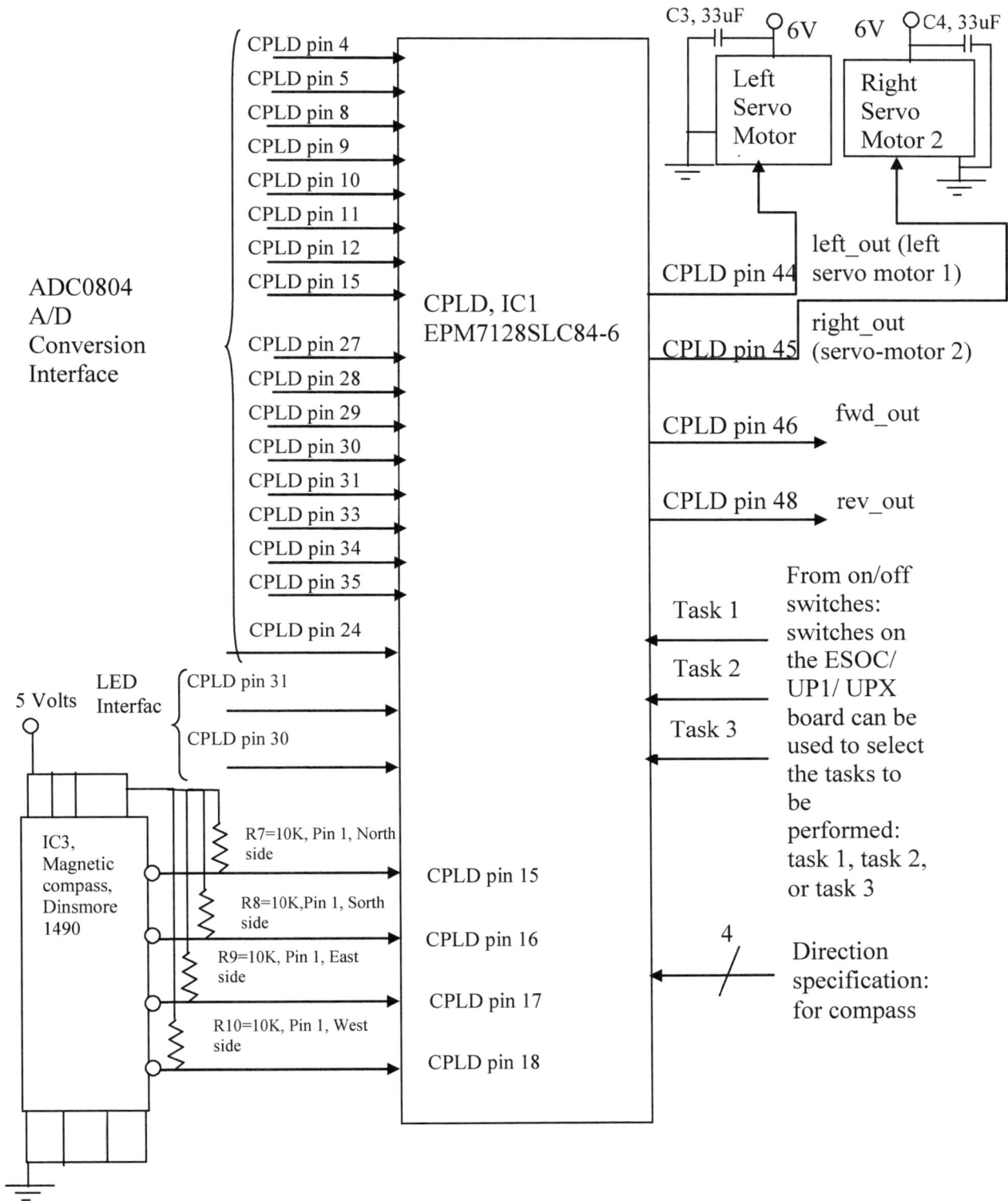

Figure 14.7. Schematic page 2: CPLD pin configuration and interfacing circuitry using the schematic I (using ADC0804 A/D converters) in Figure. Use 10K pull-up resistors for the Dinsmore 1490 compass.

14.2. Project 2: Object Avoidance and Direction using Proximity and Hall Effect Sensors

The second variation of the mobile robot uses a proximity distance sensor (Sharp GPD2D02) to determine the distance from an object (see Figure 4) and a digital magnetic compass to determine the robot's orientation. No A/D is required since the output of the both sensors are digital.

This PLD-bot has the following requirements:

Requirement 1: The PLD-bot must go north, south, east, or west at using the direction specified by input switches, task 1, task 2, and task 3.

Requirement 2: The PLD-bot must determine the closeness of objects and avoid them using an IF proximity sensor.

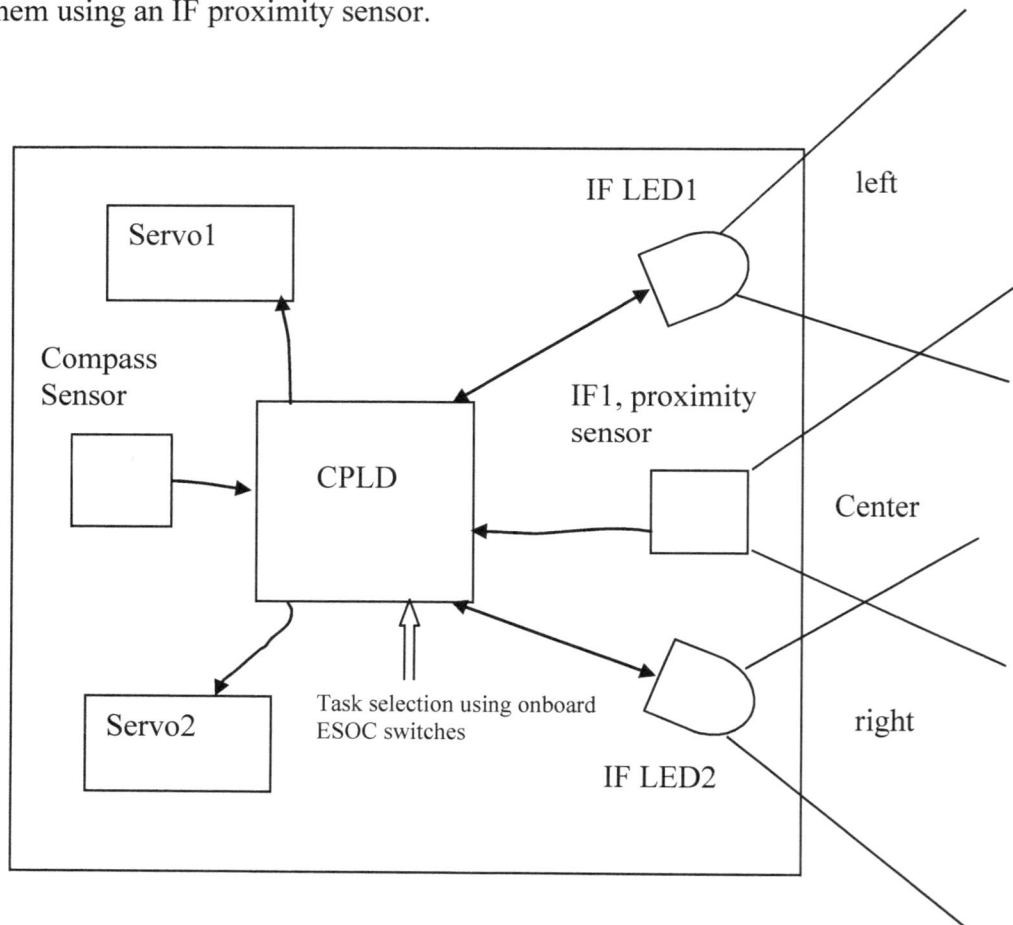

Figure 14.8. Project 2 of the PLD-bot. IF1 is the infrared proximity sensor while LED1 and LED2 are infrared LEDs.

14.2.1. Project 2 Electronic Parts List

The electronic part list for Project 2 is given in Table 14.4.

Part	Quantity	Ref designator	Description
Radio Shack or IR receiver module 276-640, IF Detector module, 276-137B (obsolete).	1	IF1	Infrared modulated detector module, bandpass center frequency=38,000Hz; Vccif=4.3 to 5.7 volts; peak wavelength = 940nm; Vout=0.5 to 4.2 volts
LED, 276-143C, Radio Shack	2	IFLED1, IFLED2	Infrared light emitting diode
Altera CPLD, EPM7128SLC84-6	1	CPLD, IC1	Devry ESOC board, Controller, see http://www.elexp.com/tst_pld2.htm
Altera CPLD, FLEX10K20 or FLEX10K70	optional	CPLD, IC1	Altera UP1 or UPX board
Magnetic Compass, Dinsmore model 1490	1	IC3	Magnetic compass, 4 bit, active low digital outputs
High Tech servos - HS300 (Futaba S3003, or Tower TS53J)	2	servo1, servo2	servomotors
Platform I	1	Platform I	Platform 1 uses two servo motors, a caster wheel, etc.
Resistor, 1000 ohms	1	R1	1 Watt resistor
Resistor, 180 ohms	2	R2, R3	1 Watt resistor

Table 14.4. Electronic part list for PLD-bot Project 2. For more details on where to order parts see Appendix D on page 577.

14.2.2. Project 2 Procedure

Acquire all the necessary parts for constructing the electronic portion of project 2. Most of these parts are listed in Table 14.4. For Project 2 the same steps are followed as in Project 1 except step 6 and 8, respectively, are replaced as follows:

6. Completely build the robot's digital circuitry for the compass sensor and IR sensor. The VHDL code for the compass sensor can be obtained from Section 9.10 on page 376. The VHDL code for the IR sensor can be obtained from Section 9.11 on page 379. Compile the design.

8. Complete the project as shown in Figure 14.8, Figure 14.9, and Figure 14.10. Lastly, program the design into the PLD or FPGA. Troubleshoot and test the robot.

14.2.3. Project 2 Schematic

The schematic for Project 2 is shown below.

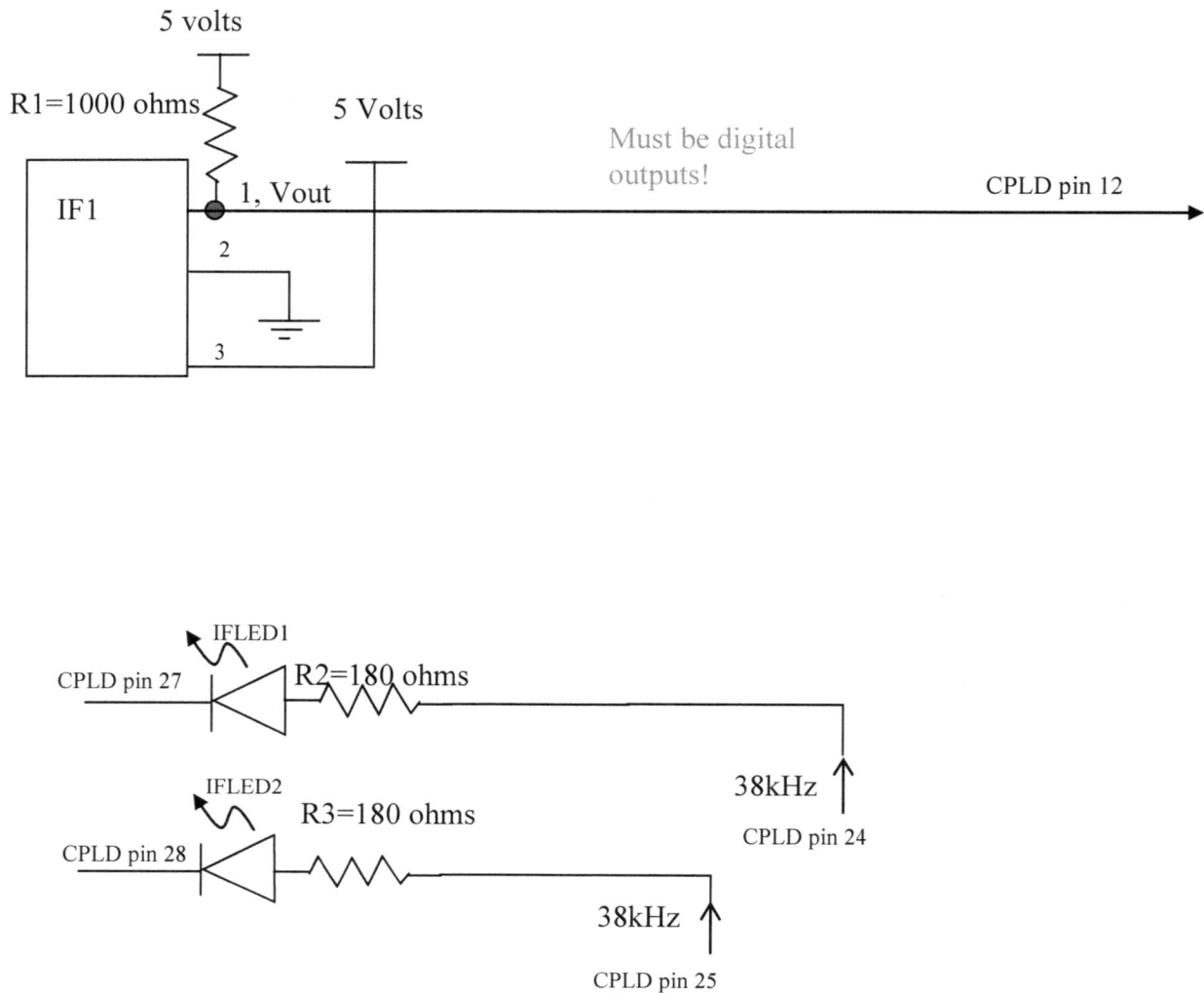

Figure 14.9. Schematic of Project 2, photo sensor input circuitry. The A/D is optional – may use for additional analog output sensors. Note. If IF1 output is not digital must use LM339 to convert it to digital.

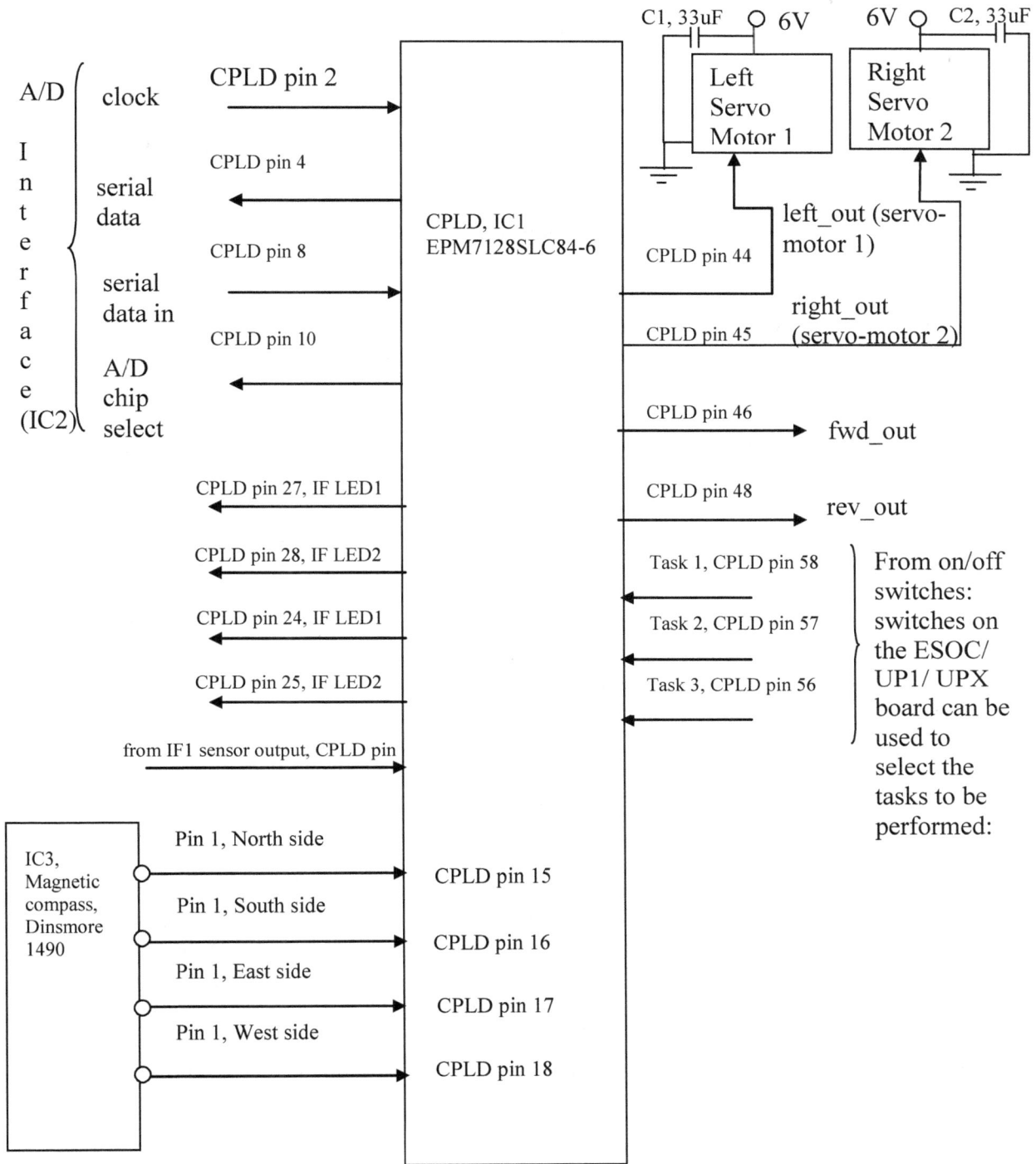

Figure 14.10. CPLD pin configuration and interfacing circuitry. Note: See Project 2 schematic to complete the Dinsmore 1490 circuitry. The 1490 require 2.2K pull-up resistors.

15. Communication Electronics and Robotics

Communication electronic robots is the focus of this Chapter. See Chapter 6 starting on page 185 for a basic description of communication systems. The following sections present descriptions, scenarios, components, etc. for building a mobile robot using a wireless transmitter and receiver. The transmitter and receiver will interface to a programmable logic device (PLD) platform or an embedded microcontroller as the contolling devices. More about the Microship microcontroller can be found in Huang (2005) and Katzen (2005) in the Reference Section on page 578. .

15.1. Project 3.0 - PLD-bot Remote Control

This project deals with sending 8 bit commands over a wireless digital transmission link to control a mobile robot and its sensors. The heart of the PLD-bot is a programmable logic device, in this case, an Altera device, EPM7128SLC84-15. The description, task, and procedure of the project are discussed next. It uses a WyJen Technologies (www.wyjen.com) transmitter and receiver to perform the wireless transmission from the controller to the robot, respectively.

15.1.1. Project 3.0 Basic Requirements

This mobile robot must meet the following requirements:

Requirement 1: The robot must respond to manual control commands. This is called manual mode. The manual control resides on the robot platform in the form of switches. These commands must include the following: 1.) turn left, 2.) turn right, 3.) go forward, 4.) go backward, and 5.) stop.

Requirement 2: The receive robot must perform the commands in requirement 1 in remote control mode. In remote control mode a transmitter must send 8 bit commands which cause the robot to perform the commands in requirement 1.

Requirement 3: The manual and remote control modes must select between sensor mode 1 or sensor mode 2, etc. If sensor 1 is a compass sensor, then if sensor mode 1 is selected the robot responds to the compass sensor. The robot must then go in the direction specified by the manual control (north, south, east, west, etc.). For example, if manual control specifies go in the direction north, the robot must turn until the north heading is found. When found the robot goes forward in the specified direction. More detailed information about this requirement can be found in section 9.10 on page 376.

Requirement 4: The project must use a wireless transmitter and receiver device. See WyJen Technologies (www.wyjen.com) transmitter and receiver module or similar devices.

15.1.2. Project 3.0 Description

This project consists of the following main components:

1. Remote control unit with a wireless transmitter and TX PLD platform.

2. Mobile robot consisting of a wireless receiver and a RX PLD platform.
3. Mobile robot with sensory circuits, e.g., compass sensor, light tracker, etc.

The mobile robot collectively is referred to as PLD-bot.

Figure 15.1 shows the robot remote control unit. The remote control unit's PLD platform receives an eight parallel bit command, converts it to serial, and then sends it to the transmitter. The transmitter sends the command data over a wireless link to a receiver on the mobile robot . These commands are then used to control the mobile robot and its sensors.   The remote control unit may optionally take command data from a PC via a parallel port or similar port.

The remote control unit has a TX PLD platform and a wireless transmitter. The transmitter sends messages over a wireless communication channel to a wireless receiver located on the robot. The receiver demodulates the message to raw data. The raw data is sent to the RX PLD platform also located on the robot. The RX PLD platform decodes the message and determines the task the robot is commanded to perform.

The TX PLD platform on the remote control unit optionally interfaces to the PC as shown in Figure 15.1. The PC in this case consists of a parallel port driver program (Dhananjay, 1998) and software to create eight bit data to send to the RX PLD platform. The default configuration is to create eight bit data using on board PLD platform switches only. The robot has the circuitry as shown in Figure 15.2.

Figure 15.2 shows the mobile robot. The main components of the mobile robot are the PLD platform, a wireless receiver, servos, and sensory circuits. The PLD platform receives the asynchronous serial digital message from the wireless receiver, decodes it, and responds by sending the appropriate frequencies to the servo motors or sensors. The square waveforms of different frequencies cause the servo motors to left, right, or stop. That is, the waveforms of different frequencies cause the servos to turn clockwise, counter clockwise, or stop. Servos typically use a pulse width modulated (PWM) waveform to cause them to turn clockwise, counter clockwise, or to be neutral (stop). Sending waveforms of different frequencies to a servo motor is seen by the servo motor to be PWM like and thus the motors respond similarly. The waveforms generated by the robot's RX PLD platform are square waveforms which have a constant pulse width. Square waves of different frequencies are selected by using a multiplexor and a counter in the receiving robot's PLD on the RX PLD platform. These waveforms of different frequencies are derived from a reference clock of 4MHz to the counter.

A compass sensor can also be included to give the robot a sense of direction. The compass sensor has four digital outputs that can be interfaced directly to the receiving robot PLD platform.

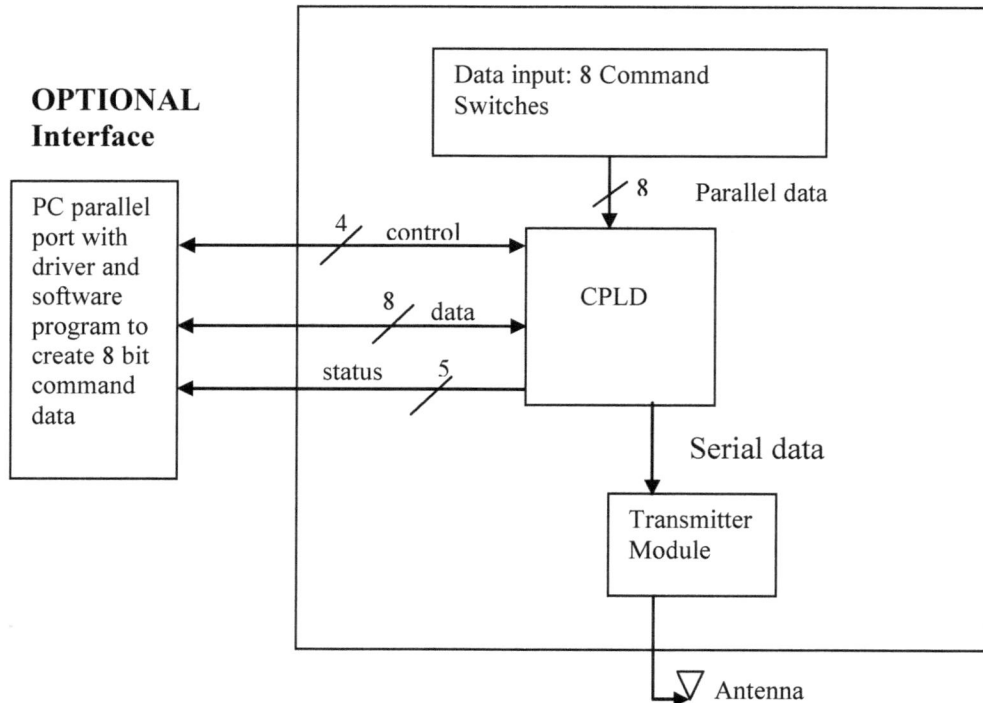

Figure 15.1. Transmit PLD platform with optional PC parallel port control. The default configuration is to use eight switches to create an eight bit command which is sent to the receive PLD platform. To control the PLD platform through a PC parallel port would require parallel port driver software and a corresponding software program to create and send command data to the PLD platform.

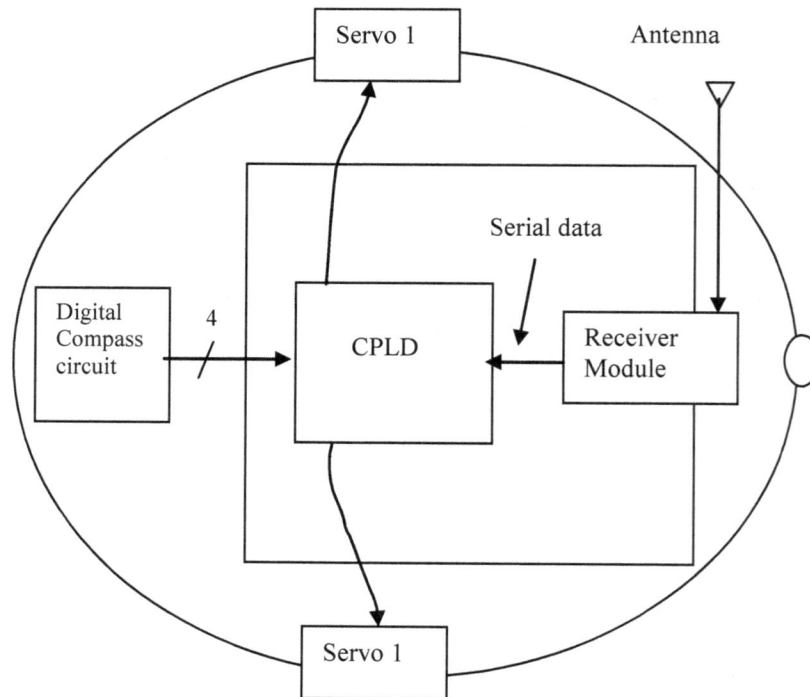

Figure 15.2. Mobile robot or PLD-bot.

15.1.3. Project 3.0 Tentative Development Schedule

The tentative schedule for developing the PLD–bot is shown in Table 15.1. You must try to follow it.

Week	Things to do
Week 1,2,3	Get ideas and teams together
Week 4,5	High-level Layout of robot complete. Devices selected (e.g., sensors, communication, etc.), start getting parts, platform selected, who will be doing what (sign off required)
Week 6,7	Parts Due, platform complete, batteries acquired (or substitute power source) (sign off required)
Week 7-12	Test all circuits, sensors, etc. (sign off required)
Week 7-10	VHDL code simulation (sign off required)
Week 9-13	Assemble robot and integrate all circuits, sensors, etc., working with VHDL code (sign off required)
Week 14	Oral presentations and reports due

Table 15.1. Tentative 15 week schedule.

15.1.1. Project 3.0 Electronic Parts List

The starter electronic parts list for Project 3.0 is given in Table 15.2.

Item	Quantity	Ref designator	Description
Altera Software	1		Software used to program the EPM7128SLC84-15. See website: http://www.altera.com/products/software/quartus-ii/switching/max-plus-ii/max-plus-ii-users.html
LED, 276-143C, Radio Shack	2	IFLED1, IFLED2	Infrared light emitting diode
Altera CPLD, EPM7128SLC84-15	2	CPLD, IC1	DeVry ESOC board, Controller. The pin connections are shown for this device. See http://www.elexp.com/tst_pld2.htm
Altera CPLD, FLEX10K20 or FLEX10K70 (optional)	2	CPLD, IC1	Altera UP1 or UPX board. The pin connections are not shown for this device
Magnetic Compass, Dinsmore model 1490	1	IC2	Magnetic compass, 4 bit, active low digital outputs
Futaba servos S3003 (Tower TS53J), or High Tech servos HS300	2	servo1, servo2	Servo motors, www.towerhobbies.com

Table 15.2. Electronic starter parts list for Project 3.0. For more details on where to order parts see Appendix D on page 577.

Item	Quantity	Ref designator	Description
Platform I	1	Platform I	Platform 1 uses two servo motors, a caster wheel, etc. See Figure 15.16 on page 485
Resistor, 1000 ohms	1	R1	1 Watt resistor
Resistor, 180 ohms	2	R2, R3	1 Watt resistor
Transceiver: GTR315M_M13	2	TXRX	315MHz transmitter/receiver, 3/5V, half duplex transmission, www.wyjen.com
Transmitter Module - GTX433M-3V-A00S06A-M30-1, etc.	1	TX	ASK modulation, WyJen Technologies, www.wyjen.com (optional)
Receiver Module - GRX433M-3V-A00S06A-M30-1, etc.	1	RX	ASK modulation, WyJen Technologies, www.wyjen.com (optional)
Circuit board	3	CB1, CB2, and CB3	Circuit board for TX, RX, and compass circuit.
1.5V AA Batteries	4	Battery	Two for transmitter and two for receiver supply voltages
MAX8510	2	IC3	Optional 3.3 V regulator, One for transmitter and One for receiver supply voltages
TPS7133QP (Digikey)	3	IC3	Optional 3.3 V regulator, One for transmitter and
20 gauge solid wire, Radio Shack part 278-1222			Black, red, green. Wire for connecting various electronic parts and components
20 gauge solid wire			Need other colors such as white, blue, yellow, etc. Wire for connecting various electronic parts and components

Table 15.2. Continued.

15.1.2. Project 3.0 Procedure

For the steps below save all Altera and VHDL files in the same directory. Preferably a directory called **work**. Follow the steps below to complete this project:

1. Acquire all the necessary parts for constructing the electronic portion of project 3. Most of these parts are listed in Table 15.2. Also see Section Appendix D on page 577 for more information on purchasing the various parts.
2. Modify the servo motors for continuous rotation by a.) clipping or removing the potentiometer, b.) clipping away the stopper.
3. It is very important that you record your results using Table 15.3 and Table 15.4 in this section for later referral! Test the servo motors using the PLD platform's 4Mhz on board clock as follows. This clock frequency must be divided down below 1000 Hz for proper operation of the servos. A divide by 2^{16} counter can achieve this. The VHDL code for a divide by 2^{16} (mod65536) counter is given in Section 9.6.2 on page 349. Send the last six outputs of the

counter to the output ports as shown in Figure 15.3. Connect each of these outputs to the signal input of the servo motor. The ground wire of the servo must be grounded to the PLD Platform ground. Similarly, the power supply wire of the servo must be connected to the PLD platform board's VCC. An example of the external test circuit is shown in Figure 15.4. Record the outputs and their respective frequencies that cause the servo to turn clockwise (forward), counter clockwise (reverse), and stop. This information may be unique for each servo and must be recorded separately for each servo motor. Use Table 15.3 and Table 15.4 below to record the effect of each output from the counter on the servos. Put a check mark in the column labeled slow speed, medium speed, and fast speed. Slow speed is when the servo turns very slow. Fast speed is the highest speed the servo will go. Medium speed is somewhere in between. You have to use your best judgment on what is a slow, medium, or fast speed.

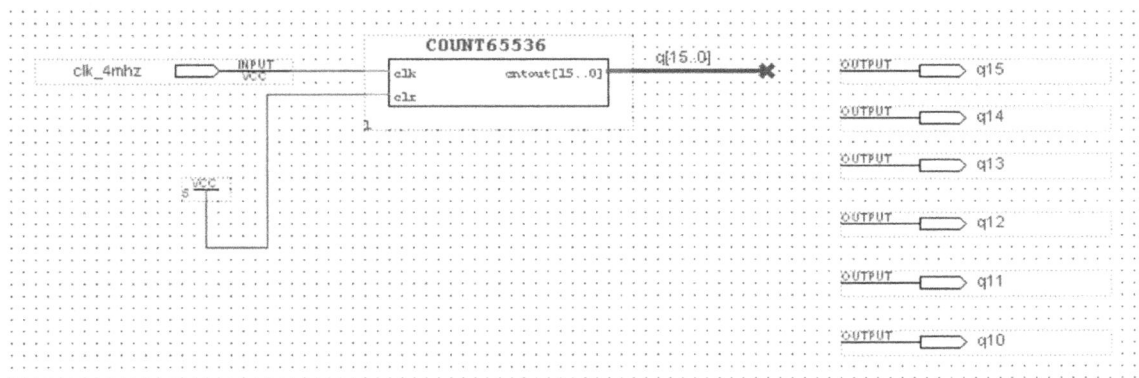

Figure 15.3. Digital test circuit for a servo motor using a mod 65536 counter to divide down a 4MHz clock signal. The output ports for q10, q11, q12, q13, q14, and q15 are individually connected to the signal input of the servo motor.

Before you begin recording the servo clock frequency verse speed and rotation direction in Table 16.4 and Table 16.5 the servos should be identified by a label as the right servo and the left servo. The servos should be positioned as shown in Figure 15.5. The shaft and tire of the servos should be pointed in opposite directions as shown.

It should be noted that clockwise is forward and counter clockwise is reverse. While each servo is positioned differently, the robot will go forward when both servos are going forward. The robot will go backward or reverse when both servos are in reverse. The robot will turn left when the left servo is stopped and the right servo is in forward and the robot will turn right when the right servo is stopped and the left servo is forward. For the robot to turn left in place the left servo needs to be in reverse and the right servo need to be in forward and for the robot to turn right in place the left servo needs to be in forward while the right servo needs to be in reverse. Keep in mind that when the robot is moving forward the left servos and the right servo are actually going in opposite directions. This is because they are positioned on opposite sides of the robot as well as faced in opposite directions.

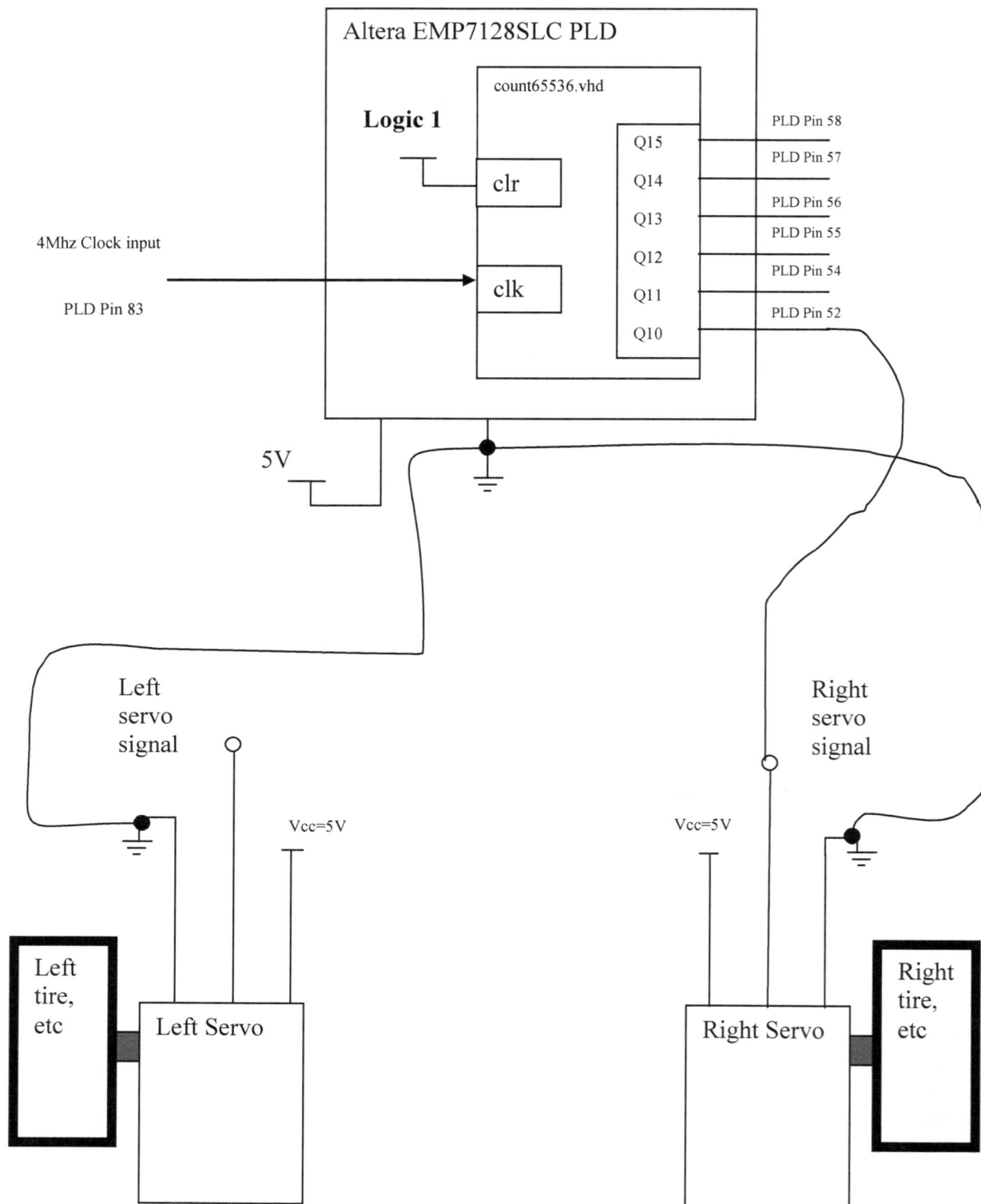

Figure 15.4. Testing of left servo and to right servo using the mod 65536 counter.

Left Servo Clock Frequency (Speed Characteristics and rotation direction)						
Counter Output	Frequency	Slow Speed	Medium Speed	Fast Speed	Forward	Reverse
Q10						
Q11						
Q12						
Q13						
Q14						
Q15						

Table 15.3. Left Servo clock frequency verses speed and rotation direction.

Right Servo Clock Frequency vs. (Speed Characteristics and rotation direction)						
Counter Output	Frequency	Slow Speed	Medium Speed	Fast Speed	Forward	Reverse
Q10						
Q11						
Q12						
Q13						
Q14						
Q15						

Table 15.4. Right Servo clock frequency verses speed and rotation direction.

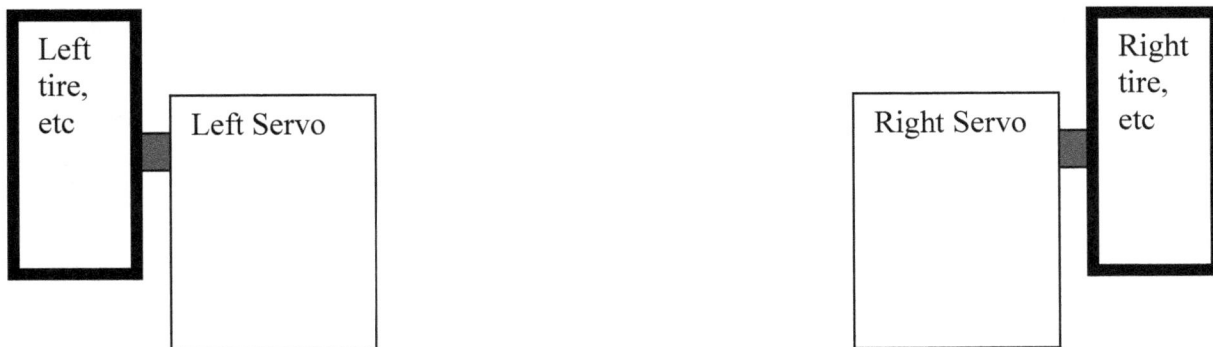

Figure 15.5. Position of left servo relative to right servo.

4. Write the VHDL code that converts parallel data (8 data bits) to serial data. This data must be in standard UART format. Standard UART requires 1 start bit, 8 data bits, and 2 stop bits. The details of this is given in Section 9.6 on page 344. Copy and compile the code in this section. The VHDL code for the mod11 counter must be written by you. This has already been done if you completed Section 9.6.1 on page 348 for count10.vhd. Connect the various VHDL files and symbols as shown in Section 9.6. Then name this graphic design file txuart.qpf. Program the PLD of the PLD platform using this design and verify the txuart digital serial data at an output port using an oscilloscope.

5. Test the transmitter and receiver modules as shown in Figure 15.6. That is, breadboard the circuit as shown in Figure 15.6. Consult your manufacturer's data sheet to connect all other inputs and outputs as required. For example, make sure the ground and power pins are connected as specified. Assuming that the TX and RX modules are wired for proper operation, now connect the output of the txuart from the PLD platform completed in procedure 4 to the transmitter module as shown. The antenna out of the TX module should be connected to the antenna input of the RX module using a short piece of wire. The ESCO, TX, and RX should share the same ground reference. Use channel A of an oscilloscope to measure the output of the txuart. Now use channel B of the same oscilloscope to measure the digital output of the RX. The data in channel A should match the data on channel B. If not, then check that the circuit is wire correctly. If all else fail, the TX or RX must be bad. Consult the distributor or manufacturer for a replacement.

6. Copy and compile the VHDL modules given in Section 9.7 on page 351 for the rxuart. Save this graphic design file as rxuart.qpf. This module will be used to convert the digital data at the output of the RX module back to 8 bit parallel data. The rxuart module consists of five main components and modules:
   a. JK flip flop
   b. clock generator
   c. serial in parallel out (SIPO) register
   d. parallel in parallel out (PIPO) register
   e. one shot

After these modules have been successfully compiled in the graphic editor. Create a default symbol from this design.

7. Open a new Quartus project file. Name this project file robottop.qpf.

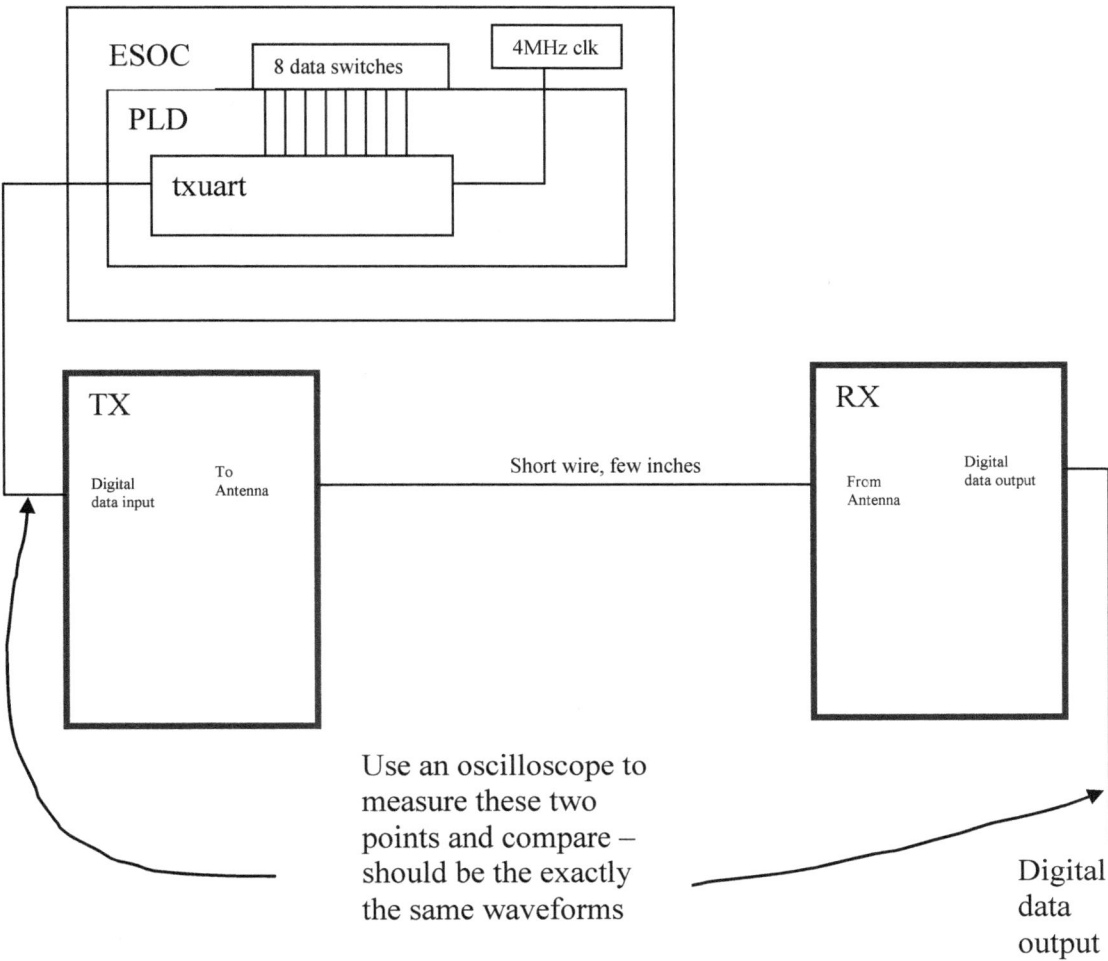

Figure 15.6. Test circuit for TX and RX.

8. Create the Quartus project file robotrx0.qpf and enter the following symbols: rxuart, servo_control, and count65536.

Now, construct the circuit shown in Figure 15.7. You must assign the frequencies determined in step 3 to the servo control module. For this particular design, w14 causes the left servo to turn clockwise, w13 causes the left servo to turn counter clockwise, w13 causes the right servo to turn clockwise, and w15 causes the right servo to turn counter clockwise. This will be different for each servo which may depend on the servo modification and the potentiometer position, etc. Assign pins to the serial data input (serialdata), clock input (clk_4Mhz), left servo output (lservo), and right servo output (rservo) signals. Compile and program the PLD on the PLD platform with this design.

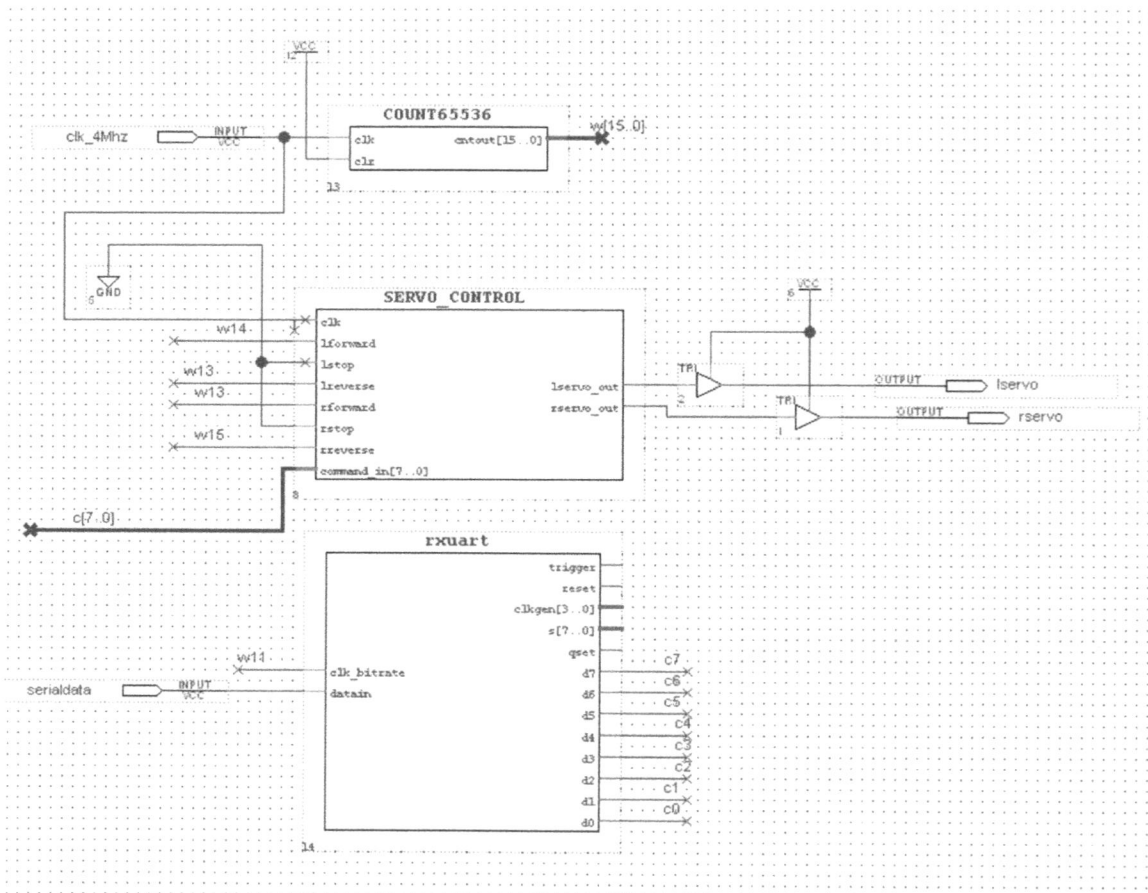

Figure 15.7. Test circuit for rxuart and servo_control.

9. Program the TX PLD platform with the txuart and the RX PLD platform with the rxuart, SERVOCONTROL, COUNT65536 (use the design in Figure 15.7). Connect the TX PLD platform directly to the RX PLD platform without the TX and RX as shown in Figure 15.9. That is, connect a wire from the output of the TXART to the input of the serial data of the rxuart. Make sure both PLD platforms share the same ground. Send various commands to RX PLD platform using the switches on the TX PLD platform. The txuart clk must be the same as the rxuart clk for this to work. For example, if the txuart clk is connected to q12 from the count65536, then the rxuart clk must also connect to q12 from count65536. The input clock to the count65536 counter in txuart and rxuart is the same clock speed. The commands should be decoded correctly by the rxuart and thus the servos should respond as commanded. If a left turn command is sent the servo should turn left. If a right turn command is sent the servo should turn right, and so on. If this step is not successful debug the code as required until the correct results are achieved. If preferred verify all parts of the design using simulation before performing this test again. See Figure 15.13 for examples of the pin assignments for the servos, etc. If the servos are not yet available test the asynchronous serial to parallel conversion using LEDs as outputs as shown in Figure 15.8 using the setup in Figure 15.10. That is, program the RX PLD platform with the design shown in Figure 15.8 instead of the design shown in Figure 15.7.

10. If step 9 is successful, perform the same procedure as given in step 9, but using the wireless link as shown in Figure 15.11. That is, include the TX and RX in the test. The transmitter and receiver circuits must be free of noise to operate optimally. One method of reducing the noise effects is to use two AA batteries in series for the transmitter 3.3V source and a separate one for the receiver 3.3V source. The two AA batteries in series produce a 3V source which is within the supply range of the WyJen transmitter and receiver. The transmitter battery ground should share the same ground as the TX PLD platform ground and the receiver battery ground should share the same ground as the RX PLD platform ground. Also locate the receiver away from the servo motors.

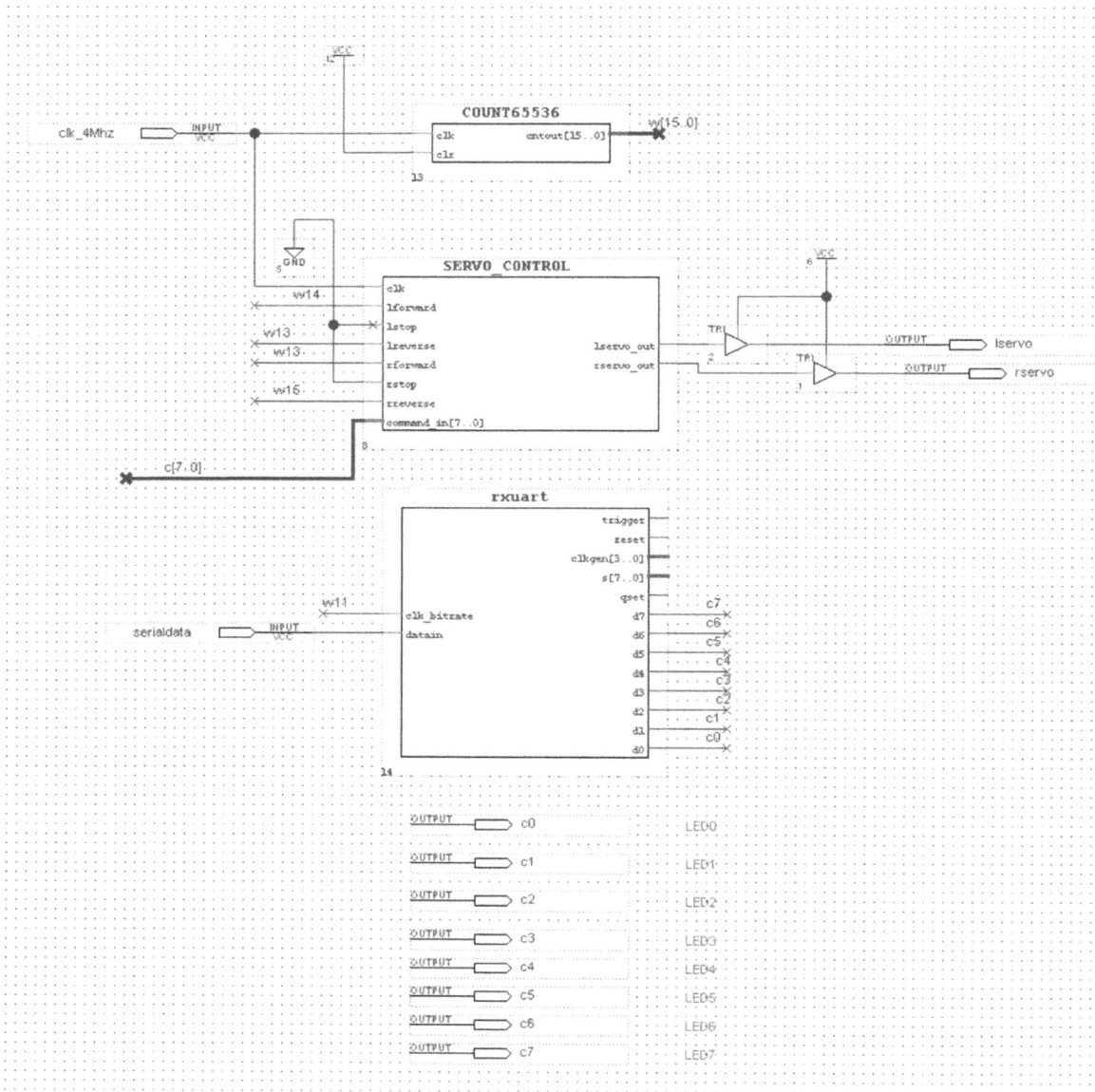

Figure 15.8. Test circuit (Quartus file: **robotrx0.qpf**) for rxuart and servo_control using LEDs.

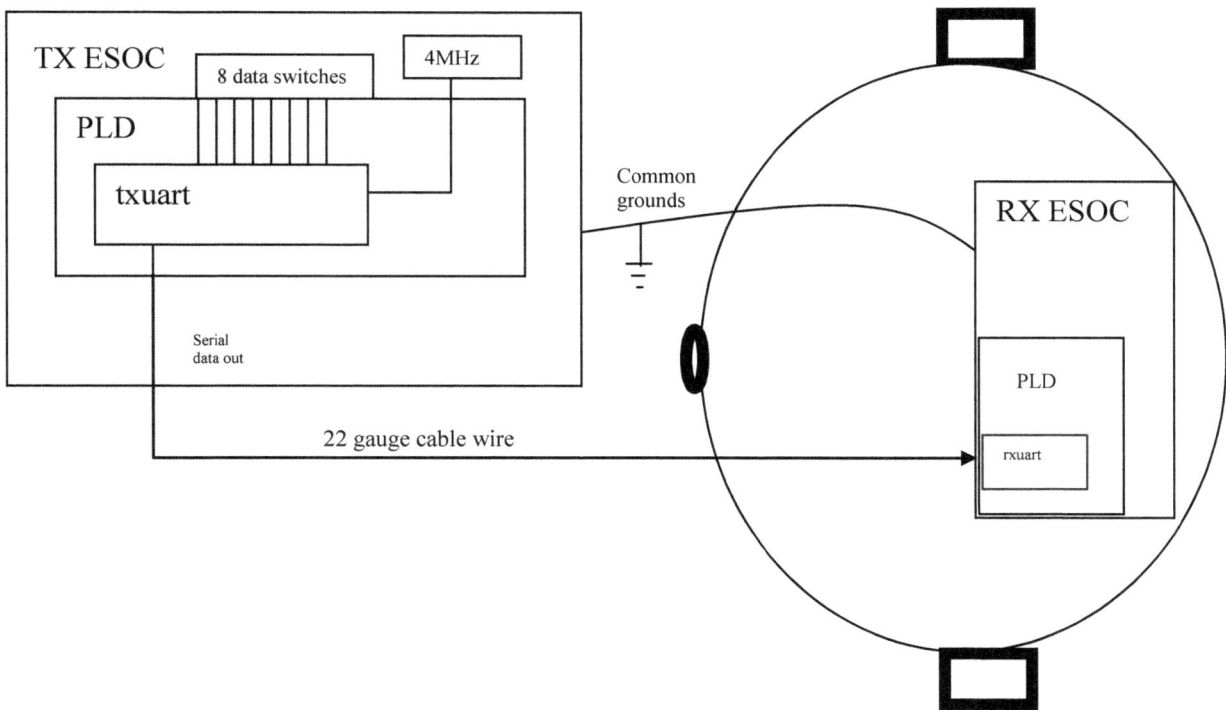

Figure 15.9. Test circuit for TX PLD platform, RX PLD platform, left and right servos, and robot platform.

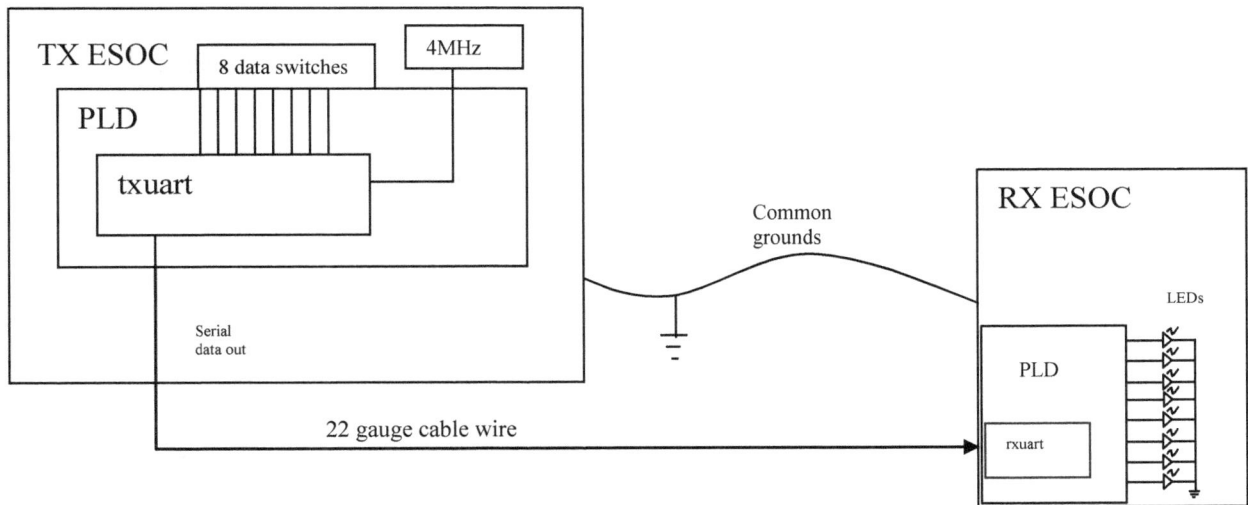

Figure 15.10. Test circuit for TX PLD platform and RX platform.

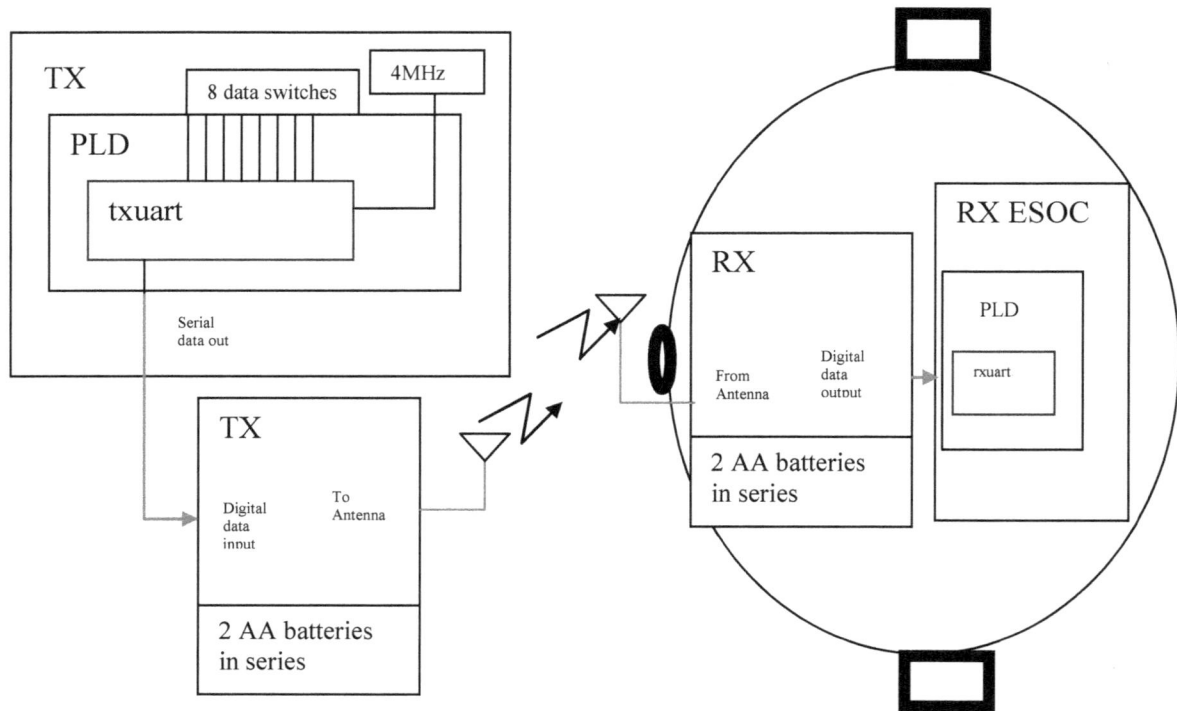

Figure 15.11. Test circuit for TX PLD platform, RX PLD platform, TX, RX, left and right servos, and robot platform.

11. Create a project called robottop in Quartus II. Now, completely build the robot's digital circuitry for the Infrared (IR) sensor and track tape sensor as shown in Figure 15.12. Compile the file. Note, you must use the VHDL modules that you agreed to do for your project. You can change this design according to your design requirements.

12. As an exercise add the compass module to the design. The VHDL code for the compass sensor can be obtained from Section 9.10 on page 376. Compile the design.

13. Acquire all the hardware components of the robot. A starting list is given in Table 15.5.

14. Construct the hardware platform for the robot as shown in Section 15.1.3 on page 484. The RX PLD platform needs to stand above the plexiglass platform. This is done here with 1 and ½ inch standoffs. The platform is about 10 inches in diameter. The weight of the ESCO board along with the other components needs to be evenly distributed to prevent the robot from easily turning over.

15. Complete the project as shown in Figure 15.13. The WyJen receiver detailed circuitry is shown in Figure 15.14. If you are using the WyJen transceiver use the circuit in Figure 15.15.

16. Lastly, program the design into the PLD or FPGA. Troubleshoot and test the robot.

17. Incorporate the battery supply into the project. The receive robot may now feed off the battery supply. The battery will be drained in a few minutes demonstration time. Make sure you have a spare battery pack or a fast charger and a power outlet nearby for uninterrupted demonstrations.

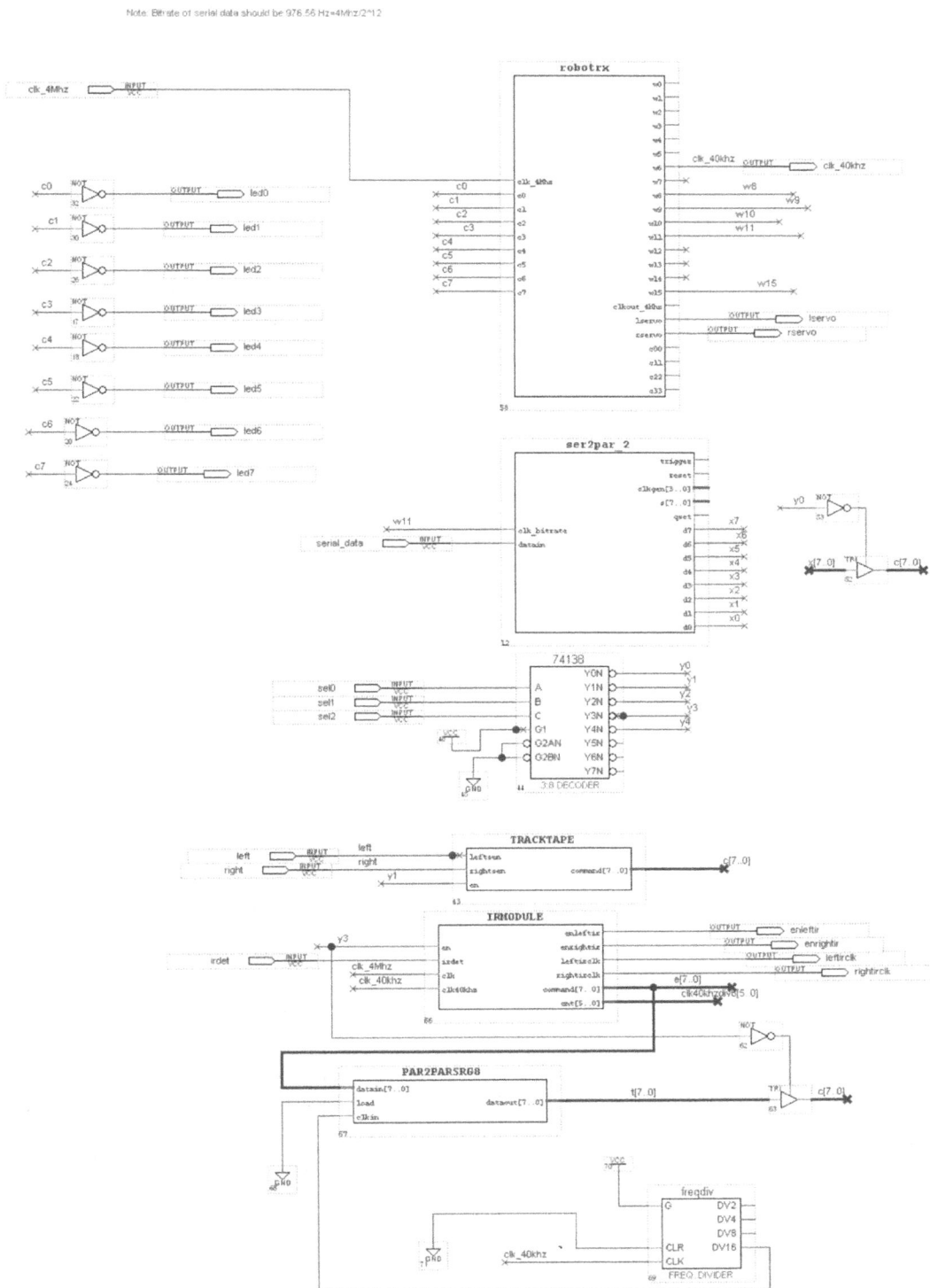

Figure 15.12. robot_top0.qpf design file.

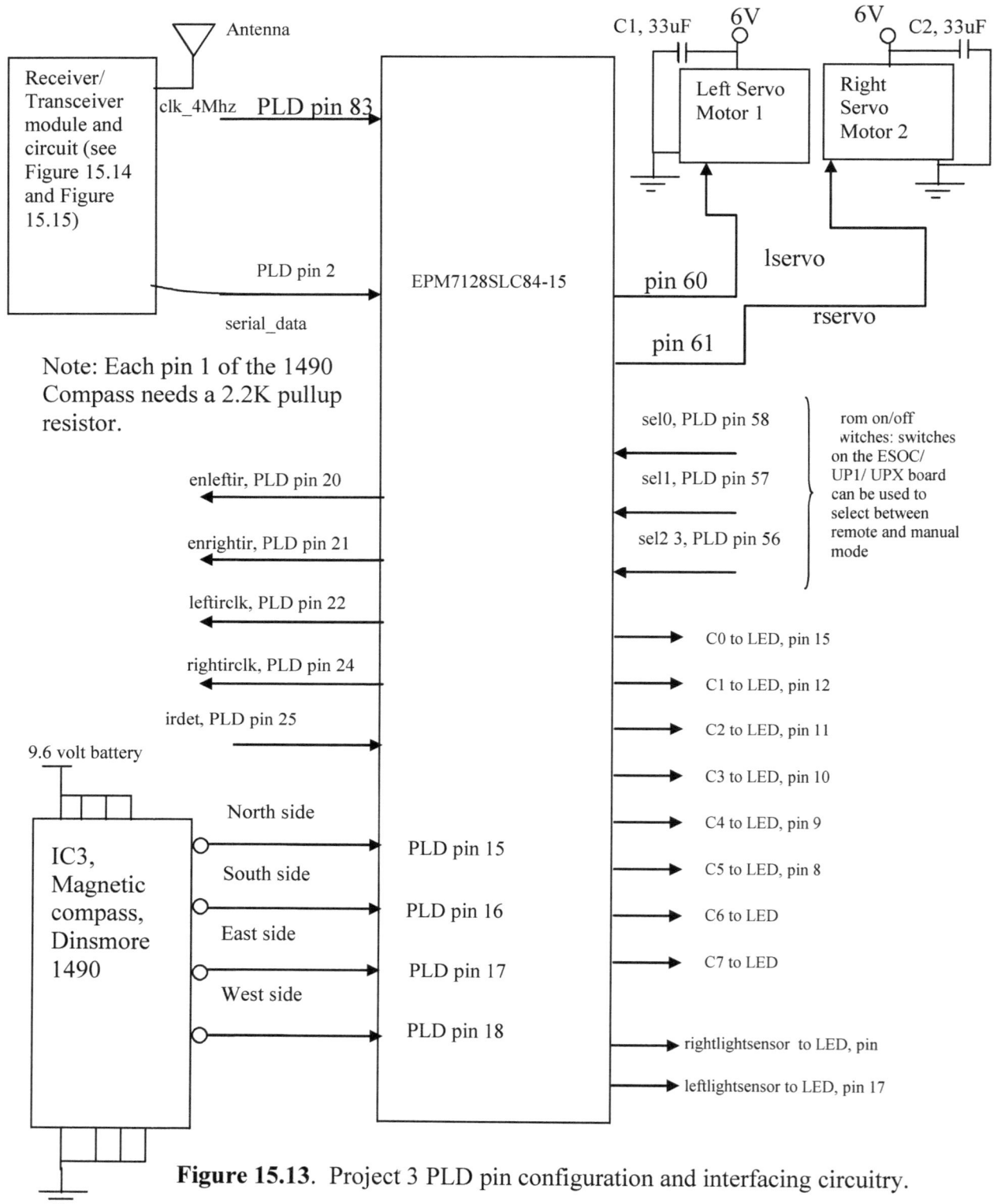

Figure 15.13. Project 3 PLD pin configuration and interfacing circuitry.

Power Supply Output: 3.3V or two
AA batteries in series at 3.0V

DeVry ESOC Board - 5V;
on robot platform

8 LEDs

3.3V

3.3V Battery Supply

Input:
serialdata
from RX

Antenna

GRX315M-3V-A00S12B-M07A or
: GRX433M-3V-A00S12B-M07A – 3V

3.3V

1 2 3 4 5 6 7 8 9 10 11 12

1
2

20

18

10

MM74HC244/SN74LVC244AN

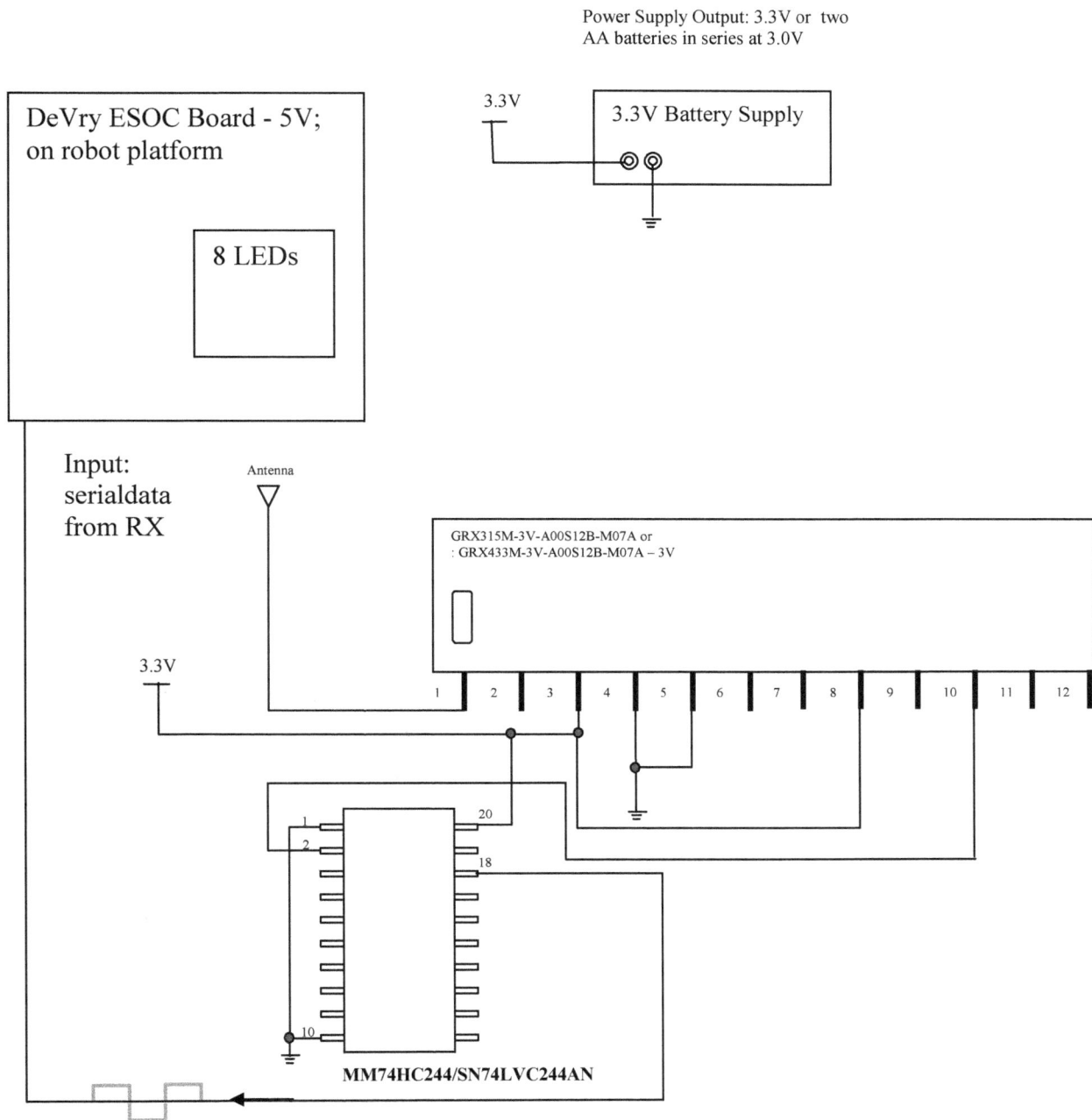

Figure 15.14. GRX315M/GRX433M Receiver circuitry. The 244 buffer provides the interface between 5V and 3.3V circuitry.

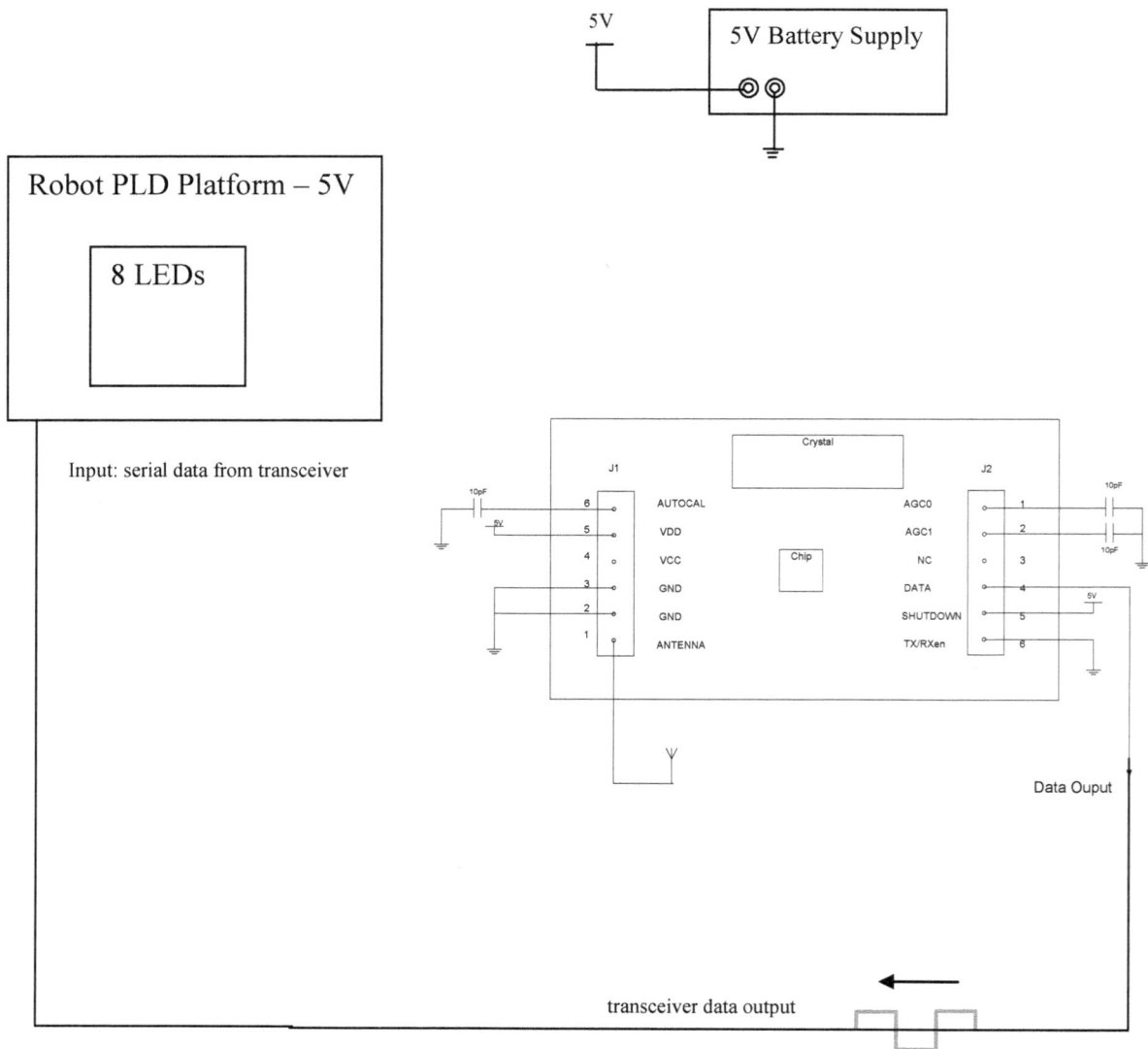

Figure 15.15. GTR315M-M13 receiver circuitry. No buffer is required since both circuits are at the same voltage levels, i.e., +5V.

15.1.3. Project 3.0 Hardware Parts List

The hardware parts list for Project 3.0 is given in Table 15.5 below. These parts can be obtained from the references listed in Section Appendix D on page 577.

Part	Quantity	Ref designator	Description
High Tech servos - HS300 (Futaba S3003, or Tower TS53J)	2	servo1, servo2	servomotors
Battery supply, radio shack, 9.6V, 1000mA, 230-0342	2	Vcc	9.6 Volt Nickel Cadmium R/C car battery pack and charger for TX ESCO and RX ESOC, Radio Shack
DC power connector, M type	2		For connecting battery packs to ESOCs, Radio Shack
Caster wheel	1		Front wheel, Menards, Lowes, Home depot, etc.
Tire	2		Rubber tire, Tower Hobbies
Wheel	2		Rim for rubber tire, Tower Hobbies
Round Platform	1	platform I	Round platform of diameter 11 inches (and thickness 3/16 inches) cut from plexiglass, etc, etc
standoffs	4	ST1, ST2, ST3, ST4	1 and ½ inch standoffs, Digikey, Radio Shack, etc
Screws, 844-712 is Menards part #	Depends on usage		Machine screws, size 4-40x ½, 30 pieces, can be used to mount caster wheel and servo if the 4-40x ¼ is not long enough.
Screws, 844-710, 844-712 is Menards part #	About 16		Machine screws, size 4-40x ¼, 34 pieces. 8 is needed for mounting the RX ESOC, 4 for the caster wheel, 2 for each servo
Nuts, 846-644 is Menards part #	About 12		Machine nuts, size 4-40, 28 pieces. 4 for the caster wheel, 4 for each servo
Washers, 154-306 is Menards part #	About 12		Washers, No. 6, 100 pieces. 4 is needed for mounting the bottom of the RX ESOC, 4 for the caster wheel, 4 for each servo
Velcro			4 inch by 2 inches, 2 sets

Table 15.5. Project 3.0 starter hardware parts list. For more details on where to order parts see Appendix D on page 577.

15.1.4. Project 3.0 Other Possible Sensors

Some other sensors are:
1. Line Tracker Sensor (Lynxmotion, digital output)
2. Infrared Proximity Detector (Sharp GP1U5 Radio Shack #276-137B, Digikey #160-1060, digital output)
3. Sonar Ranging Unit (Polaroid Sonar module)
4. IR Distance Sensor (Sharp GPD2D02, digital output)
5. Magnetic Compass Sensor (Dinsmore model 1490; digital output)
6. Other sensor – you must get approval from the professor

15.1.5. Project 3.0 Assembly of the PLD-bot: Platform I

The bottom view of the PLD-bot is shown in Figure 15.16. The PLD platform will rest on the top of the platform. The servomotors will be mounted on the bottom and attached to the wheels as shown. The control Cables run from the servomotors to the PLD platform located on the opposite side as shown. The sensors should be mounted on the top with all outputs connected to the controller (PLD platform). The switch and power cables have separate holes in the platform. The power switch is connected to the hole in the platform.

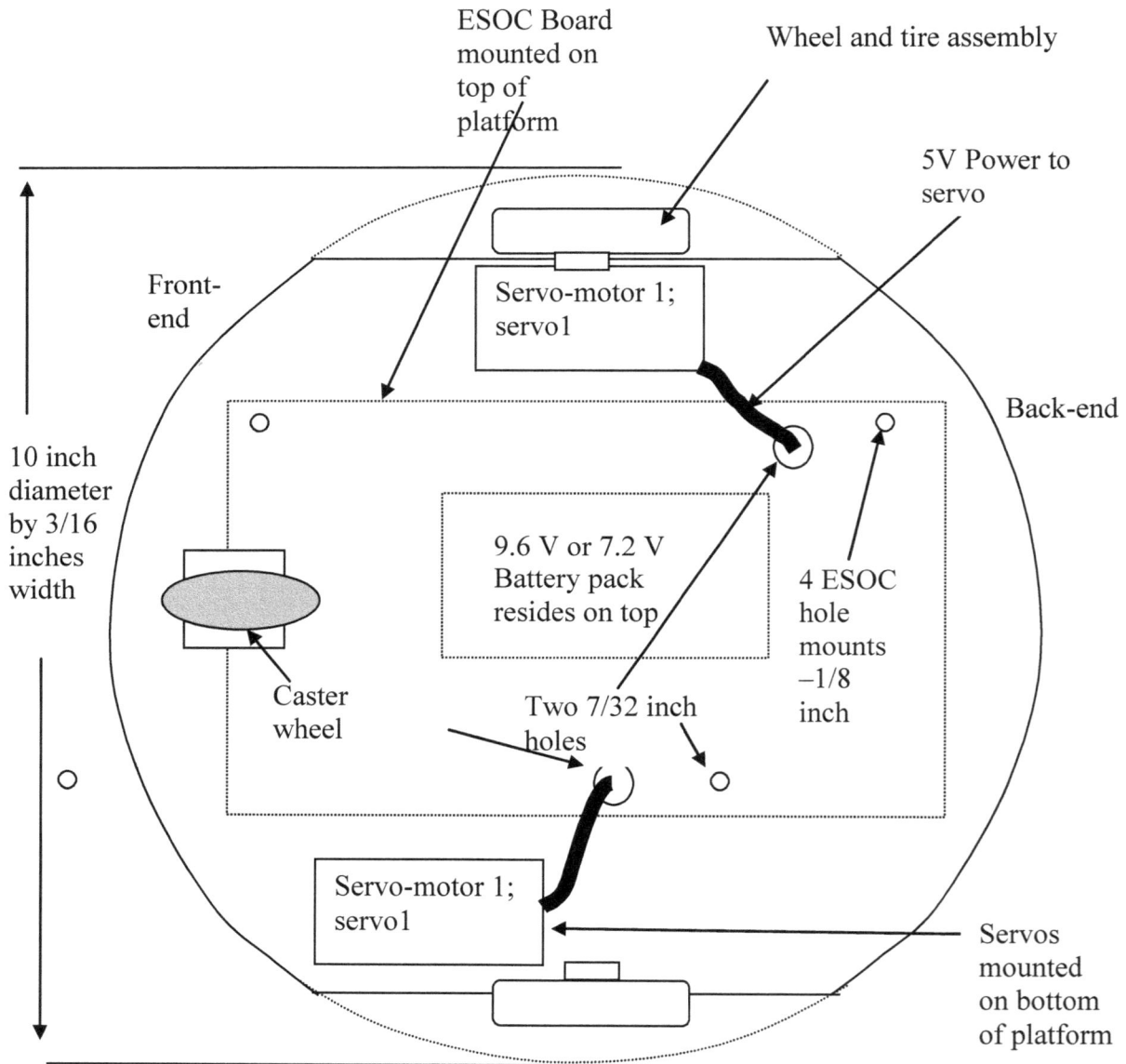

Figure 15.16. The bottom view of the PLD-bot.

15.2.   **Project 3.1 - The PIC-bot - Basic**

This project deals with sending 8 bit commands over a wireless digital transmission link to control a mobile robot, PIC-bot, and its sensors. The remote control and mobile robot consist of a PIC (originally meaning Peripheral Interface Controller) processor made by Microchip Technology. The PIC-bot system uses a WyJen Technologies (www.wyjen.com) transmitter and receiver to perform wireless transmission from the remote control unit to the mobile robot. More details on the PIC microcontroller can be referenced in Huang (2005) and Katzen (2005) in the Reference Section on page 578. Also, the MPLAB C compiler for the Peripheral Interface Controller (PIC) microcontroller can be found at:

http://www.microchip.com/stellent/idcplg?IdcService=SS_GET_PAGE&nodeId=1406&dDocName=en010014

15.2.1. Project 3.1 Basic Requirements

This mobile robot must meet the following requirements:

Requirement 1 (optional): The robot must respond to manual control commands. This is called manual mode. The manual control resides on the robot platform in the form of switches. These commands must include the following: 1.) turn left, 2.) turn right, 3.) go forward, 4.) go backward, and 5.) stop. Can use external 2 to 1 four bit multiplexer for this feature (see Section 4.11.1, on page 100.)

Requirement 2: The receive robot must perform the commands in requirement 1 in remote control mode. In remote control mode a transmitter must send 8 bit commands which cause the robot to perform the commands in requirement 1.

Requirement 3: The project must use a wireless transmitter and receiver device. See WyJen Technologies (www.wyjen.com) transmitter and receiver module or similar devices.

15.2.2. Project 3.1 Description

This project consists of the following main components:

1.   Remote control unit with a wireless transmitter, encoder, and switches.
2.   Mobile robot unit consisting of a wireless receiver and a RX PIC platform.
3.   Mobile robot with decoder circuit.

The mobile robot collectively is referred to as PIC-bot. Figure 15.26 shows the remote control unit. The TX PIC on the remote control unit receives 8 bit command data from a keypad, switches, etc., and converts it to serial data, and then sends it to a transmitter, which sends the commands over a wireless link to the receiver on the mobile robot platform. These commands are then used to control the mobile robot and its sensors. An optional PC can be used to generate the commands.

Figure 15.27 shows the mobile robot. The receiver on the mobile robot demodulates the message from the remote control unit's transmitter to raw data. The raw data is sent to a decoder, then the RX PIC. The decoder decodes the message which allows the PIC to determine the task the robot is commanded to perform.

The remote control unit optionally interfaces to the PC as shown in Figure 15.26. The PC in this case consists of a parallel port driver program (Dhananjay, 1998) and software to create eight bit data to send to the RX PLD platform. The default configuration is to create eight bit data using on board PLD platform switches only. The mobile robot has the circuitry as shown in Figure 15.27.

The main components of the mobile robot are the RX PIC platform, a wireless transceiver, servos, and decoder. The RX PIC platform receives the decoded message from the decoder and responds by sending the appropriate frequencies to the servo motors or sensors. The square waveforms of different frequencies cause the servo motors to left, right, or stop. That is, the waveforms of different frequencies cause the servos to turn clockwise, counter clockwise, or stop. Servos typically use a pulse width modulated (PWM) waveform to cause them to turn clockwise, counter clockwise, or to be neutral (stop). Sending waveforms of different frequencies to a servo motor is seen by the servo motor to be PWM like and thus the motors respond similarly. The waveforms generated by the robot's RX PIC are square waveforms which have a constant pulse width. Square waves of different frequencies are selected by using a multiplexor and a counter in the receiving robot's PIC on the RX PIC platform. On the other hand the PIC18F458 can generate the appropriate PWM signal for each motor. This is a feature of the PIC18F458.

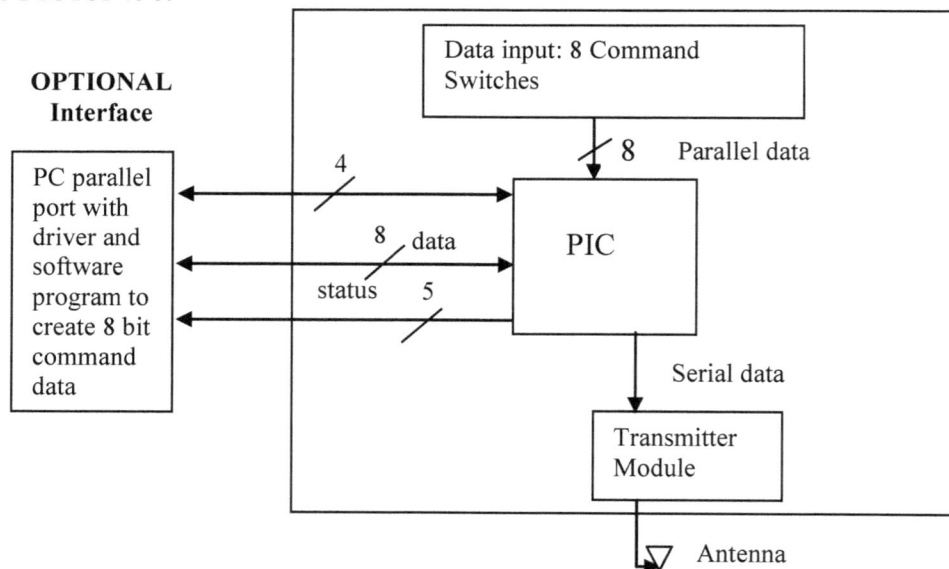

Figure 15.17. Transmit PIC with optional PC parallel port control. The default configuration is to use eight switches to create an eight bit command which is sent to the receive PIC. To control the PIC through a PC parallel port would require parallel port driver software and a corresponding software program to create and send command data to the PIC.

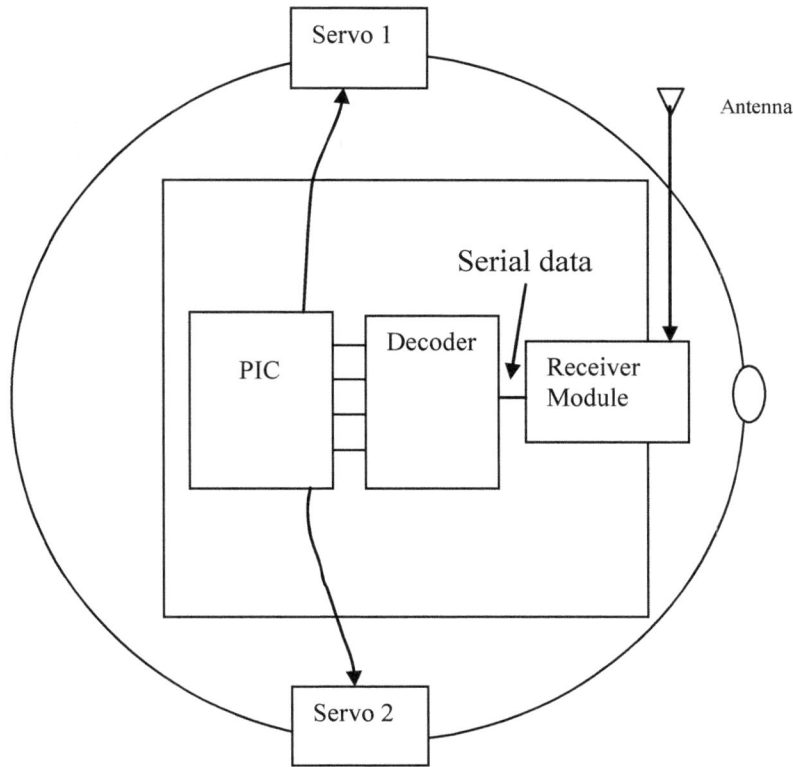

Figure 15.18. Receiving PIC-bot.

15.2.3. Project 3.1 Tentative Development Schedule

The tentative schedule for developing the PIC–bot is shown in Table 15.6. You must try to follow it.

Week	Things to do
Week 1,2,3	Get ideas and teams together
Week 4,5	High-level Layout of robot complete. Devices selected (e.g., sensors, communication, etc.), start getting parts, platform selected, who will be doing what (sign off required)
Week 6,7	Parts Due, platform complete, batteries acquired (or substitute power source) (sign off required)
Week 7-12	Test all circuits, sensors, etc. (sign off required)
Week 7-10	C code written (sign off required)
Week 9-13	Assemble robot and integrate all circuits, etc., working with C code (sign off required)
Week 14	Oral presentations and reports due

Table 15.6. Tentative 15 week schedule.

15.2.4. Project 3.1 Electronic Parts List

The starter electronic part list for Project 3.1 is given in Table 15.12.

Part	Quantity	Ref designator	Description
LED, 276-143C, Radio Shack	2	IFLED1, IFLED2	Infrared light emitting diode
Microchip PIC18F458	2	PIC, IC1	PIC board, Controller
Futaba servos S3003 (Tower TS53J), or High Tech servos HS300	2	servo1, servo2	Servo motors, www.towerhobbies.com
HT12E	1	IC4	Holtek encoder, www.holtek.com can order from www.holmate.com
HT12D	1	IC5	Holtek decoder, www.holtek.com can order from www.holmate.com
Resistor	2	Rosce, Roscd	For encoder and decoder oscillator circuits
Capacitor	6		10pF capacitor
Platform I	1	Platform I	Platform 1 uses two servo motors, a caster wheel, etc. See **Figure 15.16** on page 485
Resistor, 1000 ohms	1	R1	1 Watt resistor
Resistor, 180 ohms	2	R2, R3	1 Watt resistor
Transceiver: GTR315M_M13	2	TXRX	315MHz transmitter/receiver, 3/5V, half duplex transmission, www.wyjen.com
LEDs	4		LEDs for observing outputs
Circuit boards	3	CB1, CB2, and CB3	Circuit board for TX, RX, encoder, and decoder circuits
SPST or DPST switches	8		On/off switches for inputs
1.5V AA Batteries	4	Battery	Two for transmitter and two for receiver supply voltages
MAX8510	2	IC3	Optional 3.3 V regulator, One for transmitter and One for receiver supply voltages
TPS7133QP (Digikey)	3	IC3	Optional 3.3 V regulator, One for transmitter and
20 gauge solid wire, Radio Shack part 278-1222			Black, red, green. Wire for connecting various electronic parts and components
20 gauge solid wire			Need other colors such as white, blue, yellow, etc. Wire for connecting various electronic parts and components

Table 15.7. Electronic starter parts list for Project 3.1. For more details on where to order parts see Appendix D on page 577.

Part	Quantity	Ref designator	Description
44pF		C3, C4	Capacitors
4MHz		X	Ceramic resonator, 4.0 MHz Murata Erie CSA4.00MG ±0.5%, XT Mode Crystal

Table 15.7. Continued.

15.2.5. Project 3.1 – PIC-bot Procedure using C and an Embedded Microcontroller – the PIC18F458

For the steps below save all C files in the same directory. Preferably a directory called **cwork**. It is assumed that the reader is familiar with how to use the Microchip PIC18F458 in this project. Many examples of the PIC microcontroller programming and it applications can be found in (Huang, 2005) and (Katzen, 2005). Follow the steps below to complete this project:

1. Acquire all the necessary parts for constructing the electronic portion of project 3.1 Most of these parts are listed in Table 15.7. Also see Section Appendix D on page 577 for more information on purchasing the various parts.

2. Setup and test the serial to parallel/parallel to serial data conversion circuits with the transceivers. This can be done two ways here: 1.) encoder and decoder 2.) C code using a PIC18 UART port. In this project, the Holtek encoder (HT12E) and decoder (HT12D) are used to perform the parallel to serial and serial to parallel data conversion, respectively. The encoder will connect to the transmit transceiver (transmitter mode) input data pin as shown in Figure 11.16 (see page 430). The decoder will connect to the receive transceiver (receiver mode) output data pin as shown in Figure 11.17 (see page 431). Verify the digital serial data at the PIC's output port using an oscilloscope as indicated in Figure 15.29 or in Section 11.6.2 on page 429. The PIC that resides on the robot's platform will be called the RX PIC. The C code to control the servos is given below in Listing 15.1. Compile and program it into the Microchip PIC18F458.

3. Modify the servo motors for continuous rotation by a.) clipping or removing the potentiometer, b.) clipping away the stopper.

4. It is very important that you record your results using Table 15.8 and Table 15.9 in this section for later referral! Test the servo motors using the PICs clocking resources. Record the outputs and their respective frequencies that cause the servo to turn clockwise (forward), counterclockwise (reverse), and stop. This information may be unique for each servo and must be recorded separately for each servo motor. Use Table 15.8 and Table 15.9 below to record the effect of each output from the counter on the servos. Put a check mark in the column labeled slow speed, medium speed, and fast speed. Slow speed is when the servo turns very slow. Fast speed is the highest speed the servo will go. Medium speed is somewhere in between. You have to use your best judgment on what is a slow, medium, or fast speed.

```
#include <p18f458.h>
#include <timers.h>


#pragma config WDT = OFF
#pragma config OSC = HS
void forward(void);
void reverse(void);
void left(void);
void right(void);
void stop(void);

unsigned char control;  //value for reading timer3

float high;
char tempH;
char tempL;

void main(void)
{

TRISB = 0b11000101;     //sets port b for input
TRISC = 0x00;
TRISD = 0b11101111;
while(1)

{
PORTB = 0xFF;
tempH = 0;
tempL = 0;

if(PORTBbits.RB7 == 0)  //checks if rb7 is thrown
{
    while(PORTBbits.RB7 == 0)
    {
    forward();  //calls function until portb = 0
    }
}
```

Listing 15.1. PIC C code for servo control.

```
if(PORTBbits.RB6 == 0)  //checks if rb6 is thrown
{
      while(PORTBbits.RB6 == 0)
      {
      reverse();  //calls function until portb = 0
      }
}

if(PORTBbits.RB4 == 0)  //checks if rb1 is thrown
{
      while(PORTBbits.RB4 == 0)
      {
      stop();     //calls function until portb = 0
      }
}

if(PORTBbits.RB2 == 0)//this pin doesnt work - use
                      //one of D's
{
      while(PORTBbits.RB2 == 0)
      {
      left();     //calls function until portb = 0
      }
}

if(PORTBbits.RB0 == 0)  //checks if rb1 is thrown
      {
      while(PORTBbits.RB0 == 0)
      {
      right();    //calls function until portb = 0
      }
      }
}

}
////////////////////////////////////////////////
//
//
//unsigned char stoph = 0x03;//
//unsigned char stopl = 0xA9;//
//unsigned char cwh = 0x03; //Values for ranges of motion
//unsigned char cwl = 0x2C; //
//unsigned char ccwh = 0x04;  //
//unsigned char ccwl = 0x26;  //
////////////////////////////////////////////////
```

Listing 15.1. Continued.

```
void forward(void)
{
TRISC = 0;          //sets rc2 for output
TRISD = 0;
T3CON = 0xB9;       //intitilizes timer3
T1CON = 0xB1;
CCP1CON = 0x08;     //initializes ccp1 pin
ECCP1CON = 0x08;
ECCPR1H = 0x30;
ECCPR1L = 0xD4;
CCPR1H = 0x30;      //high byte value for count
CCPR1L = 0xD4;
TMR1H = 0;
TMR1L = 0;   //low byte value for count
TMR3H = 0;          //sets timer3 to zero
TMR3L = 0;
PIR1bits.CCP1IF = 0;
PIR2bits.ECCP1IF = 0;   //clear the interrupt flag
while(PIR1bits.CCP1IF == 0)   //loop until a match of ccpr1 and tmr3
registers
{
control = TMR3L;
control = TMR1L;
}
PIR1bits.CCP1IF = 0;     //clear flag that was just thrown
PIR2bits.ECCP1IF = 0;
CCP1CON = 0x09;                 //reinitialize ccp1 pin
ECCP1CON = 0x09;
ECCPR1H = 0x03;
ECCPR1L = 0x2C;
CCPR1H = 0x04;
CCPR1L = 0x26;
TMR1H = 0;
TMR1L = 0;
TMR3H = tempH;      //load offset of time - if any - default = 0
TMR3L = tempL;
PIR1bits.CCP1IF = 0;     //clear the interrupt flag
PIR2bits.ECCP1IF = 0;
while(PIR2bits.ECCP1IF == 0)  //loop until a match of
                             // ccpr1 and tmr3 registers
{
control = TMR3L;
control = TMR1L;
}
TMR3H = 0;
TMR3L = 0;
while(PIR1bits.CCP1IF == 0)
{
    control = TMR3L;
    control = TMR1L;
}
tempH = 0;   //store offset of extra time
tempL = 0;
}
```

Listing 15.1. Continued.

```
void reverse(void)
{
TRISC = 0;          //sets rc2 for output
TRISD = 0;
T3CON = 0xB9;       //intitilizes timer3
T1CON = 0xB1;
CCP1CON = 0x08;     //initializes ccp1 pin
ECCP1CON = 0x08;
ECCPR1H = 0x30;
ECCPR1L = 0xD4;
CCPR1H = 0x30;      //high byte value for count
CCPR1L = 0xD4;
TMR1H = 0;
TMR1L = 0;   //low byte value for count
TMR3H = 0;          //sets timer3 to zero
TMR3L = 0;
PIR1bits.CCP1IF = 0;
PIR2bits.ECCP1IF = 0;    //clear the interrupt flag
while(PIR1bits.CCP1IF == 0)   //loop until a match
                          //of ccpr1 and tmr3 registers
{
        control = TMR3L;
        control = TMR1L;
}
PIR1bits.CCP1IF = 0;     //clear flag that was just thrown
PIR2bits.ECCP1IF = 0;
CCP1CON = 0x09;              //reinitialize ccp1 pin
ECCP1CON = 0x09;
ECCPR1H = 0x04;
ECCPR1L = 0x26;
CCPR1H = 0x03;
CCPR1L = 0x2C;
TMR1H = tempH;
TMR1L = tempL;
TMR3H = 0;   //load offset of time - if any - default = 0
TMR3L = 0;
PIR1bits.CCP1IF = 0;     //clear the interrupt flag
PIR2bits.ECCP1IF = 0;
while(PIR1bits.CCP1IF == 0)   //loop until a match
                          //of ccpr1 and tmr3 registers
{
        control = TMR3L;
        control = TMR1L;
}
TMR1H = 0;
TMR1L = 0;
while(PIR2bits.ECCP1IF == 0)
{
        control = TMR3L;
        control = TMR1L;
}
tempH = 0;   //store offset of extra time
tempL = 0;
}
```

Listing 15.1. Continued.

```
void left(void)
{
TRISC = 0;           //sets rc2 for output
TRISD = 0;
T3CON = 0xB9;        //intitilizes timer3
T1CON = 0xB1;
CCP1CON = 0x08;      //initializes ccp1 pin
ECCP1CON = 0x08;
ECCPR1H = 0x30;
ECCPR1L = 0xD4;
CCPR1H = 0x30;       //high byte value for count
CCPR1L = 0xD4;
TMR1H = 0;
TMR1L = 0; · //low byte value for count
TMR3H = 0;           //sets timer3 to zero
TMR3L = 0;
PIR1bits.CCP1IF = 0;
PIR2bits.ECCP1IF = 0;   //clear the interrupt flag
while(PIR1bits.CCP1IF == 0)   //loop until a match
                             //of ccpr1 and tmr3 registers
{
     control = TMR3L;
     control = TMR1L;
}
PIR1bits.CCP1IF = 0;      //clear flag that was just thrown
PIR2bits.ECCP1IF = 0;
CCP1CON = 0x09;              //reinitialize ccp1 pin
ECCP1CON = 0x09;
ECCPR1H = 0x03;
ECCPR1L = 0x2C;
CCPR1H = 0x03;
CCPR1L = 0x2C;
TMR1H = 0;
TMR1L = 0;
TMR3H = 0;   //load offset of time - if any - default = 0
TMR3L = 0;
PIR1bits.CCP1IF = 0;      //clear the interrupt flag
PIR2bits.ECCP1IF = 0;
while(PIR1bits.CCP1IF == 0)   //loop until a match
                             //of ccpr1 and tmr3 registers
{
     control = TMR3L;
     control = TMR1L;
}


}
```

Listing 15.1. Continued.

```
void right(void)
{
TRISC = 0;           //sets rc2 for output
TRISD = 0;
T3CON = 0xB9;        //intitilizes timer3
T1CON = 0xB1;
CCP1CON = 0x08;      //initializes ccp1 pin
ECCP1CON = 0x08;
ECCPR1H = 0x30;
ECCPR1L = 0xD4;
CCPR1H = 0x30;       //high byte value for count
CCPR1L = 0xD4;
TMR1H = 0;
TMR1L = 0;   //low byte value for count
TMR3H = 0;           //sets timer3 to zero
TMR3L = 0;
PIR1bits.CCP1IF = 0;
PIR2bits.ECCP1IF = 0;    //clear the interrupt flag
while(PIR1bits.CCP1IF == 0)   //loop until a match
                             //of ccpr1 and tmr3 registers
{
     control = TMR3L;
     control = TMR1L;
}
PIR1bits.CCP1IF = 0;     //clear flag that was just thrown
PIR2bits.ECCP1IF = 0;
CCP1CON = 0x09;              //reinitialize ccp1 pin
ECCP1CON = 0x09;
ECCPR1H = 0x04;
ECCPR1L = 0x26;
CCPR1H = 0x04;
CCPR1L = 0x26;
TMR1H = 0;
TMR1L = 0;
TMR3H = 0;   //load offset of time - if any - default = 0
TMR3L = 0;
PIR1bits.CCP1IF = 0;     //clear the interrupt flag
PIR2bits.ECCP1IF = 0;
while(PIR1bits.CCP1IF == 0)   //loop until a match
                             //of ccpr1 and tmr3 registers
{
     control = TMR3L;
     control = TMR1L;
}


}
```

Listing 15.1. Continued.

```
void stop(void)
{
TRISC = 0;          //sets rc2 for output
TRISD = 0;
T3CON = 0xB9;       //intitilizes timer3
T1CON = 0xB1;
CCP1CON = 0x08;     //initializes ccp1 pin
ECCP1CON = 0x08;
ECCPR1H = 0x30;
ECCPR1L = 0xD4;
CCPR1H = 0x30;      //high byte value for count
CCPR1L = 0xD4;
TMR1H = 0;
TMR1L = 0;   //low byte value for count
TMR3H = 0;          //sets timer3 to zero
TMR3L = 0;
PIR1bits.CCP1IF = 0;
PIR2bits.ECCP1IF = 0;    //clear the interrupt flag
while(PIR1bits.CCP1IF == 0)    //loop until a match
                                //of ccpr1 and tmr3 registers
{
     control = TMR3L;
     control = TMR1L;
}
PIR1bits.CCP1IF = 0;       //clear flag that was just thrown
PIR2bits.ECCP1IF = 0;
CCP1CON = 0x09;            //reinitialize ccp1 pin
ECCP1CON = 0x09;
ECCPR1H = 0x03;
ECCPR1L = 0xA9;
CCPR1H = 0x03;
CCPR1L = 0xA9;
TMR1H = 0;
TMR1L = 0;
TMR3H = 0;   //load offset of time - if any - default = 0
TMR3L = 0;
PIR1bits.CCP1IF = 0;       //clear the interrupt flag
PIR2bits.ECCP1IF = 0;
while(PIR2bits.ECCP1IF == 0)  //loop until a match of ccpr1
                                // and tmr3 registers
{
     control = TMR3L;
     control = TMR1L;
}


}
```

Listing 15.1. Continued.

Left Servo Clock Frequency (Speed Characteristics and rotation direction)						
Counter Output	Frequency	Slow Speed	Medium Speed	Fast Speed	Forward	Reverse
Q10						
Q11						
Q12						
Q13						
Q14						
Q15						

Table 15.8. Left Servo clock frequency verses speed and rotation direction.

Right Servo Clock Frequency vs. (Speed Characteristics and rotation direction)						
Counter Output	Frequency	Slow Speed	Medium Speed	Fast Speed	Forward	Reverse
Q10						
Q11						
Q12						
Q13						
Q14						
Q15						

Table 15.9. Right Servo clock frequency verses speed and rotation direction.

Before you begin recording the servo clock frequency verse speed and rotation direction in Table 15.8 and Table 15.9 the servos should be identified by a label as the right servo and the left servo. The servos should be positioned as shown in Figure 15.19 The shaft and tire of the servos should be pointed in opposite directions as shown.

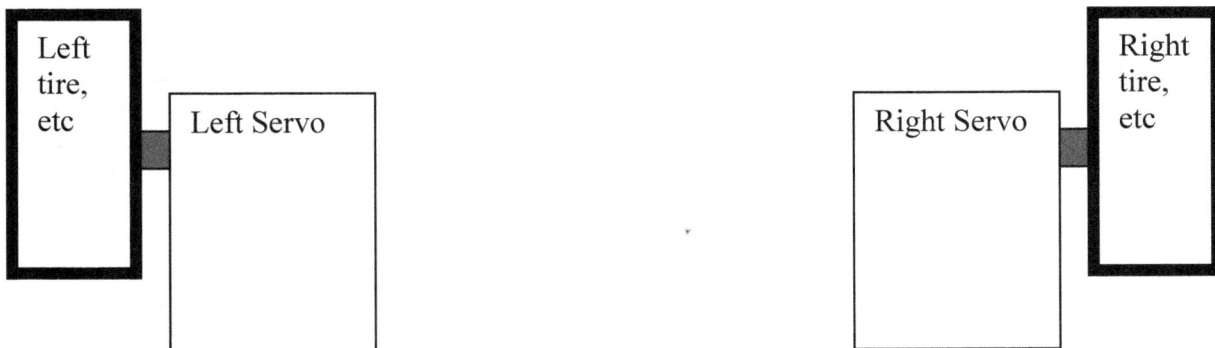

Figure 15.19. Position of left servo relative to right servo.

It should be noted that clockwise is forward and counter clockwise is reverse. While each servo is positioned differently, the robot will go forward when both servos are going forward. The robot will go backward or reverse when both servos are in reverse. The robot will turn left when the left servo is stopped and the right servo is in forward and the robot will turn right when the right servo is stopped and the left servo is forward. For the robot to turn left in place the left servo needs to be in reverse and the right servo need to be in forward and for the robot to turn right in place the left servo needs to be in forward while the right servo needs to be in reverse. Keep in mind that when the robot is moving forward the left servos and the right servo are actually going in opposite directions. This is because they are positioned on opposite sides of the robot as well as faced in opposite directions.

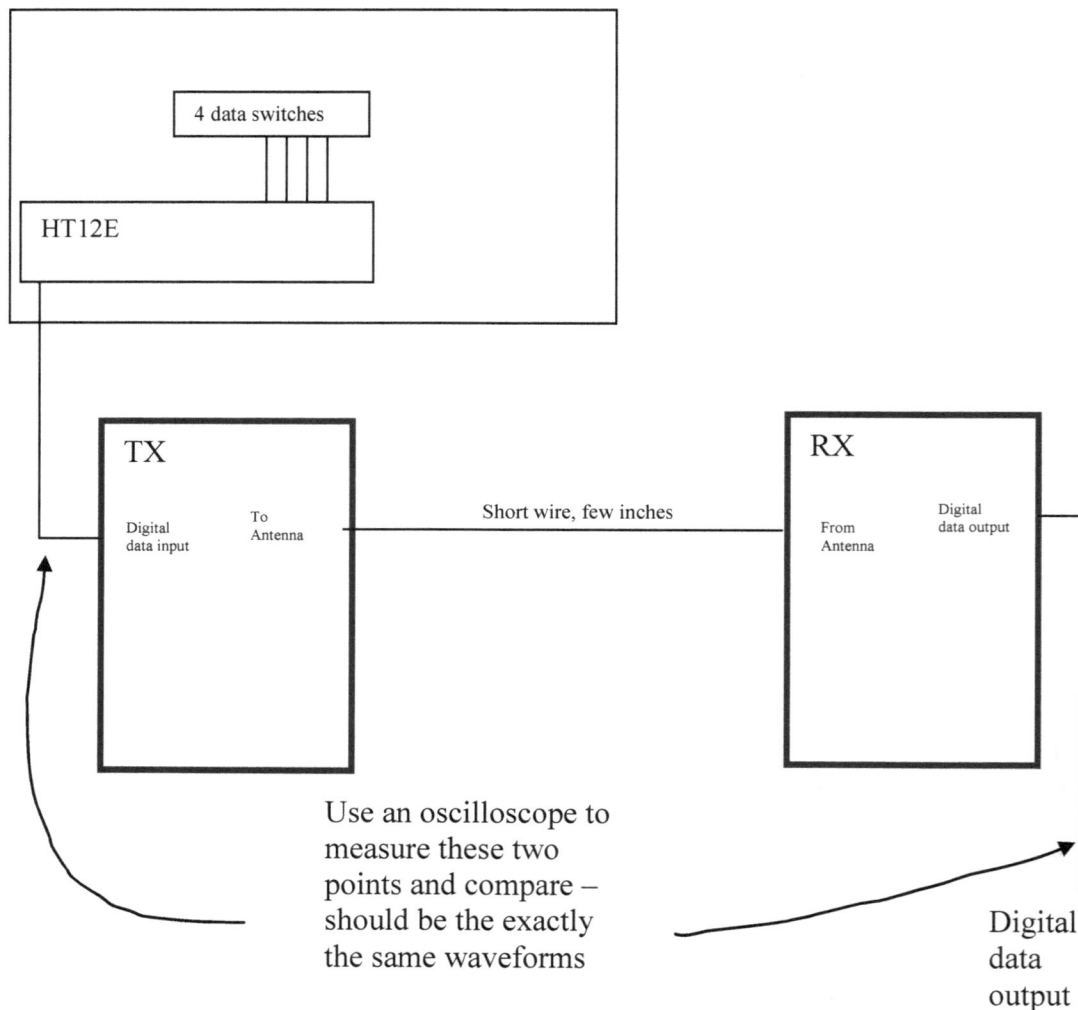

Figure 15.20. Test circuit for TX and RX.

5. If step 5 is successful perform the same procedure using the wireless link as shown in Figure 15.32. That is, include the TX and RX in the test. The transmitter and receiver circuits must be free of noise to operate optimally. One method of reducing the noise effects is to use two AA batteries in series for the transmitter 3.3V source and a separate one for the receiver 3.3V source. The two AA batteries in series produce a 3V source which is within the supply range of the WyJen transmitter and receiver. The transmitter battery ground should share the same ground as the encoder ground and the receiver battery ground should share the same ground as the RX PIC ground. Also locate the receiver away from the servo motors.

6. Acquire all the hardware components of the robot to construct the body. A starting list is given Table 15.10.

7. Construct the hardware platform for the robot as shown in Figure 15.21, Figure 15.22, Figure 15.23, Figure 15.24, and Figure 15.34. Test and complete the robot as shown. The RX PIC platform needs to stand above the plexiglass platform. This is done here with 1 and ½ inch standoffs. The platform is about 10 inches in diameter. The weight of the RX PIC board along with the other components needs to be evenly distributed to prevent the robot from easily turning over.

8. Incorporate the battery supply into the project. The receive robot may now feed off the battery supply. The battery will be drained in a few minutes demonstration time. Make sure you have a spare battery pack or a fast charger and a power outlet nearby for uninterrupted demonstrations.

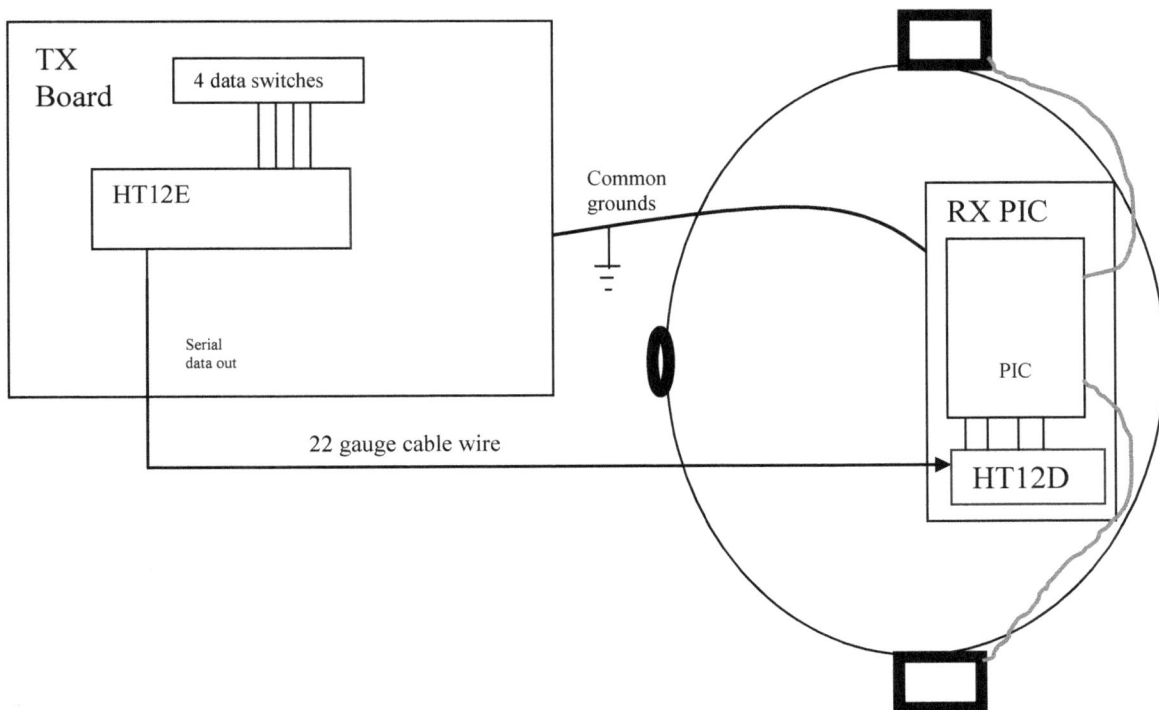

Figure 15.21. Test circuit for TX PIC, RX PIC, left and right servos, and robot platform.

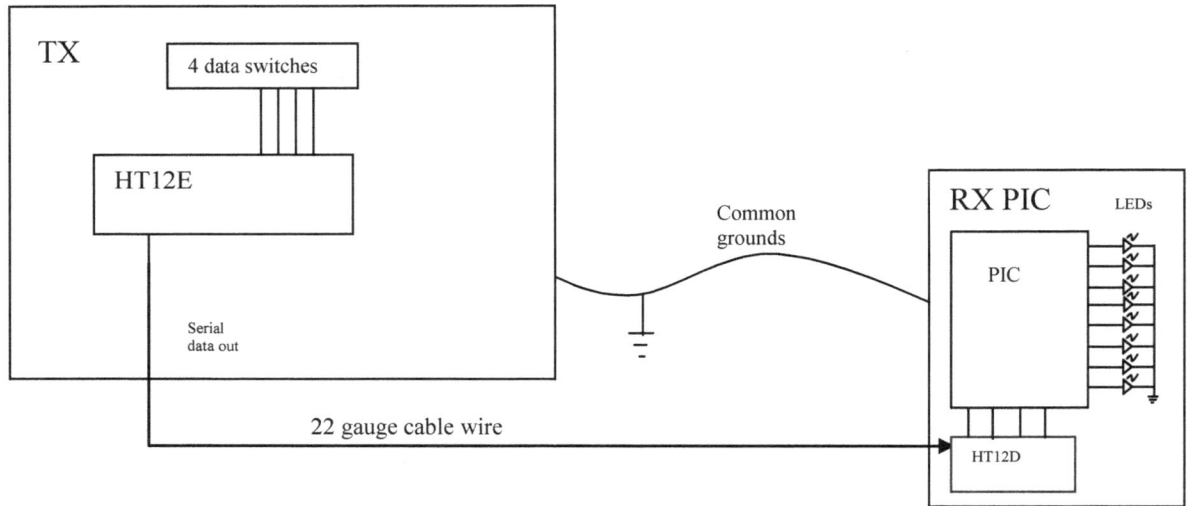

Figure 15.22. Test circuit for TX PIC and RX PIC.

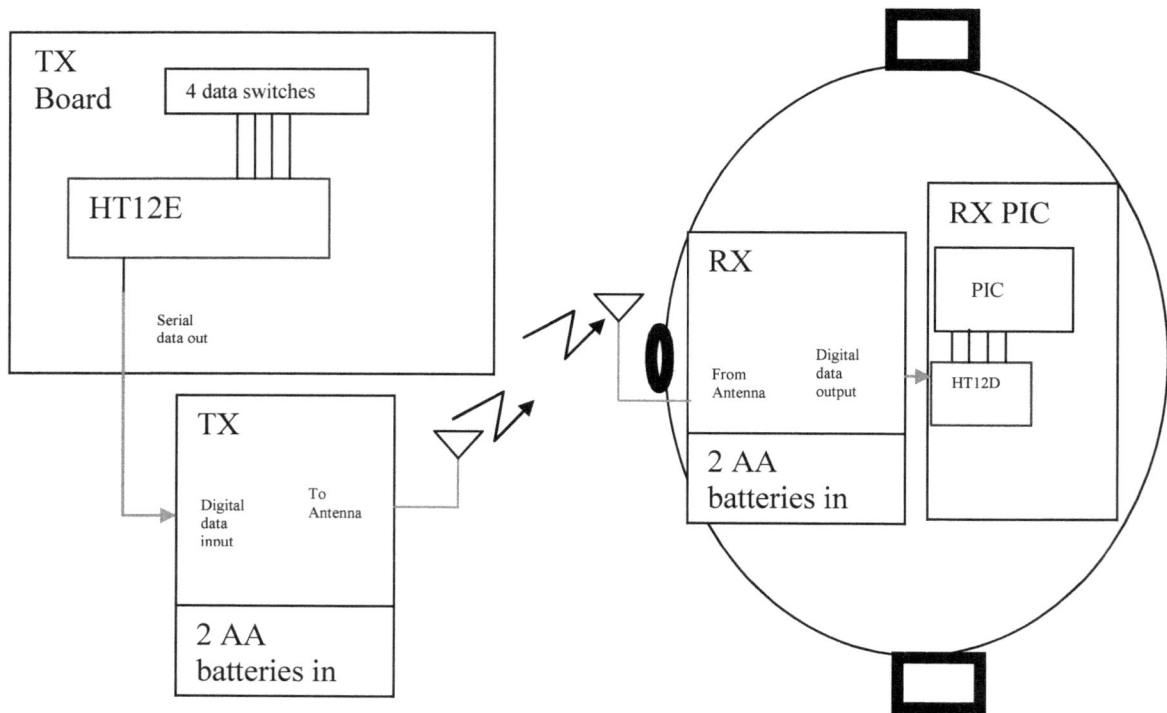

Figure 15.23. Test circuit for TX PIC, RX PIC, TX, RX, left and right servos, and robot platform.

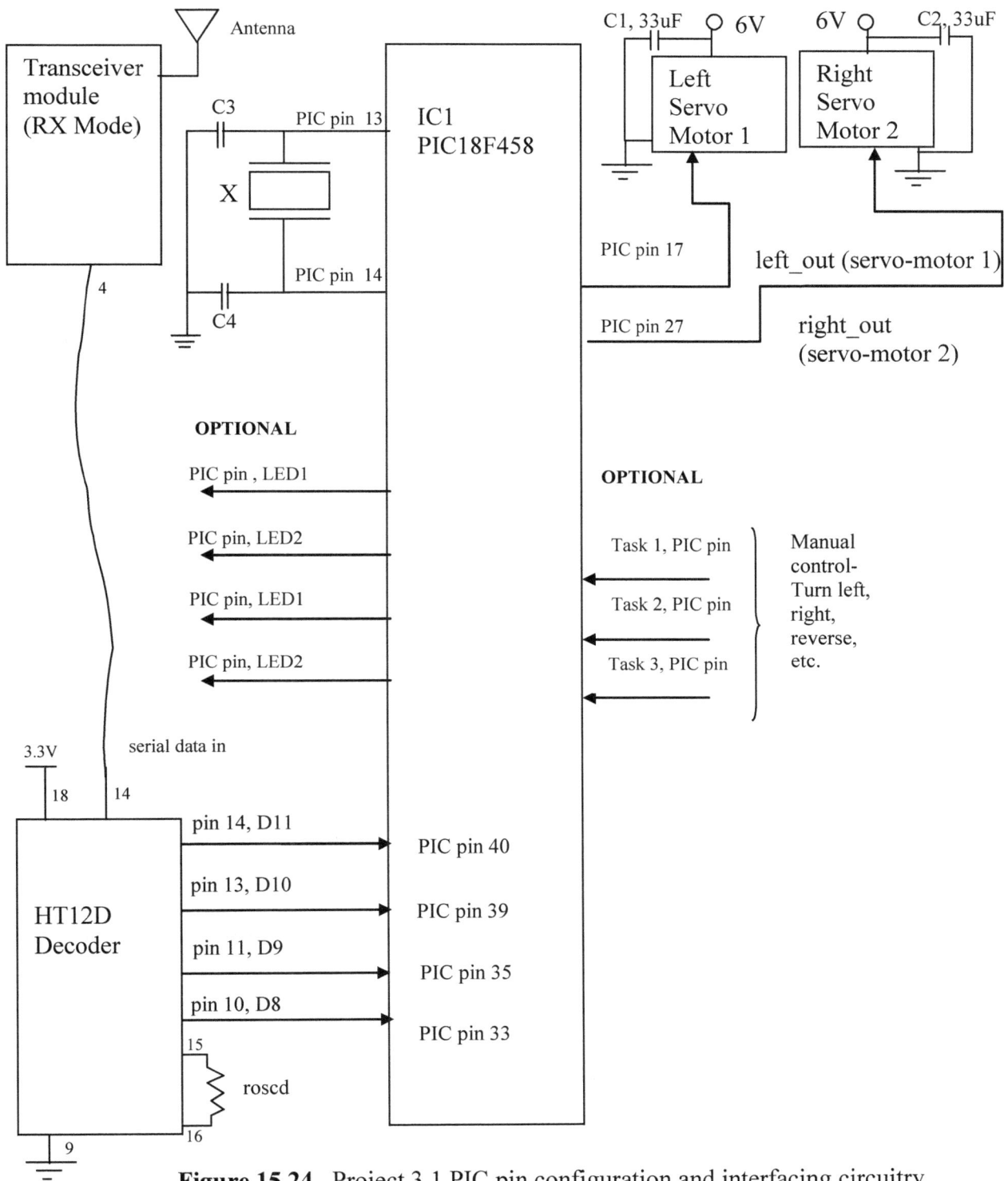

Figure 15.24. Project 3.1 PIC pin configuration and interfacing circuitry.

15.2.6. Project 3.1 Hardware Parts List

The hardware parts list for Project 3.1 is given in Table 15.15. These parts can be obtained from the references listed in Section Appendix D on page 577.

Part	Quantity	Ref designator	Description
High Tech servos - HS300 (Futaba S3003, or Tower TS53J)	2	servo1, servo2	servomotors
Battery supply, radio shack, 9.6V, 1000mA, 230-0342	2	Vcc	9.6 Volt Nickel Cadmium R/C car battery pack and charger for TX ESCO and RX ESOC, Radio Shack
DC power connector, M type	2		For connecting battery packs to ESOCs, Radio Shack
Caster wheel	1		Front wheel, Menards, Lowes, Home depot, etc.
Tire	2		Rubber tire, Tower Hobbies
Wheel	2		Rim for rubber tire, Tower Hobbies
Round Platform	1	platform I	Round platform of diameter 11 inches (and thickness 3/16 inches) cut from plexiglass, etc, etc
standoffs	4	ST1, ST2, ST3, ST4	1 and ½ inch standoffs, Digikey, Radio Shack, etc
Screws, 844-712 is Menards part #	?		Machine screws, size 4-40x ½, 30 pieces, can be used to mount caster wheel and servo if the 4-40x ¼ is not long enough.
Screws, 844-710, 844-712 is Menards part #	About 16		Machine screws, size 4-40x ¼, 34 pieces. 8 is needed for mounting the RX ESOC, 4 for the caster wheel, 2 for each servo
Nuts, 846-644 is Menards part #	About 12		Machine nuts, size 4-40, 28 pieces. 4 for the caster wheel, 4 for each servo
Washers, 154-306 is Menards part #	About 12		Washers, No. 6, 100 pieces. 4 is needed for mounting the bottom of the RX ESOC, 4 for the caster wheel, 4 for each servo
Velcro			4 inch by 2 inches, 2 sets

Table 15.10. Project 3.1 starter hardware parts list. For more details on where to order parts see Appendix D on page 577.

15.2.7. Project 3.1 Assembly of the PIC-bot: Platform I

The bottom view of the PIC-bot is shown in Figure 15.34. The PIC board will rest on the top of the platform. The servomotors will be mounted on the bottom and attached to the wheels as shown. The control Cables run from the servomotors to the PIC board located on the opposite side as shown. The sensors should be mounted on the top with all outputs connected to the controller (PIC board). The switch and power cables have separate holes in the platform. The power switch is connected to the hole in the platform.

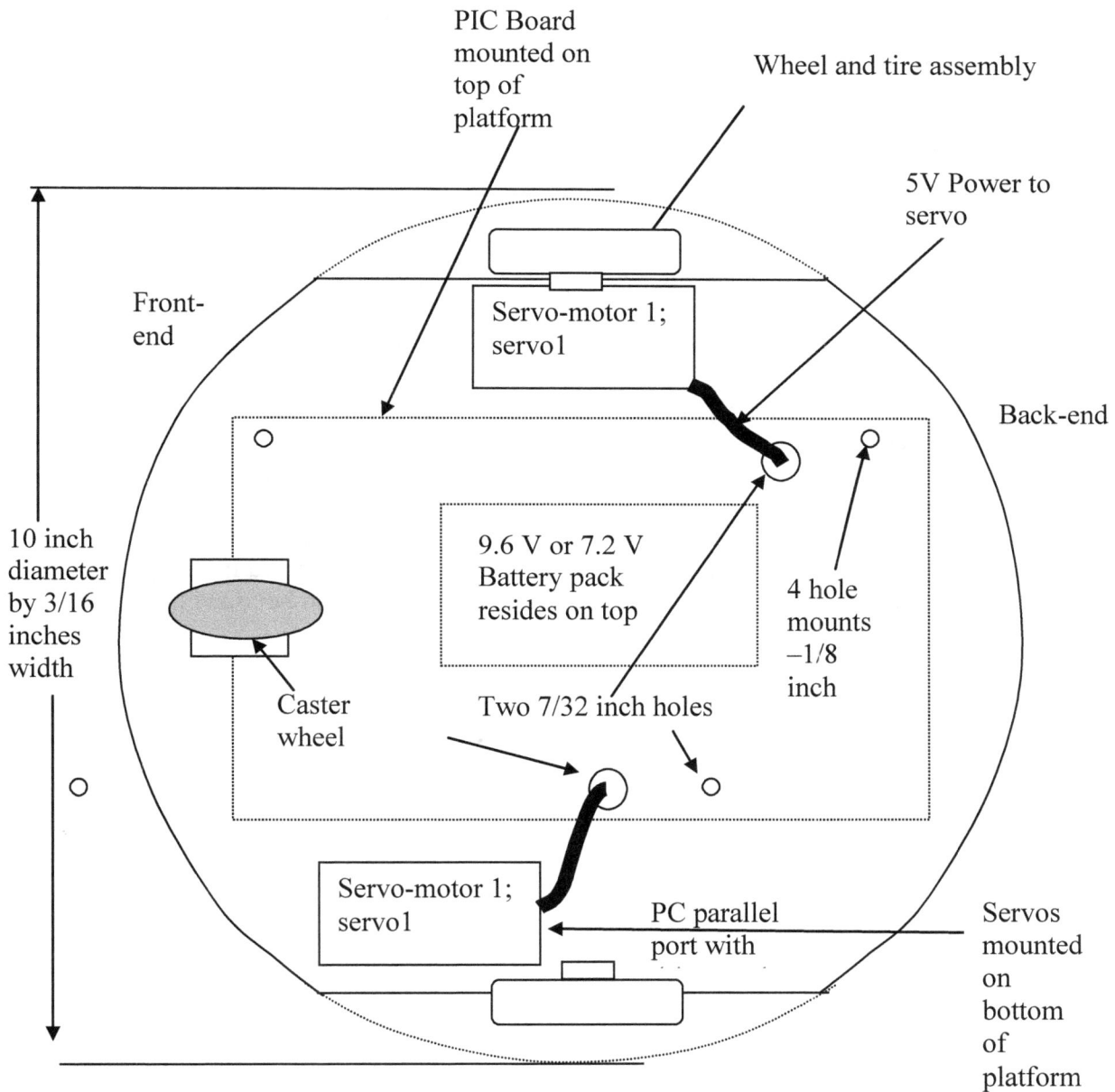

PIC Board
mounted on
top of
platform

Wheel and tire assembly

5V Power to
servo

Front-
end

Servo-motor 1;
servo1

Back-end

10 inch
diameter
by 3/16
inches
width

9.6 V or 7.2 V
Battery pack
resides on top

4 hole
mounts
−1/8
inch

Caster
wheel

Two 7/32 inch holes

Servo-motor 1;
servo1

PC parallel
port with

Servos
mounted
on
bottom
of
platform

Figure 15.25. The bottom view of the PIC bot.

15.3.   Project 3.1 - The PIC-bot – Multi Sensor

This project deals with sending 8 bit commands over a wireless digital transmission link to control a mobile robot and its sensors: compass and your choice of another sensor. The description, task, and procedure of the project are discussed next. It uses a WyJen Technologies (www.wyjen.com) transmitter and receiver to perform the wireless transmission. More details on the PIC microcontroller can be found in Huang (2005) and Katzen (2005) in the Reference Section on page 578.

15.3.1. Project 3.1 Basic Requirements

This mobile robot must meet the following requirements:

Requirement 1: The robot must respond to manual control commands. This is called manual mode. The manual control resides on the robot platform in the form of switches. These commands must include the following: 1.) turn left, 2.) turn right, 3.) go forward, 4.) go backward, and 5.) stop.

Requirement 2: The receive robot must perform the commands in requirement 1 in remote control mode. In remote control mode a transmitter must send 8 bit commands which cause the robot to perform the commands in requirement 1.

Requirement 3: The manual and remote control modes must select between sensor mode 1 or sensor mode 2, etc. If sensor 1 is a compass sensor, then if sensor mode 1 is selected the robot responds to the compass sensor. The robot must then go in the direction specified by the manual control (north, south, east, west, etc.). For example, if manual control specifies go in the direction north, the robot must turn until the north heading is found. When found the robot goes forward in the specified direction. More detailed information about this requirement can be found in section 9.10 on page 376.

Requirement 4: The project must use a wireless transmitter and receiver device. See WyJen Technologies (www.wyjen.com) transmitter, receiver, or transceiver module or similar devices.

15.3.2. Project 3.2 Description

This project consists of the following main components:

4. Remote control unit with a wireless transmitter and TX PIC platform.
5. Mobile robot unit consisting of a wireless receiver and a RX PIC platform.
6. Mobile robot with sensory circuits, e.g., compass sensor, light tracker, etc.

The mobile robot collectively is referred to as PIC-bot.

Figure 15.26 shows the remote control unit. The TX PIC on the remote control unit receives 8 bit command data from a keypad, switches, etc., and converts it to serial data, and then sends it to a transmitter, which sends the commands over a wireless link to the receiver on the mobile robot platform. These commands are then used to control the mobile robot and its sensors. An optional PC can be used to generate the commands.

Figure 15.27 shows the mobile robot. The receiver on the mobile robot demodulates the message from the remote control unit's transmitter to raw data. The raw data is sent to the RX PIC platform also located on the robot. The RX PIC platform decodes the message and determines the task the robot is commanded to perform.

The TX PLD platform on the remote control unit optionally interfaces to the PC as shown in Figure 15.26. The PC in this case consists of a parallel port driver program (Dhananjay, 1998) and software to create eight bit data to send to the RX PLD platform. The default configuration is to create eight bit data using on board PLD platform switches only. The mobile robot has the circuitry as shown in Figure 15.27.

The main components of the mobile robot are the RX PIC platform, a wireless receiver, servos, and sensory circuits such as a compass, etc. The RX PIC platform receives the asynchronous serial digital message from the wireless receiver, decodes it, and responds by sending the appropriate frequencies to the servo motors or sensors. The square waveforms of different frequencies cause the servo motors to left, right, or stop. That is, the waveforms of different frequencies cause the servos to turn clockwise, counter clockwise, or stop (refer to the C code in project 3.1). Servos typically use a pulse width modulated (PWM) waveform to cause them to turn clockwise, counter clockwise, or to be neutral (stop). Sending waveforms of different frequencies to a servo motor is seen by the servo motor to be PWM like and thus the motors respond similarly. The waveforms generated by the robot's RX PIC platform are square waveforms which have a constant pulse width. Square waves of different frequencies are selected by using a multiplexor and a counter in the receiving robot's PIC on the RX PIC platform. On the other hand the PIC18F458 can generate the appropriate PWM signal for each motor. This is a feature of the PIC18F458.

A compass sensor can also be included to give the robot a sense of direction. The compass sensor has four digital outputs that can be interfaced directly to the receiving robot PIC platform.

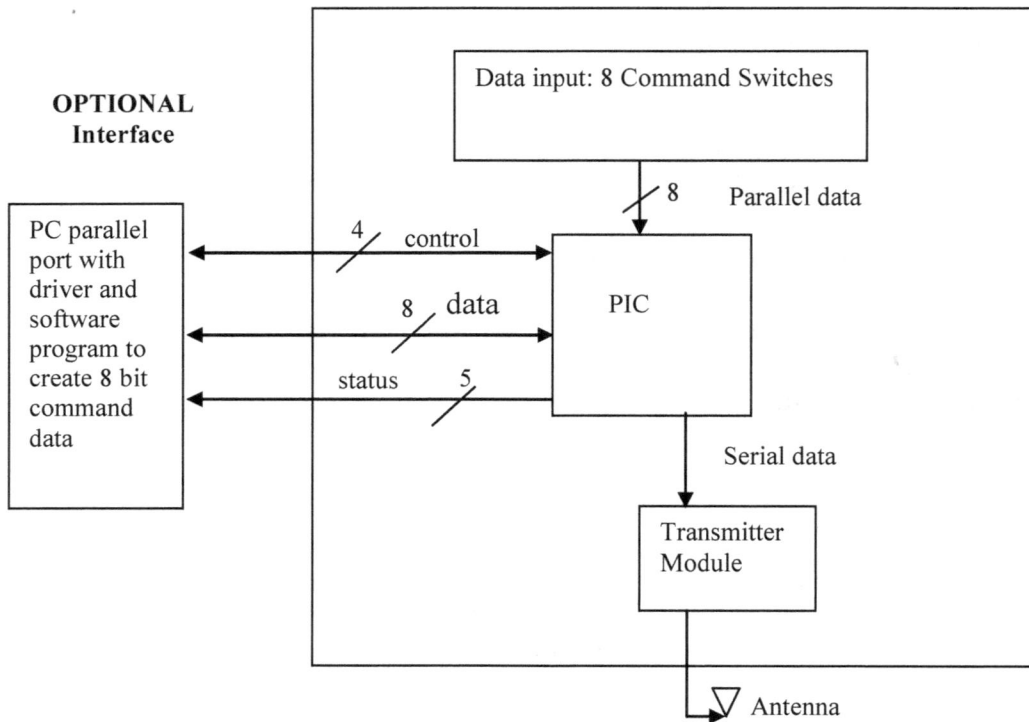

Figure 15.26. Transmit PIC with option PC parallel port control. The default configuration is to use eight switches to create an eight bit command which is sent to the receive PIC. To control the PIC through a PC parallel port would require parallel port driver software and a corresponding software program to create and send command data to the PIC.

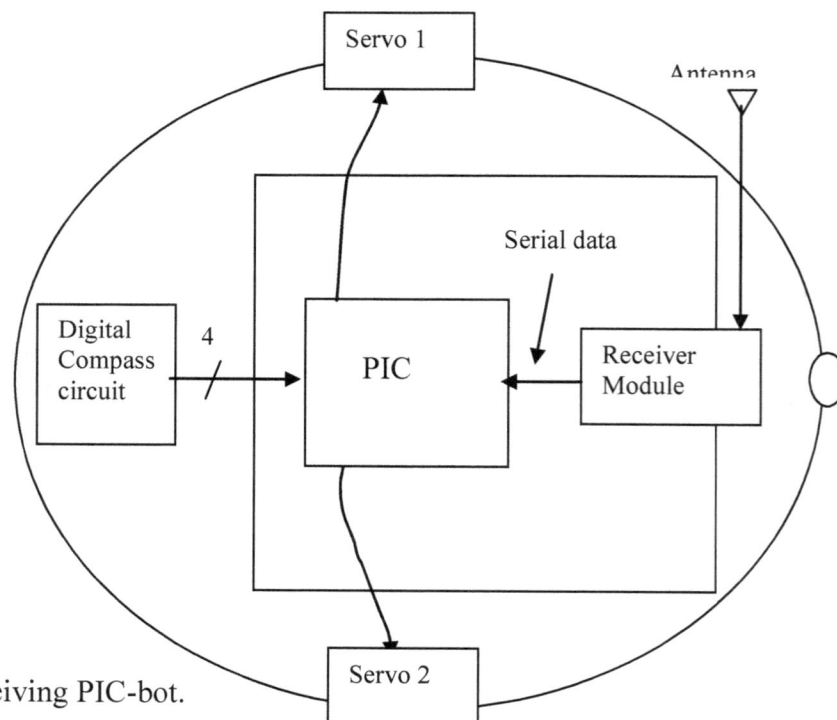

Figure 15.27. Receiving PIC-bot.

15.3.3. Project 3.2 Tentative Development Schedule

The tentative schedule for developing the PIC –bot is shown in Table 15.11. You must try to follow it.

Week	Things to do
Week 1,2,3	Get ideas and teams together
Week 4,5	High-level Layout of robot complete. Devices selected (e.g., sensors, communication, etc.), start getting parts, platform selected, who will be doing what (sign off required)
Week 6,7	Parts Due, platform complete, batteries acquired (or substitute power source) (sign off required)
Week 7-12	Test all circuits, sensors, etc. (sign off required)
Week 7-10	C code written (sign off required)
Week 9-13	Assemble robot and integrate all circuits, sensors, etc., working with C code (sign off required)
Week 14	Oral presentations and reports due

Table 15.11. Tentative 15 week schedule.

15.3.4. Project 3.2 Electronic Parts List

The starter electronic part list for Project 3.2 is given in Table 15.12.

Part	Quantity	Ref designator	Description
LED, 276-143C, Radio Shack	2	IFLED1, IFLED2	Infrared light emitting diode
Microchip PIC18F458	2	PIC, IC1	PIC board, Controller
Magnetic Compass, Dinsmore model 1490	1	IC2	Magnetic compass, 4 bit, active low digital outputs
Futaba servos S3003 (Tower TS53J), or High Tech servos HS300	2	servo1, servo2	Servo motors, www.towerhobbies.com
Platform I	1	Platform I	Platform 1 uses two servo motors, a caster wheel, etc. See **Figure 15.16** on page 485
Resistor, 1000 ohms	1	R1	1 Watt resistor
Resistor, 180 ohms	2	R2, R3	1 Watt resistor
Transceiver: GTR315M_M13	2	TXRX	315MHz transmitter/receiver, 3/5V, half duplex transmission, www.wyjen.com
Circuit board	3	CB1, CB2, and CB3	Circuit board for TX, RX, and compass circuit.
1.5V AA Batteries	4	Battery	Two for transmitter and two for receiver supply voltages

Table 15.12. Electronic starter parts list for Project 3.1. For more details on where to order parts see Appendix D on page 577.

Part	Quantity	Ref designator	Description
MAX8510	2	IC3	Optional 3.3 V regulator, One for transmitter and One for receiver supply voltages
TPS7133QP (Digikey)	3	IC3	Optional 3.3 V regulator, One for transmitter and
20 gauge solid wire, Radio Shack part 278-1222			Black, red, green. Wire for connecting various electronic parts and components
20 gauge solid wire			Need other colors such as white, blue, yellow, etc. Wire for connecting various electronic parts and components
44pF		C3, C4	Capacitors
4MHz		X	Ceramic resonator, 4.0 MHz Murata Erie CSA4.00MG ±0.5%, XT Mode Crystal

Table 15.12. Continued.

15.3.5. Project 3.2 – PIC-bot Procedure using C and an Embedded Microcontroller – the PIC18F458

For the steps below save all C files in the same directory. Preferably a directory called **cwork**. Many examples of the PIC microcontroller programming and it applications can be found in (Huang, 2005) and (Katzen, 2005). Follow the steps below to complete this project:

1. Acquire all the necessary parts for constructing the electronic portion of project 3. Most of these parts are listed in Table 15.12. Also see Section Appendix D on page 577 for more information on purchasing the various parts.

2. Modify the servo motors for continuous rotation by a.) clipping or removing the potentiometer, b.) clipping away the stopper.

3. It is very important that you record your results using Table 15.3 and Table 15.4 in this section for later referral! Test the servo motors using the PICs clocking resources. Record the outputs and their respective frequencies that cause the servo to turn clockwise (forward), counterclockwise (reverse), and stop. This information may be unique for each servo and must be recorded separately for each servo motor. Use Table 15.13 and Table 15.14 below to record the effect of each output from the counter on the servos. Put a check mark in the column labeled slow speed, medium speed, and fast speed. Slow speed is when the servo turns very slow. Fast speed is the highest speed the servo will go. Medium speed is somewhere in between. You have to use your best judgment on what is a slow, medium, or fast speed.

Left Servo Clock Frequency (Speed Characteristics and rotation direction)						
Counter Output	Frequency	Slow Speed	Medium Speed	Fast Speed	Forward	Reverse
Q10						
Q11						
Q12						
Q13						
Q14						
Q15						

Table 15.13. Left Servo clock frequency verses speed and rotation direction.

Right Servo Clock Frequency vs. (Speed Characteristics and rotation direction)						
Counter Output	Frequency	Slow Speed	Medium Speed	Fast Speed	Forward	Reverse
Q10						
Q11						
Q12						
Q13						
Q14						
Q15						

Table 15.14. Right Servo clock frequency verses speed and rotation direction.

Before you begin recording the servo clock frequency verse speed and rotation direction in Table 15.13 and Table 15.14 the servos should be identified by a label as the right servo and the left servo. The servos should be positioned as shown in Figure 15.19. The shaft and tire of the servos should be pointed in opposite directions as shown.

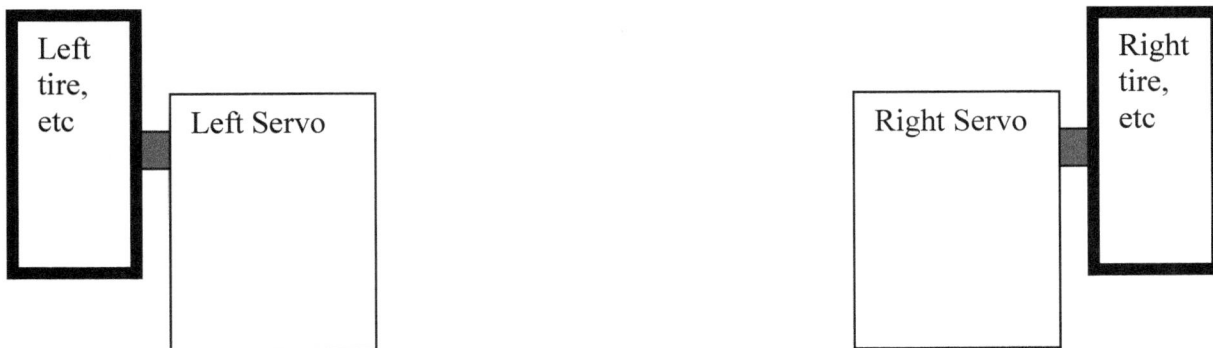

Figure 15.28. Position of left servo relative to right servo.

It should be noted that clockwise is forward and counter clockwise is reverse. While each servo is positioned differently, the robot will go forward when both servos are going forward. The robot will go backward or reverse when both servos are in reverse. The robot will turn left when the left servo is stopped and the right servo is in forward and the robot will turn right when the right servo is stopped and the left servo is forward. For the robot to turn left in place the left servo needs to be in reverse and the right servo need to be in forward and for the robot to turn right in place the left servo needs to be in forward while the right servo needs to be in reverse. Keep in mind that when the robot is moving forward the left servos and the right servo are actually going in opposite directions. This is because they are positioned on opposite sides of the robot as well as faced in opposite directions.

4. Write the C code that converts parallel data (8 data bits) to serial data. This can be done two ways here: 1.) encoder and decoder 2.) PIC18 UART port. This data must be in standard UART format. Standard UART requires 1 start bit, 8 data bits, and 1 or more stop bits. Write the C code that reads and writes the PIC18's UART port or similar. The UART port (or similar) will connect to the transmitter input data and receiver output data pins. Verify the digital serial data at the PIC's output port using an oscilloscope. The PIC that resides on the remote control unit will be referred to as the TX PIC while the PIC that resides on the robot's platform will be called the RX PIC.

5. Test the transmitter and receiver modules as shown in That is, breadboard the circuit as shown in Figure 15.29, Figure 15.30, and Figure 15.31. Consult your manufacturer's data sheet to connect all other inputs and outputs as required. For example, make sure the ground and power pins are connected as specified. Assuming that the TX and RX modules are wired for proper operation and that the PIC's UARTs are being used, now connect the output of the UART from the TX PIC completed in procedure 4 to the transmitter module as shown. The antenna out of the TX module should be connected to the antenna input of the RX module using a short piece of wire. The PICs, TX, and RX should share the same ground reference. Use channel A of an oscilloscope to measure the output of the TX PIC UART. Now use channel B of the same oscilloscope to measure the digital output of the RX. The data in channel A should match the data on channel B. If not, then check that the circuit is wire correctly. If all else fail, the TX or RX must be bad. Consult the distributor or manufacturer for a replacement.

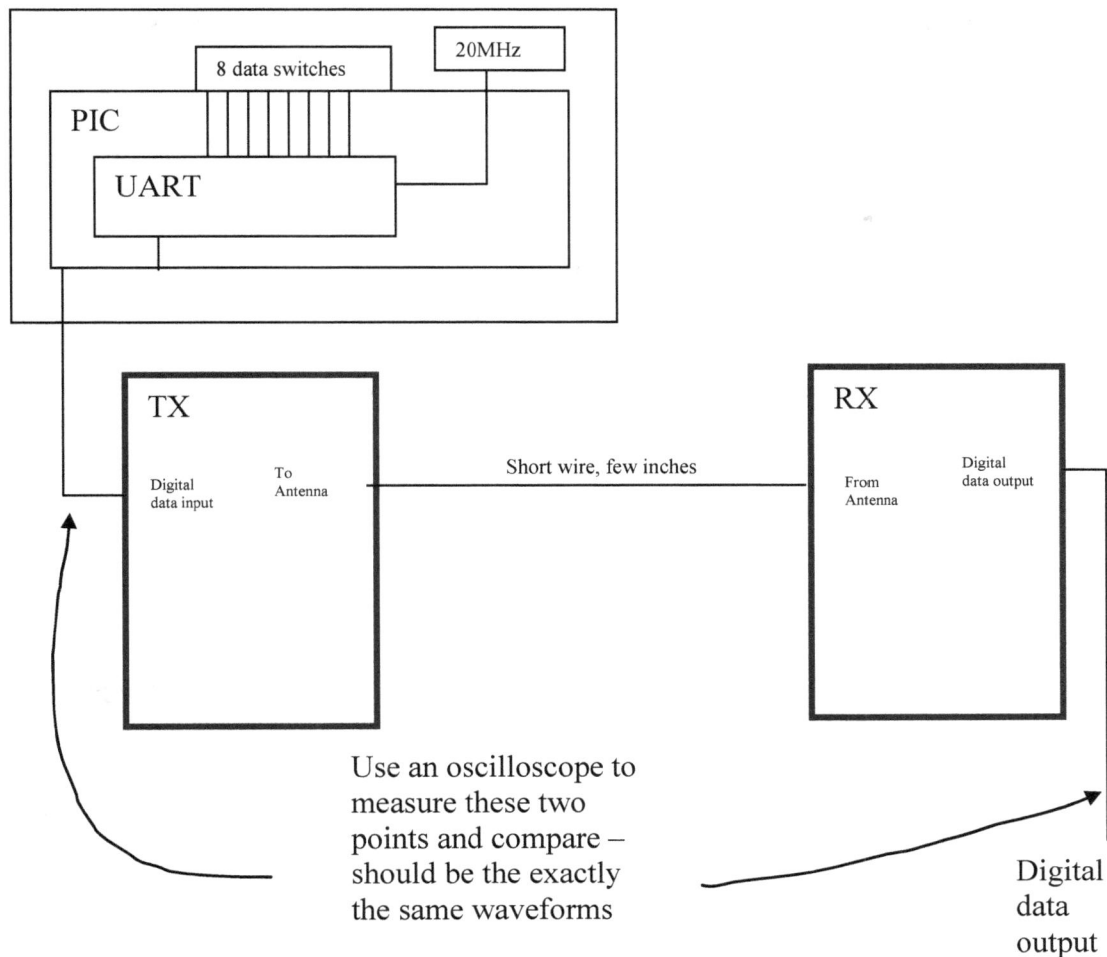

Figure 15.29. Test circuit for TX and RX.

6. If step 5 is successful perform the same procedure using the wireless link as shown in Figure 15.32. That is, include the TX and RX in the test. The transmitter and receiver circuits must be free of noise to operate optimally. One method of reducing the noise effects is to use two AA batteries in series for the transmitter 3.3V source and a separate one for the receiver 3.3V source. The two AA batteries in series produce a 3V source which is within the supply range of the WyJen transmitter and receiver. The transmitter battery ground should share the same ground as the TX PIC ground and the receiver battery ground should share the same ground as the RX PIC ground. Also locate the receiver away from the servo motors.

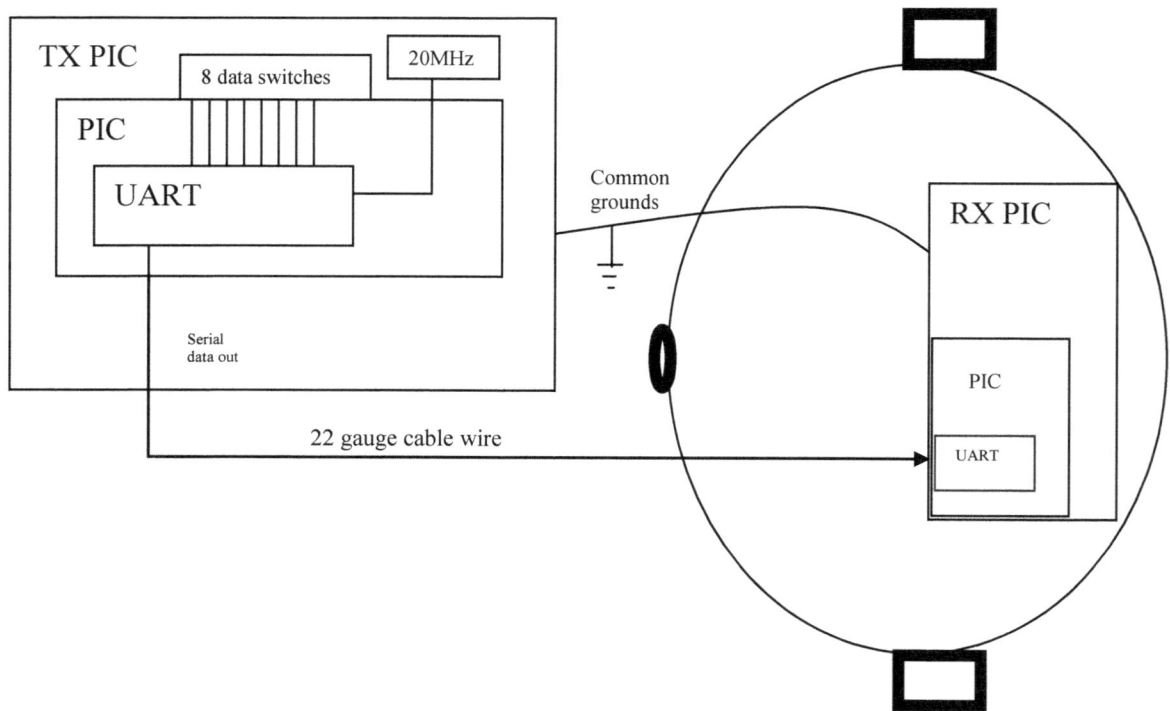

Figure 15.30. Test circuit for TX PIC, RX PIC, left and right servos, and robot platform.

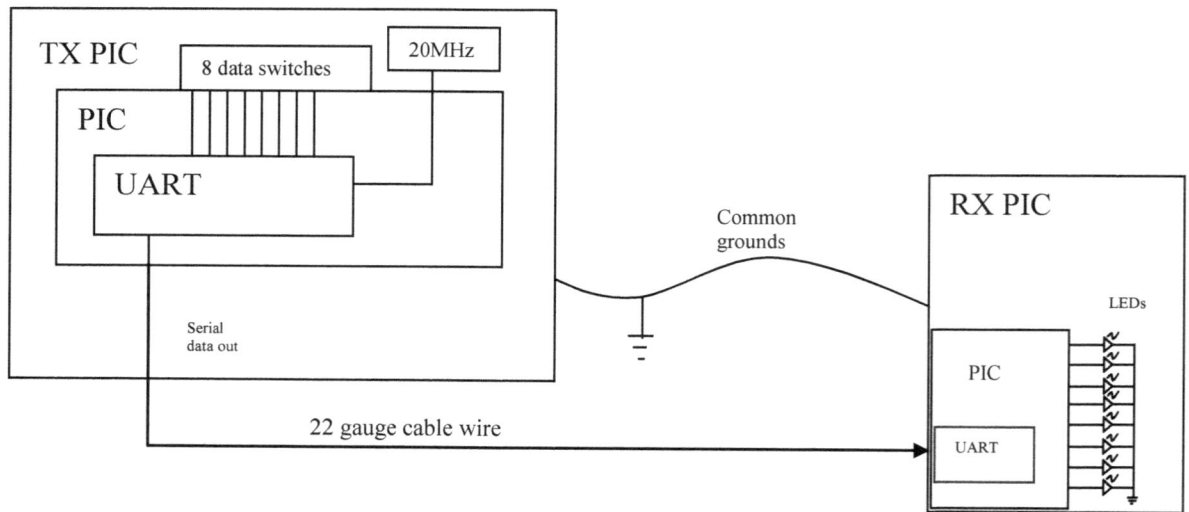

Figure 15.31. Test circuit for TX PIC and RX PIC.

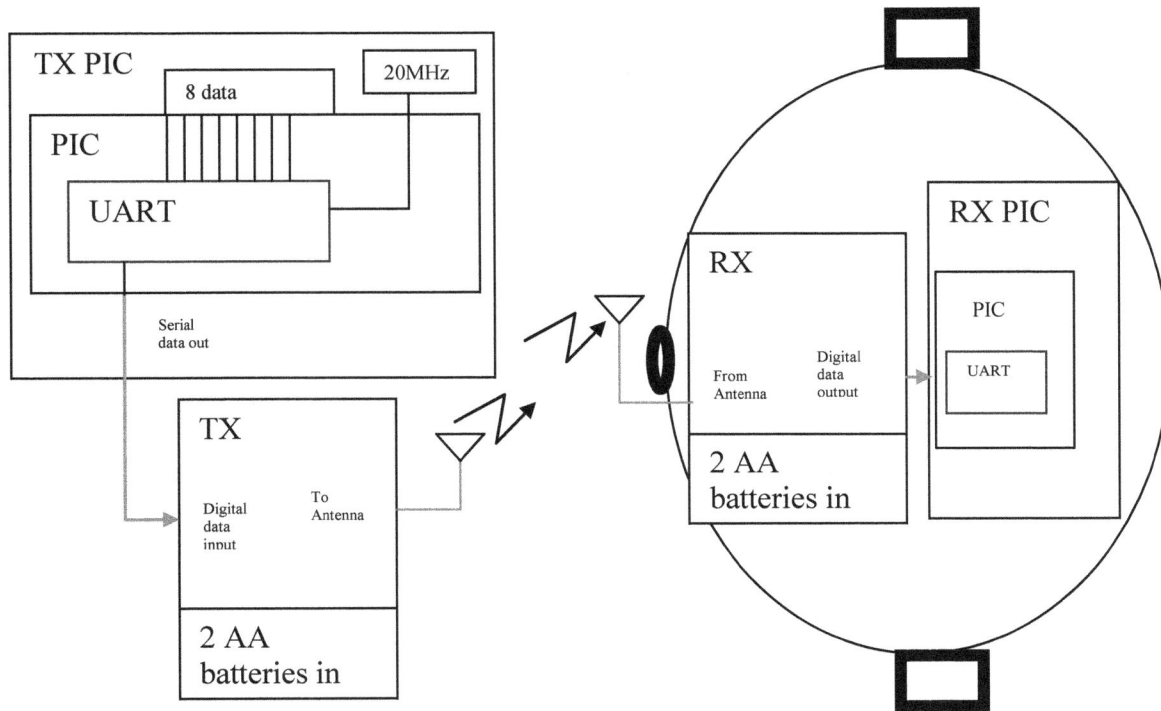

Figure 15.32. Test circuit for TX PIC, RX PIC, TX, RX, left and right servos, and robot platform.

7. Complete the project as shown in Figure 15.33 by including the compass sensor.

8. Acquire all the hardware components of the robot. A starting list is given in Table 15.15.

9. Construct the hardware platform for the robot as shown in Figure 15.34. The RX PIC platform needs to stand above the plexiglass platform. This is done here with 1 and ½ inch standoffs. The platform is about 10 inches in diameter. The weight of the RX PIC board along with the other components needs to be evenly distributed to prevent the robot from easily turning over.

10. Incorporate the battery supply into the project. The receive robot may now feed off the battery supply. The battery will be drained in a few minutes demonstration time. Make sure you have a spare battery pack or a fast charger and a power outlet nearby for uninterrupted demonstrations.

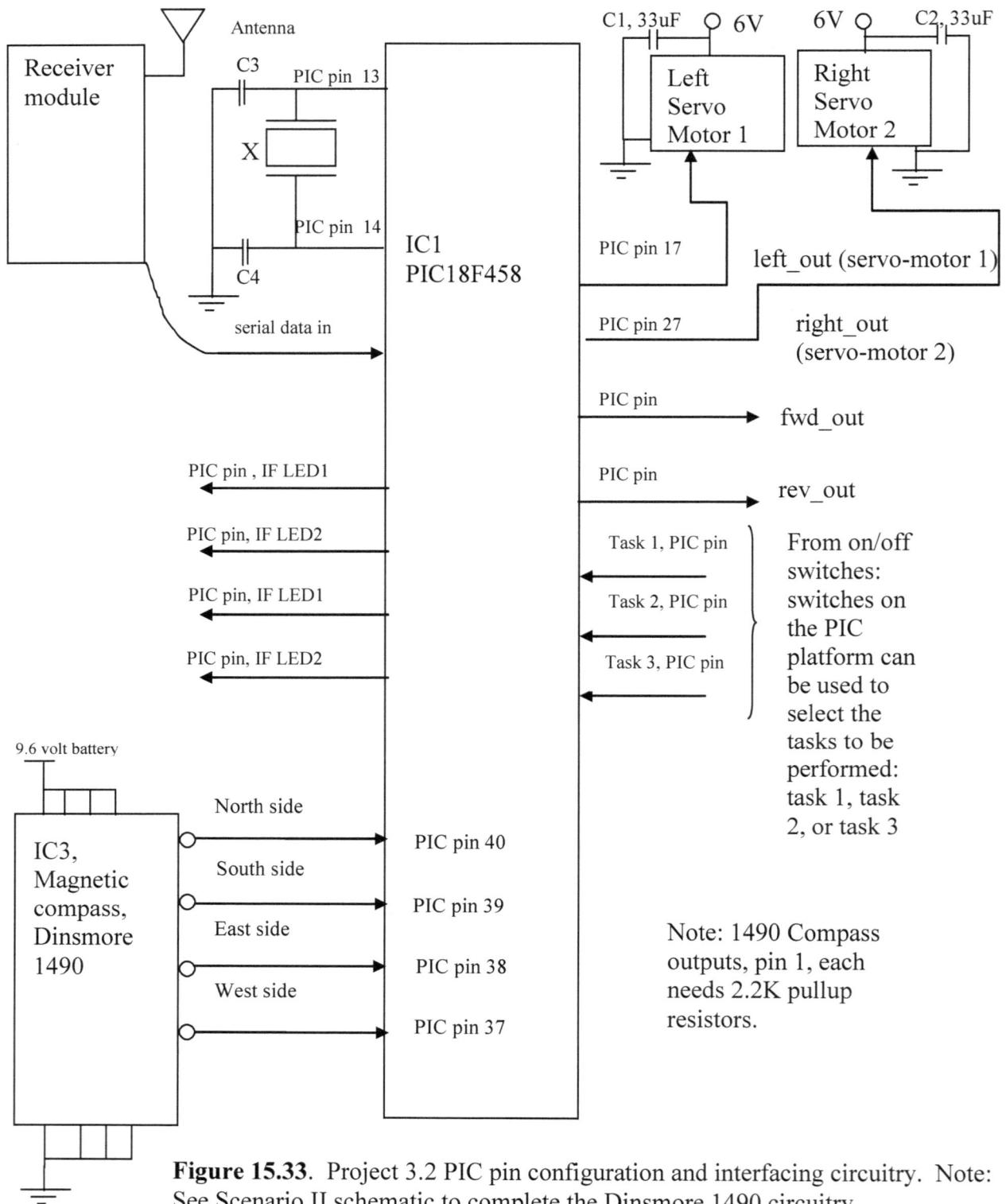

Figure 15.33. Project 3.2 PIC pin configuration and interfacing circuitry. Note: See Scenario II schematic to complete the Dinsmore 1490 circuitry.

15.3.6. Project 3.2 Hardware Parts List

The hardware parts list for Project 3.2 is given in Table 15.15. These parts can be obtained from the references listed in Section Appendix D on page 577.

Part	Quantity	Ref designator	Description
High Tech servos - HS300 (Futaba S3003, or Tower TS53J)	2	servo1, servo2	servomotors
Battery supply, radio shack, 9.6V, 1000mA, 230-0342	2	Vcc	9.6 Volt Nickel Cadmium R/C car battery pack and charger for TX ESCO and RX ESOC, Radio Shack
DC power connector, M type	2		For connecting battery packs to ESOCs, Radio Shack
Caster wheel	1		Front wheel, Menards, Lowes, Home depot, etc.
Tire	2		Rubber tire, Tower Hobbies
Wheel	2		Rim for rubber tire, Tower Hobbies
Round Platform	1	platform I	Round platform of diameter 11 inches (and thickness 3/16 inches) cut from plexiglass, etc, etc
standoffs	4	ST1, ST2, ST3, ST4	1 and ½ inch standoffs, Digikey, Radio Shack, etc
Screws, 844-712 is Menards part #	?		Machine screws, size 4-40x ½, 30 pieces, can be used to mount caster wheel and servo if the 4-40x ¼ is not long enough.
Screws, 844-710, 844-712 is Menards part #	About 16		Machine screws, size 4-40x ¼, 34 pieces. 8 is needed for mounting the RX ESOC, 4 for the caster wheel, 2 for each servo
Nuts, 846-644 is Menards part #	About 12		Machine nuts, size 4-40, 28 pieces. 4 for the caster wheel, 4 for each servo
Washers, 154-306 is Menards part #	About 12		Washers, No. 6, 100 pieces. 4 is needed for mounting the bottom of the RX ESOC, 4 for the caster wheel, 4 for each servo
Velcro			4 inch by 2 inches, 2 sets

Table 15.15. Project 3.2 starter hardware parts list. For more details on where to order parts see Appendix D on page 577.

15.3.7. Project 3.2 Other Possible Sensors

Some other sensors are:

7. Line Tracker Sensor (Lynxmotion, digital output)
8. Infrared Proximity Detector (Sharp GP1U5 Radio Shack #276-137B, Digikey #160-1060, digital output)
9. Sonar Ranging Unit (Polaroid Sonar module)
10. IR Distance Sensor (Sharp GPD2D02, digital output)

11. Magnetic Compass Sensor (Dinsmore model 1490; digital output)
12. Other sensor – you must get approval from the professor

15.3.8. Project 3.2 Assembly of the PIC-bot: Platform I

The bottom view of the PIC-bot is shown in Figure 15.16. The PIC board will rest on the top of the platform. The servomotors will be mounted on the bottom and attached to the wheels as shown. The control Cables run from the servomotors to the PIC board located on the opposite side as shown. The sensors should be mounted on the top with all outputs connected to the controller (PIC board). The switch and power cables have separate holes in the platform. The power switch is connected to the hole in the platform.

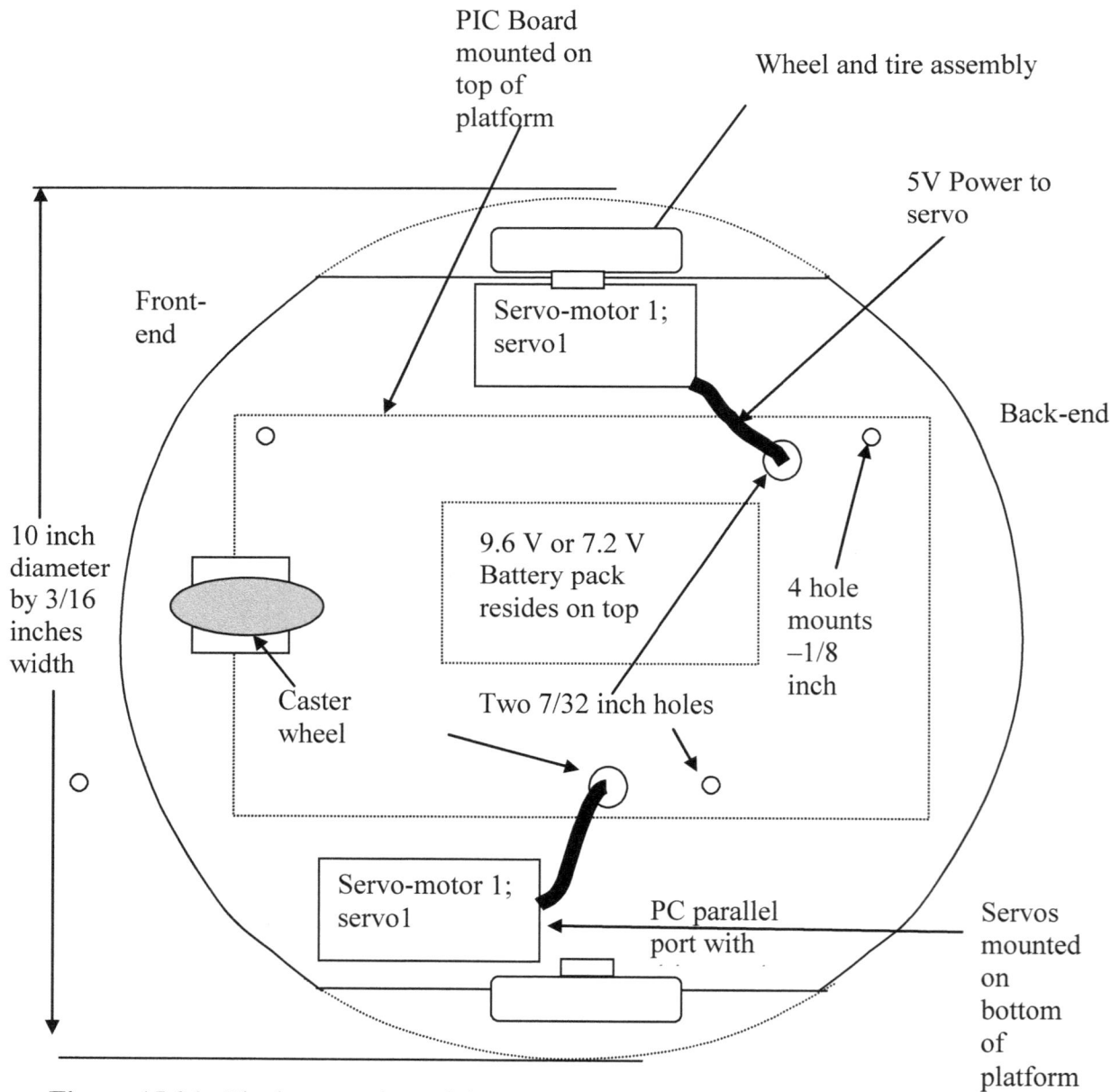

PIC Board mounted on top of platform

Wheel and tire assembly

5V Power to servo

Front-end

Servo-motor 1; servo1

Back-end

10 inch diameter by 3/16 inches width

9.6 V or 7.2 V Battery pack resides on top

4 hole mounts −1/8 inch

Caster wheel

Two 7/32 inch holes

Servo-motor 1; servo1

PC parallel port with

Servos mounted on bottom of platform

Figure 15.34. The bottom view of the PIC bot.

15.4.   **Project 4: Remote Control of the PLD-bot and Object Detection**

This PLD-bot has the following main requirements:

Requirement 1: The PLD-bot must go left, right, forward or backward at the command of the remote control unit. The PLD-bot is the receiver while the remote control unit uses another PLD platform as the transmitter. The transmitting PLD platform sends the data in UART format to the receiving PLD platform. The receiving PLD platform decodes the UART message and responds to the commands sent.

Requirement 2: The PLD-bot must perform object detect when selected from the remote control unit – the transmitting PLD platform.

Requirement 3: The PLD-bot performs direction detection when selected from the remote control unit – the transmitting PLD platform.

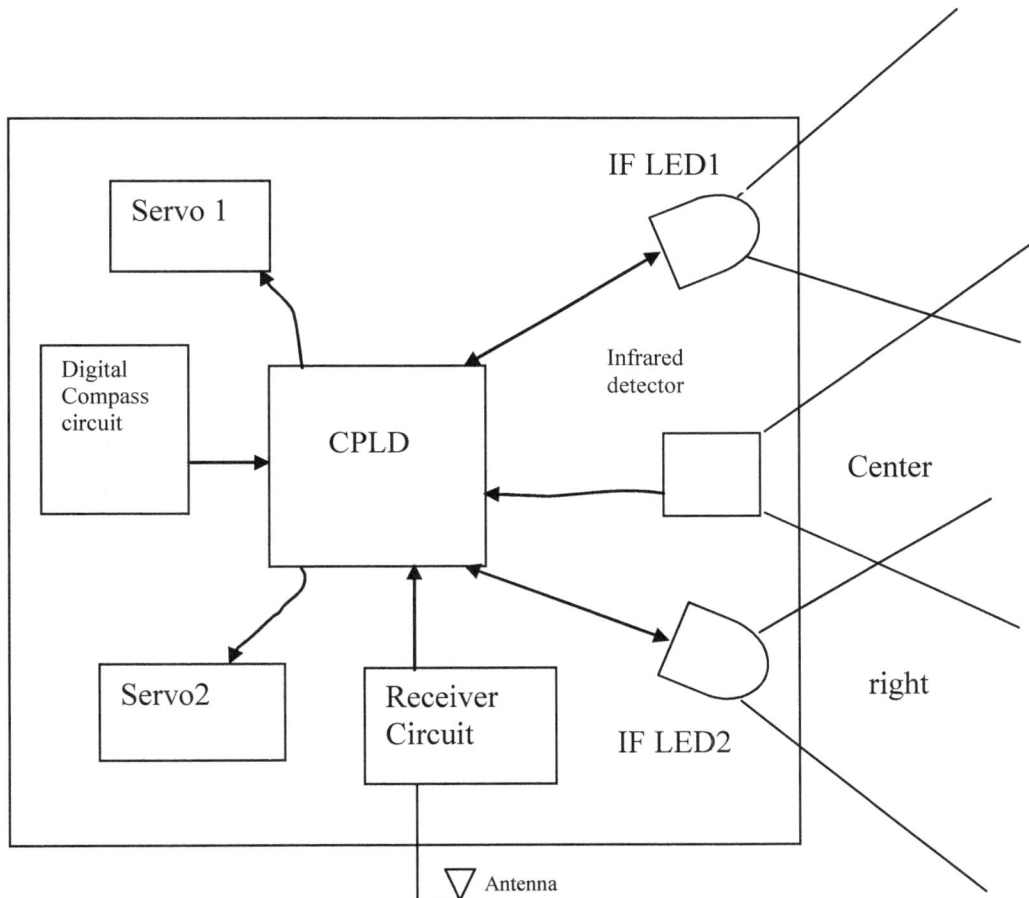

Figure 15.35. Project 4 of the receiving PLD-bot. IF1 is the infrared proximity sensor while LED1 and LED2 are infrared LEDs.

15.4.1. Project 4 Electronic Parts List

The electronic part list for Project 4 is shown in Table 15.16 below.

Part Number	Quantity	Ref designator	Description
IF Detector module, 276-640, Radio Shack	1	IF1	Infrared modulated detector module, bandpass center frequency=38,000Hz; Vcc=2.4 to 5 volts; peak wavelength = 940nm
LED, 276-143C, Radio Shack	2	IFLED1, IFLED2	Infrared light emitting diode
ESOC board	2	CPLD, IC1	Altera programmable logic device EMP7128SLC board, www.elexp.com/tst_pld2.htm
Altera CPLD, FLEX10K20 or FLEX10K70	optional	CPLD, IC1	Altera UP1 or UPX board
Magnetic Compass, Dinsmore model 1490	1	IC3	Magnetic compass, 4 bit, active low digital outputs
High Tech servos - HS300 (Futaba S3003, or Tower TS53J)	2	servo1, servo2	servomotors
Platform I	1	Platform I	Platform 1 uses two servo motors, a caster wheel, etc. See
Resistor, 1000 ohms	1	R1	1 Watt resistor
Resistor, 180 ohms	2	R2, R3	1 Watt resistor
Radio Shack – 276-1547 or 276-1429, etc.	1	J1	DB 25 male connector
Transmitter Module	1	IC	FM, FSK modules, etc. 5 volts.
Receiver Module	1	IC	FM, FSK modules, etc. 5 volts.

Table 15.16. Electronic part list for PLD-bot Project 4. For more details on where to order parts see Appendix D on page 577.

15.4.2. Project 4 Schematics

Some schematics for Project 4 are shown in Figure 15.36 and Figure 15.37.

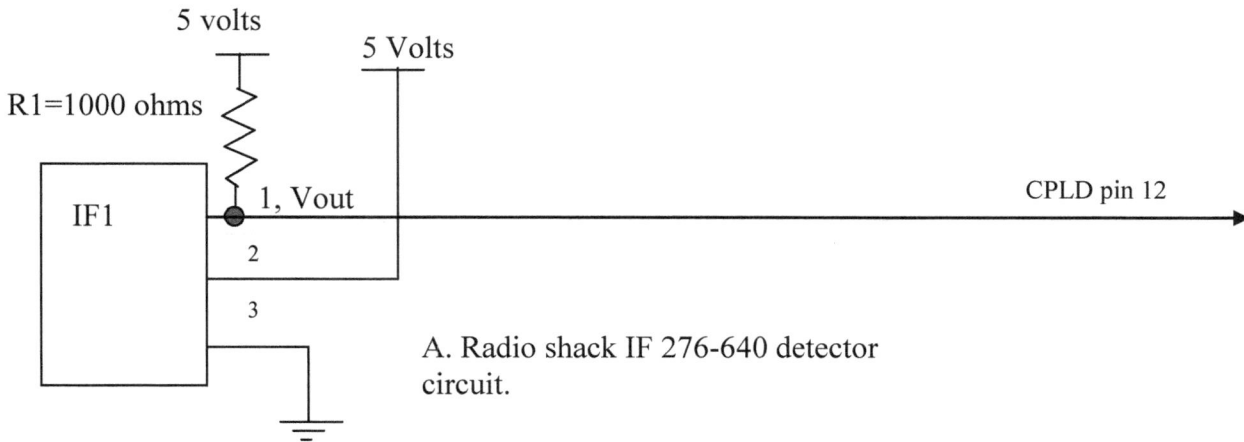

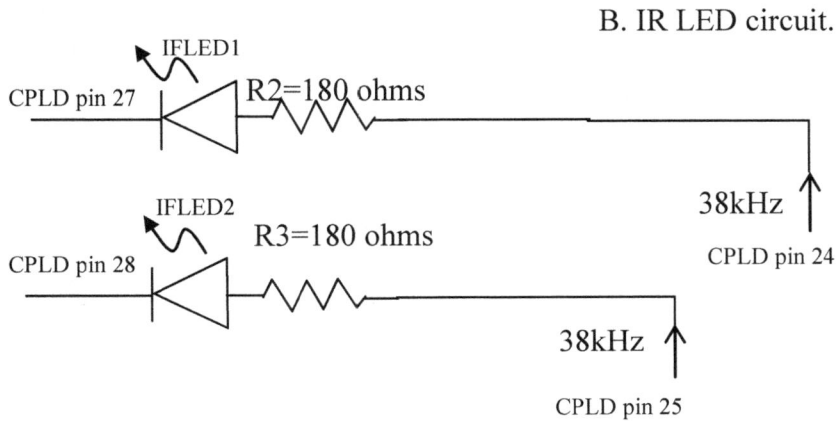

5 volts

5 Volts

R1=1000 ohms

IF1

1, Vout

2

3

CPLD pin 12

A. Radio shack IF 276-640 detector circuit.

B. IR LED circuit.

IFLED1

CPLD pin 27   R2=180 ohms

38kHz

CPLD pin 24

IFLED2

CPLD pin 28   R3=180 ohms

38kHz

CPLD pin 25

Figure 15.36. Schematic of IR Detector circuitry (A) and IR LED emitter circuitry (B).

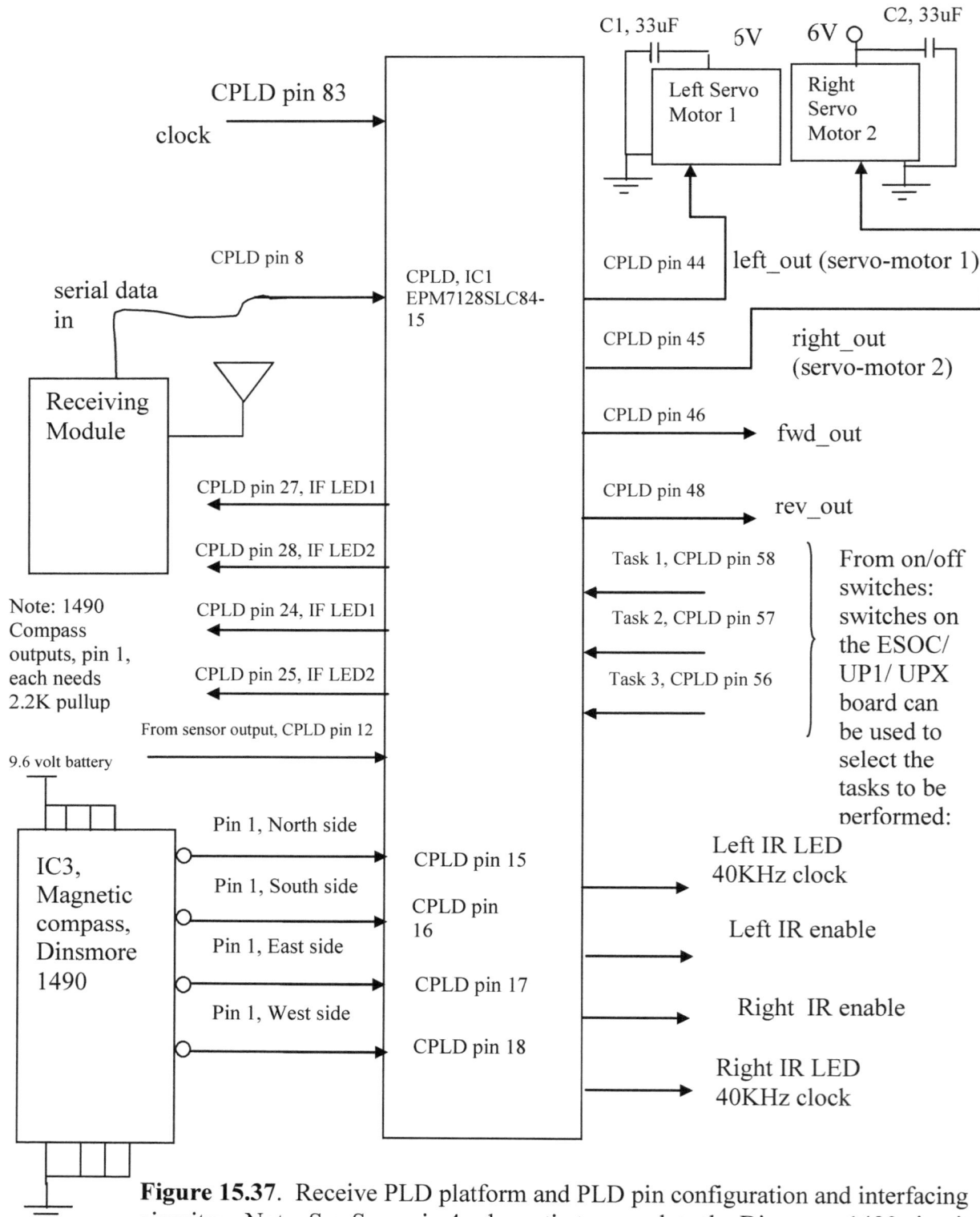

Figure 15.37. Receive PLD platform and PLD pin configuration and interfacing circuitry. Note: See Scenario 4 schematic to complete the Dinsmore 1490 circuitry.

15.4.3. Project 4 Hardware Parts List

The hardware parts list for Project 4 is given in Table 15.17 below. These parts can be obtained from the references listed in Section Appendix D on page 577.

Part	Quantity	Ref designator	Description
High Tech servos - HS300 (Futaba S3003, or Tower TS53J)	2	servo1, servo2	servomotors
Battery supply, radio shack, 9.6V, 1000mA, 230-0342	2	Vcc	9.6 Volt Nickel Cadmium R/C car battery pack and charger, for TX ESOC and RX ESOC
DC power connector, M type	2		For connecting battery packs to ESOCs
Caster wheel	1		Front wheel
Tire	2		Rubber tire
Wheel	2		Rim for rubber tire
Round Platform	1	platform I	Round platform of diameter 11 inches (and thickness 3/16 inches) cut from plastic, etc
standoffs	4	ST1, ST2, ST3, ST4	1 and ½ inch standoffs, for mounting ESOC above the platform
Screws, 844-712 is Menards part #	1		Machine screws, size 4-40x ½, 30 pieces, can be used to mount caster wheel and servo if the 4-40x ¼ is not long enough.
Screws, 844-710, 844-712 is Menards part #	About 16		Machine screws, size 4-40x ¼, 34 pieces. 8 is needed for mounting the RX ESOC, 4 for the caster wheel, 2 for each servo
Nuts, 846-644 is Menard part #	About 12		Machine nuts, size 4-40, 28 pieces. 4 for the caster wheel, 4 for each servo
Washers, 154-306 is Menard part #	About 12		Washers, No. 6, 100 pieces. 4 is needed for mounting the bottom of the RX ESOC, 4 for the caster wheel, 4 for each servo
Velcro			4 inch by 2 inches, 2 sets

Table 15.17. Project 4 hardware parts list. For more details on where to order parts see Appendix D on page 577.

16. Signal Processing Electronics and Robotics

Signal processing is a very power method of acquiring and processing analog data. Analog data is converted to digital data and then processed digitally – commonly known as digital signal processing. The following sections present descriptions, scenarios, components, etc., for building a mobile robot from a signal processing perspective. Two projects are introduced: project 5 and project 6. These projects introduce robotics with emphasis on software applications and algorithms. Project 5 is mostly digital and requires writing programs to perform most of the tasks. The programs can reside on a PC or DSP. Data is acquired from the robot's mini microphone and sent to the PC or DSP over a cable for analysis. The programs performs digital filtering and spectral analysis on the input data to determine if a specific speaker is speaking. If so, commands are returned to the robot based on the PC's or DSP's results. These commands tell the robot to perform a specific action such as move left, right, reverse, forward, etc. Project 6 is mostly analog and requires some VHDL programs to determine the commands. The analog sound from the microphone is sent to a bank of bandpass filters and comparators. The comparators convert the ringing from the bandpass filters into digital signals which can then be processed by the PLD. The PLD contains VHDL programs that count the pulses. The programs compare the count to a threshold to verify the speaker. If the speaker matches the internal threshold then the robot performs a task.

16.1. Project 5: Robotics and Signal Processing using an ADC0804 and a PLD

This PLD-bot has the following requirements:

Requirement 1: The PLD-bot must go left, right, forward, backward, reverse, etc., using the switches on the transmit PLD platform. This also serves as a test to verify communication between transmitter PLD platform (Figure 16.1) and receiving PLD platform (Figure 16.2). This task should be done before proceeding to task 2.

Requirement 2: The PLD-bot must go left, right, forward or backward upon detection of the speech phrase "left", "right", "forward", or "reverse" due to a specific speaker. The receive PLD platform is located on the robot as shown in Figure 16.2.

Requirement 3: The PLD-bot receives the speech sound from the A/D converter and sends it to the transmitting PLD platform in parallel or serial format.

Requirement 4: The PLD sends the received analog to digital data to a personal computer (PC) for speech recognition analysis.

Requirement 5: A daemon running on the PC is used to determine when the received data is available.

Requirement 6: The PC performs a speech recognition analysis on the received data. Once the analysis is done, the PC sends one command to the PLD platform for the appropriate action: turn left, turn, right, go forward, go reverse, etc.

Requirement 7: The transmitting PLD platform is interfaced to the PC's parallel port or similar as shown in Figure 16.1.

Requirement 8: Requirements 3 to 7 may be eliminated if a DSP platform (DSP-bot) on the robot in Figure 16.2 is used to perform the speech recognition. That is, a DSP can replace the PLD. The TMS320C6713 DSP is discussed briefly in Appendix A on page 561, Appendix B on page 568, and Appendix C on page 571. Also refer to Chassaing (2005) for DSP applications on the use of the C6713 Digital Signal Processing Kit (DSK) with step by step instructions. This places all the intelligence local to the robot as opposed to remotely.

Note: To do this project, if the non DSP approach is used, it is recommended that communication between platforms be done in parallel first. If enough time is left an upgrade to serial communication between boards should be considered.

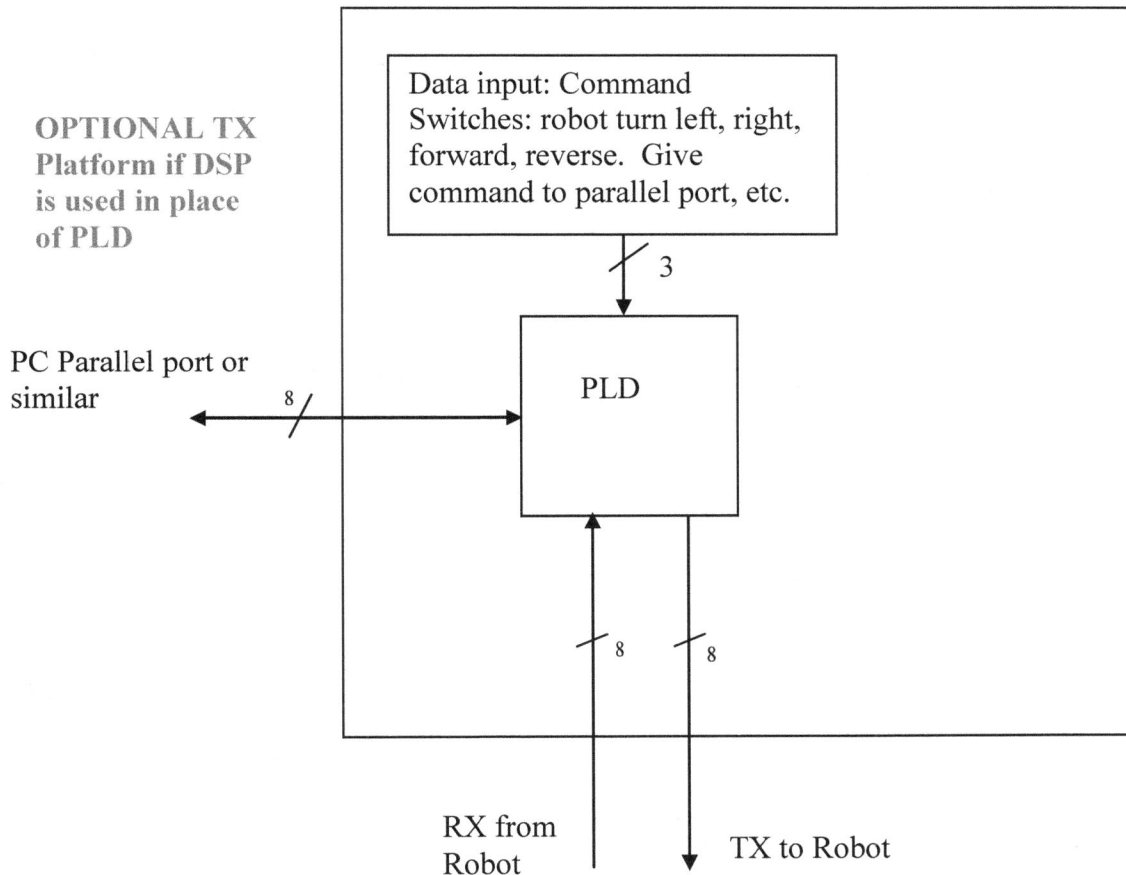

Figure 16.1. Transmit PLD Platform. Interfaces to PC.

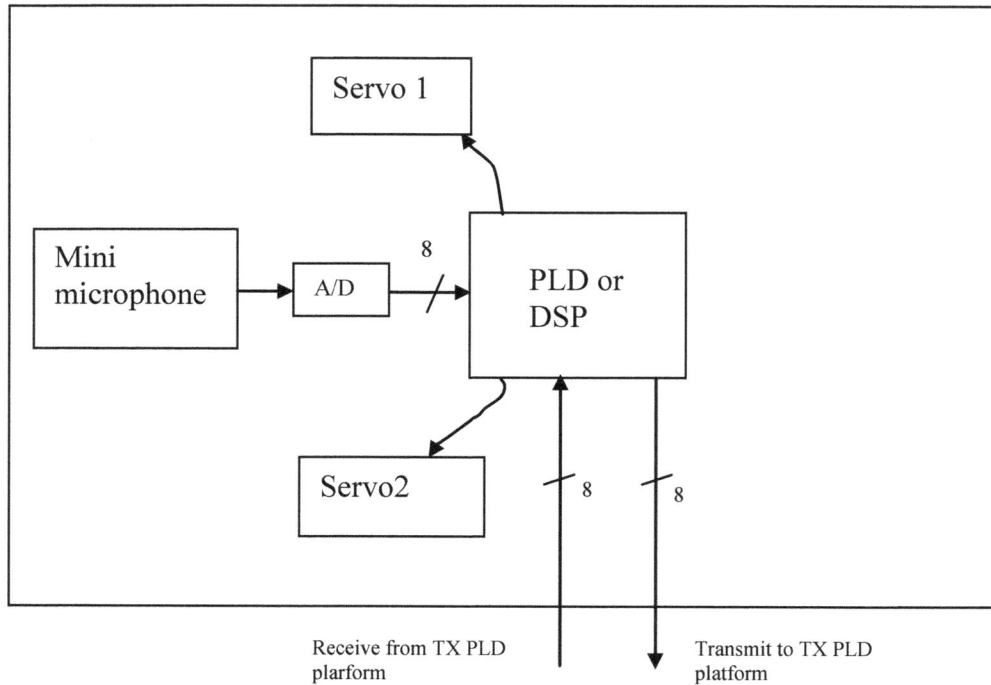

Figure 16.2. Receiving PLD platform – PLD-bot. If DSP platform is used here you may eliminate TX platform and PC interface in Figure 16.1.

16.1.1. Project 5 Electronic Parts List

The electronic parts list for Project 5 is shown in Table 16.1.

Part	Quantity	Ref designator	Description
ESOC board	2	CPLD, IC1	Altera programmable logic device EMP7128SLC board, www.elexp.com/tst_pld2.htm
Altera platform, CPLD with FPGA (FLEX10K20 or FLEX10K70) Optional	2	CPLD, IC1	Altera UP1 or UPX board
High Tech servos - HS300 (Futaba S3003, or Tower TS53J)	2	servo1, servo2	servomotors
ADC0804	1	IC	Analog to digital conversion
Radio Shack- 270-092	1	Mic	Mini Microphone, up to 15KHz input
Radio Shack – 276-1547 or 276-1429, etc.	1	J1	DB 25 male connector
Buffer HC244	5 (2 for each 8 bit cable and 1 for DB25 interface); 2 (if serial cable and DB25 interface)	IC	Buffer between cables and PLDs

Table 16.1. Electronic parts list for Project 5 PLD-bot. For more details on where to order parts see Appendix D on page 577.

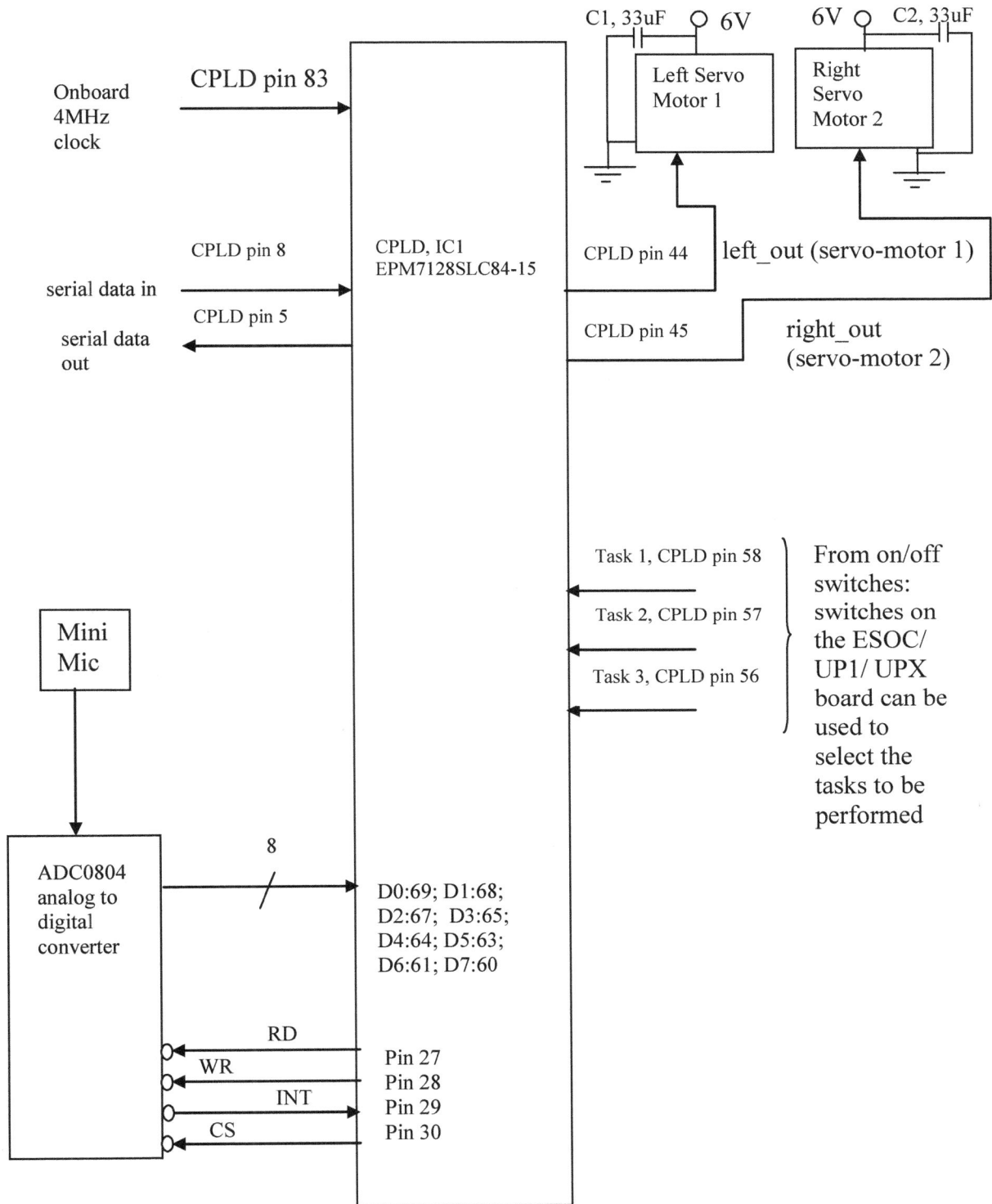

Figure 16.3. Receive PLD platform PLD pin configuration and interfacing circuitry.

16.1.2. Project 5 Hardware Parts List

The hardware parts list for platform I is given in Table 16.2.

Part	Quantity	Ref Designator	Description
High Tech servos - HS300 (Futaba S3003, or Tower TS53J)	2	servo1, servo2	servomotors
Battery supply, radio shack, 9.6V, 1000mA, 230-0342	1	Vcc	9.6 Volt Nickel Cadmium R/C car battery pack and charger
Home depot, etc., Caster wheel	1	wheel	Front wheel
Round Platform, Plexiglass/fiber glass, from Home Depot, Manards, etc.	1	platform I	Round platform of diameter 11 inches (and thickness 3/16 inches) cut from plastic, etc

Table 16.2. Project 5 hardware parts list. For more details on where to order parts see Appendix D on page 577.

16.2. **Project 6: Robotic Speaker Detection using a Zero Crossing Detector and a PLD**

The basic requirements, project description, and procedure are given for Project 6 next.

16.2.1. Project 6 Basic Requirements

This PLD-bot has the following main requirements:

Requirement 1: The PLD-bot must go left, right, forward, backward, reverse, etc., using the switches on the transmit PLD platform. This also serves as a test to verify communication between transmitter PLD platform and receiving PLD platform. This task should be done before proceeding to task 2.

Requirement 2: The PLD-bot must select a speaker in your group using 3 switches on the PLD platform.

Requirement 3: The PLD-bot must go left, right, forward or backward for the selected speaker only. This requires the PLT platform to perform a speaker identification test using the signals from the comparators for the following speech phrases: "left", "right", "forward", or "reverse".

16.2.2. Project 6 Description

The PLD platform receives the speech sound from a bank of analog filters and comparators. The output of the comparators is 5 volt digital clock signal which is fed to the PLD platform. This clock signal feeds the digital circuitry inside the PLD on the PLD platform. The digital circuitry comprises n counters in parallel, in this case mod65536 counters. The output of the counters is fed to a decoder that performs speaker identification using the threshold data from the various speakers. If a particular speaker is selected, the speaker is verified before the robot performs a task for the speaker. Otherwise the robot will not perform the task. The circuitry for project 6 is shown in Figure 16.4, Figure 16.5, and Figure 16.6.

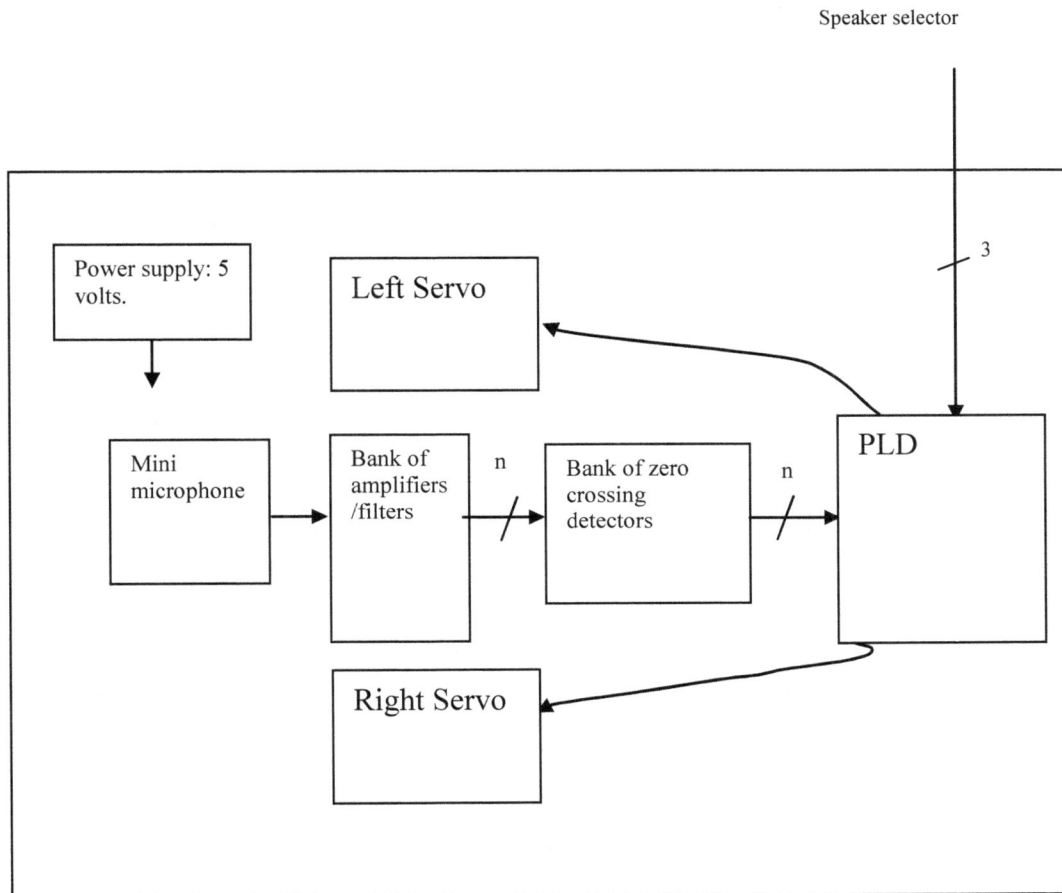

Figure 16.4. PLD-bot. n is the number of filters in parallel. The −15 and +15 volt supply is needed by the OP amps in the filter bank. 5 volts is needed by the PLD platform and remaining circuitry.

16.2.3. Project 6 Electronic Parts List

The electronic part list for Project 6 PLD platform is shown in Table 16.3 below.

Part	Quantity	Ref designator	Description
ESOC platform	2	CPLD, IC1	Altera programmable logic device EMP7128SLC board, www.elexp.com/tst_pld2.htm
Altera platform, PLD and FPGA[FLEX10K20 or FLEX10K7] (optional)	2	CPLD, IC1	Altera UP1 or UPX board (optional)
Futaba S3003 servo or Tower Hobbies TS53J pack. High Tech HS300 servos	2	servo1, servo2	servomotors
Platform I	1	Platform I	Platform 1 uses two servo motors, a caster wheel, etc. See
LM386	1	IC	Amplifier with 0 and +5volts rails
LM339	1	IC	comparators
10000 POT	1	POT	Variable resisitor, any value will do, preferably one that can be turned with fingers.
Radio Shack- 270-092	1	Mic	Mini Microphone, up to 15KHz input
Breadboard	1		For the assembly of electronic circuits

Table 16.3. Electronic part list for Project 6 PLD-bot. For more details on where to order parts see Appendix D on page 577.

16.2.1. Project 6 Procedure

The procedure for this project is given as follows:

1. Acquire all the necessary parts for constructing the electronic portion of project 6. Most of these parts are listed in Table 16.3. Also see Section Appendix D on page 577 for more information on purchasing the various parts

2. Breadboard the circuit shown in Figure 16.6 for one amplifier circuit and one zero crossing circuit. Follow the directions on the back of the Radio Shack mic for constructing the mic circuit. Use an oscilloscope to measure the output of the amplifier and the output zero crossing circuit as you speak into the microphone. The output of the amplifier should be the same as the input signal but larger. The output of the zero crossing circuit should be digital, that is, between 0 and 5 volts. The pin configuration for the LM339 and LM386 are shown in Figure 16.7 and Figure 16.8, respectively.

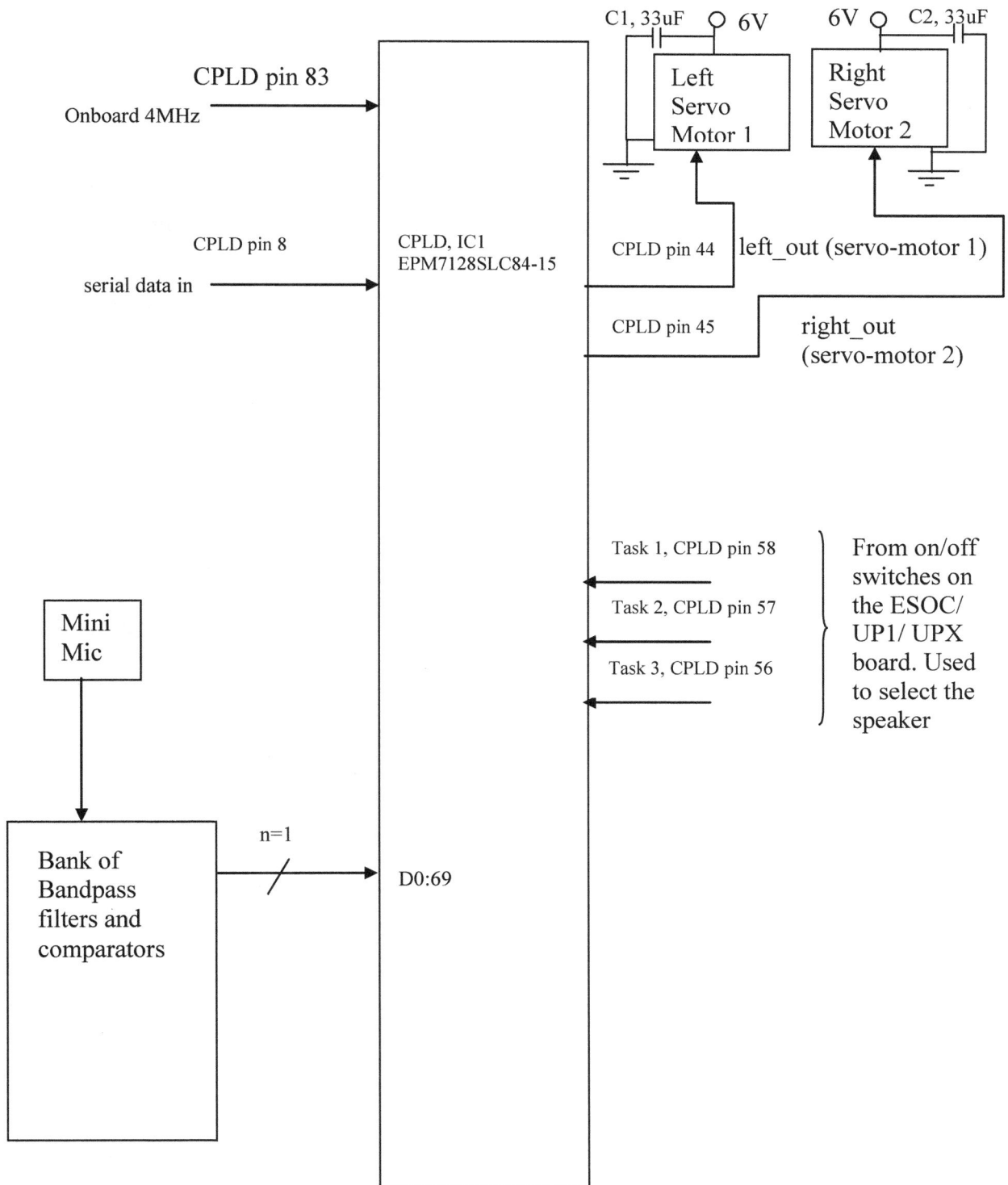

Figure 16.5. Project 6 PLD platform PLD pin configuration and interfacing circuitry.

The capacitance value for C1 should be chosen based on

$$C_1 = \frac{1}{2\pi \cdot f \cdot R_1}$$ **EQ. 16.1**

where f is the lowest voice frequency and voice has frequencies from 300 to 3800 Hz for the typical voice frequency range and R1 is the selected potentiometer resistance, in this case, R1= 10000Ω. Chose value of C using lowest voice frequency f=300 . This gives a value of C1 = 53nF. The closest standard values are 47nF or 56nF (Section 2.3.2 on page 21). Select one of the two. Note that C_1 and R_1 act like a highpass filter.

The amplifier circuit and zero crossing circuit should be tested individually before connecting the output of the amplifier to the input of the zero crossing circuit as shown in Figure 16.6. The amplifier can be tested by inputting a sinusoidal signal, e.g., 1 milli volt peak to peak and observing that the signal is amplified significantly at the output of the amplifier by a gain of 20. Higher gains can be achieved by configuring the LM386 as indicated in the application section of the datasheet for gains of 50 or 200. The zero crossing circuit can be tested by inputting a sinusoidal signal of 1000Hz, etc. and observing a square wave of the same frequency at the output of the zero crossing circuit.

3. Modify the servo motors for continuous rotation by a.) cutting off the top of the potentiometer that turns the wheel if not removable, b.) clipping and filing away the stopper, 3.) and positioning the pot midway between its two terminals. This is typically the minimum requirement for continuous rotation. Some directions specify removing the pot and replacing it with resistors. This may be un-necessary for what we have to do. However, if desired follow the directions specified for your servo motor for continuous rotation.

4. Test the servo motors using the PLD Platform's 4Mhz on board clock. This clock frequency must be divided down below 1000 Hz for proper operation of the servos. A divide by 2^{16} counter can achieve this. The VHDL code for a divide by 2^{16} (mod65536) counter is given in Section 9.6.2 on page 349. Send the last six outputs of the counter to the output ports as shown in Figure 16.9. Connect each of these outputs to the signal input of the servo motor. The ground wire of the servo must be grounded to the PLD platform's ground. Similarly, the power supply wire of the servo must be connected to the PLD platform's VCC=+5V. Before moving forward it is best to label each servo. The servo that will be on the right side of the robot should be labeled right servo and the servo that will be on the left side of the robot should be labeled left servo. Use clear tape to tape a piece of paper with left written on it on the left servo. Use clear tape to tape a piece of paper with right written on it on the right servo. Record the outputs and their respective frequencies that cause the left servo and right servo to turn clockwise (forward), counterclockwise (reverse), and stop. This information may be unique for each servo and must be recorded separately for each servo motor. Use the Table below to record the effect of each output from the counter on the servos. Put a check mark in the column labeled slow speed, medium speed, and fast speed. Slow speed is when the servo turns very slow. Fast speed is the highest speed the servo will go. Medium speed is somewhere in between. You have to use your best judgment on what is a slow, medium, or fast speed.

Mini mic

C_1 should be selected such that its reactance passes the first fundament easy but rejects DC.

Vc=5V

$R_2=10\Omega$

C_1

5V

$C_2=0.05\mu F$

$C_3=250\mu F$

Vin

LM386

$R_1=10K\Omega$
POT

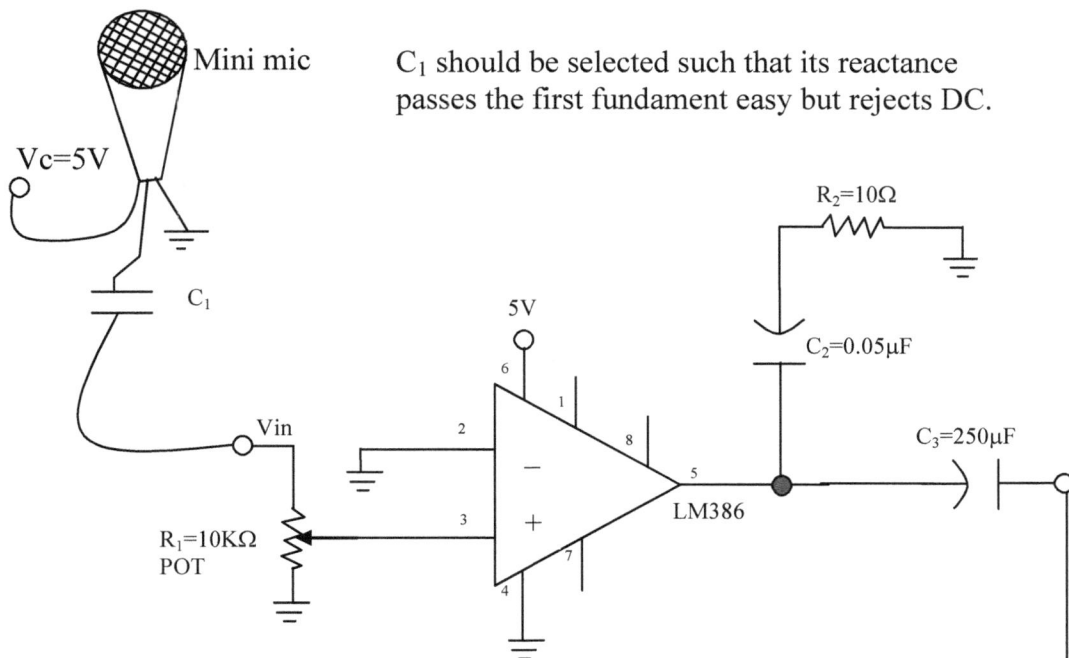

Amplifier, Gain=20, see LM386 datasheet for gains of 50 and 200

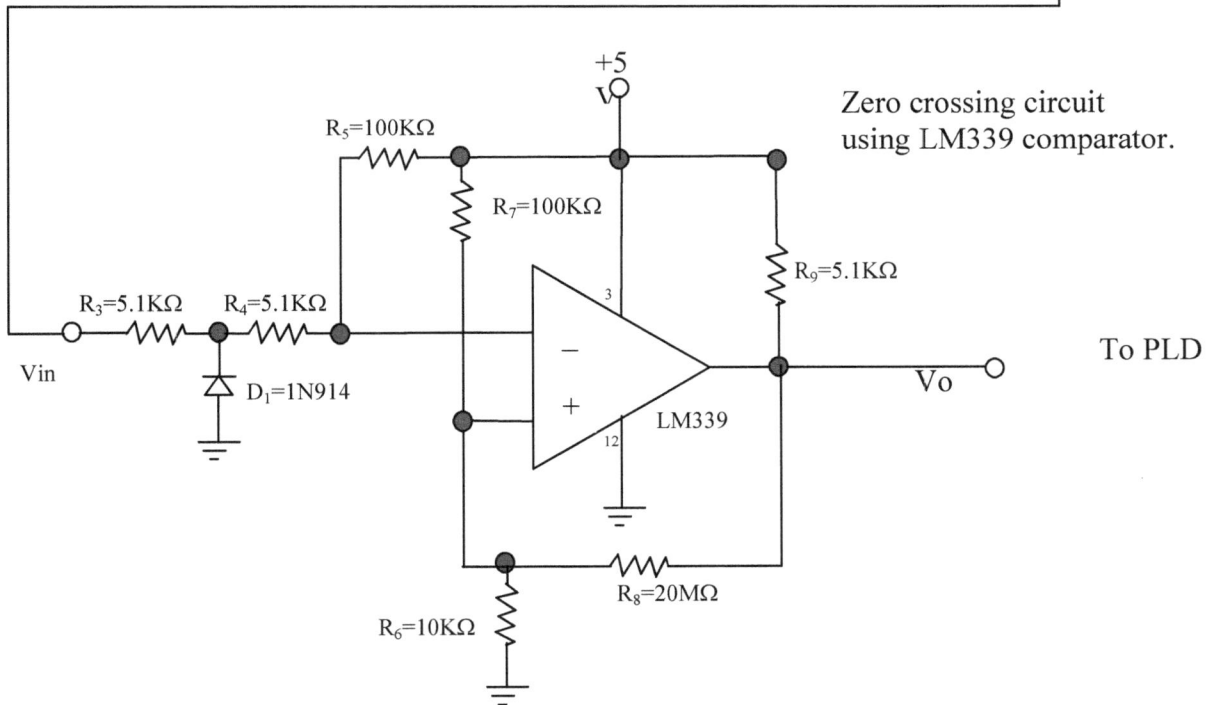

+5 V

$R_5=100K\Omega$

$R_7=100K\Omega$

Zero crossing circuit using LM339 comparator.

$R_9=5.1K\Omega$

$R_3=5.1K\Omega$ $R_4=5.1K\Omega$

To PLD

Vin

$D_1=1N914$

LM339

Vo

$R_8=20M\Omega$

$R_6=10K\Omega$

Figure 16.6. Detailed schematic of amplifier and zero crossing circuitry.

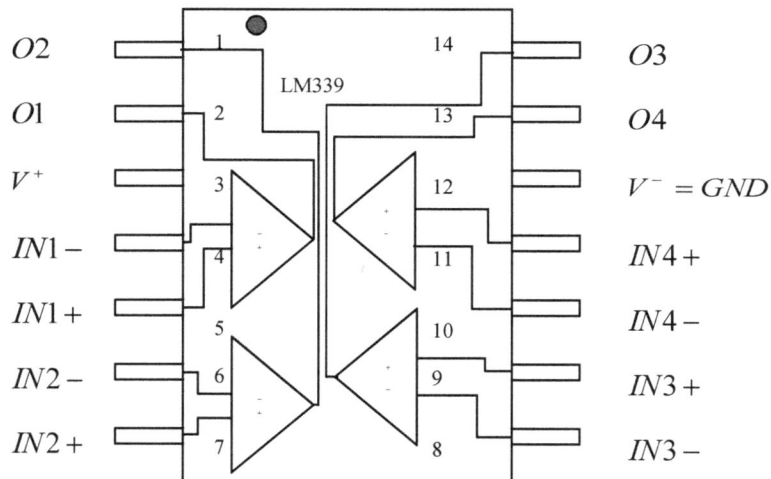

Figure 16.7. Pin configuration for the LM339 Quad comparator chip. Please see the LM339 datasheet for more information.

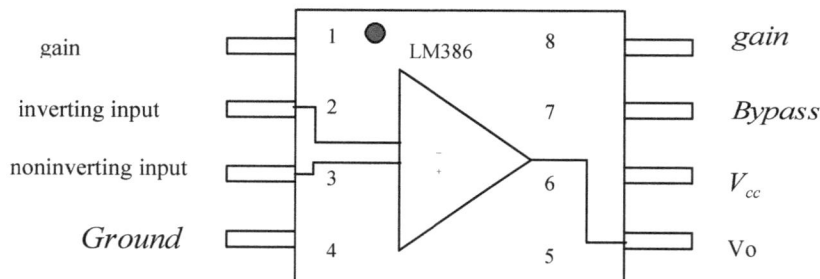

Figure 16.8. Pin configuration for the LM386 low voltage amplifier chip. Please see the LM386 datasheet for more information.

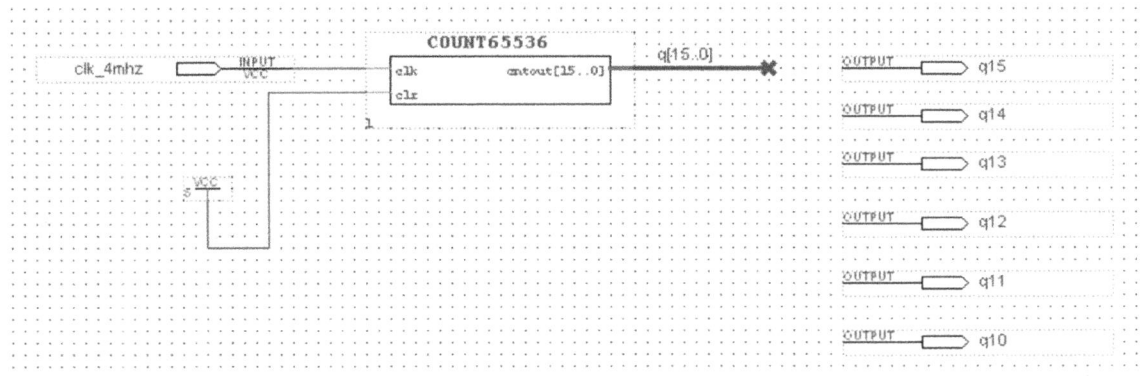

Figure 16.9. Digital test circuit for a servo motor using a mod65536 counter to divide down a 4MHz clock signal. The output ports for q10, q11, q12, q13, q14, and q15 are individually connected to the signal input of the servo motor.

Left Servo Clock Frequency (Speed Characteristics and rotation direction)						
Counter Output	Frequency	Slow Speed	Medium Speed	Fast Speed	Forward	Reverse
Q10						
Q11						
Q12						
Q13						
Q14						
Q15						

Table 16.4. Left Servo clock frequency verses speed and rotation direction.

Right Servo Clock Frequency vs. (Speed Characteristics and rotation direction)						
Counter Output	Frequency	Slow Speed	Medium Speed	Fast Speed	Forward	Reverse
Q10						
Q11						
Q12						
Q13						
Q14						
Q15						

Table 16.5. Right Servo clock frequency verses speed and rotation direction.

Before you begin recording the servo clock frequency verse speed and rotation direction in Table 16.4 and Table 16.5 the servos should be identified by a label as the right servo and the left servo. The servos should be positioned as shown in Figure 16.10. The shaft and tire of the servos should be pointed in opposite directions as shown.

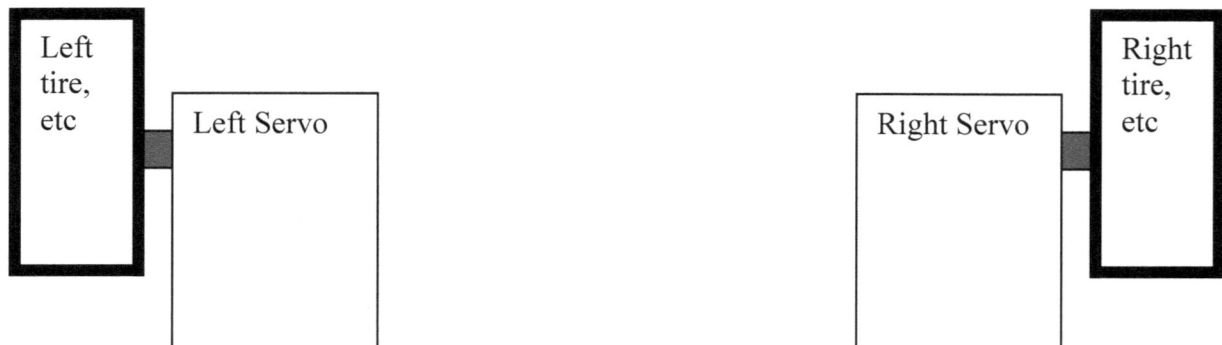

Figure 16.10. Position of left servo relative to right servo.

It should be noted that clockwise is forward and counter clockwise is reverse. While each servo is positioned differently, the robot will go forward when both servos are going forward. The robot will go backward or reverse when both servos are in reverse. The robot will turn left when the left servo is stopped and the right servo is in forward and the robot will turn right when the right servo is stopped and the left servo is forward. For the robot to turn left in place the left servo needs to be in reverse and the right servo need to be in forward and for the robot to turn right in place the left servo needs to be in forward while the right servo needs to be in reverse. Keep in mind that when the robot is moving forward the left servos and the right servo are actually going in opposite directions. This is because they are positioned on opposite sides of the robot as well as faced in opposite directions.

5. Construct the circuit shown in Figure 16.11. Figure 16.11 consist of three main VHDL modules: Servo_control.vhd, speakerid_ctl.vhd, and count65536.vhd. The servo_control VHDL module is listed in Section 9.9 on page 370. The count65536 VHDL module is listed in Section 9.6.2 on page 349. The VHDL code for speakerid_ctl VHDL module is listed in Section 9.14 on page 390. You must copy the code for each and compile each. Then created a graphic design file called robottop.qpf. Insert the symbols from the three compiled files into robottop.qpf and connect them as shown in Figure 16.11. . Lastly, assign pins as shown in Figure 16.5. sel0,sel1, and sel2 should be assigned to three switches. These inputs select a specific speaker to command the robot. Clk_4Mhz should be assigned to pin 83 which is the global clock input pin. The output of the comparator (LM339) should be assigned to the input called pulse. The output of the comparator is a clock like signal which drives a mod65536 counter. Led0, Led1, Led2, Led3, Led4, Led5, Led6, and Led7 are active low outputs that should be assigned

to LEDs on the PLD platform. Each led represent a different speaker with different thresholds. A led lights up when the speaker matches the thresholds programmed for that led. Therefore, the robot will only listen to selected speaker. When the selected speaker speaks into the microphone the robot should turn right, otherwise it goes forward. speakerreadyled is an active low output signal. It should be assigned to an LED. The speaker should speak into the microphone when this led lights up. This LED stays lit for about 2 seconds. Afterwards, the robot tries to identify the speaker. The outputs lservo and rservo should be connected to the left servo and right servo, respectively. The left servo reside on the left side of the robot. The right servo resides on the right side of the robot. You must assign the frequencies determined in step 3 to the servo control module. For this particular design, w14 causes the left servo to turn clockwise, w13 causes the left servo to turn counter clockwise, w13 causes the right servo to turn clockwise, and w15 causes the right servo to turn counter clockwise. This may be different for each servo which may depend on the servo modification and the potentiometer position, etc. Compile and program this design into the PLD platform.

6. Run the robot from the test bench in the lab to test and verify all the features. The thresholds for the speakers may need fine tuning. Connect the −12, +12, and 5 volt supply to the robot's analog circuit as needed. Connect the PLD platform to its own power supply . The ESOC board and external circuit should share the same ground reference for best performance.

7. Acquire all the hardware components for the robot. A starting list is given in Table 16.6.

8. Construct the hardware platform for the robot as shown in Section 16.2.3 on page 539. The RX PLD platform needs to stand above the plexiglass platform. This is done here with 1 and ½ inch standoffs. The platform is about 10 inches in diameter. The weight of the PLD platform along with the other components needs to be evenly distributed to prevent the robot from easily turning over.

9. Lastly, incorporate the battery supplies into the project. The receive robot may now feed off the battery supplies. If not careful, the battery pack can be drained in several minutes. Make sure you have a spare battery pack or a fast charger and a power outlet nearby for uninterrupted demonstrations.

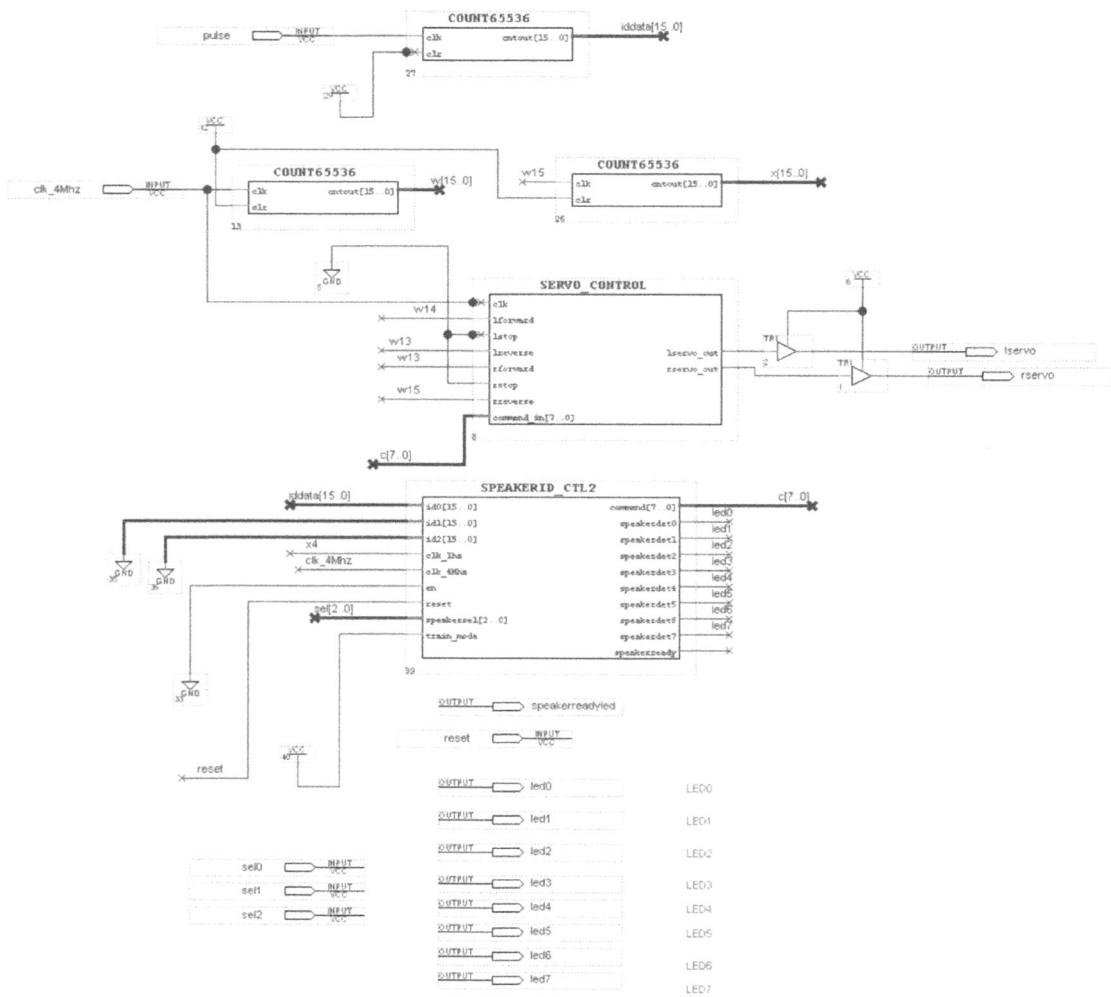

Figure 16.11. Digital Circuit to process voice signals from filterbank (Quartus Project file: speakerID_ctl3.qpf).

16.2.2. Project 6 Hardware Parts List

The hardware parts list for Project 6 is given in Table 16.6.

Part	Quantity	Ref designator	Description
Futaba S3003	2	servo1, servo2	servomotors
Battery supply, radio shack, 9.6V, 1000mA, 230-0342	1	Vcc	9.6 Volt Nickel Cadmium R/C car battery pack and charger
DC power connector, M type	1		For connecting battery pack to ESOC board
Battery supply, +12 and −12 volts.	1	-V, V+	OP177 or 741 , etc., power supplies
Home depot, etc., Caster wheel	1	wheel	Front wheel
Round Platform, plexi glass, from Home Depot, Menards, etc.	1	platform I	Round platform of diameter 11 inches (and thickness 3/16 inches) cut from fiber glass, plastic, etc
standoffs	4	ST1, ST2, ST3, ST4	1 and ½ inch standoffs
Screws, 844-712 is Menards part #	?		Machine screws, size 4-40x ½, 30 pieces, can be used to mount caster wheel and servo if the 4-40x ¼ is not long enough.
Screws, 844-710, 844-712 is Menards part #	About 16		Machine screws, size 4-40x ¼, 34 pieces. 8 is needed for mounting the RX ESOC, 4 for the caster wheel, 2 for each servo
Nuts, 846-644 is Menards part #	About 12		Machine nuts, size 4-40, 28 pieces. 4 for the caster wheel, 4 for each servo
Washers, 154-306 is Menards part #	About 12		Washers, No. 6, 100 pieces. 4 is needed for mounting the bottom of the RX ESOC, 4 for the caster wheel, 4 for each servo
Velcro			4 inch by 2 inches, 2 sets; for holding battery pack and breadboard on robot.

Table 16.6. Project 6 hardware parts list. For more details on where to order parts see Appendix D on page 577.

16.2.3. Platform for Project 5 and Project 6

Figure 15.16 shows the bottom view of the platform suggested for Project 5 and Project 6. The PLD platform and most circuitry resides on top of the platform along with the battery pack. Any other sensible platform may be used if desired.

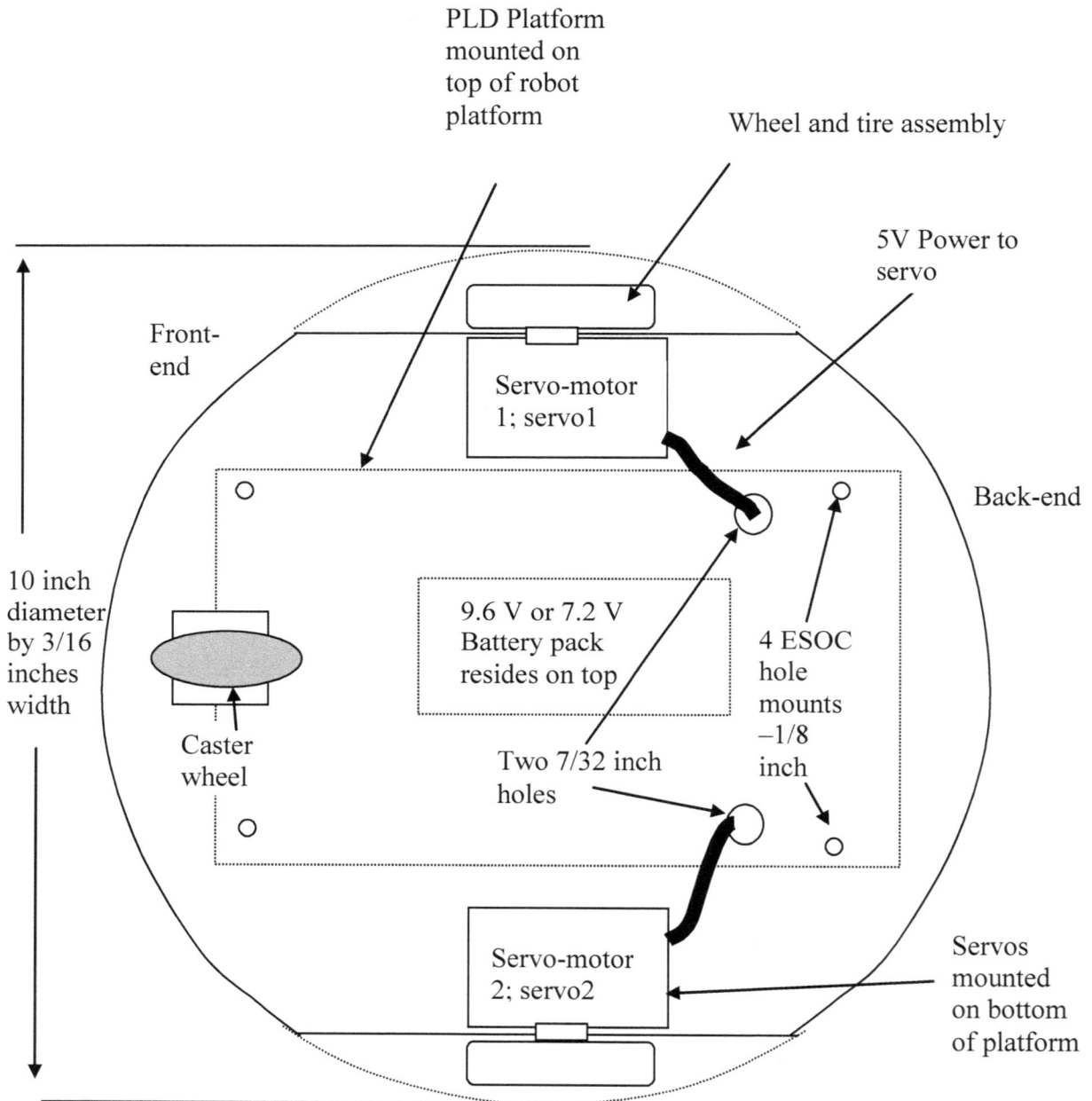

Figure 16.12. The bottom view of the PLD-bot.

17. Data Communication Projects

Data communications deal with the transmission of digital data from one device or system to another. More details can be found in Forouzan (2007) and Tomasi (2004) in the Reference Section on page 578. A wireless medium is used in Projects 7 and 8. Project 7 is a Simplex Wireless Digital Communication System using an analog to digital converter, ADC0804, and a digital to analog converter, DAC0808. On the other hand, Project 8 is a half duplex communication system. The various communication system transmission modes are reviewed in Section 6.3 on page 198. The following Sections show how to transmit a sinusoidal tone using a simplex digital communication system. The requirements followed by the procedure for building and testing the simplex wireless digital communication system are given next.

17.1. Project 7 Requirements

The requirements of this project are given as follows:

Requirement 1: Design a simplex communication system as shown in the block diagrams for the transmitter presented in Figure 17.1 and the receiver presented in Figure 17.2.

Requirement 2: A sinusoidal signal must be converted to a digital signal using an A/D converter with parallel digital outputs.

Requirement 3: The parallel digital output signal from the A/D must be converted to serial using the PLD platform. This serial data will be used to modulate the carrier frequency signal.

Requirement 4: Must use a wireless transmitter and receiver pair. Review WyJen Technologies transmitter modules which can be ordered online at www.wyjen.com.

Requirement 5: Transmitter: A small antenna may be used for the transmission of the modulated signal.

Requirement 6: Receiver: The transmitted signal must be detected by the receiver within at least 15 to 30 feet of the transmitter.

Requirement 7: The digital data recovered by the receiver may be fed to PLD platform such as the PLD platform where it is converted from serial to parallel form and sent to D/A converter which is then fed to a speaker or oscilloscope. The received D/A signal must match the analog signal sent.

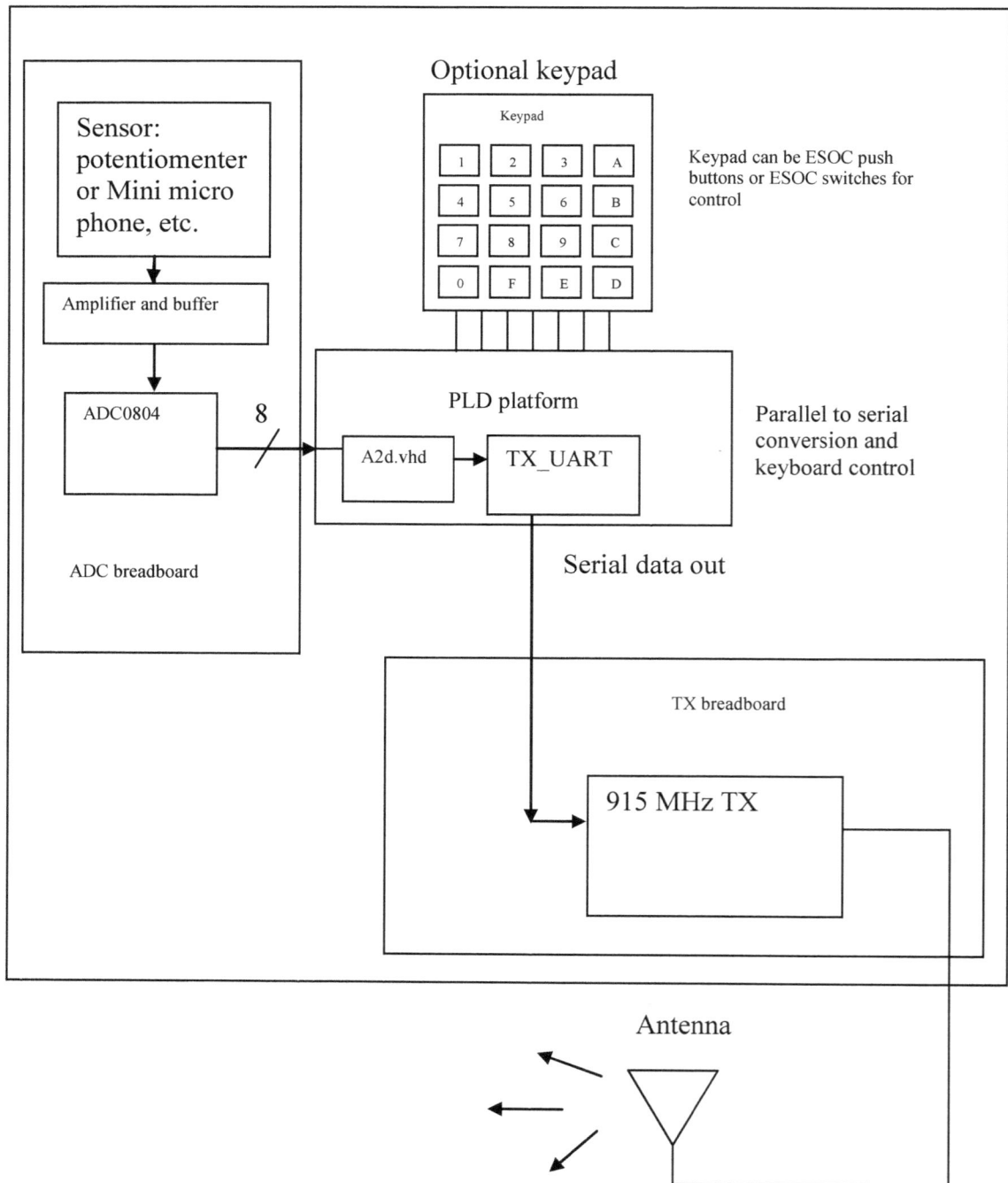

Figure 17.1. Transmitter block diagram in a simplex communication system.

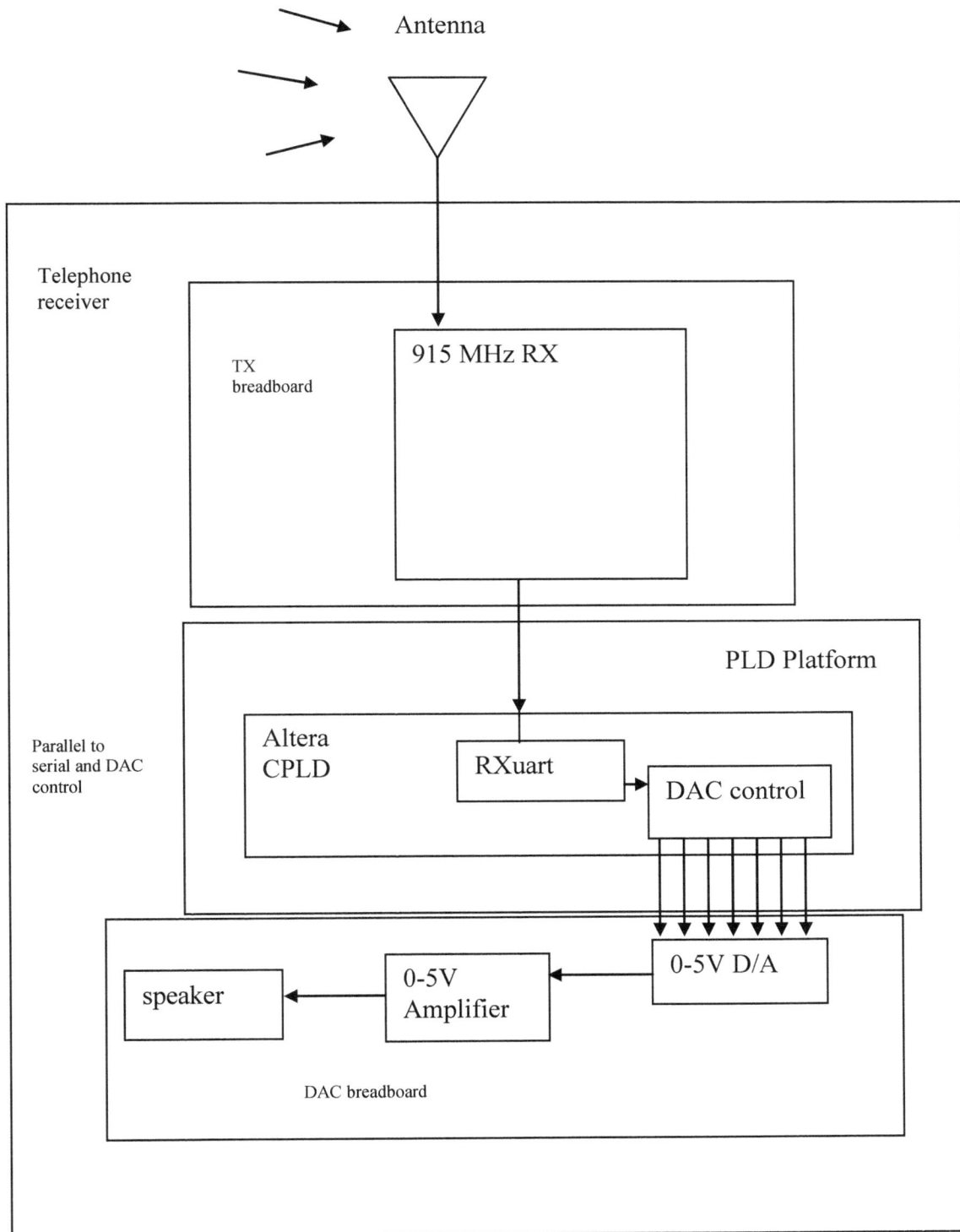

Figure 17.2. Receiver block diagram in a simplex communication system.

17.1.1. Project 7 Procedure
1. Obtain all the electronic components for the project given in Table 17.1.

Part/Equipment	Quantity	Ref Designator	Description/Purpose
Altera Software	1		http://www.altera.com/products/soft ware/quartus-ii/switching/max-plus-ii/max-plus-ii-users.html
ESOC board (or Altera PLD platform)	2	ESOC	Altera programmable logic device EMP7128SLC board, www.elexp.com/tst_pld2.htm
Function Generator	1		Sine wave generation for input to ADC0804
Oscilloscope	1		Received signal measurement on output of DAC0808
915 MHz TX	1	Module	915MHz Transmitter module or similar, www.wyjen.com
915 MHz RX	1	Module	915MHz Receiver module or similar, www.wyjen.com
Speaker	1	speaker	Small speaker, radio shack, etc
ADC0804	1	IC	Analog to digital converter, DIP
DAC0808	1	IC	Digital to analog converter, DIP

Table 17.1. Electronic components for simplex wireless phone. For more details on where to order parts see Appendix D on page 577.

2. Use the mixed digital and analog circuits from section 9.4, "Testing the ADC0804 and DAC0808 Together using a PLD", on page 336. These circuits are redrawn here for your convenience. The digital circuit (Figure 17.3) is used to program the PLD platform. The analog circuit consists of a transmit circuit (Figure 17.4) and receive circuit(Figure 17.5). These circuits are used to test the ADC0804 and DAC0808 together using one PLD platform. Mastering this step will allow you to go to next step involving two PLD platforms: The transmitting PLD platform and the receiving PLD platform.

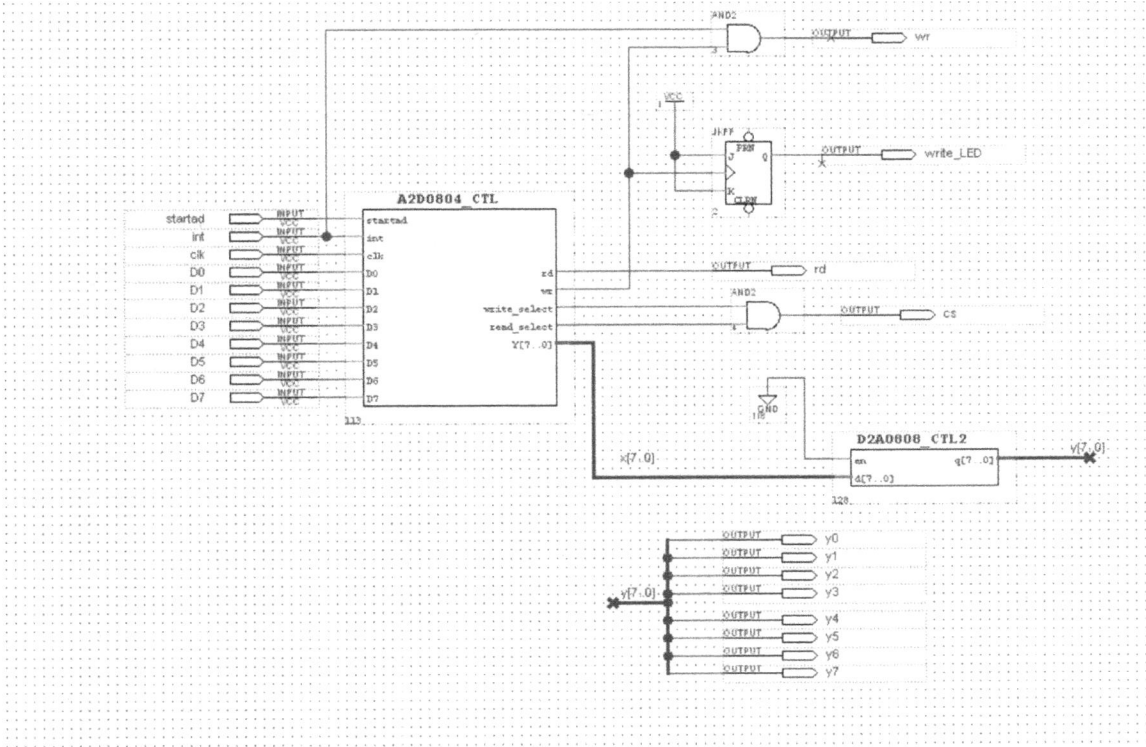

Figure 17.3. Digital circuit for the ADC0804 and DAC0808 system (Quartus file: a2d0804_2_d2a0808_ctl.qpf).

Figure 17.4. Analog transmit circuitry for ADC0804 chip.

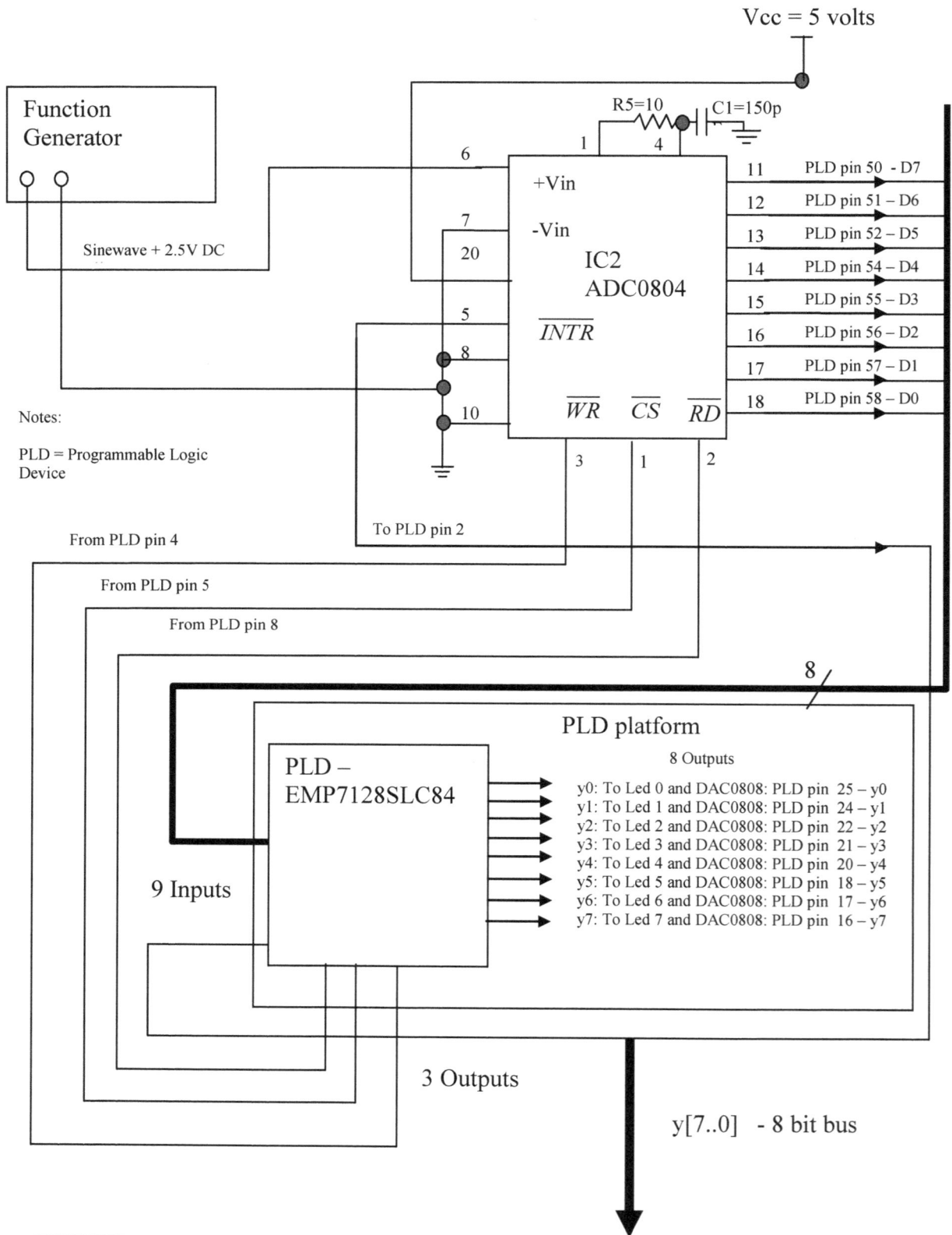

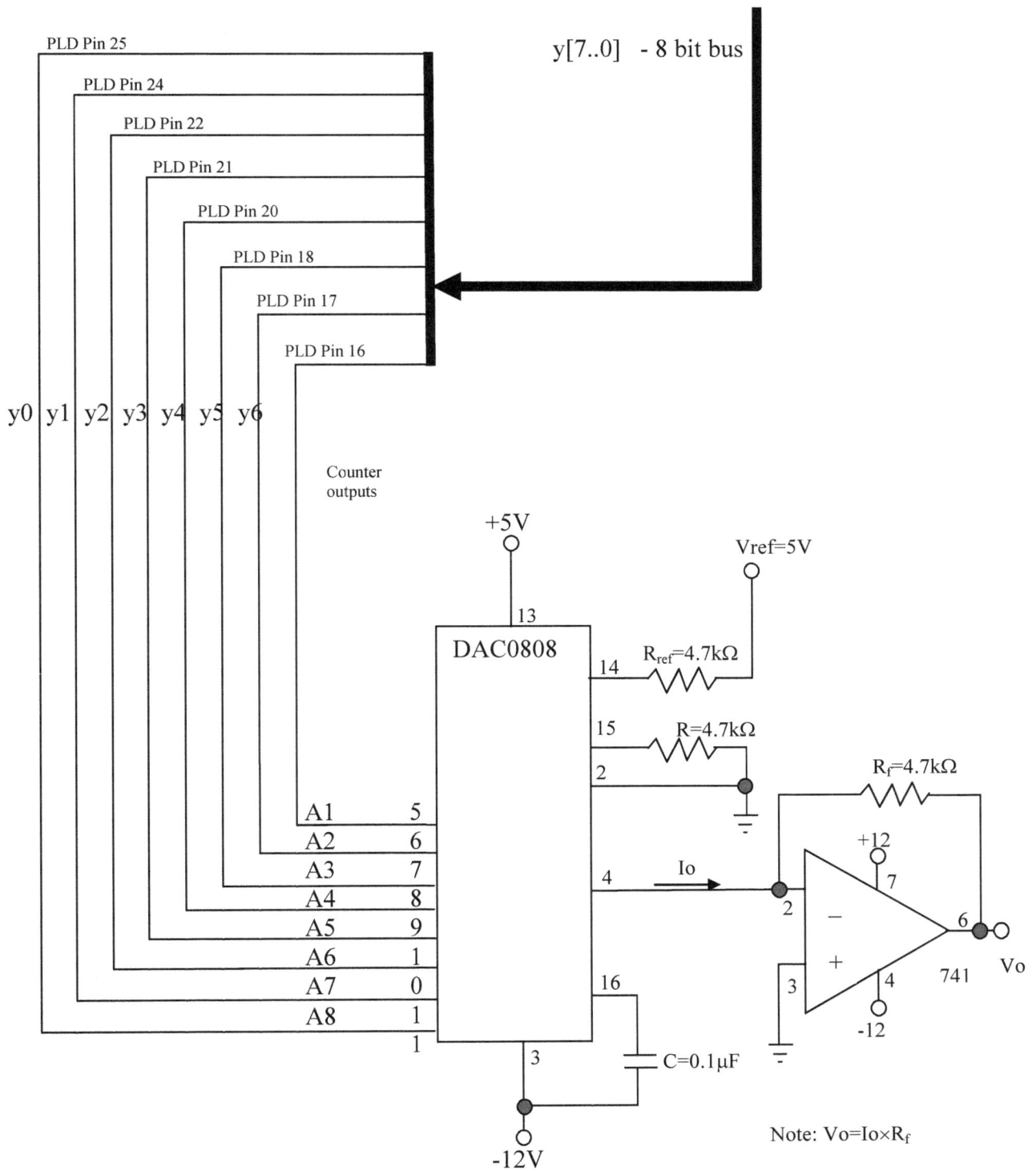

Figure 17.5. Analog receive circuitry for DAC0808 chip.

3. Now construct the transmit digital circuit (**txuart.qpf.**) for the transmitting portion of the simplex communication system as shown in Figure 17.6 and compile it. The VHDL code and details for this circuit is given in Section 9.6, The txuart, on page 344. This circuit should be programmed into the transmitting PLD platform.

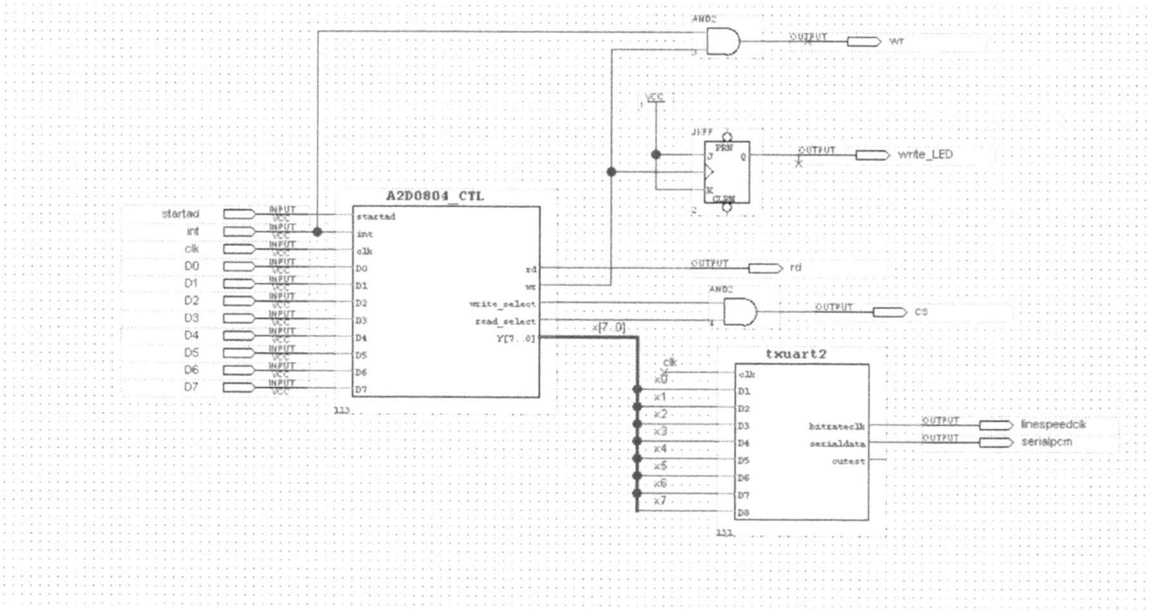

Figure 17.6. Altera digital circuit for simplex communication system showing the a2d0804_ctl and txart2 VHDL modules (Quartus file: **teletxuart.qpf**). The a2d0804_ctl controls the ADC0804 analog to digital converter and the txuart converts the parallel PCM data (see signals D0, D1, D2, D3, D4, D5, D6, D7) from the ADC0804 to serial PCM.

4. Construct the receive digital circuit (**rxuart.qpf**) for the receiving portion of the simplex communication system as shown in Figure 17.7 and compile it. The VHDL code and details for this circuit is given in Section 9.7, The rxuart, on page 351. This circuit should be program into the receiving PLD platform.

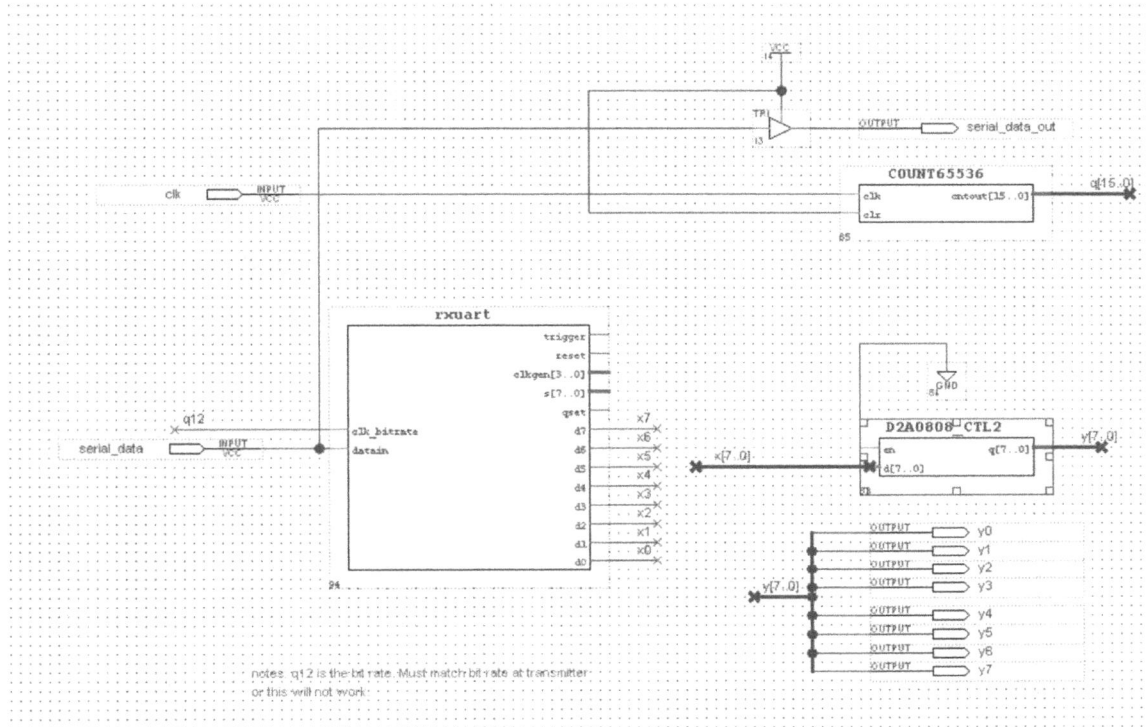

Figure 17.7. Altera digital circuit for simplex communication system showing the d2a0808_ctl and rxuart VHDL modules. The d2a0808_ctl controls the flow of data to the DAC0808 while the rxuart converts the incoming serial pcm data (serial_data) to parallel PCM data (see signals x0, x1, x2, x3, x4, x5, x6, x7). It is very important that the bit rate clk be the same as the bit rate clock at the transmitter.

5. Now test the transmitting PLD platform and receiving PLD platform using a physical wire. Connect a wire from the serial data output at the transmitter to the serial data input on the receiver. Also the two PLD platforms should share the same grounds. Yes connect them. This is shown in Figure 17.8.

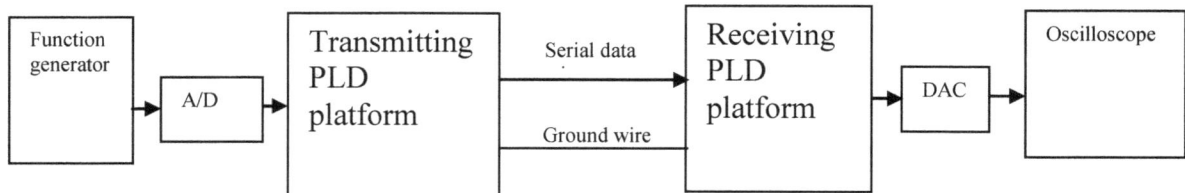

Figure 17.8. Test setup for the transmitting and receiving PLD platform using a wire as the transmission channel.

6. Now send sinusoidals of different frequencies throught the ADC0804 and transmitting PLD platform to the receiving PLD platform. Compare the transmitted signal with the received signal using an oscilloscope. They should match! Also verify the sound of the sinsusoidal using a speaker. It should sound like a tone!

7. You are now ready to perform step 6 again, only this time over a wireless channel instead of a hardwired channel. Build the circuit for the ASK transmitter and receiver then try step 6 again. This time remove the data wire and ground wire added in step 5, then connect the PLD output pin programmed for the serial data to the serial data input of the wireless 915MHz transmitter. The output of the 915MHz transmitter goes to an antenna. On the receiver end, an antenna feeds the rf input of the 915MHz receiver and the data output of the receiver connects to the input pin of the PLD programmed to receive the serial data. This is shown in Figure 17.9.

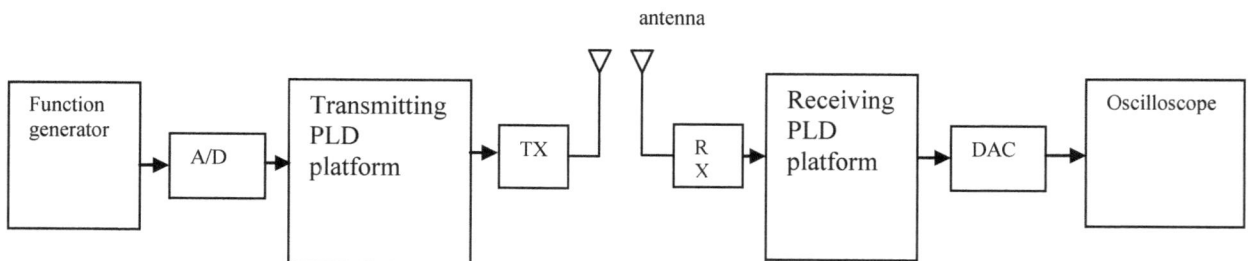

Figure 17.9. Test setup for the transmitting and receiving PLD platform using a wireless transmission channel.

17.2.   **Project 8 Half Duplex WirelessTelephony Using the Texas
Instruments Codec Chip -TP13054/TP3054**

The requirements for a half/full duplex wireless telephone system is given in the next section.

17.2.1. Project 8 Requirements

The requirements for project 8 are as follows:

Requirement 1: Must design a half duplex communication system using the block diagrams for the transmitter presented in Figure 17.10 and the receiver presented in Figure 17.11.

Requirement 2: Must successfully pass a sinusoidal signal through the half duplex system both directions. The transmitted sinusoidal must match the receiver sinusoidal.

Requirement 3: Codec - Must use a codec chip to perform the front end filtering (band limiting filter), analog to digital conversion, and parallel to serial conversion.

Requirement 4: Transceiver - Must use a 915MHz (or similar) wireless transmitter and receiver pair. Review WyJen Technologies for availability of half duplex transceiver modules which can be ordered online at www.wyjen.com.

Requirement 5: Transceiver - A small antenna may be used for both transceiver modules.

Requirement 6: Transceiver - The transmitted signal must be detected by the receiver within at least 15 to 30 feet of the transmitter.

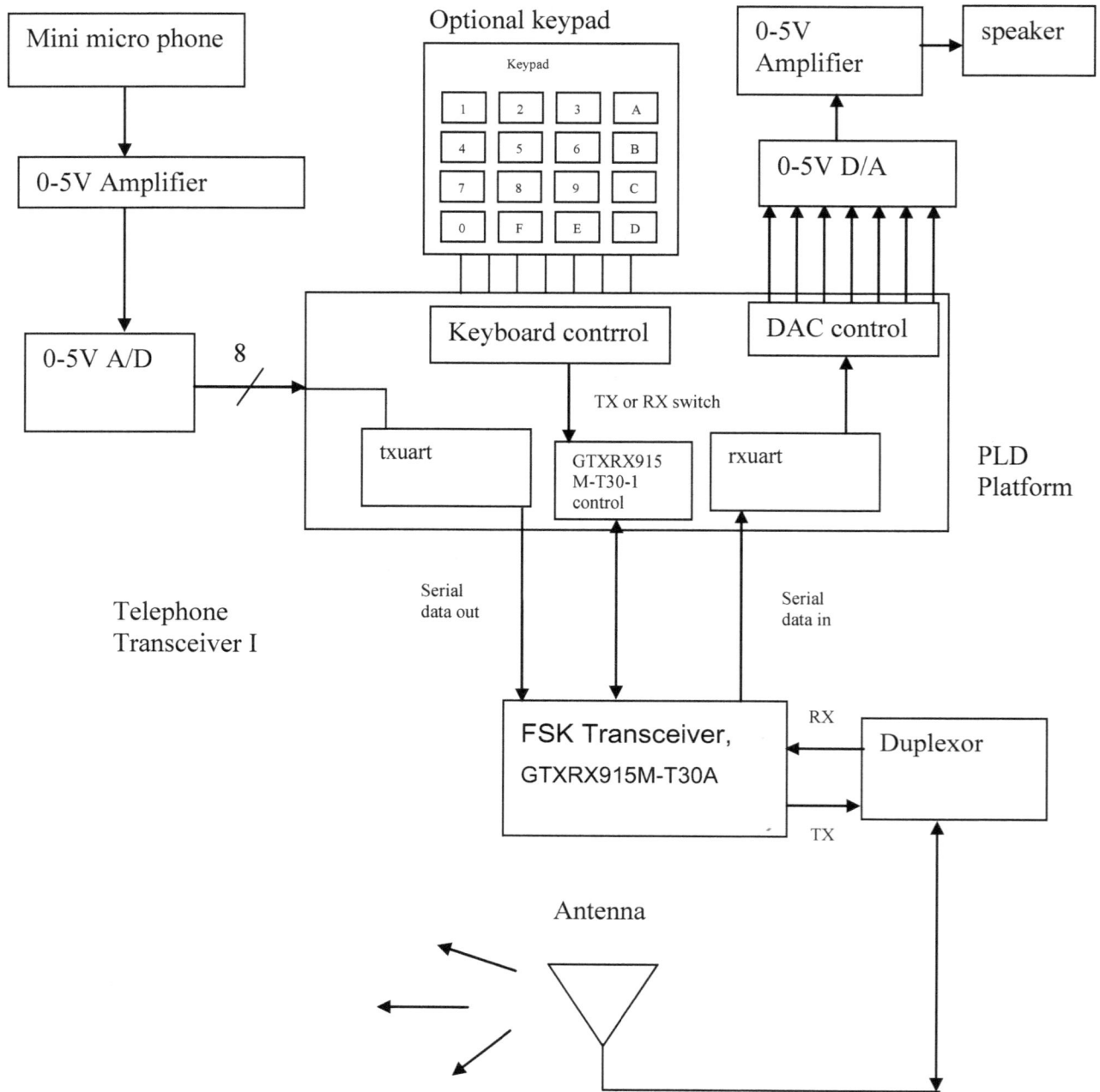

Figure 17.10. Block diagram of a transceiver in a half duplex communication system.

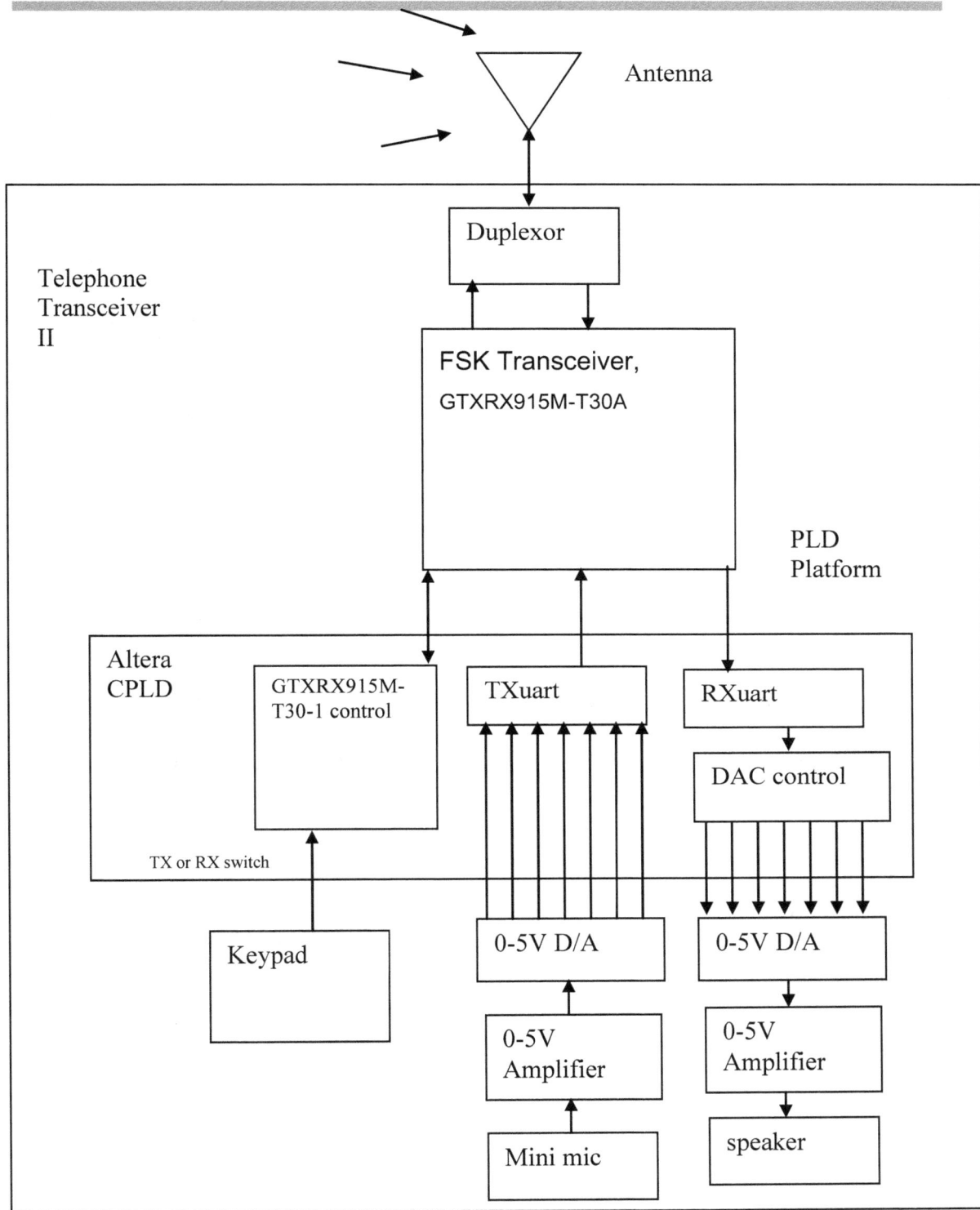

Figure 17.11. Block diagram of transceiver in a half duplex communication system.

17.2.2. Project 8 Procedure

1. Obtain all the electronic components for building the duplex communication system given in Table 17.2.

Part	Quantity	Ref designator	Description
ESOC board (or Altera PLD platform)	2	ESOC	Altera programmable logic device EMP7128SLC board, www.elexp.com/tst_pld2.htm
Battery supply, radio shack, 9.6V, 1000mA, 230-0342	2	Vcc	9.6 Volt Nickel Cadmium R/C car battery pack and charger, Radio shack
DC power connector, M type	2	Connector	For connecting battery pack to ESOC board, Radio Shack
LVM321/358/324	4	IC	Low voltage 0 to 5V rail to rail operational amplifier, National semiconductor, etc, DIP
Microphone, Radio Shack- 270-092	2	Mic	Mini Microphone, up to 15KHz input, Radio Shack
LM386	4	IC	Amplifier, 0 to 5V rails, gains=50, 200, etc
GTXRX915M-T30A	2	Module	915MHz Transceiver, SIP, www.wyjen.com or similar
Speaker	2	speaker	Small speaker
TP13054BN/TP3054AN	2	IC	Texas instrument codec and national instrument compatibles, 16 pin DIP, Digikey
6.144MHz crystal oscillator, Digikey CTX190ND or SE1226-ND	2	X	Crystal Oscillator for Baud rates, Digikey, DIP, CTS frequency controls MXO45HST-6.1440 or Epson Electronics SG-531P 6.1440MC, 8 pin PDIP

Table 17.2. Electronic components for half duplex wireless phone. Can use codec or combo chip to replace several components. For more details on where to order parts see Appendix D on page 577.

2. Write and compile the VHDL code for the TP3054/TP13054 chip. Esentially, the code derives the required clocks from the 6.144Mhz oscillator: 1.536Mhz, 64000hz, and 8000hz. The 6.144Mhz clock can be divided down as follows in the Table. The 64000 hz is derive from the 1.536Mhz clock using a MOD12 counter while the 8000hz clock is derived from the 1.536khz clock signal using a MOD192 counter.

Name of MOD65536 Counter output	Frequency of divided clock, $6.144 \times 10^6/2^n$, n=1,2,3,...16
Q_0	$6.144Mhz/2^1 = 3.072Mhz$
Q_1	$6.144Mhz/2^2 = 1.536Mhz$
Q_2	$6.144Mhz/2^3 = 768khz$
Q_3	$6.144Mhz/2^4 = 384khz$
Q_4	$6.144Mhz/2^5 = 192khz$
Q_5	$6.144Mhz/2^6 = 96khz$
Q_6	$6.144Mhz/2^7 = 48khz$
Q_7	$6.144Mhz/2^8 = 24khz$
Q_8	$6.144Mhz/2^9 = 12khz$
Q_9	$6.144Mhz/2^{10} = 6khz$
Q_{10}	$6.144Mhz/2^{11} = 3khz$
Q_{11}	$6.144Mhz/2^{12} = 1.5khz$
Q_{12}	$6.144Mhz/2^{13} = 750hz$
Q_{13}	$6.144Mhz/2^{14} = 375hz$
Q_{14}	$6.144Mhz/2^{15} = 187.5hz$
Q_{15}	$6.144Mhz/2^{16} = 93.75hz$

Table 17.3. Frequency division of the 6.144Mhz clock using a mod 65536 counter (16 bit synchronous binary counter).

3. Refer to the codec datasheet for the detailed circuitry and requirements. Then connect the data output and data input of the codec circuitry to the PLD. Then test the data transmission and decoding of the codec and PLD as shown in Figure 17.12 for one complete path. The path through the PLD should be buffered from input to output. The complete path consists of the path from the transmitting codec through the transmitting PLD through the transmission channel to the receiving PLD and receiving codec. Also the two PLD platforms should share the same grounds. In this case the transmission path is a wire instead of a wireless link. Now, using a function generator, send sinusoidals of different frequencies through the complete path. Compare the transmitted signal with the received signal using an oscilloscope. They should match! Also verify the sound of the sinsusoidal using a speaker. It should sound like a tone!

4. Now construct the digital circuit for the PLD and compile it. The digital circuit should be the same for both PLDs involved. Portion of the digital circuit is for receiving the digital data and the other is for controlling the transceiver

5. You are now ready to implement the project over a wireless link. This time remove the data wire and ground wire, then connect the PLD serial data output pin to the serial data input pin of the transceiver and connect the PLD serial data input pin to the serial data output pin of the transceiver. The rf input/output of the transceiver goes to an antenna. This is shown in Figure 17.13.

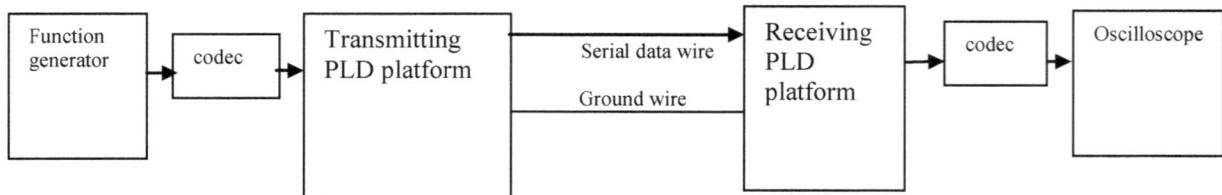

Figure 17.12. Test setup for transmitting and receiving PLD platform using a wire as the transmission channel.

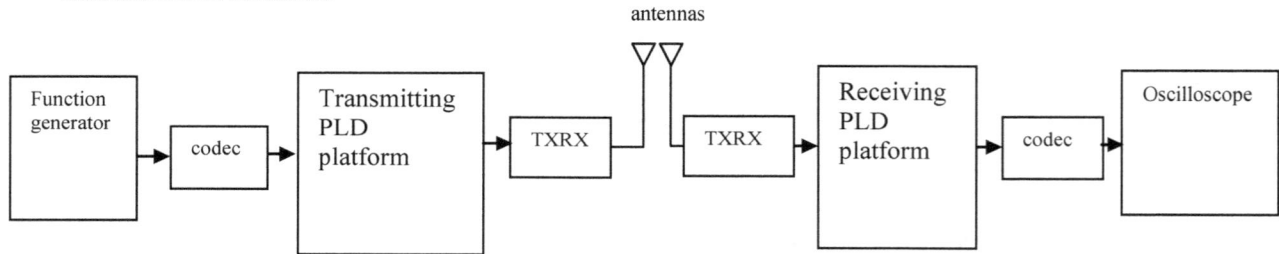

Figure 17.13. Test setup for transmitting and receiving PLD platform using a wireless transmission channel using a transceiver(TXRX).

18. Digital Filtering using PLDs or DSPs

Digital filtering in a PLD or DSP can be fun to learn. Project 9 in this section deals with digital filtering. The goal of the project is to implement a digital filter in programmable logic device (PLD) or DSP. The requirements are given next. For more on DSPs see Appendix A, Appendix B and Appendix C. Additionally, an excellent reference to many DSP applications and labs can be found in Chassaing (2005) listed in the Reference Section on page 578.

18.1.1. Project 9 Requirements

The requirements of this project are given as follows:

Requirement 1: Must create a signal of several frequencies using Matlab on a computer and send this signal to the audio port for external use.

Requirement 2: Must do a spectral analsys on the created sound and filtered sound.

Requirement 3: Must convert the audio signal to a digital signal using a analog to digital converter. This digital signal must interfaced to a Programmable Logic Device (PLD) or Digital Signal Processor (DSP) for processing.

Requirement 4: Must be able to listen to orignal and filtered sound. Can use a single-pole, double-throw (SPDT) switch to select between two speakers: one speaker for the filtered audio signal from the DSP speaker port and the other for the signal coming from the computer's audio port. That is, the switch is used to send the signal from the audio port of the computer directly to a speaker for listening or to send the audio signal to a digital processor for processing.

Requirement 5: Must use a PLD or DSP to implement the digital filter alogorithm. The PLD or DSP must remove all frequencies created in requirement 1 except 1.

Requirement 6: The digital filter implemented shall be low pass, bandpass, or highpass.

Requirement 7: Must use a digital to analog converter to convert the digital filtered signal to an analog signal.

Requirement 8: The analog filtered digital signal from requirement 6 must be sent to a speaker for listening and analysis.

18.1.2. Project 9 Procedure

Obtain all the electronic and hardware components for the project as listed in Table 18.1.

Part	Quantity	Ref designator	Description
ESOC board or PLD platform	1	ESOC	Altera programmable logic device, Devry
LVM321/358/324 (optional)	2	IC1	Low voltage 0 to 5V rail to rail operational amplifier, National semiconductor, etc, DIP
LM386	2	IC	Amplifier, 0 to 5V rails, gains=50, 200, etc
Speaker	2	speaker	Small speaker
ADC0804	1	IC	Analog to digital converter, DIP
DAC0808	1	IC	Digital to analog converter, DIP

Table 18.1. Electronic components for digital filter project.

2. Complete Section 9.4 on page 336.
3. Create and add several sinusoidals in Matlab using the frequencies specified in Table 18.2. The amplitudes of the sinusoidals representing these frequencies are calculated by $Vn=4V/n\pi$ where $n=1,3,5,7,9, \ldots$ Use $V=5$volts. The frequencies were calculated by $fn=f0\times n$. Note that $f0=f1$ in each set of frequencies in Table 18.2. Add these frequencies and send the sum to the audio port using the Matlab command **wavplay()**.

Group Number	Frequencies
1	f1=500Hz, f3=1500Hz, f5=2500Hz
2	f1=1000Hz, f3=3000Hz, f5=5000Hz
3	f1=2000Hz, f3=6000Hz, f5=10000Hz
4	f1=3000 Hz, f3=9000Hz, f5=15000Hz
5	f1=4000Hz, f3=12000Hz, f5=20000Hz

Table 18.2. Group frequencies assignment.

4. Listen to the created sound coming from the audio port using a speaker. Use the oscilloscope to record your results and measurements.

5. Perform a spectral analysis of the individual sounds and their sum from step 3. You should plot a magnitude verses frequency plot of each sound in Matlab.
6. Implement the filtering algorithm in the PLD or DSP. Use the handout given in class as a reference for the VHDL algorithm.
7. Send the filtered signal to a DAC and finally a speaker for listening and analysis.
8. Use an amplifier if necessary between the DAC and speaker.
9. Perform a spectral analysis on the filtered sound using Matlab. The output of the PLD/DSP can be connected to the audio input of the PC for recording. Then use Matlab to record the filtered data for analysis.

19. Using DC Motors Instead of Servo Motors

DC motors can be used as an alternative to servo motors. This section discusses the use of a DC motor as a replacement for the servo motor used in Section 13 through 16. If a DC motor is desired for the robot the student would need to write the VHDL code for this interface. A servo motor is controlled by a digital signal which causes it to go in reverse, forward, or stop depending on the duty cycle of the digital signal. A plain DC motor is not equipped with the control electronics as the servo motor and will need some additional circuitry to reverse and forward the motor. To control the direction of a DC motor, the current needs to be positive or negative. For example, A DC motor may require the current through it to be reversed in order to turn in a clockwise direction and forwarded to turn in a counterclockwise direction. A device called an H-bridge provides this control which itself is digitally controlled.

19.1.   The Basic Concept of a H-Bridge

The H bridge's main use here is to control a DC motor which does not have a digital interface such as the servo motors. Platform II involves modifying an existing R/C remote control car and possibly connecting the CPLD to the digital side of an H-bridge. An H-bridge is a circuit used to reverse the direction of the motor. It consists of 4 switches arranged in an H pattern. H-bridges are typically used to control the forward and reverse directions of R/C cars.

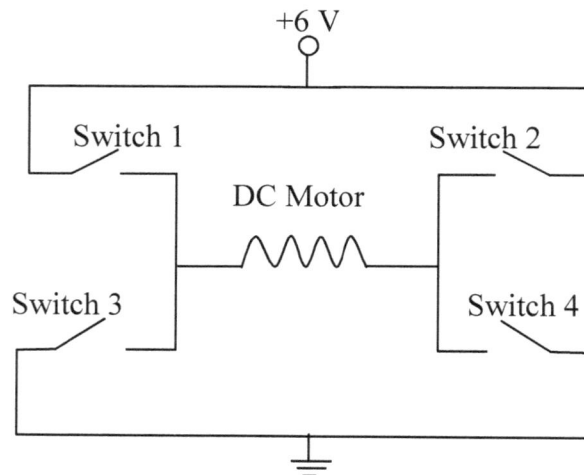

Figure 19.1. H-bridge circuit to switch direction of motor. Pressing switches 1 and 4 causes the motor to turn in one direction. Pressing switches 2 and 3 reverses the direction. Pressing switches 3 and 4 or 1 and 2 stops current flow thereby stopping the motor. The switches are typically transistors interfaced to a digital controller such as a CPLD, microprocessor, microcontroller, etc.

19.2.   Texas Instrument H-Bridge Chip – TPIC0107B

The Texas Intrument H-bridge chip, TPIC0107B, provides the following DC motor modes of operations:

1. reverse,
2. forward, and
3. brake (motor stops)

The DC motor is connected to two terminals: out1 (pins 5 & 6) and out2 (pins 15 & 16) of the chip. The out1 and out2 pins supply 3 amps of current. Two inputs, PWM (PWM control, pin 8) and DIR (direction control, pin 3), are used to control the modes of operations. The remaining pins are power grounds (pins 7, 9, 12, &14; all four must be grounded), subtrate grounds (pins 1, 10, 11, & 20; at least one must be grounded), power (pins 2, 4, 17, & 19), and diagnostic pins (13, & 18; optional connections). The modes of operation using the PWM and DIR as inputs are summarized in Table 19.1.

DIR	PWM	Mode
0	0	Brake
0	1	Counter clockwise
1	0	Brake
1	1	Clockwise

Table 19.1. Truth Table of H bridge chip (TPIC0107B).

Appendix A. Introduction to Texas Instruments Digital Signal Processors (DSPs)

As of the current date of this publication Texas Instruments currently offers the following DSP platforms:

  1.) TMS320C6000™ DSP
  2.) TMS320C5000™ DSP
  3.) TMS320C2000™ DSP

A Web Audio Video Tutorial of the various Texas Instrument DSP platform, software, etc., can be viewed at the following URL:
http://www.go-dsp.com/fet/expressdsp/index.html .

Additional, an excellent reference to many DSP applications and labs can be found in Chassaing (2005) listed in the Reference Section on page 578. I used this resource while teaching digital signal processing using the C6713 DSP while at DeVry. The students received hands on experience while bringing class room concepts to life.

A.1. The TMS320C6000 DSP Platform

The TMS320C6000 DSPs comprise fixed point and floating point DSPs. These are optimized for high performance and ease of use with high-level languages such as C. The TMS3206000 DSPs can run up to core speeds as much as 1GHz (specifically, see the TMS320C6400). A summary of information on the C6000 platform can be found at:
http://dspvillage.ti.com/docs/catalog/devtools/toolmatrix.jhtml?templateId=5121&familyId=132.

The C6000 platform consists of the following DSPs:

1. TMS320C62X for applications involving multi-channel voice, data and imaging with processor speeds between 150 MHz and 300Mhz.
2. TMS320C64X for applications involving accelerated video, data, imaging, audio with processor speeds between 300 MHz and 1000Mhz.
3. TMS320C67X for applications involving IEEE single and double-precision floating-point with processor speeds between 100 MHz and 300Mhz.

Each DSP has features that are application based. Typical applications of the TMS320C62X DSP include digital communications infrastructure and video and image processing. Typical applications of the TMS320C62X DSP include applications involving multi-channel, multi-function requirements, such as wireless base stations, multi-channel telephony systems, remote access servers (RAS), digital subscriber loop (xDSL) systems, personalized home security systems, advanced imaging/biometrics, industrial scanners, and precision instrumentation, etc. Typical applications of the TMS320C67X DSP involve demanding computational abilities such as audio, medical imaging, instrumentation, and automotive, etc.

We will be focusing on the TMS320C6713 DSP for the remainder of this publication.

A.2. Getting Started with the TMS320C6713 DSP

Before getting started it is helpful to understand what the TMS320C6713 is all about. The TI TMS320C6713 is a floating-point digital signal processor with the key features listed in Table 19.2. As shown in Table 19.2 the different TMS320C6713 DSPs are differentiated by their main clock speeds: 167Mhz, 200MHz, 225MHz, and 300MHz. After this small introduction to the TMS320C6713 you can now learn more on how to program it in the next few Sections using C67x **DSPLIB** and example programs.

DSP	TMS320C6713-167	TMS320C6713-200	TMS320C6713-225	TMS320C6713-300
Speed				
Frequency(MHz)	167	200	225	300
Peak MMACS	1336	1600	1800	2400
Memory				
RAM	256 KB	256 KB	256 KB	256 KB
On-Chip L1 Cache Memory	8 KB	8 KB	8 KB	8 KB
On-Chip L2 Cache or SRAM	64 KB	64 KB	64 KB	64 KB
External Memory Type Supported	Async: SRAM, SBSRAM, SDRAM	Async: SRAM, SBSRAM, SDRAM	Async: SRAM, SBSRAM, SDRAM	Async: SRAM, SBSRAM, SDRAM
DMA	16-Ch EDMA	16-Ch EDMA	16-Ch EDMA	16-Ch EDMA
Interfaces				
EMIF	1 16-Bit (PYP)	1 32-Bit (GDP), 1 16-Bit (PYP)	1 32-Bit (GDP/ZDP)	1 32-Bit (GDP)
McBSP	2	2	2	2
McASP	2	2	2	2
I2C	2	2	2	2
HPI	1 16-Bit	1 16-Bit	1 16-Bit	1 16-Bit
Miscellaneous				
Timers	2 32-Bit GP	2 32-Bit GP	2 32-Bit GP	2 32-Bit GP
Core Supply (Volts)	1.2 V	1.2 V/1.26 V	1.26 V	1.4 V
IO Supply (Volts)	3.3 V	3.3 V	3.3 V	3.3 V
Operating Temperature Range (°C)	-40 to 105	0 to 90 (PYP),-40 to 105 (GDP)	0 to 90	0 to 90

Table 19.2. Key features of the TMS320C6713 DSP. Each of the DSPs uses 1 c67x CPU. Notes: EDMA=Enhanced Direct-Memory-Access. McASP=Multi-Channel Audio Serial Ports, McBSP=Multi-Channel Buffered Audio Serial Ports, EMIF=External Memory Interface, DMA=Direct Memory Access, HPI=Host Port interface.

A.3. TMS320C6700 DSP Libraries

Several optimized assembly level programs have been provided by TI to speed your time to market. These assembly programs can be called by a C program or another assembly program. These assembly optimized programs are called benchmarks and are collectively called the **DSPLIB**. TI provides separate benchmarks for the following DSP generations: TMS320C62x, TMS320C64x and TMS320C67x. The benchmark programs for the TMS320C67x DSP are listed in Table 19.3. These programs are categorized as follows:

1. Adaptive filtering
2. Correlation
3. FFT
4. Filtering and convolution
5. Math
6. Matrix
7. Miscellaneous

The DSPLIB programs are free and can be downloaded from the following TI web site: http://focus.ti.com/docs/toolsw/folders/print/sprc121.html.

The ***TMS320C67x DSP Library Programmer's Reference Guide*** is a good place to start when writing C or assembly language code for the TMS320C6713 DSP. It gives instructions on how to install and use the DSPLIB relative to the C67x DSP. This reference guide is available at: http://focus.ti.com/lit/ug/spru657/spru657.pdf. (119 pages)

Furthermore, examples of how signal processing programs should use the **DSPLIB** are given in following TI reference at: http://focus.ti.com/lit/an/spra947/spra947.pdf. (18 pages)

Category	Available Programs
Adaptive Filtering	sp_lms
Correlation	sp_autocorr
FFT	sp_bitrev_cplx, sp_cfftr4_dif sp_cfftr2_dit, sp_fftSPxSP, sp_ifftSPxSP, sp_icfftr2_dif
Filtering and convolution	sp_fir_cplx, sp_fir_gen, sp_fir_r2, sp_fircirc, sp_biquad, sp_iir, sp_iirlat, sp_convol
Math	sp_dotp_sqr, sp_dotprod, sp_dotp_cplx, sp_maxval, sp_maxidx, sp_minval, sp_vecrecip, sp_vecsum_sq, sp_w_vec, sp_vecmul
Matrix	sp_mat_mul, sp_mat_trans, sp_mat_mul_cplx
MIsc	sp_blk_move, sp_blk_eswap16, sp_blk_eswap32, sp_blk_eswap64, sp_fltoq15, sp_q15tofl, sp_minerror

Table 19.3. TMS320C6700 Benchmark Programs.

Finally, if you want to develop DSP algorithms by TI's standards you should follow the TI's guide lines and rules called **eXpressDSP™- Compliant**. Several developmental references are provided at the following web address:
http://focus.ti.com/docs/toolsw/folders/print/sprc116.html

Helpful **eXpressDSP™- Compliant** user guide documents at this location include the following:

TMS320 DSP Algorithm Standard Rules and Guidelines:
http://focus.ti.com/lit/ug/spru352e/spru352e.pdf (137 pages)

TMS320 DSP Algorithm Standard Developer's Guide:
http://focus.ti.com/lit/ug/spru424c/spru424c.pdf (48 pages)

TMS320 DSP Algorithm Standard Demonstration Application:
http://focus.ti.com/lit/ug/spru361e/spru361e.pdf (137 pages)

TMS320 DSP Algorithm Standard API Reference:
http://focus.ti.com/lit/ug/spru360c/spru360c.pdf (117 pages)

Getting started with Code Composer

Texas Instruments provides several references for getting started. One is the ***Code Composer Studio v3.0 Getting Started Guide*** located at the following web address:
http://focus.ti.com/lit/ug/spru509e/spru509e.pdf **(151 pages)**

This document and several others can found at the following TI site:
http://focus.ti.com/docs/toolsw/folders/print/ccstudio.html#technicaldocuments

A.4. Setting up the TMS3206713 DSK and installing DSPLIB

This Section introduces the TMS320C6713-225 DSK and shows how to setup the DSK kit and how to install the benchmark C67x DSP library.

A.5. Introduction to the TMS320C6713 DSK

The TMS320C6713 DSK from Spectrum Digital Incorporated has the interfaces list in Table 19.4. These are shown in Figure 19.2 relative to the TMS320C6713 DSK board.

Category	Quantity	Items
Expansion connectors	3	J4, Memory expansion connector J3, Peripheral expansion connector J1, HPI exapansion connector
Audio connectors	4	J301, microphone in connector J303, Audio stereo line in connector J304, Audio stereo line out connector J302, headphone out connector
Power connectors	2	J5, +5 volt connector J6, optional 12 volt power connector
Miscellaneous connectors		J201, USB connector J8, External JTAG connector JP3, PLD programming connector System LEDS: LED7, LED8, LED9, LED10 SW2, Reset switch

Table 19.4. TMS320C6713 DSK interfaces.

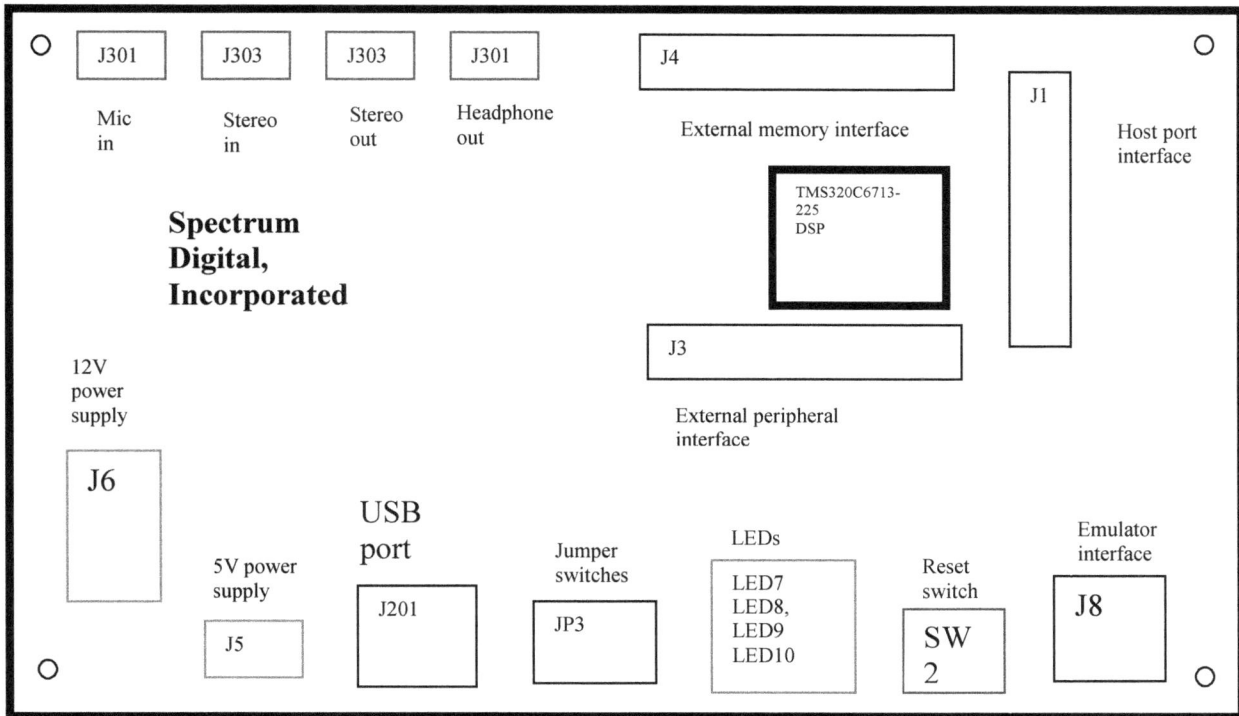

Figure 19.2. TMS320C6713 DSK layout showing hardware interfaces.

A.6. Procedure for setting up the TMS320C6713 DSK with DSPLIB

The procedure for setting up the the TMS320C6713 DSK is:

1. Acquire the items listed inTable 19.5 before beginning the TMS320C6713 configuration.

Equipment	Description
TMS320C6713 DSK board and acessories to connect to PC, powers cords, etc	TMS320C6713-225 DSP Development board, spectrum Digital, $395.00
Code composer	Software development software, comes with DSK
C67x DSPLIB	TMS320C6713 DSP library, see Section 0 on page 563 on where to download
C67x Software Developer guide	Texas Instrument software development guide
Computer or Lap Top	Computer for running code composer software and for interfacing to TMS32C6713 DSK via USB cable, etc.

Table 19.5. List of items required for setting up the TMS320C6713 DSK.

2. Power up the DSK kit and connect it to the computer using the required cable. Install code composer on the computer. When the installing is complete the software will ask whether you would like to test the DSK board. Select Yes to verify the board and its interfaces are working properly.
3. Download the C67X DSPLIB from the Texas Instrument Website given in Section 0 on page 563. Save this file, called **c67xdsplib.exe** into a directory.
4. Open the file, C67xDSPLIB.exe. a.) Click Yes to install the library. b.) Click Next to continue with the Install Shield Wizard. c.) Read the Software Licenses, and choose either "I accept" or "I don't accept." d.) if the click "I accept" the software installation will continue otherwise it will end.
5. Choose the location where you would like to install the library. The wizard will install the library into the c6700 sub-directory of the parent directory you choose. The default parent directory created is c:\ti. Now click next. Therefore, by default the directory structure created should be c:\ti\c6700\.
6. If the library has already been installed, you will be prompted to decide whether to replace the files or not. Click Yes to update the library. The Install Shield will complete the installation. When the installation is complete, click Finish.
7. In summary, the completed installation should have installed Code Composer Studio (CCS), DSK drivers, CCS and DSK test desktop icons, C67x DSPLIB, and tested your TMS320C6713 DSK board.

Appendix B.   Getting Started with Reference Frameworks for the TMS320C6713 DSP

Reference framework 3 is prewritten software to help the developer get started quickly. You will learn how to use this software to get started in this section. There are currently three reference frameworks available:

1. Reference Framework 1 (RF1)
2. Reference Framework 3 (RF3)
3. Reference Framework 5 (RF5)

The documentation for reference framework 3 is located at:
https://www-a.ti.com/downloads/sds_support/targetcontent/RF/RF%202.2/spra793d.pdf.

Additional, an excellent reference to many DSP applications and labs is (Chassaing, 2005).

B.1. Procedure for installing and using RF3 for the TMS320C6713 DSP

The procedure for implementing a FIR filter using Reference Framework 3 is given as follows:

1. Acquire the items listed in Table 19.6.

Equipment	Quantity	Description
TMS320C6713 DSK board and acessories to connect to PC, powers cords, etc	1	TMS320C6713-225 DSP Development board, spectrum Digital, $395.00
Code Composer 2.2 or greater	1	Software development software, comes with DSK(code composer getting started guide: http://focus.ti.com/lit/ug/spru509e/spru509e.pdf (151 pages)
Download Reference Framework 3 software to c:\ti\myprojects	1	https://www.ti.com/downloads/sds_support/targetcontent/RF/index.html
Computer or Lap Top	2	Computers for running code composer and Matlab software and for interfacing to TMS32C6713 DSK via USB cable, etc.
Matlab software	1	Digital signal processing software for designing and analyzing filters, signal generation, etc
Audio jack cable	1	Audio jack cable for interfacing the computers's audio output port to the TMS320C6713 DSK's audio input port
Computer speakers	1	Speakers to listen to Matlab generated sound
Computer microphone	1	Microphone for recording sounds

Table 19.6. List of items required for implementing a FIR filter in the TMS320C6713 DSK board using Reference Framework 3.

1 Power up the DSK kit and connect it to the computer using the required cable. Install code composer on the computer. When the installation is complete the software will ask whether you would like to test the DSK board. Select Yes to verify that the DSK board and its interfaces are working properly.

2 Download the reference framework 3 (RF3) executable file, **setup_rf_v2_20_win32.exe**, from the following URL:

https://www.ti.com/downloads/sds_support/targetcontent/RF/index.html

Once downloaded, install the RF3 in the following directory under the code composer directory: **myprojects/** For example, if your code composer studio is installed under c:\ti, then the RF3 files should be installed in the following path: **c:\ti\myprojects**. This should install the following files in your directory:
rf_v2_20_00_08.zip,
relnotes_rf_v2_20_00_08.htm,
filediff_rf_v2_20_00_08.txt,
_jvm,
_uninst.

Now, Unzip the zip file **rf_v2_20_00_08.zip** into the **c:\ti\myprojects** directory.

2. Open an MS-DOS command prompt and run **dosrun.bat** from the CCS for your DSK installation. Look in **c:\ti**.

3. Run the top-level build configuration script **c:\ti\myprojects\referenceframeworks\buildConfig.bat**. This will create DSP/BIOS configuration files (cdb's) compatible with your DSK release, for all Reference Frameworks levels (RF1, RF3, RF5, etc). Another way is to just run **makeConfig.bat** in the individual RF Level board port of interest (e.g. **c:\referenceframeworks\apps\rf3\dsk6713**).

4. Now start CCS, open the RF project for your board by selecting **Project->Open c:\ti\myprojects\referenceframeworks\apps\rf3\dsk6713\app .pjt**

5. Go to **Project -> Build Options**. Add the following include path: **-i"$(Install_dir)\c6000\xdais\include"**.

6. Now, build the application for your board by choosing **Project->Build**

7. Load the program into your DSP board by selecting: **File -> Load Program.**

8. Choose **Debug-> Run.**

9. Prepare your TMS320C6713 board as shown in Figure 19.3 for sending voice data. Connect your microphone to the PC's input audio port as shown. Connect the USB cable from your computer to your DSK board. Connect a computer speaker to your DSK's board audio output port. Connect an audio cable from your PC to the DSK's input audio port.

10. Record 5 seconds of voice speech to the file **sp0dat**. Send this to the DSK board. Play it back using speakers connected to the PCs audio output port. Now send it to the DSK board.

11. As an alternative you can also play a tutorial from TI's code composer install on your PC to test the code. Go to the Help menu and select tutorials.

12. While the speech file or tutorial is playing choose **File->Load GEL** and select **app.gel** which is located in the project folder. This file activates a graphical user interface (GUI) which can be used to send control data to the DSP's memory causing one of two channels to be active or inactive. The volume can also be controlled.

13. Now choose **GEL->Application Control->Set Active Channel**. The down position of the slider select channel 0 and the up position of the slider selects channel 1. Channel 0 is filtered using a low pass filter while channel 1 is filtered using a high pass filter. Listen to the sound coming from the speakers as you raise and lower the controls.

14. Now choose **GEL->Application Control->Set_Channel_0_Gain**. Change the position of the slider for various values ranging from 0 to 200 and listen to the sound coming from the speakers.

15. Repeat the previous step for channel 1. That is, choose **GEL->Application Control->Set_Channel_1_Gain**.

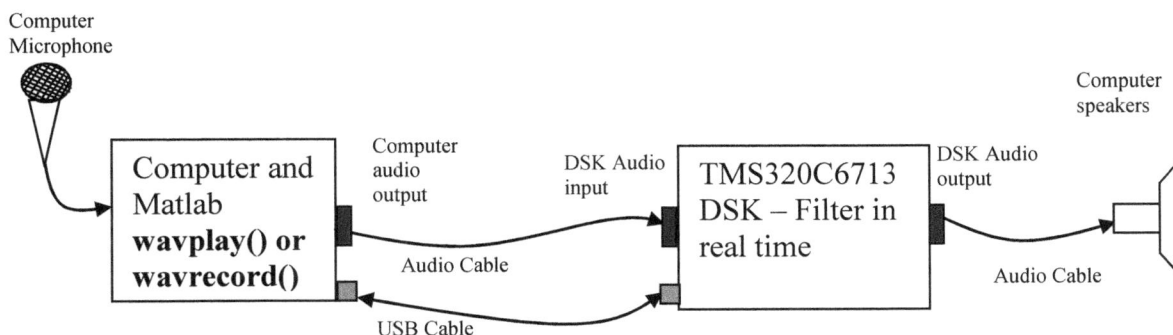

Figure 19.3. Illustration of how two computers are connected to DSK board for testing reference framework.

Appendix C. TMS320C6713 DSK Experiments

This section gives experiments utilizing the TMS320C6713 DSK board. A finite impulse response (FIR) filter design experiment is given next. Follow the procedures to complete the FIR experiment.

Additional, an excellent reference to many DSP applications and labs is (Chassaing, 2005).

C.1. Procedure for Designing a FIR Filter in Matlab.

The procedure for design an FIR filter is as follows:

1. Open Matlab.
2. Design a 40 tap FIR filter, h[n], using the required specifications in the Figure 19.4. The pass-band must be maximally flat. One algorithm to do this is the **remez()** algorithm. Save the coefficients to the variable **hn0dat**. Save as ascii data using the **save** command.
3. Plot h[n] using the **stem()** function. Print results.
4. Plot the magnitude and phase response verses frequency. Save the magnitude of h[n] to the variable **hn0datmag** and its phase to **hn0datang**. Print results. Display the magnitude in dB.

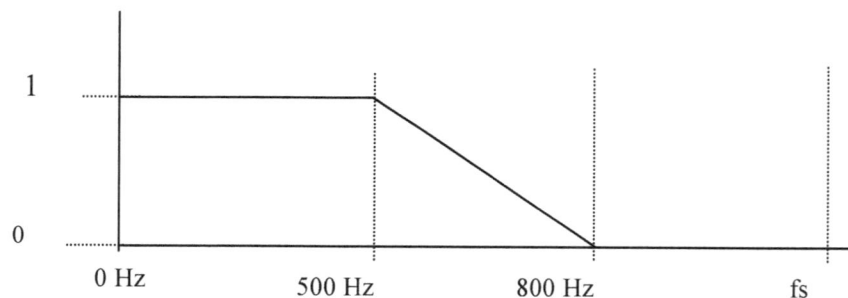

Figure 19.4. Lowpass filter specifications.

C.2. Procedure for Implementing a FIR Filter in the TMS320C6713 DSP
The procedure for implementing the FIR filter in the previous section is provided in the following steps:

1. Acquire the items listed in Table 19.7.

Equipment	Quantity	Description
TMS320C6713 DSK board and acessories to connect to PC, powers cords, etc	1	TMS320C6713-225 DSP Development board, spectrum Digital, $395.00
Code composer	1	Software development software, comes with DSK
C67x DSPLIB	1	TMS320C6713 DSP library, see Section 0 on page 563 on where to download
C67x Software Developer guide	1	Texas Instrument software development guide
Computer or Lap Top	2	Computers for running code composer and Matlab software and for interfacing to TMS32C6713 DSK via USB cable, etc.
Matlab software	1	Digital signal processing software for designing and analyzing filters, signal generation, etc
Audio jack cable	2	Audio jack cable for interfacing the computers's audio output port to the TMS320C6713 DSK's audio input port
Computer speakers	1	Speakers to listen to Matlab generated sound
Computer microphone	1	Microphone for recording sounds

Table 19.7. List of items required for designing and implementing a FIR filter in the TMS320C6713 DSK board.

2. Acquire the lowpass FIR filter coefficients h[n] from Section 0 on page 571. Prepare them for implementation inside the DSP.
3. Use Matlab to generate a sum of sinusoidals as follows:

$$x(t)=5\sin(2\pi200t) + 2\sin(2\pi1500t) + \sin(2\pi3200t)$$
EQ C.1

Save this signal to the Matlab variable **xndat.** One way to generate the signal x(t) in Matlab is more C like as follows:

```
fs= 44100;   % sample rate in samples per second
Ts=1/fs;  % sampling period
M=100000;  %number of desired samples
for i =0:M
    t(i+1)=Ts*i;
    xndat(i+1)= 5*sin(2*pi*200/fs * i) + 2*sin(2*pi*1500/fs * i) + sin(2*pi*3200/fs * i);
end;
plot(t,xndat);  title('x(t)=5sin(2π200t) + 2sin(2π1500t) + sin(2π3200t)');
xlabel('t (time)' );
ylabel ('Amplitude');
```

Another way to generate the signal x(t) in Matlab is more Matlab script like as follows:

```
fmin=200;  % minimum signal frequency
Tmax=1/fmin;  % maximum period
fs= 44100;   % sample rate in samples per second
Ts=1/fs;  % sampling period
t=0:Ts:5*Tmax;   % generate 5 periods of the signal x(t)
xndat= 5*sin(2*pi*200.*t) + 2*sin(2*pi*1500.*t) + sin(2*pi*3200.* t);
plot(t,xndat);  title('x(t)=5sin(2π200t) + 2sin(2π1500t) + sin(2π3200t)');
xlabel('t (time)' );
ylabel ('Amplitude');
```

4. Plot a 3ms window of the signal x(t) in Matlab and save for your records. One way to do this is to determine the number of samples required for a 3 ms window. The sampling rate is fs with the unit's samples per second. Using the window size and seconds it is easy to determine the number of samples given by:

$$samples = \left(\frac{samples}{\sec onds} \right) \sec onds \qquad\qquad \textbf{EQ C.2}$$

$$= fs \times \sec onds$$

Therefore, to plot 3ms of the signal x(t) requires $44100 \times 3 \times 10^{-3} = 132$ samples. The Matlab code to do this is:

```
plot(t(0:131),xndat(0:131) );  title('x(t)=5sin(2π200t) + 2sin(2π1500t) + sin(2π3200t)');
xlabel('t (time)' );
ylabel ('Amplitude');
```

To plot ½ second requires $44100 \times 1/2 = 22050$ samples. The Matlab code to plot a ½ second window is:

```
plot(t(0:22049),xndat(0:22049) );  title('x(t)=5sin(2π200t) + 2sin(2π1500t) + sin(2π3200t)');
xlabel('t (time)' );
ylabel ('Amplitude');
```
Any window in the signal will do not just from 0 to samples −1 as shown in these examples.

5. Use **waveplay**() to play this sound to the computer's audio port and listen to it using your computer speakers. Explain what it sounds like.

6. Now, plot the magnitude and phase plot of 3 seconds of x[n] verses frequency. Save the magnitude to the variable **xndatmag** and the phase to **xndatang**. Plot and print the Figure generated. Explain what you see.

7. Now filter the **xndat** using the filter h[n]. Save the filtered **xndat** to the variable **xnfilt0dat**.

8. Test the DSK hardware using the installed desktop icon called "6713 DSK Diagnostics Utility". Open code composer studio using the desktop icon called "C6713 DSK CCS". Setup the DSK to with the programs to read data from the stereo audio port, filter the data, and send it back to the stereo and headphone output ports. Now connect the computer's audio output port to the audio input port of the DSK and play x[n] using **wavplay**(). This is illustrated in Figure 19.5. Use the DSPLIB filter to implement the filter and filter the sound coming from the computer's audio port. Play the filtered signal in real time back to the DSK's audio output port and listen to it using your computer speaker. Explain what it sounds like.

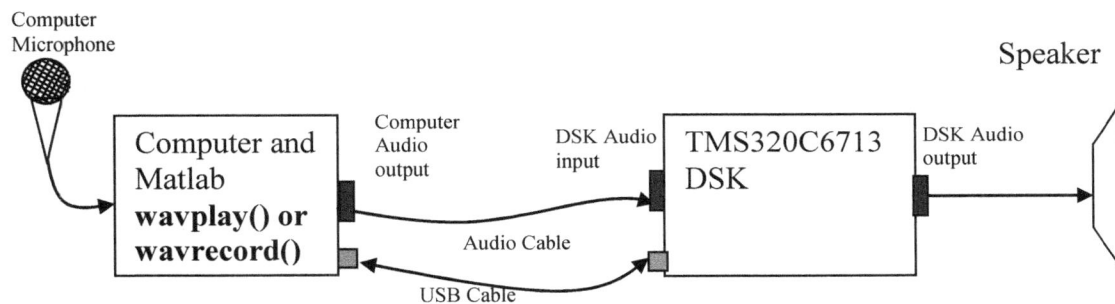

Figure 19.5. Illustration of computer connected to DSK board and DSK connected to speaker.

9. Now while filtering the sound in real time using the DSK record the sound in Matlab to the variable **xndskfilt0dat**. This is illustrated in Figure 19.6.

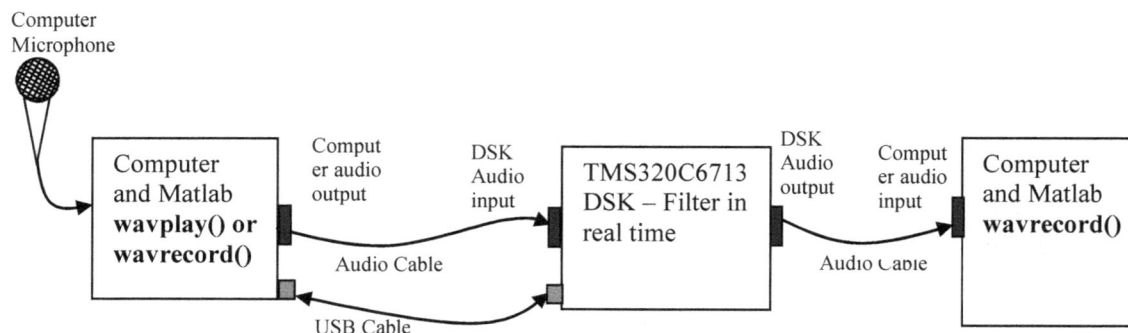

Figure 19.6. Illustration of how two computers are connected to DSK board.

10. Plot the magnitude and phase plot of **xndskfilt0dat**. Save the magnitude in the variable **xndskfilt0datmag** and the phase in **xndskfilt0datang**.

11. Now, using Matlab record the phrase "What Has God Wrought" using your computer's microphone. Record the phrase to the variable **s0dat**. Print the whole phrase from Matlab. Now plot and print a 300 ms window of the word "God" in the recorded phrase. Preferably select any 300ms in the "G".

12. Now plot the magnitude and phase of the 300 ms window. Save the magnitude in the variable **s0datmag** and the phase in **s0datang**.

13. Filter the phrase **s0dat** in Matlab using h[n]. Save the filtered phrase to the Matlab variable **s0filt0dat**. Plot the same 300 ms window used in step 10. Write down what you observed.

14. Now , using the filtered **s0filt0dat**, plot the magnitude and phase of the same 300 ms window in step 11. Save the magnitude to the variable **s0filtdatmag** and the phase to the variable **s0filtdatang**.

15. Connect the audio cable from the computer audio output to the DSK audio input port. Play the **s0dat** file to the computer's audio port while using the DSK to filter (using h[n]) the sound in real time while playing it to the speaker. Record how it sounds. This is illustrated in Figure 19.5.

16. Now connect the DSK's audio output port back to the computer's audio input port as shown in Figure 19.6. Now play the **s0dat** file from Matlab while recording it on another computer. Save the recorded data to **s0dskfilt0dat**.

17. Plot the magnitude and phase plot of **s0dskfilt0dat** using the same 300 ms window from step 11. Save the magnitude in the variable **s0dskfilt0datmag** and the phase in the variable **s0dskfilt0datang**.

18. **You are done! Have the professor verify your work.**

Project 10 –Speaker Identification using the Texas Instrument TMS320C6713 Digital
Signal Processor (DSP)
  The requirements of this project are given next.

C.3. Requirements
  The requirements for this project are as follows:
  1. Modify Reference Framework 3 where applicable for the requirements of this
     project.
  2. Record 21 seconds of the following speech phrase for each member in the group:

We Were a Way a Year Ago

Use 1 second breaks between phrases. You must repeat this phrase until at-least 21
seconds are complete.

  3. Must send recorded speech phrase for each group member to the DSK for speaker
     identification.
  4. Must use two channels in the DSP to process the speech from the codec.
  5. Implement a bandpass filter and volume control in each channel.
  6. Implement a level crossing analysis after each bandpass filter in each channel.
  7. Implement a threshold detection analysis after each level crossing analysis in each
     channel.
  8. Implement an algorithm that uses the threshold analysis from each channel to
     determine who the current speaker is.
  9. After the current speaker is determined display a message indicating the speaker
     using one of the following methods:

a. Using an LED on the DSP board that is unique to each speaker that lights up when
speaker is identified.
b. Storing a phrase of the speaker's voice in memory and sending the speaker's voice
to a speaker when identified.
c. Sending a sinusoidal sound to the speaker which is unique to each speaker
d. Displaying the speaker's name on an LCD display.

Appendix D. Robotic Electronic Component Stores
A list of stores for purchasing parts is provided in Table 19.8.

Name and phone	Website
Altera PLD Programming Software	http://www.altera.com/products/software/quartus-ii/switching/max-plus-ii/max-plus-ii-users.html
DeVry PLD Platform	http://www.elexp.com/tst_pld2.htm
WyJen Technologies, 800-490-4165	http://www.wyjen.com
Radio Shack, 800-843-7812	http://www.radioshack.com
Tech America, 800-877-0072	http://www.techam.com
Digikey, 800-344-4539	http://www.digikey.com
Mouser, 800-346-6873	http://www.mouser.com
Small Parts, Inc, 800-220-4242	http://www.smallparts.com
Robot Store, 800-374-5764	http://www.robotstore.com
Tower Hobbies, 800-637-6050	http://www.towerhobbies.com/
Robson Company	http://www.robson.com
Home Depot	Local Store.
Menards	Local Store.
Lowes	Local Store.

Table 19.8. List of distributors.

The various distributors in Table 19.8 provide the following items as described:
1. Altera: Quartus Software used to program the EPM7128SLC PLD device.
2. Electronix Express: DeVry PLDT-2 Board, 01PLDT2, and supporting items.
3. WyJen Technologies: wireless transmitters and receivers, Wireless devices, etc.
4. Tower Hobbies: Tires, wheels, different types of servo motors, etc.
5. Digikey: Many electronic components, including transmitters and receivers, connectors, etc.
6. Radio Shack: solder, tools, electronic components, wire, electronic components, battery packs, connectors, etc.
7. Mouser: Many electronic components including transmitters, receivers, connectors, etc
8. Home Depot, Marnard, Lowes, etc: hardware parts such as plexiglass, screws, nuts, washers, caster wheels, etc.
9. Robson Company : 1490 Dinsmore compass.

References

Blake, Roy. 2002. *Electronic Communication Systems.* Albany : Delmar Learning, 2002. 0-7668-2684-8.

Chassaing, Rulph. 2005. *Digital Signal Processing and Applications with the C6713 and C6416 DSK.* Hoboken : A John Wiley and Sons, Inc, 2005. 0-471-69007-4.

Coughlin, R and Driscoll. 2001. *Operational Ampfliers and Linear Integrated Circuits.* s.l. : Prentice Hall, 2001. 978-0130149916.

Dhananjay, Gadre V. 1998. *Programming the Parallel Port: Interface the PC for Data Acquisition and Process Control.* Berkeley : Publisher Group West, 1998. 0-87930-513-4.

Dueck, Robert K. 2001. *Digital Design with CPLD Applications and VHDL.* Albany, NY : Delmar Thompson Learning, 2001. 0-7668-1160-3.

Forouzan, Behrouz A. 2007. *Data Communications and Networking.* New York : McGraw Hill, 2007. 978-0-07-296775-3.

Frenzel, Louis E. 1998. *Principles of Electronic Communication Systems.* Columbus : McGraw Hill, 1998. 0-02-800409-4.

Furman, James O. Hamblen and Michael D. 2001. *Rapid Prototype of Digital Systems - A Tutorial Approach.* Norwell, Massachusetts : Kluwer Academic Publishers, 2001. 0792374398.

Gadre, Dhananjay V. 1998. *Programming the Parallel Port - Interfacing the PC for Data Acquisition and Process Control.* Emeryville : Miller Freeman, 1998. 0-87930-513-4.

Huang, Han-Way. 2005. *PIC Microcontroller: An Introduction to Software and Hardware Interfacing.* Clifton Park : Delmar Learning, 2005. 1-4018-3967-3.

Ifeachor, Emmanuel C. and Jervis, Barrie W. 2002. *Digital Signal Processing - A Practical Approach.* Upper Saddle River : Pearson Education, 2002. 0-201-59619-9.

Jacob, Michael. 1989. *Industrial Control Electronics: Applications and Designs.* s.l. : Prentice Hall, 1989. 978-0134593067.

Katzen, Sid. 2005. *The Quintessential PIC Microcontroller.* London : Springer-Verlag, 2005. 1-85233-942-X.

Tomasi, Wayne. 2004. *Electronic Communication Systems - Fundamentals through Advanced.* Upper Saddle River : Pearson Education, Inc, 2004. 0-13-049492-5.

Vegte, Joyce Van de. 2002. *Fundamentals of Digital Signal Processing.* Upper Saddle River : Pearson Education, 2002. 0-13-016077-6.

Waterman, Steve. 2003. *Digital Logic Simulation and CPLD Programmimg wuth VHDL.* Upper Saddle River : Pearson Education, 2003. 0-13-096760-2.

Index

U

V

W

Z

www.ingramcontent.com/pod-product-compliance
Lightning Source LLC
Chambersburg PA
CBHW060944210326
41598CB00031B/4716